Parasites of the Colder Clin

Parasites of the Colder Climates

Edited by
Hannah Akuffo, Ewert Linder,
Inger Ljungström, and
Mats Wahlgren

CRC Press
Taylor & Francis Group
Boca Raton London New York

CRC Press is an imprint of the
Taylor & Francis Group, an **informa** business
A TAYLOR & FRANCIS BOOK

First published 2003 by Taylor & Francis

Published 2019 by CRC Press
Taylor & Francis Group
6000 Broken Sound Parkway NW, Suite 300
Boca Raton, FL 33487-2742

© 2003 by Taylor & Francis Group, LLC
CRC Press is an imprint of Taylor & Francis Group, an Informa business

First issued in paperback 2019

No claim to original U.S. Government works

ISBN 13: 978-0-367-45464-7 (pbk)
ISBN 13: 978-0-415-27584-2 (hbk)

Visit the Taylor & Francis Web site at
http://www.taylorandfrancis.com

and the CRC Press Web site at
http://www.crcpress.com

Typeset in Baskerville MT by
Newgen Imaging Systems (P) Ltd, Chennai, India

British Library Cataloguing in Publication Data
A catalogue record for this book is available from the British Library

Library of Congress Cataloging in Publication Data
A catalog record for this book has been requested

Contents

Contributors

Hannah Akuffo
Microbiology and Tumor Biology Centre,
Karolinska Institutet and
Swedish Institute for Infectious
Disease Control, Box 280,
SE-171 77 Stockholm, Sweden
hannah.akuffo@mtc.ki.se

Kjell Alestig
Department of Infectious Diseases,
SU/Östra,
SE-416 85 Göteborg, Sweden
kjell.alestig@infect.gu.se

Jan Andersson
Division of Infectious Diseases,
Karolinska Institutet,
Huddinge University Hospital,
SE-141 86 Stockholm, Sweden
jan.andersson@medhs.ki.se

Tomas Bergström
Department of Clinical Virology,
University of Göteborg,
Guldhedsgatan 10 B,
SE-413 46 Göteborg, Sweden
tomas.bergstrom@microbio.gu.se

Bjørn Berland
Zoologisk Institutt, University of Bergen,
Allégt. 41, N-5007 Bergen, Norway
berland@zoo.uib.no

Sven F. F. Britton
Department of Infectious Diseases,
Karolinska Hospital,
SE-171 76 Stockholm, Sweden
sven.britton@medks.ki.se

Ib C. Bygbjerg
Department of Infectious Diseases,
Rigshospitalet, Copenhagen, Denmark
i.bygbjerg@pubhealth.ku.dk

B. Göran Bylund
Institute of Parasitology, Åbo Akademi
University, BioCity, Artillerigatan 6,
20520 Åbo, Finland
goran.bylund@abo.fi

Johan Carlson
National Board of Health and Welfare,
SE-106 30 Stockholm, Sweden

Kerstin Elvin
Department of Clinical Immunology and
Transfusion Medicine,
Karolinska Hospital,
SE-171 76 Stockholm, Sweden
kerstin.elvin@ks.se

Birgitta Evengård
Division of Infectious Diseases,
Huddinge University Hospital,
SE-141 86 Stockholm, Sweden
birgitta.evengard@infect.hs.sll.se

Matthías Eydal
Institute for Experimental Pathology,
Keldur, University of Iceland,
IS-112 Reykjavik, Iceland
m.eydal@hi.is

Theresa W. Gyorkos
Division of Clinical Epidemiology,
Montreal General Hospital,
Room L 10-420 1650 Cedar Avenue,
Montreal, Quebec, Canada H3G 1A4
theresa.gyorkos@mcgill.ca

Thomas G. T. Jaenson
Medical Entomology Unit, Department of
Systemic Zoology, Evolutionary Biology
Centre, Uppsala University, Norbyvägen
18d, SE-752 36 Uppsala, Sweden
thomas.jaenson@zoologi.uu.se
thomas.jaensen@ebc.uu.se

Ants Jõgiste
Health Protection Inspectorate,
Paldiski mnt. 81,
10614 Tallinn, Estonia
Ants.Jogiste@tervisekaitse.ee

Ludmila Jurevica
Public Health Agency,
Klijanu iela 7,
Riga LV-1012, Latvia
jurevica@baltcom.lv

Agneta Aust Kettis
Department of Infectious Diseases,
Karolinska Institute, Danderyd Hospital
Medical Products Agency, Box 26,
SE-751 03 Uppsala, Sweden
agneta.aust-kettis@mpa.se

Johan Landegren
Värtavägen 16, 5 tr, SE-115 24 Stockholm,
Sweden

Maija Lappalainen
Helsinki University Central Hospital,
Department of Virology,
PO Box 400,
00029 Helsinki, Finland
maija.lappalainen@hus.fi

J. I. Ronny Larsson
Department of Cell and
Organism Biology,
University of Lund,
SE-223 62 Lund, Sweden

Marianne Lebbad
Swedish Institute for Infectious Disease
Control
SE-171 82 Solna, Sweden
marianne.lebbad@smi.ki.se

Ewert Linder
Swedish Institute for Infectious Disease
Control
SE-171 82 Solna, Sweden
ewert.linder@smi.ks.se

Johan Lindh
Swedish Institute for Infectious Disease
Control
SE-171 82 Solna, Sweden
johan.lindh@smi.ki.se

Inger Ljungström
Swedish Institute for Infectious Disease
Control
SE-171 82 Solna, Sweden
inger.ljungstrom@smi.ki.se

Bettina Lundgren
Department of Clinical Microbiology,
Hvidovre University Hospital, 2650
Hvidovre, Denmark
bettina.lundgren@hh.hosp.dk

J. Dick MacLean
McGill University Centre for
Tropical Diseases,
Montreal General Hospital,
Room D7-153 1650 Cedar Avenue
Montreal, Quebec
dick.maclean@mcgill.ca

Ats Metsis
Faculty of Mathematics and
Natural Sciences,
Tallinn Pedagogical University,
Chair of Biology, Tallinn, Estonia
ats@globalgenomics.con
ats@kbfi.ee

Inga-Lill Gustavsson Moringlane
Department of Infectious Diseases,
Karolinska Institutet, Huddinge Hospital,
SE-141 86 Stockholm, Sweden

Nancy P. Nenonen
Department of Clinical Virology,
University of Göteborg,
Guldhedsgatan 10 B,
SE-413 46 Göteborg, Sweden
nancy.nenonen@microbio.gu.se

Sigvard Olofsson
Department of Clinical Virology,
University of Göteborg,
Guldhedsgatan 10 B,
SE-413 46 Göteborg, Sweden
sigvard.olofsson@microbio.gu.se

Mats Olsson
Virologigruppen Octapharma AB,
SE-112 75 Stockholm, Sweden
mats.olsson@biovitrum.com

N. N. Ozeretskovskaya
Martinovsky Institute of Medical
Parasitology and Tropical Medicine,
Moscow, 117830, Russia
gvi@ioc.ac.ru

Eskild Petersen
Department of Gastrointestinal and
Parasitic Infections,
Statens Serum Institut,
2300 Copenhagen, Denmark
epi@ssi.dk

Sigurður H. Richter
Institute for experimental Pathology,
Keldur, University of Iceland,
IS-112 Reykjavik, Iceland
shr@hi.is

V. P. Sergiev
Martinovsky Institute of Medical
Parasitology and Tropical Medicine,
Moscow, 117830, Russia

Bouchra Serhir
Départment de sérologie-virologie,
Laboratoire de santé publique du Québec,
20045 Chemin Ste Marie,

Ste Anne de Bellevue, Québec
Canada H9X 3R5
bserhir@lspq.org

Karl Skírnisson
Institute for Experimental Pathology,
Keldur, University of Iceland,
IS-112 Reykjavik, Iceland
karlsk@hi.is

Elda Sparrelid
Division of Infectious Diseases,
Karolinska Institutet,
Huddinge University Hospital,
SE-141 86 Stockholm, Sweden
elda@hem.utfors.se

Babill Stray-Pedersen
Department of Gynecology and Obstetrics,
National Hospital, University of Oslo,
Oslo, Norway
Babill.stray-pedersen@klinmed.uio.no

Staffan Svärd
Swedish Institute for Infectious Disease
Control
SE-171 82 Solna, Sweden
staffan.svard@mtc.ki.se

Cecilia Thors
Swedish Institute for Infectious Disease
Oslo, Control
SE-171 82 Solna, Sweden
cecilia.thors@smi.ki.se

Regina Virbaliené
Vilnius Public Health Centre,
Laboratory of Parasitology,
Geliu 9 Vilnius,
Lithuania
vilnvscparazlab@takas.lt

Mats Wahlgren
Swedish Institute for Infectious Disease
Control and Microbiology and Tumor
Biology Center, Karolinska Institutet,
SE-171 82 Solna, Sweden
mats.wahlgren@smi.ki.se

Brian Ward
Division of Infectious Diseases,
Department of Medicine,
Montreal General Hospital,
Room R3-133 1650 Cedar Avenue
Montreal, Quebec,
Canada H3G 1A4
brian.ward@mcgill.ca

Michael Willcox
Formely Gävle Hospital,
SE-801 87 Gävle, Sweden

Jadwiga Winiecka-Krusnell
Swedish Institute for Infectious Disease
Control
SE-171 82 Solna, Sweden
jadwiga.winiecka-krusnell@smi.ki.se

Pål Wölner-Hanssen
Department of Obstetrics and Gynecology,
University Hospital of Lund,
SE-221 85 Lund, Sweden
pal.wolner-hanssen@gyn.lu.se

Preface

Parasites have been our follower and foe during the history of *Homo sapiens*. The prehistoric man was infected and we are still at risk, despite all efforts to eradicate parasitic diseases.

There are various publications of parasitology available on the market for reference. Most of these books tend to concentrate on parasitic infections in the Tropics. There is, however, a dearth of publications that specifically addresses parasites of the Northern hemisphere. At present, there is no publication which addresses the parasites indigenous to the North, given the impression to many that parasites are not a problem in this part of the world. Parasitology is only recently becoming an academic subject in its own right in several northern countries and it would be very useful for medical students and others to have the opportunity to have a concise text book describing the parasites in their own part of the world. Such a text book, concentrated on an area other than the Tropics, would be expected to be of interest to the international community as well.

While many of parasites of the North are present as a zoonosis in the animal population there are also occasional water contamination leading to outbreaks. Many visitors to the cold north are not prepared for and are taken aback by the extent of the micro-predators/ectoparasite population in the beautiful forests and around the picturesque waterways. Such irritating arthropods include mosquitoes, gnats (black-flies), and ticks which dominate the summer months, while lice (the head louse) remain an all-year-round menace in the day care centres. Search through the literature show that parasitic infections of a more exotic nature may also be indigenous to the North including myiasis. The habit of eating uncooked fish in conjunction with the high rate of infection of the fish population with such parasites as *Diphyllobotirum latum* (40% of the pike in the Lake Mälaren of Stockholm are carriers) makes the potential of such infections high. Toxoplasmosis is one of the most important parasitic infections in the North, found both as the congenital form and in immunosuppressed patients. The increase in this latter group of patients is responsible for the increased incidence in a number of other parasites including *Toxoplasma gondii*, Microsporidia, *Pneumocystis carinii*, *Cryptosporidia* spp., and *Acantamoeba* spp.

This volume gives a broad overview of the parasites present in the North and the diseases they may cause. The first part of the book presents the general epidemiology in different countries of the North, past and present, followed by the biology, pathogenesis, and the specific epidemiology of important parasites in this part of the world. The clinical presentation and laboratory investigations of the diseases are given in separate sections together with methods for laboratory diagnosis. The book concludes with a historical review named Linnaeus, Armauer Hansen and Nordic Research on some neglected diseases, in memory of Elias Bengtsson, a pioneer among the persons introducing clinical parasitology in Sweden.

Part 1

Introduction

1 Parasites indigenous to the North (tables)

Cecilia Thors, Inger Ljungström, and Thomas Jaenson

Nematodes, cestodes, and trematodes

Phylum/ subphylum	Class	Species	Disease	Transmission	Distribution	References
Nematoda		*Anisakis simplex*	Anisakidosis (anisakiasis)	Ingestion of larva in raw salt-water fish	For example Alaska, Canada, Iceland, Japan, western Europe (Holland), Baltic countries	[6, 17, 25]
		Ascaris lumbricoides	Ascaridosis	Ingestion of eggs	World-wide	[29]
		Capillaria aerophila	Capillariosis	Ingestion of eggs (from dogs, cats, foxes or other carnivors' faeces in the soil)	Russia	[14, 41]
		Enterobius vermicularis	Enterobiosis	Ingestion of eggs	World-wide	[14, 20]
		Pseudoterranova decipiens (*Phocanema*)	Anisakidosis	Ingestion of larva in raw salt-water fish	Japan, US	[6, 24, 36, 43]
		Toxocara canis	Toxocariosis visceral larva migrans	Ingestion of eggs (from dog faeces in the soil)	World-wide	[21, 27, 28, 37]
		Toxocara cati	Toxocariosis	Ingestion of eggs (from cat faeces in the soil)	World-wide	[49]
		Trichinella britova	Trichinellosis	Ingestion of larvae in meat	Palaearctic region (from western Europe to eastern Asia)	[32]
		Trichinella nativa	Trichinellosis	Ingestion of larvae in meat	Holarctic region	[7, 32, 33, 38, 52]
		Trichinella spiralis	Trichinellosis	Ingestion of larvae in meat	World-wide except Oceania	[8, 13, 19, 32, 33, 35, 41, 50]

Class	Species	Disease	Transmission	Distribution	References
Cestoda	*Diphyllobothrium latum*	Diphyllobothriosis	Ingestion of plerocercoid (sparganum) in flesh of infected fish	Scandinavian, Baltic countries, northern Russia, Japan, North America, Chile, Uganda	[1, 20, 31, 42, 45]
	Echinococcus granulosus	Echinococcosis (or hydatide disease)	Ingestion of eggs (dogs)	World-wide	[2, 16, 46, 51]
	Echinococcus multilocularis	Echinococcosis (or hydatide disease)	Ingestion of eggs (cats, foxes, dogs)	Central Europe	[47]
	Hymenolepis nana	Hymenolepiasis	Ingestion of embryonated eggs (in fleas)	World-wide (Baltic countries, Russia)	[44]
	Taenia saginata	Taeniosis saginata	Ingestion of cysticercus in beef (uncooked)	World-wide	[10, 18]
Trematoda	*Fasciola hepatica*	Fasciolosis	Ingestion of metacercariae in fresh-water plants	Europe, Latin America, North Africa, Asia, Western Pacific	[3, 15]
	Metorchis conjunctus	*Metorchis conjunctus*-infection	Ingestion of metacercariae in fish (sucker)	Canada	[22]
	Nanophyetus salmincola (Troglotrema salmincola)	Nanophyetiasis	Ingestion of metacercariae in salmon	Eastern Siberia, US Pacific Northwest	[11, 12, 23, 26, 40]
	Opisthorchis felineus	Opisthorchiosis	Ingestion of metacercariae in fresh-water fish	Russia, Central and eastern Europe, India	[15, 48]
	Trichobilharzia spp.	Swimmer's itch	Skin penetration of cercariae	World-wide	[4, 5, 9, 30]

References

[1] Adams, A. M. *et al.* (1997). Parasites of fish and risks to public health. *Revue Scientifique et Technique*, **16**, 652–660.

[2] Ammann, R. and Eckert, J. (1996). Cestodes Echinococcus. *Gastroenterology Clinics of North America*, **25**, 655–689.

[3] Arjona, R. *et al.* (1995). Fascioliasis in developed countries: a review of classic and aberrant forms of the disease. *Medicine*, **74**, 13–23.

[4] Beer, S. and German, S. (1994). The ecological prerequisites for the spread of schistosomal dermatitis (cercariasis) in Moscow and the Moscow area (in Russian). *Meditsinkaia Parazitologiia i Parazitarnye Bolezni*, **1**, 16–19.

[5] Berg, K. and F. Reiter (1960). Observations on schistosome dermatitis in Denmark. *Acta Dermato-Venereologica*, **40**, 369–380.

[6] Bouree, P. *et al.* (1995). Anisakidosis: report of 25 cases and review of the literature. *Comparative Immunology, Microbiology & Infectious Diseases*, **18**, 75–84.

[7] Capo, V. and Despommier, D. D. (1996). Clinical aspects of infection with *Trichinella* spp. *Clinical Microbiology Reviews*, **9**, 47–54.

[8] Clausen, M. *et al.* (1996). *Trichinella* infection and clinical disease. *Quarterly Journal of Medicine*, **89**, 631–636.

[9] Cort, W. (1950). Studies on schistosome dermatitis XI. Status of knowledge after more than twenty years. *American Journal of Hygiene*, **52**, 251–307.

[10] Craig, P. *et al.* (1995). Hydatidosis and cysticercosis – larval cestodes. *Medical parasitology. A practical approach.* S. Gillespie and P. Hawkey. Oxford, IRL press, pp. 209–237.

[11] Eastburn, R. *et al.* (1987). Human intestinal infection with *Nanophyetus salmincola* from salmonid fishes. *American Journal of Tropical Medicine and Hygiene*, **36**(3), 586–591.

[12] Fritsche, T., Eastburn, R. *et al.* (1989). Praziquantel for treatment of human *Nanophyetus salmincola* (*Troglotrema salmincola*) infection. *Journal of Infectious Diseases*, **160**, 896–899.

[13] Gamble, H. (1997). Parasites associated with pork and pork products. *Revue Scientifique et Technique*, **16**, 496–506.

[14] Grencis, R. K. and Cooper, E. S. (1996). Enterobius, trichuris, capillaria, and hookworm including ancylostoma caninum. *Gastroenterology Clinics of North America*, **25**, 579–597.

[15] Harinasuta, T. *et al.* (1993). Trematode infections. Opisthorchiasis, clonorchiasis, fascioliasis, and paragonimiasis [published erratum appears in *Infect. Dis. Clin. North. Am.* 1994 Mar. **8**(1), following table of contents]. *Infectious Disease Clinics of North America*, **7**, 699–716.

[16] Huldt, G. *et al.* (1973). Echinococcosis in northern Scandinavia. Immune reactions to *Echinococcus granulosus* in Kautokeino Lapps. *SO – Archives of Environmental Health*, **26**, 36–40.

[17] Ishikura, H. *et al.* (1993). *Anisakidae* and anisakidisis. *Progress in Clinical Parasitology*, **3**, 43–102.

[18] Ito, A. *et al.* (1998). Novel antigens for neurocysticercosis: simple method for preparation and evaluation for serodiagnosis. *American Journal of Tropical Medicine & Hygiene*, **59**(2), 291–294.

[19] Kapel, C. M. (1997). *Trichinella* in arctic, subarctic and temperate regions: Greenland, the Scandinavian countries and the Baltic States. *Southeast Asian Journal of Tropical Medicine & Public Health*, **28**(Suppl 1), 14–19.

[20] Kyrönseppä, H. (1993). The occurrence of human intestinal parasites in Finland. *Scandinavian Journal of Infectious Diseases*, **25**, 671–673.

[21] Lewis, J. W. and Maizels, R. M. (1993). *Toxocara and Toxocariasis.* London, Institute of Biology and the British Society for Parasitology.

[22] MacLean, J. D., Arthur J. R. *et al.* (1996). Common-source outbreak of infection due to the North American liver fluke *Metorchis conjunctus*. *Lancet*, **347**, 154–158.

[23] Millemann, R. and S. Knapp (1970). Biology of *Nanophyetus salmincola* and 'salmon poisoning' disease. *Advances in Parasitology*, **8**, 1–41.

[24] Myers, B. (1959). *Phocanema*, a new genus for the anisakid nematode of seals. *Canadian Journal of Zoology*, **37**, 459–465.

[25] Myjak, P. *et al.* (1994). Anisakid larvae in cod from the southern Baltic Sea. *Archive of Fishery and Marine Research*, **42**, 149–161.

[26] Odhner, T. (1914). Die Verwaudschaftsbeziehungen der Trematodengattung Paragonimus Brn. *Zoologiska Bidrag från Uppsala*, **3**, 231–246.

[27] Overgaauw, P. A. (1997). Aspects of *Toxocara* epidemiology: human toxocarosis. *Critical Reviews in Microbiology*, **23**, 215–231.

[28] Overgaauw, P. A. (1998). *Toxocara* infections in dogs and cats and public health implications. *Veterinary Quarterly*, **20**(Suppl 1), S97–S98.

[29] Pawlowski, Z. S. (1982). Ascariasis: host–pathogen biology. *Reviews of Infectious Diseases*, **4**, 806–814.

[30] Prilä, P. and Wikgren, B.-J. (1957). Cases of swimmers itch in Finland. *Acta Dermatovenerologica*, **37**, 140–148.

[31] Pliushcheva, G. L. et al. (1997). The spread of *Diphyllobothrium* having medical importance on the territory of Russia. *Meditsinskaia Parazitologiia i Parazitarnye Bolezni* (2), 55–60.

[32] Pozio, E. et al. (1992). Taxonomic revision of the genus *Trichinella*. *Journal of Parasitology*, **78**, 654–659.

[33] Pozio, E. et al. (1995). Concurrent infection with sibling *Trichinella* species in a natural host. *International Journal for Parasitology*, **25**, 1247–1250.

[34] Pozio, E. et al. (1996). Environmental and human influence on the ecology of *Trichinella spiralis* and *Trichinella briovi* in Western Europe. *Parasitology*, **113**(Pt 6), 527–533.

[35] Pozio, E. et al. (1998). Distribution of sylvatic species of *Trichinella* in Estonia according to climate zones. *Journal of Parasitology*, **84**, 193–195.

[36] Sakanari, J. A. and McKerrow, J. H. (1989). Anisakiasis. *Clinical Microbiology Reviews*, **2**, 278–284.

[37] Schantz, P. M. (1989). *Toxocara* larva migrans now. *American Journal of Tropical Medicine & Hygiene*, **41**(Suppl 3), 21–34.

[38] Shaikenov, B. (1995). The distribution of *Trichinella nativa*, *T. nelsoni* and *T. pseudospiralis* in Eurasia. *Meditsinskaia Parazitologiia i Parazitarnye Bolezni* (3), 20–24.

[39] Skrjabin, K. et al. (1970). Investigations on trichinellosis in the U.S.S.R. *Wiadomosci Parazytologiczne*, **16**, (Suppl 67–8), 21–34.

[40] Skrjabin, K. and W. Podjapolskaja (1931). *Nanophyetus schikhobalowi* n. sp. Ein neuer Trematode aus dem Darm des Menschen. *Zentralblatt für Bakteriologie, Parasitologie und Infektionskrankheiten*, **119**, 294–297.

[41] Skrjabin, K. et al. (1957, 1970). Essentials of nematodology. 6. *Trichocephalidae and Capillariidae* of animals and man and diseases caused by them, Academy of Sciences of USSR, Moscow, pp. 420–423, 1957 (in Russian). *Israel Program for Scientific Translation, Jerusalem*: 416–419, 513–515.

[42] Suvorina, V. I. and Simonova, N. F. (1993). The epidemiological aspects of diphyllobothriasis in Yakutia. *Meditsinskaia Parazitologiia i Parazitarnye Bolezni*, (4), 23–26.

[43] Suzuki, H. et al. (1972). *Terranova* (Nematoda: Anisakidae) infection in man. I. Clinical features of five cases of *Terranova* larva infection. *Japanese Journal of Parasitology*, **21**, 252–256.

[44] Tamblyn, S. (1975). Intestinal parasites in an institution – Ontario: Preliminary report. *Epidemiological Bulletin*, **19**, 53–54.

[45] Tanowitz, H. B. et al. (1993). Diagnosis and treatment of intestinal helminths. I. Common intestinal cestodes. *Gastroenterologist*, **1**, 265–273.

[46] Thompson, R. C. A. and Lymbery, A. J. (1995). *Echinococcus and Hydatid Disease*. Oxon, Cab International.

[47] Tornieporth, N. G. and Disko, R. (1994). Alveolar hydatid disease (*Echinococcus multilocularis*) – review and update. *Progress in Clinical Parasitology*, **4**, 55–76.

[48] Tsybina, T. N. (1994). The ecological–epidemiological characteristics of opisthorchiasis in Sverdlovsk Province. *Meditsinskaia Parazitologiia i Parazitarnye Bolezni* (3), 45–50.

[49] von Reyn, C. F. et al. (1978). Infection of an infant with an adult *Toxocara cati* (Nematoda). *Journal of Pediatrics*, **93**, 247–249.

[50] Wakelin, D. (1993). *Trichinella spiralis*: immunity, ecology and evolution. *Journal of Parasitology*, **79**, 488–494.

[51] Wilson, J. F. et al. (1995). Alveolar hydatid disease. Review of the surgical experience in 42 cases of active disease among Alaskan Eskimos. *Annals of Surgery*, **221**, 315–323.

[52] Zarnke, R. L. et al. (1995). Prevalence of *Trichinella nativa* in lynx (*Felis lynx*) from Alaska, 1988–1993. *Journal of Wildlife Diseases*, **31**, 314–318.

Protozoa

Phylum / subphylum	Species	Disease	Transmission	Distribution	References
Ciliata	*Balantidium coli*	Balantidiasis	Faecal–oral	World-wide	[4, 24, 45]
Mastigophora (flagellates)	*Chilomastix mesnili*	Uncertain	Faecal–oral	World-wide, more often in warmer climate	[5, 45]
	Dientamoeba fragilis	Dientamoeba infection	Uncertain	World-wide	[4, 51]
	Enteromonas hominis	Uncertain	Faecal–oral	World-wide	[45]
	Giardia intestinalis	Giardiasis	Faecal–oral	World-wide	[13, 16, 26, 33, 45, 47]
	Trichomonas hominis (*Pentatrichomonas hominis*)	Uncertain	Faecal–oral	World-wide	[14, 19]
	Trichomonas tenax	Respiratoric infection thoracic abscesses	Uncertain	World-wide	[17, 25]
	Trichomonas vaginalis	Trichomoniasis	Sexual	World-wide	[23, 40]
Microspora	*Microsporidia* spp.	Microsporidiosis	Unknown	World-wide	[9, 50]
Sarcodina (amoeba)	*Acanthamoeba* spp.	Acanthamoeba keratitis	Poor lens care	World-wide	[1, 20, 43, 46]
	Blastocystis hominis	Uncertain	Faecal–oral	World-wide	[4, 45, 52]
	Endolimax nana	Non-pathogenic	Faecal–oral	World-wide	[22, 45]
	Entamoeba coli	Non-pathogenic	Faecal–oral	World-wide	[4, 27, 45]
	Entamoeba dispar	Non-pathogenic	Faecal–oral	World-wide	[12, 21, 49]
	Entamoeba gingivalis	Non-pathogenic	Unknown	World-wide	[8, 11, 32]
	Entamoeba hartmanni	Non-pathogenic	Faecal–oral	World-wide	[22, 42]

	Entamoeba histolytica	Amoebiasis, amoebic dysentery, extraintestinal infections	Faecal–oral	World-wide	[30, 34, 39, 44, 45]
	Iodamoeba bütschlii	Non-pathogenic	Faecal–oral	World-wide	[22, 42, 45]
Sporozoa	*Babesia divergens*	Babesiosis	Tick borne (Ixodes)	Northern Europe, the British isles	[7, 15, 48]
	Cryptosporidium parvum	Cryptosporidiosis	Faecal–oral	World-wide	[3, 10, 16, 34, 35, 41]
	Sarcocystis spp.	Sarcocystosis	Oral	World-wide	[2, 6]
	Toxoplasma gondii	Toxoplasmosis	Oral, placental, tranplantation	World-wide	[18, 28, 38]
Fungi	*Pneumocystis carinii*	*Pneumocystis pneumoni*	Uncertain	World-wide	[29, 31, 36, 37]

References

[1] Aasly, K. and Bergh, K. (1992). *Acanthamoeba keratitis*; report of the first Norwegian cases. *Acta Ophthalmologica*, **70**, 698–701.

[2] Ackers, J. P. (1997). Gut *Coccidia – Isospora, Cryptosporidium, Cyclospora* and *Sarcocystis*. *Seminars in Gastrointestinal Disease*, **8**, 33–44.

[3] Atterholm, I., Castor, B., and Norlin, K. (1987). Cryptosporidiosis in southern Sweden. *Scandinavian Journal of Infectious Diseases*, **19**, 231–234.

[4] Aucott, J. N. and Ravdin, J. I. (1993). Amebiasis and "nonpathogenic" intestinal protozoa. *Infectious Disease Clinics of North America*, **7**, 467–485.

[5] Barnham, M. (1977). Is *Chilomastix* harmless? *Lancet*, **2**, 1077–1078.

[6] Beaver, P. C., Gadgil, R. K., and Morera, P. (1979). *Sarcocystis* in man: a review and report of five cases. *American Journal of Tropical Medicine and Hygiene*, **28**, 819–844.

[7] Christensson, D. A. (1989). *Babesia* of cattle and sheep in Sweden. Thesis (ISBN 91-576-3712-1).

[8] Clark, C. G. and Diamond, L. S. (1997). Intraspecific variation and phylogenetic relationships in the genus *Entamoeba* as revealed by riboprinting. *Journal of Eukaryotic Microbiology*, **44**, 142–154.

[9] Conteas, C. A., Didier, E. S., and Berlin, O. G. W. (1997). Workup of gastrointestinal microsporidiosis. *Digestive Diseases*, **15**, 330–345.

[10] Current, W. L. and Garcia, L. S. (1991). Cryptosporidiosis. *Clinical Microbiology Reviews*, **4**, 325–358.

[11] Dao, A. H., Robinson, D. P., and Wong, S. W. (1983). Frequency of *Entamoeba gingivalis* in human gingival scrapings. *American Journal of Clinical Pathology*, **80**, 380–383.

[12] Diamond, L. S. and Clark, C. G. (1993). A redescription of *Entamoeba histolytica* Schaudinn 1903 (Emended Walker, 1911) separating it from *Entamoeba dispar* Brumpt, 1925. *Journal of Eukaryotic Microbiology*, **40**, 340–344.

[13] Farthing, M. J. (1996). Giardiasis. *Gastroenterology Clinics of North America*, **25**, 493–515.

[14] Felleisen, R. S. J. (1997). Comparative sequence analysis of 5-center-dot-8S rRNA genes and internal transcribed spacer (ITS) regions of trichomonadid protozoa. *Parasitology*, **115**, 111–119.

[15] Gorenflot, A., Moubri, K., Precigout, E., Carcy, B., and Schetters, T. P. M. (1998). Human Babesiosis. *Annals of Tropical Medicine & Parasitology*, **92**, 489–501.

[16] Hansen, A. and Stenström, T. A. (1998). Kartläggning av *Giardia* och *Cryptosporidium* i svenska ytvattentäkter, (in Swedish), ISBN 1400-3473.

[17] Hersh, S. M. (1985). Pulmonary trichomoniasis and *Trichomonas tenax*. *Journal of Medical Microbiology*, **20**, 1–10.

[18] Holliman, R. E. and Greig, J. R. (1997). Toxoplasmosis in immunocompromised patients. *Current Opinion in Infectious Diseases*, **10**, 281–284.

[19] Honigberg, B. M. (ed.) (1989). *Trichomonads parasitic in humans*. Springer-Verlag, New York. (ISBN 3 540 96903 9).

[20] Illingworth, C. D. and Cook, S. D. (1998). *Acanthamoeba* keratitis. *Survey of Ophthalmology*, **42**, 493–508.

[21] Jackson, T. F. H. G. (1998). *Entamoeba histolytica* and *Entamoeba dispar* are distinct species – clinical, epidemiological and serological evidence. *International Journal for Parasitology*, **28**, 181–186.

[22] Jokipii, L., Sargeaunt, P. G., and Jokipii, A. M. (1989). Coincidence of deficient delayed hypersensitivity and intestinal protozoa in homosexual men. *Scandinavian Journal of Infectious Diseases*, **21**, 563–571.

[23] Kent, H. L. (1991). Epidemiology of vaginitis. *American Journal of Obstetrics & Gynecology*, **165**, 1168–1176.

[24] Khamtsov, V. G. (1969). The distribution of balantidiasis (Review of the national and foreign literature for 110 years, 1857–1966 (in Russian) *Meditsinskaia Parazitologiia i Parazitarnye Bolezni*, **38**, 598–606.

[25] Kikuta, N., Yamamoto, A., Fukura, K., and Goto, N. (1997). Specific and sensitive detection of *Trichomonas tenax* by polymerase chain reaction. *Letters in Applied Microbiology*, **24**, 193–197.

[26] Kyrönseppä, H. (1993). The occurence of human intestinal parasites in Finland. *Scandinavian Journal of Infectious Diseases*, **25**, 671–673.

[27] Kyrönseppä, H. and Pettersson, T. (1976). The occurence of human intestinal parasites in Finland. *Scandinavian Journal of Infectious Diseases*, **8**, 199–202.

[28] Lebech, M., Joynson, D. H., Seitz, H. M., Thulliez, P., Gilbert, R. E., Dutton, B. N. *et al.* (1996). Classification system and case definitions of *Toxoplasma gondii* infection in immunocompetent pregnant women and their congenially infected offspring. *European Journal of Clinical Microbiology & Infectious Diseases*, **15**, 799–805.

[29] Lee, C. H., Helmeg-Larsen, J., Tang, X., Jin, S., Li, B., Bartlett, M. S. *et al.* (1998). Update on *Pneumocystis carinii* f. sp. hominis typing based on nucleotide sequence variations in internal transcribed spacer regions of rRNA genes. *Journal of Clinical Microbiology*, **36**, 734–741.

[30] Li, E. and Stanley, S. L., Jr. (1996). Protozoa. Amebiasis. *Gastroenterology Clinics of North America*, **25**, 471–492.

[31] Lidman, C., Olsson, M., Bjorkman, A., and Elvin, K. (1997). No evidence of nosocomial *Pneumocystis carinii* infection via health care personnel. *Scandinavian Journal of Infectious Diseases*, **29**, 63–64.

[32] Linke, H. A., Gannon, J. T., and Obin, J. N. (1989). Clinical survey of *Entamoeba gingivalis* by multiple sampling in patients with advanced periodontal disease. *International Journal for Parasitology*, **19**, 803–808.

[33] Ljungström, I. and Castor, B. (1992). Immune response to *Giardia lamblia* in a water-borne outbreak of giardiasis in Sweden. *Journal of Medical Microbiology*, **36**, 347–352.

[34] Marshall, M. M., Naumowitz, D., Ortega, Y., and Sterling, C. R. (1997). Waterborne protozoan pathogens. *Clinical Microbiology Reviews*, **10**, 67–85.

[35] Meng, J. and Doyle, M. P. (1997). Emerging issues in microbiological food safety. *Annual Review of Nutrition*, **17**, 255–275.

[36] Olsson, M., Elvin, K., Linder, E., Loïdahl, S., and Wahlgren, M. (1999). *Pneumocystis carinii* is still a dangerous opportunist. The infection is continuously a threat to immuno-compromised patients (in Swedish) *Läkartidningen*, **96**, 328–331.

[37] Olsson, M., Lidman, C., Latouche, S., Bjorkman, A., Roux, P., Linder, E., and Wahlgren, M. (1998). Identification of *Pneumocystis carinii* f. sp. hominis gene sequences in filtered air in hospital environments. *Journal of Clinical Microbiology*, **36**, 1737–1740.

[38] Pavesio, C. E. and Lightman, S. (1996). *Toxoplasma gondii* and ocular toxoplasmosis: pathogenesis. *British Journal of Ophtalmology*, **80**, 1099–1107.

[39] Petri, W. A., Jr. (1996). Recent advances in amebiasis. *Clinical Reviews in Clinical Laboratory Sciences*, **33**, 1–37.

[40] Petrin, D., Delgaty, K., Bhatt, R., and Garber, G. (1998). Clinical and microbiological aspects of *Trichomonas vaginalis*. *Clinical Microbiology Reviews*, **11**, 330.

[41] Pohjola, S., Oksanen, H., Jokipii, L., and Jokipii, A. M. (1986). Outbreak of cryptosporidiosis among veterinary students. *Scandinavian Journal of Infectious Diseases*, **18**, 173–178.

[42] Sargeaunt, P. G. and Williams, J. E. (1979). Electrophoretic isoenzyme patterns of the pathogenic and non-pathogenic intestinal amoebae of man. *Transactions of the Royal Society of Tropical Medicine & Hygiene*, **73**, 225–227.

[43] Skarin, A., Floren, I., Kiss, K., Miorner, H., and Stenevi, U. (1996). *Acanthamoeba* keratitis in the south of Sweden. *Acta Ophthalmologica Scandinavica*, **74**, 593–597.

[44] Stanley, S. L., Jr. (1997). Progress towards development of a vaccine for amebiasis. *Clinical Microbiology Reviews*, **10**, 637–649.

[45] Svensson, R. (1935). Studies on human intestinal protozoa. *Acta Medica Scandinavica*, suppl. LXX.

[46] Szenasi, Z., Endo, T., Yagita, K., and Nagy, B. (1998). Isolation, identification and increasing importance of free-living amoebae causing human disease. *Journal of Medical Microbiology*, **47**, 5–16.

[47] Thompson, R. C. and Meloni, B. P. (1993). Molecular variation in *Giardia*. *Acta Tropica*, **53**, 167–184.

[48] Uhnoo, I. Cars, O., Christensson, D., and Nyström-Rosander, C. (1992). First documented case of human babesiosis in Sweden. *Scandinavian Journal of Infectious Diseases*, **24**, 541–547.

[49] Walderich, B., Weber, A., and Knobloch, J. (1997). Differentiation of *Entamoeba histolytica* and *Entamoeba dispar* from German travelers and residents of endemic areas. *American Journal of Tropical Medicine and Hygiene*, **57**, 70–74.

[50] Weber, R., Bryan, R. T., Schwartz, D. A., and Owen, R. L. (1994). Human microsporidial infections. *Clinical Microbiology Reviews*, **7**, 426–461.

[51] Yang, J. and Scholten, T. (1977). *Dientamoeba fragilis*: a review with notes on its epidemiology, pathogenicity, mode of transmission and, diagnosis. *American Journal of Tropical Medicine and Hygiene*, **26**, 16–22.

[52] Zierdt, C. H. (1991). *Blastocystis hominis* – past and future. *Clinical Microbiology Reviews*, **4**, 61–79.

Parasitic arthropods

Common name	Scientific name	Disease(s) transmitted[a] or other harm caused
Insects	Insecta	
Flies	Diptera	Myiasis[b]
Fleas	Siphonaptera	Plague, murine typhus
Bedbugs	Cimex	Anaemia
Lice	Anoplura	Pediculosis capitis, pediculosis corporis, epidemic typhus, relapsing fever, trench fever
Mites and ticks	Acari	
Chigger mites	Leptotrombidium	Scrub typhus (Tsutsugamushi disease)
Scabies mites	Sarcoptes scabiei	Scabies
Soft ticks	Argasidae	Tick-borne relapsing fever
Hard ticks	Ixodidae	Viral and rickettsial diseases, lyme disease, babesioses

Notes
a Only the most important diseases are included.
b Myiasis, the infestation by fly larvae in tissues or organs of living vertebrates.

References

See Chapter 23.

2 Parasites and parasitism

Mats Wahlgren and Hannah Akuffo

This chapter summarizes the concepts of parasitism and parasitology. It describes the main characteristics of the different groups of parasites, the protozoa, the helminths, and the arthropods. Their organization into different kingdoms, sub-kingdoms, and phyla, as well as their developmental relationships are discussed.

What is a parasite?

Any organism which lives, either temporarily or permanently, in, on, or, with another living organism from which it also obtains its nourishment is, broadly speaking, a parasite. The host is usually the larger and stronger of the two. This definition is a wide one and it could obviously include most micro-organisms except those that are free-living. Yet, only eukaryotic microbes such as protozoa, helminths, and the arthropods, the latter of the animal kingdom, are defined as parasites. Fungi, which occupy their own kingdom, bacteria, which are prokaryotes, and viruses, which lack a complete cellular machinery, are obviously not covered by the definition. Nevertheless, some parasitic protozoa, such as *Pneumocystis carinii*, have only recently been found to be fungi (Cushion *et al.* 1990), and some protozoa, for example, the microspora, have been suggested to be, but still remain to be proven, to be fungi. The regrouping of these into the kingdom of fungi depends on studies of the sequences of small-sub-unit ribosomal RNA. Researchers have today compiled a rather robust map of the evolutionary relationship of microbes using these sequences. Three primary domains of organisms have been identified, including Archea (prokaryotes with a eukaryote-like transcriptional system), Bacteria, and Eukarya (eukaryotes). It is interesting to find the highly parasitic protozoa *Giardia* and *Trichomonas* to be the most primitive of all eukaryotes (Figure 2.1).

Animal associations

There are no living organisms, besides maybe viruses, that do not carry their own parasitic organisms. Yet, parasitism is a question of specificity, and the unravelling of the complex interactions of species to host, species to vector, and sometimes vector to host, continues to captivate many biologists. Three types of parasitic associations can be contemplated. *True parasitism* is where the parasites live at the expense of their hosts and always have the potential to harm them, either directly, by depriving them of essential material or by destroying tissues, or indirectly, by damaging their hosts by the liberation of toxins. This is the case with pathogenic parasites. (Sometimes the host–parasite interaction does not initially induce a pathogenic state due to the capacity of the host immune system to hold it at bay.) In the event where this balance is tipped against the host it may lead to the parasite gaining

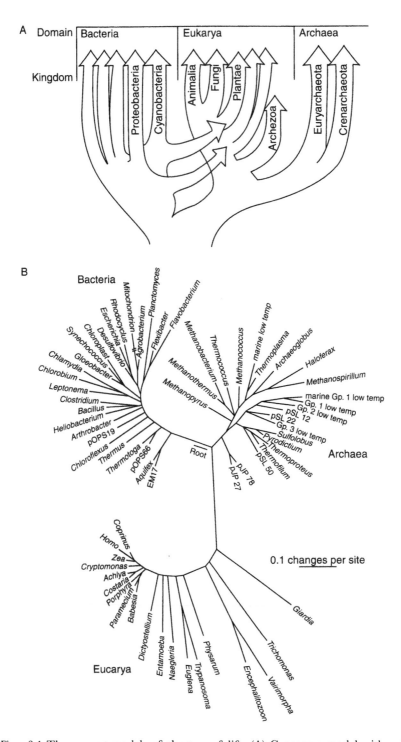

Figure 2.1 The current models of the tree of life. (A) Consensus model with only a few of the 'kingdoms' of the bacteria shown. Mitochondrial and chloroplast endosymbioses are indicated by lower and upper diagonal arrows, respectively (from Doolittle 1999). (B) Molecular tree of life based on ssRNA sequences and modified by trimming with other sequences (from Pace 1997). Reprinted with permission of the editors.

the upper hand.) *Commensalism* is when the eukaryote acquires nourishment from the host but does not hurt it and is not metabolically dependent on it. Good examples of this are the usually harmless intestinal amoebae, *Entamoeba dispar* or *Entamoeba coli*, which probably do not remove anything of value to man. *Symbiosis*, called by some as mutualism, is an interaction where both hosts benefit. A good example is the role of the prokaryotic bacterium *Wolbachia* which has been shown to infect filarial worms such as *Brugia malayi* which parasitizes man. The interaction of the bacterium with the filaria seems important for the development of the latter, as suggested by the lack of sexual development of the worm when the *Wolbachia* are eradicated. Thus, the interaction between the human host and the filarial worm is parasitic, while that between the bacterium and the worm is a form of symbiosis. This fascinating example is one of many interactions of parasites with their hosts.

Those parasites that harm the host are called pathogens, while those that do not are non-pathogenic (commensals). Those that do not depend on the host during the entire life-cycle but also live outside of it are facultative parasites (e.g. *Strongyloides* spp.), while obligate parasites are unable to survive without their hosts. Some parasites are ectoparasites as they live on the body surfaces. This definition is applicable no matter how brief the period of time, as occurs, for example, with arthropods such as ticks or mosquitoes. The latter form of parasitism is called infestation. Helminths or protozoa living in the body of the host cause infections. They are endoparasites. Hosts are also placed into different categories depending on where the parasite reaches sexual maturity. A definitive host is one in which sexual maturity and reproduction occur, while an intermediate host is one in which they do not; the mosquito is unexpectedly, therefore, the definitive host of the malaria parasite, while man is the intermediate host.

Medical parasitology includes the study of parasites, their vectors, definitive and intermediate hosts, and factors of ecological or epidemiological importance relevant to a disease. The delicate balance of these interactions is what determines whether a parasitic disease remains endemic or is eradicated. Thus, climatic and ecological changes that affect the vector may drastically alter the incidence of a given disease. In the same way, a break at any point in the complex chain may be enough to eradicate some parasitic diseases. It is worth noting that diseases such as malaria, which are often defined as being 'tropical' in nature, were prevalent in northern countries prior to the 1930s. However, changes, including those in the social structure, apparently broke the chain and thus the adequate transmission of *Plasmodium*. A narrow definition of parasitology, such as that used for medical parasitology, may be preferred by those studying parasitic infections, while population geneticists may favour, for example, a more comprehensive definition. Parasitism, in the broader sense, embraces every gradation of a host–parasite relationship, from interactions in which the partners are mutually and equally beneficial to cases in which the parasite is pathogenic or even lethal to its host.

Protozoa

Protozoa are the most basic form of animal life carrying nuclei; most have a Golgi, an endoplasmatic reticulum, and at least a rudimentary mitochondrion. Only 73 of the approximately 40 000 protozoa found in nature are of medical importance and parasitize man (Ashford and Crewe 1998). For further details of protozoa indigenous to the North, see Chapters 3–7.

The protozoa (protos = first; zoon = animal; greek) form their own sub-kingdom within the kingdom of single-cell eukaryotes, the Protista (Figure 2.2). They comprise the phyla Apicomplexa, Ciliata, Microspora, and Sarcomastigophora. *Apicomplexan* parasites include,

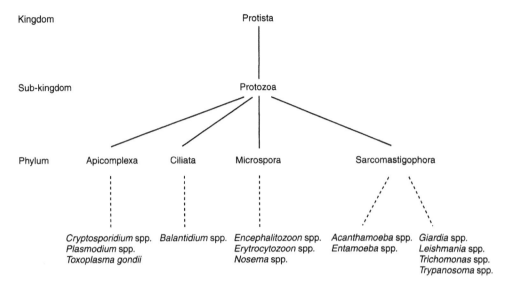

Figure 2.2 Classification of protozoa according to Levine *et al.* (1980).

for example, the malaria parasites (*Plasmodium falciparum, P. malariae, P. ovale, P. vivax*) and *Toxoplasma gondii.* The Apicomplexa derive their name from the presence of an apical complex composed of rhoptries, micronemes, and other structures used by the parasite during the invasion of cells. Most of them also have a cytostome (micropore) for the acquistion of nutrients and carry a rudimentary organelle called the apicoplast, a modified chloroplast, which in plants is involved in the generation of chlorophyll. It lacks genes of importance for photosynthesis but still carries those encoding proteolytic enzymes as well as molecules involved in fatty-acid synthesis. The phylum Ciliata does not include any parasites pathogenic to man but the rather uncommon ciliated *Balantidium coli* which causes diarrhoea (first identified in 1851 in Stockholm by Malmsten). The phylum Microspora includes intracellular spore-forming parasites which infect nearly all major animal groups. Microsporan parasites only (or mainly) cause gastro-intestinal disease in immunosuppressed individuals, such as those with AIDS. In contrast, a large number of pathogeneic parasites that affect the immunocompetent host are included in the phylum Sarcomastigophora. Protozoa of the order Kinetoplastida, for example, trypanosomes and leishmania parasites, typically carry a flagellum and a DNA-containing kinetoplast located near the flagellar basal body (kinetosome). Some of the kinetoplastids (trypanosomes) have additional organelles called glycosomes for storing carbohydrates. *Giardia lamblia*, which causes diarrhoea, belongs to the order diplomonida and are characterized by multiple flagella, the absence of a complete mitochondrion, and their capacity to form cysts. Protozoa of the sub-phylum Sarcodina are uninucleated and possess mitochondria; many have the capacity to form cysts, for example, *Entamoeba histolytica.* The order Thrichomonadida, which comprises trichomonas and dientamaoebas, frequently carry flagella and lack a mitochondrion but have hydrogenosomes which generate free H-ions (trichomonatids).

Helminths

Around 100 000 different worm species exist, many of which are free-living; some are parasitic, but fewer of them parasitize man. Helminthic infections are, nonetheless, common on a global scale with an estimated 4–5 billion individuals infected. A total of 319 different species infect man (Ashford and Crewe 1998), for example, pinworm and ascaris infections, both of which are endemic to the North and to the rest of the world. For further details of helminths indigenous to the North, see Chapters 3–7.

The word helminth is derived from Greek where 'helmins' means parasitic worm. Helminths belong to sub-kingdom Metazoa of the animal kingdom. They are divided into nematodes (roundworms) and platyhelminths (flatworms). The flatworms can be further divided into cestodes (tapeworms) and trematodes, the latter of which comprises flukes of different kinds. Most flatworms are hermaphrodites, except the group of schistosomes. A few human infections have been documented to be caused by helminths belonging to the phylum Acantocephala.

Helminths are complex multicellular organisms similar to *Homo sapiens* in many aspects as they posses, for example, neuronal tissues, muscle fibres, digestive canals, and sexual organs. They can be very small worms that can hardly be seen with the naked eye (e.g. schistosome or pinworm) or very large ones, like some of the cestodes. *Taenia saginata* or *Diphyllobtrium latum* can reach of length of several meters.

Arthropods

The phylum Arthropoda, which includes insects and their like, has over four times as many species of organisms as the rest of the animal kingdom. There are three particular characteristics which distinguish them from the lower eukaryotes: at some time in their life-cycle they all have a chitinized exoskeleton, a hemocele, and jointed appendages.

The arthropods play a major role in the transmission of disease-producing agents such as protozoa and helminths, and of course of bacteria, viruses, and rickettsiae. Although their roles as vectors are important, it must be remembered that arthropods may also parasitize man. The bites of ticks and mosquitoes and of a large number of other insects are unfortunately familiar to us. Forty-five different arthropods belonging to five classes are of global medical importance, including centipedes (Chilopoda), crabs, crayfish and copepods (Crustaceas), insects (Hexopoda), and tongueworms (Pentastomida). For further details of arthropods indigenous to the North, see Chapters 3–7.

References and selected references for further reading

Ashford, R. W. and Crewe, W. (1998). *The parasites of Homo sapiens*. Liverpool School of Tropical Medicine, Liverpool.

Boothroyd, J. C. and Komuniecki, R. (eds) (1995). *Molecular approaches to parasitology*. Wiley-Liss, New York.

Cushion, M. *et al.* (1990). Molecular biology of *Pneumocytis carinii*. *Ann. New York Acad. Sci.*, **616**, 415.

Doolittle, W. F. (1999). Phylogenetic classification and the universal tree. *Science*, **284**, 2124–2128.

Levine, N. D., Corliss, J. O., Cox, F. E., Deroux, G., Grain, J., Honigberg, B. M., Leedale, G. F., Loeblich, A. R. III, Lom, J., Lynn, D., Merinfeld, E. G., Page, F. G., Poljansky, G., Sprague, V., Vavra, J., and Wallace, F. G. (1980). A newly revised classification of the protozoa. *J. Protozool.*, **27**, 37–58.

Pace, N. R. (1997). A molecular view of microbial diversity and the biosphere. *Science*, **276**, 734–740.

Schmidt, G. and Roberts, L. S. (1989). *Foundations of parasitology*, 4th edn. Times Mirror/Mosby, College Publishing, St Louis, MO.

Trager, W. (1986). *Living together: the biology of animal parasitism*. Plenum Press, New York.

Part 2

General epidemiology in countries in the North

Past and present

3 Prevalence of human parasites and parasitic infections in Denmark, Finland, Norway, and Sweden

Staffan Svärd, Inger Ljungström, and Mats Wahlgren

Here, we summarize the presence and prevalence of different parasites and parasitic infections in Denmark, Finland, Norway, and Sweden. Helminthic infections such as pinworm and protozoal infections caused by *Giardia intestinalis* and *Toxoplasma gondii* are prevalent. Other parasites appear to infect mainly the immunosuppressed (e.g. Microsporidia and *Pneumocystis carinii*). Immigrants and travellers bring into the country parasites of the south but there are no known cases where transmission has been established. The most common infestations yet are those caused by arthropods such as gnats, mosquitoes, and ticks during the relatively warm summers but they are not included in this chapter.

Introduction

Denmark, Finland, Norway, and Sweden are in the northernmost part of Europe with a total population of about 20 million. The distribution of people is uneven, with the largest groups (≈80%) living in Denmark and in the southernmost parts of Finland, Norway, and Sweden and only few inhabiting the huge areas of the very north. The populations are ethnically relatively homogenous, but for groups of immigrants of different ethnic and religious backgrounds and a small number of Lapps in the north of Finland, Norway, and Sweden.

The southern part of the region is temperate with a continental climate. There are four distinct seasons: summer is between June and August–September and winter is between December and March (3–4 months below zero). The very north is, in contrast, characterized by a sub-arctic climate without tundra but with long, dark winters of 4–6 months below zero, and short but relatively warm summers with long days and light nights. Parasites which are present in these northern parts of Europe will be discussed here (excluding arthropods), focussing on those that are the most frequent or are the most important in the view of their morbidity and mortality. Iceland, which also belongs to the Nordic countries, has a very distinct climate and geographical location. A separate chapter has therefore been dedicated to the parasites of Iceland (see Chapter 4).

The overall endemicity is very similar in Denmark, Finland, Norway, and Sweden but the surveillance systems used for detecting different pathogens vary, a fact that complicates the comparisons of their prevalence and incidence between the countries. Each of the different countries has a national agency which follows the epidemiology of infectious diseases and provides diagnostic help and advice to national health authorities (The State Serum Institute in Denmark, National Public Health Institute in Finland, National Institute of Public Health in Norway, and Swedish Institute for Infectious Disease Control in Sweden).

Helminthic infections

Cestodes

Diphyllobothrium

There is, today, no reliable estimate of the prevalence of *Diphyllobothrium latum*, but it seems that prevalence has decreased during the last few decades. Human infections have been followed in eastern Finland from 1960 to 1978, and during that time the infection rate dropped from 18 to about 4.5% (Bonsdorff and Bylund 1982). During 1950–1970 the total infection rate in the Finnish population decreased from about 20 to 1.8% (Wikström 1972), and by the 1980s the prevalence of *D. latum* had decreased further to 1.4% (Kyrönseppä 1993).

In the beginning of the 1970s Ronéus and Dalborg (1975) investigated 65 lakes and 12 coastal areas of the Baltic Sea in Sweden. Infected fish were present in 75% of the lakes and in 58% of the coastal areas. In Lake Mälaren, 74% of the fish were infected, and in 1999 the figure has decreased to around 30% (Arvid Uggla, personal communication). The results from coastal areas show that most of the infected fish were found in the northern part of the Baltic Sea. Experience from Sweden has demonstrated that through adequate sanitary programs it is possible to eradicate the infection from an endemic centre within 10 years. In such cases the worm carriers were identified, dewormed, and provisional water-purification plants were replaced by modern and more efficient plants (Almer 1974).

Other cestode infections

Taenia saginata is found in cattle, but it is rarely transmitted to humans. There are no cases of *Taenia solium* reported since the beginning of the twentieth century.

Echinococcus

There is evidence that the cervid strain of *Echinococcus granulosus* occurs in Norway, Sweden, and Finland in a dog/reindeer cycle (Eckert and Thompson 1988). This strain can cause human infections, and autochthonous cases have been reported in reindeer breeders in northern Sweden (Lindholm and Lantto 1968; Huldt *et al.* 1973). Till date prevalence figures of echinococcosis in the Nordic countries are not available, but over the years studies of diagnosed human cystic echinococcosis in Sweden and Norway have been presented. However, the prevalence has fallen drastically and most cases are today imported.

Echinococcus granulosus cysts were found in the lungs of 0.3–2.1% of slaughtered reindeer in northern Sweden, according to reports from 1960 to 1972 (Ronéus 1974). In 1969, 1.2% of reindeer were reported to be infected with cysts in Finland and in areas bordering northern Sweden (Stössel 1989). In 1978, northern Norway introduced one annual praziquantel treatment of dogs and their exclusion from slaughter places, which resulted in a decrease in the prevalence of *E. granulosus* cysts in the lungs of reindeer from 1.45 to 0.17%. (Kummeneje 1982). Today very few infected reindeer are recognized in the north.

Human *E. multilocularis* infections have not yet been reported from this northern part of Europe. However, there are reports showing that the infection is spreading northwards from Central Europe (Eckert and Deplazes 1999). In the beginning of 2000, the first *E. multilocularis*-infected fox (and one mouse) was reported from the Nordic countries, the fox was killed in a traffic accident in Copenhagen, Denmark (Petersen and Kapel 2000).

Trematodes

Trichobilharzia *spp.*

Swimmers' itch or cercarial dermatitis is endemic in northern Europe (see Chapter 22). The disease is caused by the penetration of the skin by cercaria of non-human schistosomes (often *Trichobilharzia* spp. and *Bilharziella polonica*) with birds and rodents as primary hosts. The penetration of cercariae into the skin produces a short-lived dermatitis or even urticarial and vesicular eruptions in the case of heavy and repeated exposures. It is a problem mainly during warm summer months when the number of cases increase drastically due to bathing in shallow lakes containing high levels of cercariae and in some coastal areas where the salt level is low. The distribution is not well understood but it is often focal, and 50% of 248 Swedish districts reported problems with swimmers' itch during the summer of 1999 (Cecilia Thors, personal communication). Cases of swimmers' itch have also been reported from Norway (Thune 1994).

Nematodes

Ascaris

Ascaris infections are found in the northern part of Europe as well as in the rest of the world, but the incidence of indigenous infections is not well defined. There are indications that most of the individuals have contracted the infection outside the Nordic countries, but 35 cases of indigenous *Ascaris* were reported in Denmark in 1997 (Petersen 1998). An outbreak of ascariasis, probably due to imported vegetables, has been reported from Finland (Raisanen *et al.* 1985).

Asymptomatic carriers

Pinworm or *Enterobius vermicularis* is the most common human helminth infection. Indeed, the Guinness Book of Records claims that it is the second most-common human infection, after the common cold. Interestingly, the prevalence is higher in temperate than in tropical climates and it is more common in developed than in underdeveloped countries.

In Finland, 344 children were studied for the presence of pinworm and 5.2% of the children were found positive (Kyrönseppä 1993), while two studies in Swedish children showed higher prevalence figures of 27 (Kjellberg and Heyman 1993) and 21% (Herrström *et al.* 1997). The latter study showed that more children than previously known, are asymptotic carriers (36/41 positive children) and that finger sucking is strongly associated with a positive tape-test.

Toxocara

Toxocara canis has a cosmopolitan distribution. The parasite is found both in countries with severe winters and in those with dry summers, although it seems more common in wet tropical areas. In Sweden the first serologically confirmed case of toxocarosis was published in 1979 (Carlson *et al.* 1979), and in 1996 the first report of visceral larval migrans (VLM) in an adult Norwegian was published (Lund-Tønnesen 1996). A seroepidemiological survey of young healthy Swedish adults showed 7% prevalence, indicating that subclinical toxocarosis occurs in healthy Swedes. In the sera of patients suspected of having contracted toxocarosis, 25% were seropositive suggesting that clinically covert toxocarosis also exists in Sweden (Ljungström and van Knapen 1989). In Stockholm, 86 sandboxes were investigated for *Toxocara* eggs and 32% were found to be infected (Christensson 1983, 1988). A similar figure was also observed in Oslo, where of 13 sandpits sampled, 38.5% were found infected with *Toxocara* spp.

Toxocara canis infection has also been reported in animals, and 28% of wild red foxes collected from all over Sweden were found to be positive. Fourteen per cent of foxes above latitude 67°N were infected (Christensson 1983, 1988). In the metropolitan area of Copenhagen, stools from red foxes were collected and *Toxocara* eggs could be detected in 23.5% and *T. canis* worms were recovered from 81% of these foxes (Willingham *et al.* 1996). An autopsy of 230 stray cats gave further evidence of the high endemicity of these worms in the north as 79% of them were infected with *Toxocara cati* (Engbaek *et al.* 1984).

During the years 1971–1981 the National Veterinary Institute of Sweden analyzed the occurrence of *T. canis* in Swedish dogs by autopsy and routine stool examination ($n = 19\,044$). The prevalence was found to be 5 and 6.5%, respectively (Christensson 1983). In routine stool samples from dogs sent to the laboratory of The Norwegian College of Veterinary Medicine during 1981–1982, *Toxocara* was diagnosed in 4% of the samples. In hospitalized dogs, during the five-year period of 1978–1982, only 0.7% had *Toxocara* eggs based on stool sample examination (Tharaldsen 1983).

Trichinella

The epidemiology of trichinellosis is characterized by two main cycles: a synanthropic-domestic cycle (*Trichinella spiralis*) and a sylvatic cycle (all species and phenotypes). *Trichinella spiralis*, *T. britovi*, and *T. nativa* are found in the north.

Over the years, several outbreaks have occurred in Sweden (Table 3.1) with the last outbreak occurring in 1984. Since then only a few cases have been diagnosed, and these

Table 3.1 *Trichinella* outbreaks in Sweden during the twenty-first century

Locality	Year	No. of cases	Source of infection	References
Kiruna, Stockholm, and Karlskrona	1917	≥18	Bear meat	Berglund (1917)
Lidköping	1937	≥5	Pork	Lundmark (1937)
Lindesberg	1937	≈50	Pork	Hallén (1937, 1938)
Kristianstad	1941	7	Home slaughter	Bergwall (1941)
Kristianstad	1943	7	Home slaughter	Bergwall (1943)
Alingsås	1944	≥13	Pork	Norup and Roth (1944)
Borås	1946	35	Uninspected pork	Wird (1946)
Östergötland	1953	14	Unknown	Arvidsson (1954)
Blekinge	1961	338	Sausage	Ringertz *et al.* (1962)
Vadstena	1969	15	Unknown	Odelram (1973); Ljungström (1974)
Västergötland	1969	9	Uncooked sausage meat	Ljungström (1974)
Kristianstad	1984	4	Imported pork	Personal communication

patients have contracted the infection outside the Nordic countries. However, *Trichinella* infection occurs in wildlife such as foxes and wild boar.

There are few documented outbreaks in Norway. The first reported outbreak was in 1881, with four cases (Bull 1882), and the last ones in 1953–1954. These last outbreaks were in a farming family who all became infected at five different occasions (Hauge 1969). The largest outbreak occurred in Halden in 1940, when 681 German soldiers and at least eight civilian Norwegians became infected. The cause of infection was a cold smoked sausage 'Braunschweiger Wurst' or 'Mettwurst' containing 50% pork and 50% beef (Sannum 1980). The same year an outbreak occurred in Akerhus County with seven reported cases. In addition, other people from the same area showed symptoms suggestive of trichinella infection, but the diagnosis was never verified (Kobro and Owren 1941).

At the compulsory slaughterhouse control of pigs in Norway, trichinellosis has been found only occasionally. The last two cases were in 1994. The sylvatic trichinellosis (*T. nativa* and *T. britovi*) are found in wildlife and *T. spiralis* seems very rare (Tharaldsen 1997). The compulsory control of pigs in Sweden has shown that very few pigs are infected with trichinellosis, and the same holds true for Finland.

Both domestic and sylvatic trichinellosis have increased dramatically during the last decade in Finland. The raccoon dog population has increased over the years, and they might be an important reservoir of sylvatic trichinellosis in Finland (Oksanen *et al.* 1998). The raccoon dog has also been suspected as one of the links in transferring *Trichinella* between wildlife and farm animals. Fortunately, very few human cases have been reported, mainly in bear hunters (Nurmi 1983).

Denmark is regarded as *Trichinella* free with respect to domestic trichinellosis and sylvatic trichinellosis is very rare (Kapel 1997).

Protozoal infections

Toxoplasma

Toxoplasma gondii is a globally widespread parasite that is common in the North. The sero-prevalence in sheep is high both in Sweden and in Norway (10–45%; Lundén *et al.* 1994) while it is much lower in swines ($\approx 3\%$; Hirvela-Koski 1992; Nielsen and Wegener 1997). The prevalence is also high in many other groups of animals: 42% of cats are sero-positive, 31% of red foxes, 23% of dogs, 21% of wild rabbits, 12% of red deer, 1% of horses, and 3 out of 732 small rodents (0.4%) were found sero-positive for toxoplasmosis while reindeer were always sero-negative (Kapperud 1978; Uggla *et al.* 1990).

The first case reports occurred during the 1950s when $\approx 50\%$ of pregnant women were sero-positive (Forsgren *et al.* 1991). In one study it was reported that antibodies were more common in females than in males. Even small girls were more often infected than boys of the corresponding age. The difference became marked during puberty and was significant in adults (Huldt *et al.* 1979). The main risk factors of contracting toxoplasmosis during pregnancy according to a Norwegian study was found to be the consumption of undercooked minced meat products, unwashed raw vegetables or fruits, and raw or undercooked mutton or pork. Cleaning cat litter was also associated with the higher incidence of toxoplasmosis (Kapperud *et al.* 1996). The sero-prevalence in pregnant women is today 25–30% in the southernmost parts of Sweden (Evengård, personal communication) and in Denmark (Lebech *et al.* 1999). The sero-prevalence in pregnant Finns in the Helsinki area is about 20% (Lappalainen *et al.* 1992). It is lower in the northern parts of Sweden and Norway

(7.5 versus 6.7%) while an intermediate sero-prevalence level is found in Stockholm (14%; Evengård *et al.* 1999) and in Oslo (13.2%; Jenum *et al.* 1998).

Some recent incidence figures suggest that congenital spread of the parasite to the foetus occurs in 1 in 10 000 pregnancies in Sweden (Evengård, personal communication) while it is higher in other countries (0.9–5/1 000 sero-negative women; Jenum *et al.* 1998; Lebech *et al.* 1999; see also Chapter 14).

A remarkably high prevalence of anti-toxoplasma antibodies has been found among slaughterhouse and greenhouse workers in Denmark (Lings *et al.* 1994). Other risk groups are the immunosuppressed who have been transplanted or those who have AIDS. A few cases of toxoplasmosis have been described where patients acquire the infection from heart transplants (Andersson *et al.* 1992), and the sero-positivity of Swedish homosexual men was found to be higher than the rest of the population (37%; Bergqvist *et al.* 1984). A longitudinal comparison of the sero-prevalence over the last four decades in Stockholm, Sweden, has shown the prevalence to be on a steady decrease from 47.7% in 1957 to about 10–15% in 1997 (Forsgren *et al.* 1991; Evengård *et al.* 1999). Thus, the incidence of toxoplasmosis has decreased in the Nordic countries over the last few decades.

Malaria

Malaria does not occur in northern Europe today, but for around 500 cases imported each year; 463 cases were, for example, imported in 1998 whereof 181 were to Denmark, 55 to Finland, 74 to Norway, and 153 to Sweden.

Malaria is known to have occurred for several 100 years in the north; 'ague' and 'quartana' were already well known during the Middle Ages. At this time, malaria was caused by a *Plasmodium vivax* parasite, which had a prolonged incubation time, so that the individual was sometimes infected the year before the clinical symptoms appeared. Transmission was also secured by winter-hibernating female *Anopheles* mosquitoes infected during the previous year, as well as by relapses of malaria in individuals infected previously. Although the parasite present in the north has always been *P. vivax* the death rates were quite high. This provokes the thought that *P. vivax* previously caused malignant infection.

An early record of malaria is in the thesis of Carolus Linnaeus presented to the University of Haderwijk, Holland, in the year 1735 (Linnaeus 1735) where he describes the epidemiology of malaria and his thoughts on current treatment of the disease. He suggests that malaria is frequent around lakes and water holes and argues that malaria is caused by the clay in the water. It should also be remembered that the malarial parasite was used worldwide to treat cerebral syphilis until the 1940s when penicillin became generally available. For example, *P. vivax* was used in the treatment of syphilis until 1945 at dermatology and venerology clinics, such as St Görans Sjukhus in Stockholm.

Five large epidemics of malaria occurred in Europe, including the Nordic countries, during the nineteenth century (1812–1816, 1819–1821, 1830–1832, 1846–1848, and 1853–1862). The most extensive epidemic in Denmark occurred during 1830–1831 with its centre at Lolland and Falster (Andersen 1976; Bengtsson 1992). Åland was severely affected by malaria during the 1850s when half of the population was infected and 149 died. There were about 500 cases of malaria per year during the First World War in Finland that had spread from the Cossacks in southern Russia. The incidence decreased between the wars; but malaria was re-introduced in Finland during the Second World War with 1 252 patients who had contracted the disease on Carelian isthmus during 1944 (Figure 3.1).

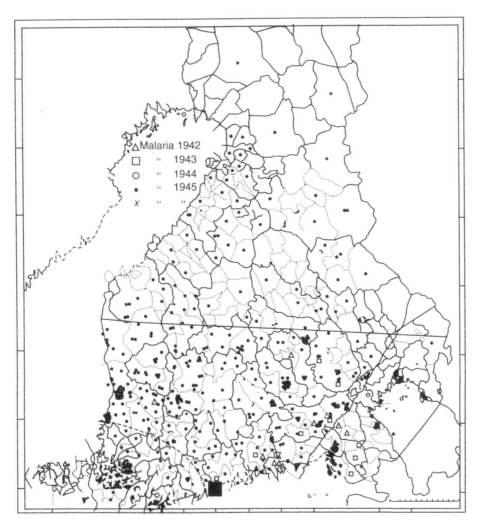

Figure 3.1 Malaria cases in Finland during 1945. Each point represents five cases of malaria (from Hernberg and Tuomela 1948).

A few cases of malaria remained in Finland into 1948 suggesting that the incubation time for the disease could be very long (Hernberg 1947; Hernberg and Tuomela 1948; Bengtsson 1992).

Some small epidemics are known to have hit Norway around Oslo and Fredriksstad during the early nineteenth century. A malaria epidemic also appeared later during the century with 143 cases recorded in Oslo (previously Christiania) during 1870. This was due to a receding water level at the seashore, which created novel marshlands where mosquitoes could breed (Grelland 1965; Bengtsson 1992).

Malaria was endemic in Sweden during the nineteenth century along the southeast coast and around Lake Mälaren, with an epidemic spread throughout southern Sweden (Figure 3.2). The last record from Sweden dates back to 1930 in Oxelösund.

Figure 3.2 The endemic and the epidemic spread of malaria in Sweden during the 1850s (from
    Bengtsson 1992).

Babesia divergens

Babesia divergens is endemic in cattle in Denmark and in the southern parts of Sweden, and
a few cases of accidental transmission to humans occur now and then (Wahlgren *et al.* 1984;
Uhnoo *et al.* 1992).

Pneumocystis carinii

Pneumocystis carinii was earlier considered a protozoan parasite but recent data suggest that it
is a primitive fungus (Miller and Wakefield 1996). However, since it is on the border line
between fungal and parasitic diseases it is included here. *Pneumocystis carinii* gives severe pneumonia (PCP) in immunosuppressed patients (see Chapter 12). It became known as one
of the major opportunistic diseases affecting HIV patients during the early 1980s but the
number of cases in HIV-infected patients has decreased since then due to prophylactic
PCP treatment and anti-virus drugs. However, pneumocystis is still detected in 50% of
the samples from HIV patients investigated in Sweden; 165 cases were reported in 1998
(Olsson *et al.* 1999). Similar numbers were obtained in Denmark. Among these cases are also

other immunocompromised patients, that is, malignancy patients, transplant recipients, and patients receiving immunosuppressive therapy.

Intestinal protozoa

Giardia

Giardia intestinalis is the most common cause of parasite-induced diarrhoea in Scandinavia. Lambl described *Giardia* from humans under the name *Cercomonas intestinalis* in 1859, and Müller (1889) described the parasite in Sweden the first time. His publication describes flagellates with the characteristic *Giardia* morphology attached, in several layers, to the mucosa of the small intestine of an executed murderer. The prevalence of *Giardia* was found to be 6% in Stockholm, Sweden and Helsinki, Finland during the 1930s (Svensson 1935). The most recent survey of the occurrence of intestinal pathogenic protozoa performed in Finland identified *G. intestinalis*, and the prevalence of *Giardia* was found to be 1% in healthy adults and in children (Kyrönseppä 1993). Nowadays, the incidence of *Giardia* infection per 100 000 habitants is 3.6 in Sweden. A study of entropathogens in 851 adult patients in Stockholm with diarrhoea identified *Giardia* in 2% of the cases, whereas none of the 200 healthy controls were positive for *Giardia* (Svenungsson *et al.* 2000). *Giardia* was found in 2% of samples in a Danish study of 109 children with gastrointestinal symptoms (Hjelt *et al.* 1987) and 1 500 cases were reported in Denmark in 1997 (Petersen 1998). The parasite is mainly found in travellers and immigrants: of a total number of 1 475 Swedes infected by *Giardia* during 1998, only 315 were infected in Sweden, and similar figures were obtained in Denmark (Petersen 1998). *Giardia* was found in 10% of faecal samples during routine screening of refugees into Sweden (Benzeguir *et al.* 1999). However, the percentage of infected in Sweden has increased during the 1990s, although symptomatic cases are predominantly investigated, which likely underestimates the prevalence rates in the general population. *Giardia* cysts were recently found in 26% of investigated Swedish raw water samples and similar percentages have been found in Norwegian water supplies (Hansen and Stenström 1998). One of the largest outbreaks of water-borne giardiasis reported in Europe occurred in Sweden in 1986 with more than 1 400 cases being diagnosed (Ljungström and Castor 1992), and several other water-borne outbreaks have been reported recently (Hansen and Stenström 1998). The occasional transmission of the parasite among children in day care centres and among institutionalized geriatric patients has also been reported (Svensson 1935; Christenson *et al.*, 1992). *Giardia* is a potential zoonotic disease and a Danish study revealed that 7.6% of faecal samples from asymptomatic dairy cows and calves contained *Giardia* cysts (Iburg *et al.* 1996). *Giardia* have also been detected in faeces of Swedish dogs (Castor and Lindquist 1990), roe deer, lynx, sheep, and moose (Dan Christensson, personal communication).

Entamoeba

Entamoeba histolytica was first described in 1875 by Lösch in Russia in the stool of a dysenteric patient. It was recently discovered, however, that a morphologically identical parasite *E. dispar* exists as a separate non-pathogenic species. This makes it possible that not all present and previously reported cases are/were due to *E. histolytica* as we still lack a simple and highly specific assay system to separate them.

Sievers reported 10 cases of amoebiasis in Finland in 1906; the prevalence was 9% in Helsinki in 1927 (Svensson 1935). The number of reported amoebiasis cases in Sweden has widely varied during the 1990s, mainly due to changes in the number of immigrants. In Sweden 3 830 cases of amoebiasis were reported in 1992. The majority of these were diagnosed in the faeces of individuals from the former Yugoslavia. In 1998, in Sweden, 445 cases were reported and only 27 of those were infected in Sweden. During a water-borne outbreak in Sweden 106 became infected by *E. histolytica* (Ljungström and Castor 1992). The domestic incidence of amoeba infection during 1999 in gastroenteritis cases was 0.3 per 100 000 inhabitants. In a recent survey, 1% of Swedish adults with enteritis had *Entamoeba* infection while none of the controls was infected (Svenungsson *et al.* 2000). As a result the number of people infected with *E. histolytica* is most likely overestimated. Very little data exists regarding the prevalences of *Entamoeba* in other Scandinavian countries, but *Entamoeba* was found in 0.3% of investigated travellers in Finland (Kyrönseppä 1993) and 300 imported cases were reported from Denmark in 1997 (Petersen 1998).

Other intestinal protozoa

There is a large uncertainty as to the prevalence of other intestinal protozoa since it is not obligatory that they are reported. Microsporidia spp. occur relatively frequently in AIDS patients but only 13 cases were, for example, reported to the Swedish Institute for Infectious Disease Control between 1994 and 1995. *Cryptosporidium parvum* is yet another parasite which, at the beginning of the 1980s, was believed to only affect immunosuppressed individuals, when serious outbreaks were reported. This is no longer the case. Spread from animals to humans has been reported and outbreaks have occurred at day care centres in Scandinavia (Hjelt *et al.* 1987). *Cryptosporidium* was identified in 3% of diarrhoeal cases in three Swedish studies (Atterholm *et al.* 1987; Svantesson *et al.* 1988; Svenungsson *et al.* 2000) and in 6% of faecal samples from neonatal calves (Viring *et al.* 1993). None of the 519 healthy individuals excreted cryptosporidia. Similar results have been obtained in Denmark (Holten-Andersen *et al.* 1984). *Balantidium coli*, the only ciliate known to be pathogenic to humans, was described for the first time by Malmsten in Stockholm in 1856, and Sievers reported 140 cases in Finland in 1906. However, transmission of *Balantidium* is not known to occur today. *Blastocystis hominis* and *Dientamoeba fragilis* are frequently detected in faecal samples from children at day care centres in Sweden but their pathogenicity is uncertain (Weiss and Keohane 1997; Marianne Lebbad, personal communication). *Entamoeba coli* was found in 15% of investigated faecal samples in Finland in 1935 (Svensson 1935) and odd cases of *E. coli*-associated diarrhoea has been reported to be associated with gastroenteritic symptoms (Wahlgren 1991).

References

Almer, B. (1974). Controlling the broad tapeworm, *Diphyllobothrium latum* (L.) (in Swedish). *Limnological Surveys from Swedish Environmental Agency SNV PM*, **460**, 1–30.

Andersen, O. (1976). A malaria epidemic in Denmark. *Ugeskrift for Laeger*, **138**, 3228–3230.

Andersson, R., Sandberg, T., Berglin, E., and Jaensson, S. (1992). Cytomegalovirus infections and toxoplasmosis in heart transplant recipients in Sweden. *Scand. J. Infect. Dis.*, **24**, 411–417.

Arvidsson, A. O. (1954). En trichinosepidemi. *Svenska Läkartidningen*, **51**, 1547.

Atterholm, I., Castor, B., and Norlin, K. (1987). Cryptosporidiosis in southern Sweden. *Scand. J. Infect. Dis.*, **19**, 231–234.

Bengtsson, E. (1992). Malaria och andra Parasitoser. In: *Epidemiernas historia och framtid* (B. Evengård, ed.). Natur och Kultur, Stockholm, pp. 199–217.

Benzeguir, A. K., Capraru, T., Aust-Kettis, A., and Björkman, A. (1999). High frequency of gastrointestinal parasites in refugees and asylum seekers upon arrival in Sweden. *Scand. J. Infect. Dis.*, **31**, 79–82.

Berglund, H. (1917). Meddelande angående den senaste trikinepidemin. *Svenska Läkarsällskapets förhandlinga*, **13**(2), 133–138.

Bergqvist, R., Morfeldt-Mansson, L., Pehrson, P. O., Petrini, B., and Wasserman, J. (1984). Antibody against *Encephalitozoon cuniculi* in Swedish homosexual men. *Scand. J. Infect. Dis.*, **16**, 389–392.

Bergwall, Å. (1941). Några trichinosfall. *Svenska Läkartidningen*, **33**, 1702–1708.

Bergwall, Å. (1943). Ytterligare trichinosfall. *Svenska Läkartidningen*, **40**, 72–79.

Bonsdorff, B. and Bylund, G. (1982). The ecology of *Diphyllobothrium latum*. *Ecol. dis.*, **1**, 21–26.

Bull, C. (1882). Trichinosis i Evje. *Norsk Magazin for Lægevidenskapen*, **12**, 229–249.

Carlson, M. G., Grabell, I., Lindahl, L., and Tonell, U. (1979). Toxocariosis – A neglected diagnosis. *Läkartidningen*, **76**, 2691–2692.

Castor, S. B. and Lindqvist, K. B. (1990). Canine giardiasis in Sweden: no evidence of infectivity to man. *Trans. R. Soc. Trop. Med. Hyg.*, **84**, 249–250.

Christensson, D. (1983). *Toxocara* and related ascarid roundworms in some Swedish animals – occurrence and prevention. In *Proceedings of the XI Symposium of the Scandinavian Society for Parasitology. Information*, Vol. 17. Åbo Akademi, Finland, pp. 14–15.

Christensson, D. (1988). High-latitude *Toxocara*. *Parasitol. Today*, **4**, 322.

Christenson, B., Jacobson, B., Ryd, G., and Bergström, I. T. (1992). Giardiasis among children in day care centers in 2 northern communities of Stockholm. *Läkartidningen*, **89**, 763–764.

Eckert, J. and Deplazes, P. (1999). Alveolar echinococcosis in humans: the current situation in Central Europe and the need for countermeasurements. *Parasitol. Today*, **15**, 315–319.

Eckert, J. and Thompson, R. C. A. (1988). *Echinococcus* strains in Europe: a review. *Trop. Med. Parasitol.*, **39**, 1–8.

Engbaek, K., Madsen, H. and Larsen, S.O. (1984). A survey of helminths in stray cats from Copenhagen with ecological aspects. *Zeitschrift für parasitenkunde*, **70**, 87–94.

Evengård, B., Lilja, G., Capraru, T., Malm, G., Kussofsky, E., Oman, H., and Forsgren, M. (1999). A retrospective study of seroconversion against *Toxoplasma gondii* during 3,000 pregnancies in Stockholm. *Scand. J. Infect. Dis.*, **31**, 127–129.

Forsgren, M., Gille, E., Ljungström, I., and Nokes, D. J. (1991). *Toxoplasma gondii* antibodies in pregnant women in Stockholm in 1969, 1979 and 1987. *Lancet*, **337**, 1413–1414.

Grelland, R. (1965). Malaria transmitted in Norway. *Tidskrift for den Norske Laegeforeningen*, **85**, 1147–1149.

Hallén, L. E. (1937). En trikinosepidemi I Lindesberg under slutet av maj och början av juni 1937. *Svenska Läkartidningen*, **28**, 1020–2026.

Hallén, L. E. (1938). Beobachtungen über eine Trichinoseepidemie in Lindesberg (Schweden). *Acta Med. Scand.*, **96**, 355–365.

Hansen, A. and Stenström, T.-A. (1998). *Giardia* and *Cryptosporidium* in Swedish surfacewaters. *SMI-tryck*, **118**, 1–44.

Hauge, S. (1969). Trikinosins epidemiologi i Norge. *Medlemsblad Norsk Veterinær-Förening*, **81**, 325–331.

Hernberg, C. A. (1947). Clinical observations on malaria tertiana in Finland, and on the difference between autumn and spring malaria. *Acta Med. Scand.*, **128**, 428–451.

Hernberg, C. A. and Tuomela, A. (1948). Incubation time of malaria tertiana during an epidemic in Finland in 1945. *Acta Med. Scand*, **130**(Suppl.), 534–539.

Herrström, P., Friström, A., Karlsson, A., and Högstedt, B. (1997). *Enterobius vermicularis* and finger sucking in young Swedish children. *Scand. J. Primary Health Care*, **15**, 146–148.

Hirvela-Koski, V. (1992). The prevalence of toxoplasma antibodies in swine sera in Finland. *Acta Vet. Scand.*, **33**, 21–25.

Hjelt, K., Paerregaard, A., Nielsen, O. H., Grauballe, P. C., Gaarslev, K., Holten-Andersen, W., Tvede, M., Orskov, F., and Krasilnikoff, P. A. (1987). Acute gastroenteritis in children attending day-care centers with special reference to rotavirus infections. I. Aetiology and epidemiologic aspects. *Acta Paed. Scand.*, **76**, 754–762.

Holten-Andersen, W., Gerstoft, J., Henriksen, S. A. and Pedersen, N. S. (1984). Prevalence of *Cryptosporidium* among patients with acute enteric infection. *J. Infect.*, **9**, 277–282.

Huldt, G., Johansson, G. O. and Lantto, S. (1973). *Echinococcus* in northern Scandinavia. Immune reactions to *Echinococcus granulosus* in Kautokeino Lapps. *Arch. Environ. Health*, **26**, 36–40.

Huldt, G., Lagercrantz, R., and Sheehe, P. R. (1979). On the epidemiology of human toxoplasmosis in Scandinavia especially in children. *Acta Paed. Scand.*, **68**, 745–749.

Iburg, T., Gasser, R. B., and Henriksen, S. A. (1996). First record of Giardia in cattle in Denmark. *Acta Vet. Scand.*, **37**, 337–341.

Jenum, P. A., Stray-Pedersen, B., Melby, K. K., Kapperud, G., Whitelaw, A., Eskild, A., and Eng, J. (1998). Incidence of *Toxoplasma gondii* infection in 35,940 pregnant women in Norway and pregnancy outcome for infected women. *J. Clin. Microbiol.*, **36**, 2900–2906.

Kapel, C. M. (1997). *Trichinella* in arctic, subarctic and temperate regions: Greenland, the Scandinavian countries and the Baltic states. *Southeast Asian J. Trop. Med. Public Health*, **28**(Suppl. 1), 14–19.

Kapperud, G. (1978). Survey for toxoplasmosis in wild and domestic animals from Norway and Sweden. *J. Wildlife Dis.*, **14**, 157–162.

Kapperud, G., Jenum, P. A., Stary-Pedersen, B., Melby, K. K., Eskild, A. and Eng, J. (1996). Risk factors for *Toxoplasma gondii* infection in pregnancy. Results of a prospective case-control study in Norway. *Am. J. Epidemiol.*, **144**, 405–412.

Kjellberg, G. and Heyman, B. (1993). Springmaskinfektion hos förskolebarn. *Läkartidningen*, **90**, 2993–2995.

Kobro, M. and Owren, P. (1941). Trikinosepidemien på Nedre Romerike (Akershus) *Tidsskrift for Norske Lægeforbundet*, **61**, 223–227.

Kummeneje, K. (1982). Control of echinococcosis (hydatidosis) in reindeer and dogs. Abattoir statistics from Kautokeino, West Finland, northern Norway. *Norsk Veterinaertidsskrift*, **94**, 419.

Kyrönseppä, H. (1993). The occurrence of human intestinal parasites in Finland. *Scand. J. Infect. Dis.*, **25**, 671–673.

Lappalainen, M., Koskela, P., and Hedman, K. (1992). Incidence of primary toxoplasma infections during pregnancy in southern Finland: a prospective cohort study. *Scand. J. Infect. Dis.*, **24**, 97–104.

Lebech, M., Andersen, O., Christiansen, N. C., Hertel, J., Nielsen , H. E., Peitersen, B., Rechnitzer, C., Larsen, S. O., Norgaard-Pedersen, B., and Petersen, E. (1999). Feasability of neonatal screening for toxoplasma infection in the absence of prenatal treatment. Danish Congenital Toxoplasmosis Study Group. *Lancet*, **353**, 1834–1837.

Lindholm, A. and Lantto, S. (1968). Hydatid disease in Northern Sweden. An orientation of hydatid disease in Northern Sweden and an evaluation of the Casoni and complement fixation tests. *Arch. Environ. Health*, **17**, 685–688.

Lings, S., Lander, F., and Lebech, M. (1994). Antimicrobial antibodies in Danish slaughterhouse workers and greenhouse workers. *Int. Arch. Occup. Environ. Health*, **65**, 405–409.

Linnaeus, C. (1735). Hipothesis nova de febrium intermittentium causa. Harderwijk Dissertatio Medica (in Swedish by Gustaf Drake) (1933). *En ny hypotes om orsaken till frossa*, Almqvist & Wiksells Bocktryckeri, Uppsala.

Ljungström, I. (1974). Antibody response to *Trichinella spiralis*. In: *Trichinellosis*. (C. W. Kim, ed.). pp. 449–460. Intext Educational, New York.

Ljungström, I. and Castor, B. (1992). Immune response to *Giardia lamblia* in a water-borne outbreak of giardiasis in Sweden. *J. Med. Microbiol.*, **36**, 347–352.

Ljungström, I. and van Knapen, F. (1989). An epidemiological and serological study of *Toxocara* infection in Sweden. *Scand. J. Infect. Dis.*, **21**, 87–93.

Lund-Tønnesen, S. (1996). Visceral larva migrans. An unusual cause of eosinophilia in adults. *Tidsskrift for Den Norske Lægeforening*, **116**, 2660–2661.

Lundén, A., Nasholm, A., and Uggla, A. (1994). Long-term study of *Toxoplasma gondii* infection in a Swedish sheep flock. *Acta Vet. Scand.*, **35**, 273–281.

Lundmark, F. (1937). Fem fall av trichinos. *Svenska Läkartidningen*, **34**, 867–876.

Miller, R. F. and Wakefield, A. E. (1996). Pneumocystis carinii: molecular taxonomy and epidemiology. *J. Med. Microbiol.*, **45**, 233–235.

Müller, E. (1889). Ett fynd av *Cercomonas intestinalis* i jejunum från människa. *Nordiskt Medicinskt Arkiv*, **XXI**, 1–21.

Nielsen, B. and Wegener, H. B. (1997). Public health and pork products: regional perspectives of Denmark. *Rev. Sci. Technol.*, **16**, 513–524.

Norup, E. B. and Roth, H. (1944). Något om trikinos, med anledning av en epidemi i Alingsås-trakten. *Svensk Veterinär-tidskrift*, **49**, 370–382.

Nurmi, E. (1983). Trichinosis in Finland. *Wiadomosci Parazytologiczne*, **29**, 617–619.

Odelram, H. (1973). A trichinosis epidemic. *Scand. J. Infect. Dis.*, **5**, 293–298.

Oksanen, A., Lindgren, T., and Tunkkari, P. (1998). Epidemiology of trichinellosis in lynx in Finland. *J. Helminthol.*, **72**, 47–53.

Olsson, M., Elvin, K., Linder, E., Löfdahl, S., and Wahlgren, M. (1999). *Pneumocystis carinii* fortfarande en svår opportunist. *Läkartidningen*, **96**, 328–331.

Petersen, E. (1998). Parasitaere intestinale infektioner. *Månedsskrift for praktisk laegegerning*, **76**, 1371–1375.

Petersen, E. and Kapel, C. M. (2000). *Echinococcus multicularis* – a new zoonotic infection in Denmark and a new differential diagnosis in hepatic tumors. *Ugeskrift for Laeger*, **162**, 1242.

Raisanen, S., Ruuskanen, L., and Nyman, S. (1985). Epidemic ascariasis—evidence of transmission by imported vegetables. *Scand. J. of Primary Health Care*, **3**, 189–191.

Ringerts, O. *et al.* (1962). Trichinosis in Sweden in 1961. *Acta Pathol. Microbiol. Scand.*, **54**, 351.

Ronéus, O. (1974). Prevalence of echinococcosis in reindeer (*Rangifer tarandus*) in Sweden. *Acta Vet. Scand.*, **15**, 170–178.

Ronéus, O. and Dalborg, G. (1975). Larvae of the human tapeworm in fishes from Swedish waters and the infection potential of some lakes. *Swedish Environmental Protection Agency SNV PM*, **616**, 1–51.

Sannum, A. (1980). Utbruddet av trikinose I Halden 1940. *Norsk Veterinærtidskrift*, **92**, 309–312.

Stössel, T. (1989). Literaturubersicht zur Haufigheit und geographischen Verbreitund der Echinokokkose in Landern der EG und EFTA. Medical dissertation, University of Zürich.

Svanteson, B., Thoren, A., Castor, B., Barkenius, G., Bergdahl, U., Tufvesson, B., Hansson, H. B., Möllby, R., and Juhlin, I. (1988). Acute diarrhoea in adults: aetiology, clinical appearance and therapeutic aspects. *Scand. J. Infect. Dis.*, **20**, 303–314.

Svensson, R. M. (1935). Studies of human intestinal protozoa. Medical thesis. *Acta Med. Scand.*, **LXX**, 1–115.

Svenungsson, B., Lagergren, å., Ekwall, E., Evengård, B., Hedlund, K.-O., Kärnell, A., Löfdahl, S., Svensson, L., and Weintraub, A. (2000). Enteropathogens in adult patients with diarrhea and healthy control subjects: a 1-year prospective study in a Swedish clinic for infectious diseases. *Clin. Infect. Dis.*, **30**, 770–778.

Tharaldsen, J. (1983). On the prevalence of *Toxocara* in Norway. In: *Proceedings of the XI Symposium of the Scandinavian Society for Parasitology. Information*, Vol. 17. Åbo Akademi, Finland, pp. 12–13.

Tharaldsen, J. (1997). *Trichinella* ssp in wild red fox (*Vulpes vulpes*) in Norway. *Proc. WAAVP*, **16**, 1–82.

Thune, P. (1994). Cercarial dermatitis or swimmer's itch-a little known but frequently occuring disease in Norway. *Tidsskrift for Norske Laegeforening*, **114**, 1694–1695.

Uggla, A., Mattson, S., and Juntti, N. (1990). Prevalence of antibodies to *Toxoplasma gondii* in cats, dogs and horses in Sweden. *Acta Vet. Scand.*, **31**, 219–222.

Uhnoo, I. Cars, O., Christensson, D., and Nyström-Rosander, C. (1992). First documented case of human babesiosis in Sweden. *Scand. J. Infect. Dis.*, **24**, 541–547.

Viring, S., Olsson, S. O., Alenius, S., Emanuelsson, U., Jacobsson, S. O., Larsson, B., Linde, N., and Uggla, A. (1993). Studies of enteric pathogens and gamma-globulin levels in neonatal calves in Sweden. *Acta Vet. Scand.*, **34**, 271–279.

Wahlgren, M. (1991). *Entamoeba coli* as cause of diarrhoea? *Lancet*, **337**, 675.

Wahlgren, M. Christensson, D., Bergquist, R., Björkman, A., Pehsson, P. O., and Rombo, L. (1984). Babesios, en risksjukdom för immundefekta i Sverige. OPMEAR, **29**, 26–28.

Weiss, L. M. and Keohane, E. M. (1997). The uncommon gastrointestinal protozoa: Microsporidia, Blastocystis, Isospora, Dientamoeba, and Balantidium. *Curr. Clin. Topics Infect. Dis.*, **17**, 147–187.

Wikström, M. (1972). The incidence of fish tapeworm, *Diphyllobothrium latum*, in the human population of Finland. *Commentationes Biologicae Societas Scientiarum Fennica*, **58**, 3–11.

Willingham, A. L., Ockens, N. W., Kapel, C. M., and Monrad, J. (1996). A helminthological survey of wild red foxes (*Vulpes vulpes*) from the metropolitan area of Copenhagen. *J. Helminthol.*, **70**, 259–263.

Wird, K. (1946). A trichinosis epidemic in the Borås district, its clinical and epidemiological aspects. *Acta Med. Scand.*, **126**, 1–16.

4 Prevalence of human parasites in Iceland

Past and present status

Karl Skírnisson, Sigurður H. Richter, and Matthías Eydal

The origin of human parasites in Iceland is discussed and historical information reviewed. Furthermore, the past and present status of endemic and imported human endo- and ectoparasites, recorded in Iceland so far, is reviewed. Several zoonotic parasites which have not been confirmed in humans and non-host specific species, which are capable of causing irritations on human skin, are also mentioned. Most endemic parasites are shortly reviewed. The reference list reflects research on human parasites in Iceland.

Origin of human parasites in Iceland

Iceland is an isolated island situated in the North Atlantic Ocean, south of the polar circle, between Norway and Greenland. In 1997, Iceland had approximately 275 000 inhabitants.

First to live in Iceland were Irish hermits. It is generally believed that they were living on the island during the eighth and ninth centuries but disappeared when Norse people started to arrive in Iceland in 874. In 930 Iceland is believed to have been fully settled with a rural population of at least 20 000 inhabitants. The majority of the settlers came from Norway but some came from Sweden, Ireland, and Scotland. Already, during the settlement period livestock (sheep, horses, cattle, pigs, and poultry), pets (dogs and cats), and accidentally probably also house mouse *Mus musculus* and field mouse *Apodemus sylvaticus* were imported from one or more of these countries. Certainly, both the settlers and the animals carried their parasites to Iceland.

Until the twentieth century most foreign connections (travel and trade) were with countries in northern Europe. Therefore, it is expected that the Icelandic parasite fauna has, through the centuries, resembled the endemic fauna of Scandinavia and the British Isles.

Historical information

In 825, the Irish clergyman Dicuil described the land of Thule (Iceland), where there was no daylight in winter, but on summer nights 'whatever task a man wishes to perform, even picking lice from his shirt, he can manage as well as in clear daylight' (Tierney 1967). Echinococcosis, fleas, and lice of humans are already mentioned in the Sagas, which were written in Iceland in the twelfth and thirteenth centuries (Magnússon 1913; Norðdal 1944; Benediktsson 1967). Furthermore, lice have been found in medieval and post-medieval excavations (Sveinbjarnardóttir and Buckland 1983; Amorosi *et al.* 1992; Buckland *et al.* 1992).

Until the middle of the eighteenth century, however, little was known about human parasites in Iceland. Ólafsson and Pálsson, who travelled in the country from 1752 to 1757 and systematically collected contemporary information on the natural history, reported the presence of some human parasites (Ólafsson 1981). Somewhat later, Mohr (1786) also reported some human parasites. Also, the General of Health in north Iceland, from 1775 to 1801, included some information on human parasites in his 'Therapy book for the general public' (Pétursson 1834). However, during most of the nineteenth century, scarce additional information was published on human parasites in Iceland.

Research on human parasites in Iceland

Since 1881, annual reports on public health, written by local medical doctors, have been compiled and published by the Directorate General of Health in Iceland. These reports not only contain valuable information on the occurrence of sarcoptidosis and echinococcosis, the only notable human parasitic diseases in the country, but also on some other endemic parasites.

During 1939–1962, a checklist of most parasitic groups occurring in Iceland was compiled by several foreign scientists and published, together with some taxonomic and zoogeographical information, in the series 'Zoology of Iceland'. Human parasites are mentioned in the issues on Cestoda (Baer 1962), Nematoda (Kreis 1958), Hemiptera (Fristrup 1945), Siphonaptera (Henriksen 1939), and Mallophaga and Anoplura (Overgaard 1942). However, the occurrence of protozoan parasites does not seem to have been reported until the second half of the twentieth century.

In recent decades human parasites have been identified and examined at different institutions in Iceland. Ectoparasites and parasites in blood and tissues, have mainly been identified by specialists working at Icelandic hospitals or health centres. During 1973 to May 1999, gastrointestinal parasites (protozoan oocysts, helminth eggs, or larvae) were routinely surveyed at the Institute for Experimental Pathology at Keldur, mainly by studying faecal samples using the formalin–ethylacetate concentration method (Richter *et al.* 1990a,b). In this period approximately 12 750 faecal samples from 8 650 individuals were examined (Table 4.1). On an average, parasites have been found in *c.*10% of these individuals. In recent years approximately three-fourths of the samples (72% in 1997) originated from Icelandic citizens but the remaining part was from guest workers and immigrants. Since June 1999, faecal samples have been examined at the National University Hospital in Reykjavík.

Endemic endoparasites

Trichomonas vaginalis has not yet been a subject of a systematic investigation in Iceland. However, it is sometimes seen in cervical samples screened at The Cancer Detection Center of The Icelandic Cancer Society. During 1986–1997 the incidence of *T. vaginalis* decreased more or less continuously from 1.2 to 0.25% in 22 369–29 129 samples annually screened (Margrét Snorradóttir, personal communication).

Cryptosporidium parvum was first searched for and detected in human faeces in 1986 (Eydal *et al.* 1990). Until 1997 its prevalence in 5 945 faecal samples examined at Keldur was 0.8%. Most of the infected individuals were hospitalized children. *Cryptosporidium parvum* is a common zoonotic species in Iceland. In a survey carried out on single faecal samples during 1990–1992, 50% of 40 foals examined were infected, so were 49.2% of 65 calves, 19.1% of 115 lambs, 11.1% of nine kittens and 5.5% of 55 pigs (Skírnisson *et al.* 1993b).

Toxoplasma gondii was already reported in humans in the 1950s (Feldman and Miller 1956; Björnsson 1958). Jónsdóttir and Árnadóttir (1988) reviewed five antibody prevalence studies carried out on 1 084 humans during 1956–1987. The overall prevalence was 10.1%.

Table 4.1 Endemic (end), probably endemic (end?), zoonotic (zoon), eradicated (erad), and imported (imp) human endoparasites found in Iceland. Occurrence score is based on the number of cases which have been reported by faecal examinations at the Institute for Experimental Pathology at Keldur. 0: not reported during 1973–1999, 1: one or few cases reported during 1971–1999, 2: annually reported during 1973–1999

	Endemic (score)	Zoonotic	Eradicated	Imported[a] (score)
Protozoa				
Entamoeba histolytica	(0)			imp (2)
Entamoeba hartmanni	(0)			imp (2)
Entamoeba coli	(0)			imp (2)
Entamoeba polecki	end[b] (0)	zoon		imp (1)
Endolimax nana	(0)			imp (2)
Iodamoeba buetschlii	end[b] (0)	zoon		imp (2)
Acanthamoeba spp.	end?			
Dientamoeba fragilis	end?			
Giardia lamblia	end? (2)			imp (2)
Chilomastix mesnili	(0)			imp (2)
Trichomonas vaginalis	end			
Balantidium coli	end[b] (0)	zoon		(0)
Blastocystis hominis	(0)			imp (1)
Cryptosporidium parvum	end (2)	zoon		imp (1)
Toxoplasma gondii	end	zoon		
Plasmodium spp.				imp
Leishmania spp.				imp
Pneumocystis carinii	end			
Encephalitozoon cuniculi	end[b]	zoon		
Trematoda				
Clonorchis sinensis and/or related species	(0)			imp (2)
Fasciolopsis buski and/or related species	(0)			imp (1)
Schistosoma spp.	(0)			imp
Cestoda				
Taenia saginata	(0)			imp (1)
Taenia solium	(0)			imp (1)
Hymenolepis nana	(0)			imp (1)
Dipylidium caninum	(0)		erad	(0)
Echinococcus granulosus			erad	
Nematoda				
Enterobius vermicularis	end (2)			(0)
Ascaris lumbricoides	(0)		erad	imp (2)
Ascaris suum	end (1)	zoon		
Trichuris trichiura	(0)			imp (2)
Trichostrongylus spp.	(0)			imp (1)
Necator americanus/				
Ancylostoma duodenale	(0)			imp (2)
Strongyloides stercoralis	(0)			imp (1)
Toxocara cati	end[b]	zoon		
Toxocara canis	end[b]	zoon		
Pseudoterranova decipiens	end[b]	zoon		
Anisakis simplex	end[b]	zoon		
Wuchereria bancrofti				imp (1)
Oncocerca volvulus				imp (1)

Notes
a Infections of tourists and immigrants acquired abroad.
b Not reported from Icelandic humans but a known zoonotic species.

Furthermore, a *T. gondii*-induced abortion as well as ocular and congenital toxoplasmosis have been reported (Björnsson 1958; Jónsdóttir and Árnadóttir 1988; Thorarensen *et al.* 1992). A recent survey showed that 30.2% of 149 domestic cats were seropositive (Finnsdóttir 1997). Also, *T. gondii* oocysts have been detected in cat faeces collected in playground sandboxes (Smáradóttir and Skírnisson 1996).

Pneumocystis carinii is detected in one or two immunocompromised patients in Iceland every year. Records from the National University Hospital indicate that a stable incidence of *P. carinii*-caused pneumonia over the past 15 years (Ingibjörg Hilmarsdóttir and Jónas Hallgrímsson, personal communications).

Enterobius vermicularis was first reported by Mohr (1786) but since 1881 its occurrence has been reported frequently (Health Reports 1881–1990; Matthíasson 1917, 1921; Sambon 1925; Hansen 1926; Kreis 1958). A recent survey on pinworm infections of 184 children in play schools in Iceland indicated that infections were rare in 2- and 3-year-old children but 13.2% of 4-year-old and 7.1% of 5-year-old children were infected (Jónsson and Skírnisson 1998; Skírnisson 1998). Another survey among 186 children, 6- to 8-years-old, in four primary schools in the capital area revealed an overall infection prevalence of 14.5%. No pinworm infections were detected in two of the 12 schoolrooms included in the survey, but the highest prevalence found in a schoolroom was 38.5% (Skírnisson and Stefánsdóttir, unpublished data).

Additional, possibly endemic, endoparasites

Three other protozoans are probably also endemic in humans in Iceland. However, low prevalence and limited research makes their endemic status uncertain.

In the 1970s and the 1980s *Giardia lamblia* infections were supposed to occur exclusively in people who had been staying abroad. In recent years, however, the parasite could possibly have become endemic, at least locally. This view is based on the fact that temporarily, restricted local epidemics have been confirmed and that *G. lamblia* has occasionally been found in humans who have not been abroad.

At least three ocular infections caused by *Acanthamoeba* have been diagnosed in Iceland since 1990 (Ólafur Grétar Guðmundsson, personal communication).

Dientamoeba fragilis has been detected in an Icelandic child who had never been abroad (Ingibjörg Hilmarsdóttir, personal communication).

Eradicated endoparasites

Three previously endemic helminth species have been eradicated in Iceland.

Echinococcus granulosus is definitely the most serious parasite to have ever affected humans in Iceland. Since 1863, extensive research has been carried out on the echinococcosis problem in the country (for reviews see, e.g. Krabbe 1865; Magnússon 1913; Dungal 1946, 1957; Pálsson *et al.* 1953, 1971; Jónsson 1962; Beard 1973; Ólafsson 1979; Pálsson 1984; Arinbjarnar 1989).

Symptoms that can be related to human echinococcosis are already described in the Icelandic literature written in the twelfth and thirteenth centuries (Magnússon 1913). In the nineteenth century, Iceland had the highest prevalence of human hydatid disease ever recorded anywhere in the world (Dungal 1946, 1957). In 1863 *E. granulosus* was found in 28% of 100 dogs, which were examined from various parts of Iceland (Krabbe 1865). A campaign against echinococcosis has been in effect since 1869. It has primarily been based on education and general information to the public on the nature of the disease and the life cycle of the parasite. Access of dogs and foxes to hydatids of sheep has been prevented by systematic removal of slaughterhouse offal and dogs that have been treated with anticestodals.

Reports from necropsies performed during 1932–1982 strongly indicate that the spread of human hydatid disease was practically brought to an end by the end of the nineteenth century (Ólafsson 1979; Pálsson 1984). Several surveys and reports from meat inspectors indicate that echinococcosis in sheep and cattle was apparently brought under control 30–40 years later (Pálsson *et al.* 1953, 1971; Pálsson 1984). Studies on dogs and the arctic fox *Alopex lagopus* in the latter half of the twentieth century have not revealed the parasite (Pálsson *et al.* 1953, 1971; Baer 1962; Skírnisson *et al.* 1993a). At present all evidence indicates that the parasite has been eradicated.

Dipylidium caninum usually occurs in dogs but occasionally human infections can also be found. In 1863 the parasite was found in 57% of 100 dogs examined (Krabbe 1865). The infection prevalence was reduced to 1% in 200 dogs examined during 1950–1960 (Pálsson *et al.* 1971). Since then it has never been detected and is regarded to be extinct. Although never reported in Icelandic humans, the high prevalence of the tapeworm in Icelandic dogs in the last century indicates that human infections might occasionally have also occurred.

Ascaris lumbricoides, first reported in the eighteenth century by Mohr (1786), is supposed to have been endemic in Iceland at least until early in the twentieth century. Health reports from 1890 to 1910 report tens of human ascariosis cases, mainly in children. The infected persons were living in different parts of the country. The endemic status was also confirmed by Sambon (1925), Matthíasson (1928), and Kreis (1958). However, since 1973, as routine search for human endoparasites started at Keldur, no endemic cases have been confirmed.

Zoonotic endoparasites

Several zoonotic protozoans and helminths, which have not (yet) been reported from humans (except probably for *Ascaris suum*, see subsequently), are known to be endemic in Iceland.

Recently *Entamoeba polecki*, *Iodamoeba buetschlii*, and *Balantidium coli* were reported in domestic pigs (Eydal and Konráðsson 1997). However, indigenous human infections have never been reported. The fourth protozoan, *Encephalitozoon cuniculi*, occasionally reported from humans abroad, has been confirmed in some wild mammals and in farmed foxes (Hersteinsson *et al.* 1993).

Already in the nineteenth century *Toxocara cati* and *Toxocara canis* were reported from cats and dogs (Krabbe 1865). Recent surveys on endoparasites of cats (Ágústsson and Richter 1993) and dogs (Richter and Elmarsdóttir 1997) showed that both species are still common. Also, *T. canis* has been reported from the native arctic fox (Skírnisson *et al.* 1993a). However, human toxocariosis has not been reported in Iceland and none of 307 healthy blood donors, examined in 1981 for *Toxocara* antibodies, appeared to be seropositive (Woodruff and Savigny 1982). Infectious eggs of both species have been detected in playground sandboxes in Iceland (Smáradóttir and Skírnisson 1996).

Pseudoterranova decipiens, *Anisakis simplex*, and other anisakine larvae are common in marine fish around Iceland. Potentially, they are capable of infecting humans but as traditional preparation of fish in Iceland does not include the consumption of raw fish the risk of getting infected is regarded to be minimal.

Ascaris suum is an endemic parasite of domestic pigs in Iceland (Richter 1991; Eydal and Konráðsson 1997). As already mentioned the morphologically similar *A. lumbricoides* is not regarded to be endemic anymore. A case is known where an *Ascaris* infection was acquired on a pig farm. In general, however, human *A. suum* infections seem to be exceptional.

Endemic ectoparasites

Three host-specific human ectoparasites are endemic in Iceland (Table 4.2).

Sarcoptes scabiei, first reported by Pétursson (1834), has probably been endemic since Iceland was settled. Since 1881 sarcoptidosis has been a notable disease. The number of annually registered cases has markedly changed. As an example, only eight cases were reported in 1961 but most cases in this period, 1 569, were recorded in 1941. In the 1980s, on an average, 426 cases were annually registered (Health Reports 1881–1990).

Phthirus pubis has been endemic in Iceland for centuries, but its incidence has probably always remained low. First confirmed reports on its occurrence are from post-medieval archaeological excavations from Reykholt (Buckland *et al.* 1992). By the middle of the eighteenth century Ólafsson (1981) reported a species with the name *Pediculus ferus* and mentioned that infestations were sometimes observed on foreigners. Pétursson (1834) discussed methods to get rid of *P. pubis*, which he obviously regarded as a problem in Iceland in the late eighteenth century. Health Reports from the former half of the twentieth century occasionally report phthiriasis but Overgaard (1942) and Gígja (1944) state that *P. pubis* hardly occurs outside harbor areas in Iceland. According to Health Reports the number of phthiriasis cases increased during the Second World War. At present, dermatologists regularly confirm the presence of *P. pubis* (Jón Hjaltalín Ólafsson, personal communication).

Table 4.2 List of endemic (end), possibly endemic (end?), eradicated (erad), and imported (imp) human ectoparasites known in Iceland. Species capable of causing skin irritation are also included. Further information and references are given in the text

Species	Endemic	Eradicated	Imported	Comments
Hirudinea	end			Three freshwater leeches occur in Iceland
Avian schistosome cercariae	end			'Swimmers itch' has been reported since 1997
Ixodes uriea	end			Very common on wild birds
Ixodes ricinus	end?		imp	Regularly imported on migratory birds
Ornithonyssus bacoti			imp	Few rat and gerbil infestations reported
Sarcoptes scabiei	end			Occurs rarely on humans
Cheyletiella spp.	end			Occurs rarely on cats and dogs
Phthirus pubis	end			Occurs rarely on humans
Pediculus humanus capitis	end			Occurs rarely on humans
Pediculus humanus humanus		erad		Eradicated decades ago
Cimex lectularius		erad	imp	Occasionally imported after eradication
Simulium vittatum	end			Common in running water ecosystems
Pulex irritans		erad	imp	Eradicated decades ago, occasionally imported
Nosopsyllus fasciatus	end			Common on rodents
Ctenophthalmus agyrtes	end			Common on rodents
Ceratophyllus gallinae	end			Common on wild bird species
Ceratophyllus garei	end			Common on wild bird species

Pediculus spp. have been detected in medieval and post-medieval archaeological excavations from Iceland (Sveinbjarnardóttir and Buckland 1983; Amorosi *et al.* 1992; Buckland *et al.* 1992). Written sources report *Pediculus* spp. by the middle of the eighteenth century (Ólafsson 1981). The body louse *P. h. humanus* Linnaeus had already been eradicated but the head louse, *P. h. capitis* de Geer still survives. Both subspecies are regarded to have been very common in Iceland until the twentieth century.

Pediculus humanus capitis was very common until approximately the middle of the twentieth century (Health Reports 1881–1990). In the 1920s, health examination of school children revealed *P. h. capitis* prevalence as high as 82% in some schools. Quite often prevalence values ranged between 20 and 50% (Sigurðsson 1928). In the 1930s and 1940s its occurrence dropped markedly. Obviously, it disappeared somewhat earlier in urban areas than in rural districts. During 1980–1982, on an average, 184 *P. h. capitis* cases were reported annually (Magnússon 1983). At present *P. h. capitis* is occasionally observed, mainly on school children.

Eradicated ectoparasites

Three ectoparasites have already been eradicated in Iceland (Table 4.2).

Pediculus humanus humanus (syn. *P. h. corporis*) was frequently mentioned in Health Reports until the middle of the twentieth century. Quite often the typical cutaneous lesions, which follow a *P. h. humanus* infestation, were described. Medical examination of school children carried out in the health district of Flatey in 1926 revealed that 50% of the children were infested with *P. h. humanus* and 54% with *P. h. capitis* (Health Report 1926). Overgaard (1942) also mentions that the two subspecies occurred at a similar prevalence.

Health Reports (1930, 1934) state that bedbugs, *Cimex lectularius*, were imported to Iceland by Norwegian whalers, who established a whaling station in Dýrafjörður, NW Iceland in 1890. In 1898, bedbugs were already found on adjacent farms. In the following decades the parasite not only colonized most farms around the fjord but had also spread to adjacent districts and to other parts of the country. *Cimex lectularius* was reported in Reykjavík in 1923 and soon became a real pest, which had to be eradicated from hundreds of flats in the forthcoming decades (Health Reports 1930–1947). Gígja (1944) and Fristrup (1945) state that *C. lectularius* was the worst pest that had ever occurred in human dwellings in Iceland. The parasite survived locally until the 1970s. Since then *C. lectularius* has only been reported on rare occasions and is always regarded to have been imported from abroad.

Pulex irritans, first mentioned by Mohr (1786), was probably very common in Iceland; so common and widespread that its occurrence does not seem to be mentioned. Sometimes, however, *P. irritans* might have been confused with endemic bird and rodent fleas (Table 4.2) (Henriksen 1939; Skírnisson and Richter 1992; Skírnisson 1995). In the nineteenth century fleas were common on most farms in Iceland (Jónasson 1934). Matthíasson (1920) advises on how to get rid of fleas from turf houses where they could easily complete their life cycles, for example, in beds and hay mattresses (Gígja 1944). Still, in the 1930s fleas were abundant in houses. Thus, Health Report (1934) informs about such a flea epidemic on most farms in the district of Dalir that the local physician regarded it as impossible for humans to stay on infested farms if the individuals had previously not become used to bites of human fleas. *P. irritans* seems to have disappeared from Iceland during the middle of the twentieth century. Surprisingly *P. irritans* has not been found in medieval and post-medieval archaeological excavations in Iceland, whereas in Greenland *P. irritans* has been found in Viking-age deposits from several farmsteads that date from the initiation of the Viking settlement in 986 to its final demise around 1350 (Sadler 1990). The settlers in Greenland mainly originated from Iceland.

Non-host-specific ectoparasites

Some non-host-specific, endemic parasites, which can cause irritation on human skin, are listed in Table 4.2.

Three freshwater Hirudinea (*Theromyzon tessulatum*, *T. garjaewi*, and *T. maculosum*) are known to attack humans but normally they parasitize on waterfowl (Bruun 1938; Hallgrímsson 1979).

Recently cercariae of an unknown avian schistosome species caused 'swimmers itch' on humans wading in a pond in the Family Park of Reykjavík (Kolarova *et al.* 1999).

Acarina ectoparasites, sometimes attacking humans, include the blood-sucking ticks *Ixodes uriea* which is common on several sea-bird species and *Ixodes ricinus* which is probably frequently imported on migratory birds but might be endemic in some localities (Richter 1981). In the 1970s the tropical rat mite *Ornithonyssus bacoti* was reported to annoy humans living close to nests of the common rat *Rattus norvegicus* (Richter 1977) and in 2001 the species was again reported to attack owners of Mongolian gerbils (*Meriones unguiculatus*), which had been kept as pets (Skírnisson 2001). Further species are *Cheyletiella parasitovorax*, which has caused skin irritation on Persian cat owners (Skírnisson *et al.* 1997) and *C. yasguri* (Karl Skírnisson, unpublished data).

Among the insects, bird fleas, mainly *Ceratophyllus gallinae*, and to a lesser extent *Ceratophyllus garei*, and the rodent fleas *Nosopsyllus fasciatus* and *Ctenoptalmus agyrtes* sometimes attack humans who live in close proximity to infested birds and rodents (Richter 1977; Bengtson *et al.* 1986; Skírnisson 1995). The black fly *Simulium vittatum* frequently sucks blood from humans and sometimes the louse-fly *Ornithomya chloropus* also bites humans. Both species are common in Iceland (Ólafsson 1991). None of the insects are known to be vectors of transmittable human diseases in Iceland.

Non-endemic human parasites confirmed in Iceland

During the last decades travels to and from Iceland have rapidly increased. Icelandic citizens commonly travel abroad and foreigners visit or settle in the country. Some of these persons have carried parasites to the country and therefore, at present most of the common human endoparasites have been observed in Iceland (Table 4.1).

Final remarks

As already mentioned it is expected that the Icelandic human parasite fauna, which probably remained stable until the twentieth century, has, through the centuries, resembled the endemic fauna of Scandinavia and the British Isles. The absence of certain species might be a result of the isolation of the country or due to the lack of necessary intermediate hosts. During the twentieth century, however, the metazoan parasite fauna in Iceland has markedly changed. Once very common parasites like, for example, *E. granulosus* and *P. irritans* have been eradicated, most other species occur so rarely that they are generally not regarded as a problem anymore. As regards the protozoan parasites, no information is available on their presence in Iceland until recent decades. However, most of the protozoans may have been endemic in Iceland for a long time.

High standard of hygiene and medical services, use of effective drugs and insecticides, good general education, and construction of modern houses with closed effluent system are among the factors that have made it difficult for most human parasites to establish themselves or sustain in Iceland.

Acknowledgements

Ingibjörg Hilmarsdóttir, Department of Microbiology, and Jónas Hallgrímsson, Department of Pathology, National University Hospital provided unpublished data on the occurrence of *D. fragilis, Plasmodium* spp., *Leishmania* spp., *P. carinii*, and *Schistosoma* spp. Margrét Snorradóttir, Icelandic Cancer Society, provided data on *T. vaginalis*, Jón Hjaltalín Ólafsson, Dermatology Center in Kópavogur, informed about the present status of *P. pubis*, and Ólafur Grétar Guðmundsson, Landakot Medical Clinic, provided unpublished data on *Acanthamoeba* spp.

References

Amorosi, T., Buckland, P. C., Ólafsson, G., Sadler, J. P., and Skidmore, P. (1992). Site status and the palaeoecological record: a discussion of the results from Bessastaðir, Iceland. In: *Norse and later settlement and subsistence in the north Atlantic* (C. D. Morris and D. J. Rackham, eds). Department of Archaeology, Glasgow, pp. 169–191.

Arinbjarnar, G. (1989). Four cases of Hydatid disease treated at Akureyri Regional Hospital during 1984–1988. *Icelandic Med. J.*, **75**, 399–403 (in Icelandic with English summary).

Ágústsson, Þ., and Richter, S. H. (1993). Endo- and ectoparasites of domestic cats in SV-Iceland. *Icelandic Vet. J.*, **8**, 24–29. (in Icelandic with English summary).

Baer, J. G. (1962). Cestoda. In: *The zoology of Iceland*, Vol. II, Part 12. Ejnar Munksgaard, Copenhagen.

Beard, T. C. (1973). The elimination of echinococcosis from Iceland. *Bull. World Health Org.*, **48**, 653–660.

Benediktsson, J. (ed.) (1967). Íslensk Fornrit. In: *Íslendingabók-Landnámabók* (Book of Icelanders, Book of Settlement). Hið Íslenska Fornritafélag, Reykjavík (in Icelandic).

Bengtson, S.-A., Brinck-Lindroth, G., Lundquist, L., Nilsson, A., and Rundgren, S. (1986). Ectoparasites on small mammals in Iceland: origin and population characteristics of a species-poor insular community. *Holarctic Ecol.*, **9**, 143–148.

Björnsson, G. (1958). Toxoplasmosis. *Icelandic Med. J.*, **42**, 118–128 (in Icelandic with English summary).

Bruun, A. (1938). Freshwater Hirudinea. In: *The zoology of Iceland*, Vol. II, Part 22. Ejnar Munksgaard, Copenhagen.

Buckland, P. C., Sadler, J. P., and Sveinbjarnardóttir, G. (1992). Palaeoecological investigations at Reykholt, Western Iceland. In: *Norse and later settlement and subsistence in the north Atlantic* (C. D. Morris and D. J. Rackham, eds). Department of Archaeology, Glasgow, pp. 149–167.

Dungal, N. (1946). Echinococcosis in Iceland. *Am. J. Med. Sci.*, **212**, 12–17.

Dungal, N. (1957). Eradication of hydatid disease in Iceland. *NZ Med. J.*, **56**, 212–222.

Eydal, M., Richter, S. H., and Skírnisson, K. (1990). Infections caused by the coccidian *Cryptosporidium* in humans in Iceland. *Icelandic Med. J.*, **76**, 264–266 (in Icelandic with English summary).

Eydal, M. and Konráðsson, K. (1997). The prevalence of *Balantidium coli* and other zoonotic parasites in Icelandic pigs. *Bull. Scand. Soc. Parasitol.*, **7**, 66.

Feldman, H. A. and Miller, L. T. (1956). Serological study of toxoplasmosis prevalence. *Am. J. Hyg.*, **64**, 320–335.

Finnsdóttir, H. (1997). Tíðni mótefna gegn bogfrymlasótt í blóði katta á Íslandi (Prevalence of antibodies against *Toxoplasma* in sera of Icelandic cats). *Icelandic Vet. J.*, **1997**, 16–21 (in Icelandic).

Fristrup, B. (1945). Hemiptera. 1. Heteroptera and Homoptera Auchenorhyncha. In: *The zoology of Iceland*, Vol. III, Part 51. Ejnar Munksgaard, Copenhagen.

Gígja, G. (1944). *Meindýr í húsum og gróðri og varnir gegn Þeim* (Pests in houses and on plants and their control). Jens Guðbjörnsson, Reykjavík (in Icelandic).

Hallgrímsson, H. (1979). *Veröldin í vatninu* (The World in the Water). Bókagerðin Askur, Reykjavík (in Icelandic).

Hansen, H. (1926). Njálgur (oxyuriasis) (Pinworm). *Icelandic Med. J.*, **12**, 182–184 (in Icelandic).

Health Reports (1881–1990). Annual reports compiled and published by the Directorate General of Health, Iceland. Gutenberg, Reykjavík (in Icelandic with English summary).

Henriksen, K. L. (1939). Siphonaptera. In: *The zoology of Iceland*, Vol. III, Part 47. Ejnar Munksgaard, Copenhagen.

Hersteinsson, P., Gunnarsson, E., Hjartardóttir, S., and Skírnisson, K. (1993). Prevalence of *Encephalitozoon cuniculi* antibodies in terrestrial mammals in Iceland 1986 to 1989. *J. Wildlife Dis.*, **29**, 341–344.

Jónasson, J. (1934). *Íslenskir þjóðhættir* (Icelandic folk customs). Ísafoldarprentsmiðja, Reykjavík (in Icelandic).

Jónsdóttir, K. E. and Árnadóttir, Þ. (1988). Mælingar á mótefnum gegn bogfrymlum í nokkrum hópum Íslendinga (Test for antibodies against *Toxoplasma* in some groups of Icelanders). *Icelandic Med. J.*, **74**, 279–284 (in Icelandic).

Jónsson, B. (1962). The last *Echinococcus*? *Icelandic Med. J.*, **46**, 1–13 (in Icelandic with English summary).

Jónsson, B. and Skírnisson, K. (1998). Pinworm infections in children in playschools in Iceland. *Icelandic Med. J.*, **84**, 215–218 (in Icelandic with English summary).

Kolarova, L., Skirnisson, K. and Horák, P. (1999). Schistosome cercariae as the causative agent of swimmer's itch in Iceland. *J. Helminthol.*, **73**, 215–220.

Krabbe, H. (1865). Helmintologiske Undersøgelser i Danmark og paa Island med særligt Hensyn til Blæreormlidelserne paa Island. *Det Kongel. Danske Videnskabernes Selskabs Skrifter*, **5**, 1–71.

Kreis, A. (1958). Parasitic nematoda. In: *The zoology of Iceland*, Vol. II, Part 15b. Ejnar Munksgaard, Copenhagen.

Magnússon, G. (1913). *Yfirlit yfir sögu sullaveikinnar á Íslandi* (Review of the echinococcosis history in Iceland). Prentsmiðjan Gutenberg, Reykjavík (in Icelandic).

Magnússon, G. (1983). Höfuðlús – Höfuðverkur á hverju hausti (Head-lice. A headache every autumn). *Heilbrigðismál*, **31**, 8–9 (in Icelandic).

Matthíasson, S. (1917). Appendicitis og oxyuriasis. *Icelandic Med. J.*, **3**, 97–101.

Matthíasson, S. (1920). *Heilsufræði* (Hygiene), 2nd edn. Gutenberg, Reykjavík (in Icelandic).

Matthíasson, S. (1921). Oxyures og önnur corpora aliena í appendix (Oxyures and other corpora aliena in the appendix). *Icelandic Med. J.*, **7**, 145–147 (in Icelandic).

Matthíasson, S. (1928). Ritsjá og eigin reynsla (Literature review with reports from the author's praxis). *Icelandic Med. J.*, **14**, 112–114 (in Icelandic).

Mohr, N. (1786). *Forsøg til Islandsk Naturhistorie*. Christian Friederich Holm, København.

Norðdal. S. (ed.) (1944). *Flateyjarbók*. Flateyjarútgáfan hf, Prentverk Akraness (in Icelandic).

Ólafsson, E. (1981). *Ferðabók Eggerts Ólafssonar og Bjarna Pálssonar um ferðir peirra á Íslandi árin 1752–1757* (Travel in Iceland 1752–1757). 2. Bindi. Örn og Örlygur, Reykjavík (in Icelandic).

Ólafsson, E. (1991). A checklist of Icelandic insects. *Fjölrit Náttúrufræðistofnunar*, Vol. **17**. Náttúrufræðistofnun Íslands, Reykjavík (in Icelandic with English summary).

Ólafsson, G. (1979). Has echinococcosis been eradicated from Iceland? *Icelandic Med. J.*, **65**, 139–142 (in Icelandic with English summary).

Overgaard, C. (1942). Mallophaga and Anoplura. In: *The zoology of Iceland*, Vol. III, Part 42. Ejnar Munksgaard, Copenhagen.

Pálsson, P. A. (1984). Echinococcosis and its elimination in Iceland. *Hist. Med. Vet.*, **1**, 4–10.

Pálsson, P. A., Sigurðsson, B., and Hendriksen, K. (1953). Sullaveikin á undanhaldi (Echinococcosis diminishing). *Icelandic Med. J.*, **37**, 1–13 (in Icelandic with English summary).

Pálsson, P. A., Vigfússon, H., and Henriksen, K. (1971). Heldur sullaveikin velli? (Will echinococcosis persist?). *Icelandic Med. J.*, **57**, 39–51 (in Icelandic with English summary).

Pétursson, J. (1834). *Lækningabók fyrir almúga* (Therapy Book for the General Public). Möller, Kaupmannahöfn (in Icelandic).

Richter, S. H. (1977). Humans bitten by a bird flea, a rat flea and a rat mite. *Icelandic Med. J.*, **63**, 107–110 (in Icelandic with English summary).

Richter, S. H. (1981). *Ixodes ricinus* in Iceland. *Icelandic Vet. J.*, **2**, 14–17 (in Icelandic with English summary).

Richter, S. H. (1991). Prævalensundersægelse i Island. Parasitære infektioner hos svin. *NKJ-Projekt*, **59**, 63–65.

Richter, S. H. and Elmarsdóttir, Á. (1997). Intestinal parasites in dogs in Iceland: the past and present. *Icelandic Agric. Sci.*, **11**, 151–158.

Richter, S. H., Eydal, M., and Skírnisson, K. (1990a). A survey of endoparasites found in humans in Iceland. *Icelandic Med. J.*, **76**, 287–293 (in Icelandic with English summary).

Richter, S. H., Eydal, M., and Skírnisson, K. (1990b). Endoparasites in humans in Iceland found in studies during 1973–1988. *Icelandic Med. J.*, **76**, 224–225 (in Icelandic with English summary).

Sadler, J. P. (1990). Records of ectoparasites on humans and sheep from Viking-age deposits in the former western settlement of Greenland. *J. Med. Entomol.*, **27**, 628–631.

Sambon, L. W. (1925). Researches on the epidemiology of cancer made in Iceland and Italy (July–October, 1924). *J. Trop. Med. Hyg*, **28**, 39–71.

Sigurðsson, P. (1928). Burt með lúsina (Eradicate the louse). *Icelandic Med. J.*, **16**, 9–13. (in Icelandic).

Skírnisson, K. (1995). On the fauna of Icelandic fleas. In: *Insects. Proceedings of the 14th Symposium of the Society of Icelandic Biologists*, 28 and 29 October, 1995. Abstract p. 12 (in Icelandic).

Skírnisson, K. (1998). On the biology of the pinworm. *Icelandic Med. J.*, **84**, 208–213 (in Icelandic with English summary).

Skírnisson, K. (2001). The tropical rat mite *Ornithonyssus bacoti* attacks humans in Iceland. *Icelandic Med. J.*, **87**, 991–993 (in Icelandic with English summary).

Skírnisson, K., Eydal, M., Gunnarsson, E., and Hersteinsson, P. (1993a). Parasites of the arctic fox *Alopex lagopus* in Iceland. *J. Wildlife Dis.*, **29**, 440–446.

Skírnisson, K., Eydal, M., and Richter, S. H. (1993b). *Cryptosporidium* spp. in animals in Iceland. *Icelandic Vet. J.*, **8**, 4–13 (in Icelandic with English summary).

Skírnisson, K., Ólafsson, J. H., and Finnsdóttir, H. (1997). Dermatitis in cats and humans caused by *Cheyletiella* mites reported in Iceland. *Icelandic Med. J.*, **83**, 30–33 (in Icelandic with English summary).

Skírnisson, K. and Richter, S. H. (1992). Óværa á köttum (Ectoparasites of cats). *Icelandic Vet. J.*, **7**, 3–8 (in Icelandic).

Smáradóttir, H. and Skírnisson, K. (1996). Zoonotic parasites of cats and dogs found in play-school sandboxes in the Reykjavík area, Iceland. *Icelandic Med. J.*, **82**, 627–634 (in Icelandic with English summary).

Sveinbjarnardóttir, G. and Buckland, P. C. (1983). An uninvited guest. *Antiquity*, **58**, 127–130.

Tierney, J. J. (ed.) (1967). Dicuil: Liber de Mensura orbis terrae. In: *Scriptores Latini Hiberniae*, Vol. 6. Dublin.

Thorarensen, Ó., Júlíusson, P. B., Jónsson Ó. G., and Laxdal, Þ. (1992). Congenital toxoplasmosis. *Icelandic Med. J.*, **78**, 411–417 (in Icelandic with English summary).

Woodruff, A. W. and Savigny D. H. (1982). Toxocaral and toxoplasmal antibodies in cat breeders and in Icelanders exposed to cats but not to dogs. *Br. Med. J.*, **284**, 309–310.

5 Prevalence of parasites in the Baltic states (Estonia, Latvia, and Lithuania)

Epidemiology past and present

Ats Metsis, Ants Jõgiste, Regina Virbalienė, and Ludmila Jurevica

The history of prevalence studies of parasitic infections in the Baltic countries goes back to the nineteenth century. At the Tartu University in Estonia, at least four dissertations dedicated to the epidemiology of helminthosis have been defended during the last century (Erdmann 1833; Lieven 1834; Karamarenkow 1841; Szydlowski 1879). The data available from Szydlowski (1879) indicate the prevalence of helminthosis among the population of Tartu: *Ascaris lumbricoides*, 25%; *Diphyllobothrium latum*, 10%; *Trichocephalus trichiurus*, 4%; *Taenia solium*, 1%.

The present review of parasitic infections among the inhabitants of the three Baltic states is mainly based on the analysis of the available official statistical data of case reports registered by the healthcare officials of the three countries over different periods of times:

- For Estonia, official publications of the National Board for Health Protection (Jõgiste *et al.* 1995, 1996) were used, where the available statistics about all communicable diseases have been generalized for the period 1919–1995.
- For Latvia, the data have been obtained from the Latvian National Environmental Health Centre and cover the period 1967–1996.
- For Lithuania, the data were made available by the Communicable Diseases Prevention and Control Centre and they cover the period from 1949 to 1996.

Introduction

The comprehensive data available for Estonia, Latvia, and Lithuania begin from the Soviet period (post-Second World War), the pre-war data about parasitic infections being fragmentary. We will thus, in this review, concentrate on the period beginning from the end of the 1940s to the present.

The helminth fauna of the three Baltic states localized in the temperate climate zone is represented mainly by parasites of animals. Humans are hosts for eight main local helminthosis: two geohelminthosis [ascariosis due to *A. lumbricoides*, and trichiuriosis due to *T. trichiurus* (also known as *Trichuris trichiura*)], four biohelminthosis [taeniosis due to *T. solium* and *Taeniarhynchus saginatus* (also known as *Taenia saginata*)], diphyllobothriosis due to *D. latum*, and trichinellosis due to *Trichinella spiralis*), and two contact helminthosis (hymenolepiosis due to *Hymenolepis nana*, and enterobiosis due to *Enterobius vermicularis*).

In the end of the 1940s and during the 1950s, helminthosis screening and control programmes were introduced in the territory of all Baltic states of the former Soviet Union. From the 1960s to the end of the 1980s, between 30 and 50% of the total population in all three countries was tested yearly for helminth eggs by coprological analysis or perianal scrapings. For instance, in 1949 in Lithuania only 17 103 persons were tested for helminthosis. The following year the number was already 76 572 and the number of people tested increased every year reaching about 1.5 million in the 1970s. The largest number of people were screened at the end of the 1980s, reaching 1.85 million in 1988 in Lithuania and more than 1.5 million in Latvia in 1986, constituting about 60% of the population. The individuals tested positive for helminths were treated free of charge. So for instance in Lithuania every fourth (520 096) person was tested for helminths and every eighth was dehelminthized in 1955 (Biziulevicius, 1958). After restoration of independence of the Baltic states in 1991, the proportion of the population screened has declined to 15–30%.

The statistics of parasitoses from the three Baltic states embrace mostly the eight helminthoses mentioned earlier, so these form the main theme of the current review, supplemented with a brief overview of the data available about other parasites.

Parasitosis in the Baltic states

Geohelminthosis

Ascaris lumbricoides *(Figure 5.1)*

The highest prevalence of ascariosis in population in the past has been characteristic for Lithuania where in 1949, 67.3% prevalence was registered. During the 1950s, the prevalence began to decline from 55.9% in 1950 to 30.4% in 1960 due to the introduction of control measures.

For Estonia, there are data for the period from 1948 to 1951 but the numbers available for these years seem to be inadequately low (0.3–0.7%) compared with the numbers beginning from 1957 (2.58%), when the long-scale control programme was introduced. It should also

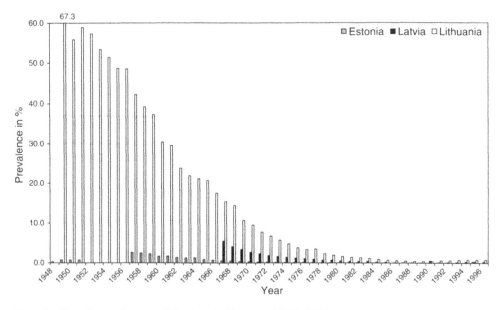

Figure 5.1 Prevalence of ascariasis in three Baltic states (1948–1996).

be indicated here that from the end of the 1950s to the beginning of the 1960s, there are some contradictory data available for Estonia. An independent survey of distribution of helminthosis among the population was carried out between 1959 and 1962, covering most of the territory (Kondratjeva *et al.* 1963). In that survey, the prevalence of ascariosis given is 5.2%, indicating a possibly higher prevalence of the infection among the population.

For Latvia, the available data begin from 1967 with 5.5% prevalence although the control programme was introduced in 1948.

The control measures over these last 50 years have had significant effect on the prevalence of ascariosis in all three countries. In 1995, the following prevalence were reported: 0.029% for Estonia, 0.3% for Latvia, and 0.79% for Lithuania. As the initial level of prevalence in Lithuania was very high, it took about 35 years to achieve prevalence lower than 1% of the population. In Estonia, with the lowest initial proportion of the population infected, it took only eight years. The situation stabilized in Estonia to lower than 0.1% in mid-1970s and has been around 0.03% in the 1990s. The situation in Latvia is apparently similar, although the initial prevalence seems to have been higher. In Lithuania, the prevalence declined below 1% in mid-1980s and the lowest level was recorded in 1990 (0.37%). Beginning from 1991, a mild rise, both in Latvia and Lithuania has taken place: 0.3% in Latvia and 0.79% in Lithuania in 1996. This change might be attributed to the significant decrease in the overall number of people tested and paying more attention to the risk groups.

Trichocephalus trichiurus (Trichuris trichiura)

When looking at the prevalence of trichuriosis among the inhabitants of Baltic states we see the same picture as with ascariosis: with a high (12.6%) prevalence in Lithuania in 1949, which declined to 9% by 1959. For Estonia, the first data are from 1959 indicating a low prevalence of 0.21%. The results of an independent study by Kondratjeva *et al.* (1963) give the prevalence of 1.1% in 1961 in Estonia. By the middle of the 1960s, the prevalence in Estonia was below 0.1% and has been falling further since then, with only single cases being registered during the last years. In Latvia, 0.9% prevalence was registered in 1967 and 18 cases registered in 1996 (0.005% prevalence). In Lithuania, the prevalence of trichuriosis reached lower than 1% in 1981, and has remained at 0.14% for the last years with a little more than 750 cases.

As with the significant decrease in the prevalence of ascariosis and trichuriosis in humans, a similar tendency has been observed for the contamination of soil with the eggs of corresponding helminths (Jõgiste and Barotov 1993). In the 1960s, about 4% of the soil probes in Estonia were contaminated with the helminth eggs, while in the 1980s only 0.9% was reported. The percentage of vegetables found to be contaminated decreased during the same period from 1.7% to 0.7%.

Biohelminthosis

Taenia solium *and* Taeniarhynchus saginatus (= Taenia saginata)

Taeniosis has been registered in Lithuania from 1949 till 1970 as a single infection for both *T. solium* and *T. saginatus*. In 1949, 0.9% of the people tested were infected with taeniosis in Lithuania, and by 1959 it decreased to 0.6%. In the same year in Estonia, the prevalence was 0.015% for *T. solium* and 0.006% for *T. saginatus*. These numbers have been decreasing in all three countries and in 1990s there have been single cases registered in some years. The decrease can be followed also for the prevalence of cysticerci in cattle and pigs. In 1964 in

Lithuania, 1083 cases or 0.19% prevalence was registered in slaughtered cattle and 183 cases or 0.02% prevalence in pigs. By 1982, 152 cases were documented in cattle (0.02%) and one case in pigs. Beginning from 1986, only two cases of *T. solium* infection have been registered, and this was in 1993 in Estonia, while the veterinary control did not identify any infected pigs (Jõgiste and Barotov 1993). The prevalence of *T. saginatus* in cattle has been higher, but that too is declining. During the 1970s, veterinary control found 868 infected animals, while in the 1980s the number was 303; of them 230 between 1980 and 1984, and 73 from 1985 to 1989. This situation is reflected in the cases of human infection as well: one to five cases being registered yearly since 1980.

The improvement in the situation with taeniosis infections in all three countries could be attributed to the changes in cattle and pig farming throughout the Soviet period: concentration of animals had taken place in big farms, or so called factories in the case of pigs, enabling more efficient veterinary control. After the breakdown of the collective farm system, cattle breeding is decentralizing, and an increase in the prevalence of taeniosis can be anticipated as has happened with trichinellosis, if the veterinary control measures are not adapted to the new situation.

Diphyllobothrium latum *(Figure 5.2)*

In contrast to the infections reviewed so far, the prevalence of diphyllobothriosis has been the highest in Estonia and is more than 10 times higher even nowadays than in the other two Baltic countries. Based on the data of Kondratjeva *et al.* (1963), it could have been even significantly higher in the end of the 1950s and beginning of the 1960s (3.7%) than indicated in the official publication of the National Board of Health Protection (between 0.43 and 0.3%). Since then, the prevalence has decreased significantly to 0.037% in 1995 with 548 cases. It should be pointed out that historically there have been two main foci of infection in

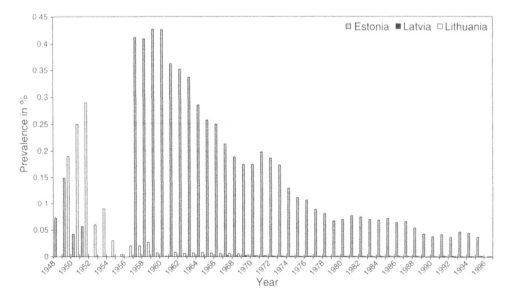

Figure 5.2 Prevalence of diphyllobothriasis in three Baltic states (1948–1996).

Estonia: the coastal regions of the Lake Peipus in the east of Estonia and on the island of Saarema (Ösel). In the coastal regions of Lake Peipus the prevalence in the late 1950s and beginning of the 1960s was extremely high, constituting 53.3% of the inhabitants tested (Kondratjeva *et al.* 1963). In 1971, the prevalence of diphyllobotriasis in that region had decreased to between 11.2 and 2.6% in different villages thanks to the control measures (Vasiljeva, 1973). The high prevalence of *D. latum* in that region has been attributed to the high prevalence of plerocercoids in fish in Peipus (Jõgiste and Barotov 1993). Although the data for 1960s may not be concrete, it has been reported that nearly all the fish caught were infected. Between 1985 and 1991, the infection was found in 26.4% of burbots, 14.4% of pikes, and 3.3% of perches tested. The overall level of infection of the Peipus fish (12.5%) was much higher than in the second biggest lake of Võrtsjärv (4.7%) in Estonia. Such a high prevalence in fish has been attributed to the high wastewater pressure on that lake from the city of Tartu in Estonia and the city of Pskov on the Russian side. The positive effect on the decrease of prevalence in Estonia has been associated with the freezing of fish caught from Lake Peipus since the 1980s. For the 1990s, the number of cases reported have been between 715 in 1993 and 548 in 1995.

For Latvia and Lithuania, the prevalence has been constantly significantly lower. In Latvia, the prevalence among humans has decreased from 0.002% (46 cases) in 1967 to 0.0007% (18 cases) in 1996. The cases have been registered all over the country. The number of cases has been higher in the northeastern part of the country in the regions closest to the Estonian border. A significant number of cases have been attributed to fishing in, and consumption of the fish from, Lake Peipus. The data from veterinary services indicate a low incidence of fish infected with plerocercoids in rivers and lakes of Latvia, but the geographical distribution of cases make that data dubious. Complementary studies are needed.

In Lithuania, the highest prevalence of 0.25% was recorded in 1951 (273 cases). In the 1990s, this indicator has been 0.0013% in 1995 (seven cases from 540 893 people tested) and 0.0003% in 1990 and 1993 (five and two cases, respectively). In Lithuania, the foci of diphyllobothriosis have been in the region of the lagoon of Kuršiu Marios and the Trakai Lake (Biziulevicius and Krotas 1961; Sangaila and Federene 1973). In 1973, it was stated that these two foci were in the state of elimination.

Trichinella spiralis *(Figure 5.3)*

If the parasitic infections described previously have shown the tendency of declining prevalence in all three countries, then the situation has been contrary with trichinellosis. In Estonia, there have been the lowest number of cases through out the period reviewed. In the 1920s four cases were reported between 1924 and 1926. Beginning from 1945, when monitoring of *T. spiralis* was started, the first episode of four cases was reported in 1969. From 1985 to 1994, there have been 3 years when no trichinellosis was reported. The highest number of individuals infected (43) was in 1993. The only source of infection identified so far has been the meat of wild boars (Jõgiste and Barotov 1993). In domestic pigs, for the last years only one case has been reported on the island of Hiiumaa (Dagö) and the carcass was not used for food.

Since 1967, the first three cases of human infection were reported in Latvia in 1973. Beginning from 1991, the number of cases has been rising every year with 81 cases (two lethal) in 1996. Beginning from 1994, the role of domestic pigs as the source of infection is rising: more than half of the patients being infected from the meat of domestic pigs.

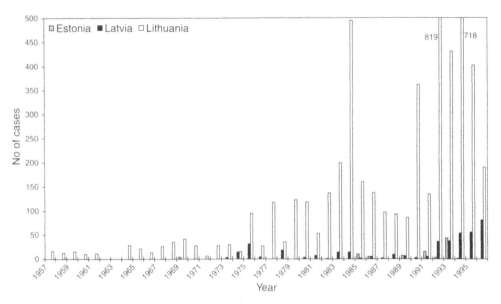

Figure 5.3 Incidence of trichinellosis in three Baltic states (1957–1996).

The trichinellosis has been stated the most serious parasitic disease in Lithuania (Rockiene 1994). The epidemiology of trichinellosis in Lithuania has been thoroughly reviewed previously (Rockiene 1994). The foci of trichinellosis began to spread from the beginning of the 1970s from the southeast of Lithuania, and are now registered all over the country. The highest number of foci (101) was identified in 1994 with 718 people infected. The highest number of cases was in 1992 (819) from 76 foci. Since 1982, there have been 11 lethal cases of infection in Lithuania.

The ways of infection in all three countries are similar, several people being infected from the meat of one animal that had not gone through adequate veterinary control. With wild boars, it is often the case of poaching. Since 1990, licences on boar hunting have been abolished in Lithuania so that people could check the meat in veterinary laboratories without legal consequences. This has resulted in significant decrease in the numbers of infection due to consuming meat of wild boars (Rockiene 1994). The situation with domestic pigs can be associated with the decentralization process of animal breeding because of the reforms in the agricultural sector. It seems that the veterinary control measures have not kept in touch with all the changes, at least where trichinellosis is concerned. There are also case reports when farmers have sold meat without veterinary licence.

In Latvia and Lithuania, the main source of infection has changed during the 1990s: if in the 1970s and the 1980s people were infected mainly from wild boars, then in the last years domestic pigs have become the predominant source of infection.

Contact helminthosis

Hymenolepis nana

The prevalence data concerning hymenolepiasis are available only for Estonia and Lithuania, while for Latvia the data available were presented in the form of registered case reports. The situation with hymenolepiasis in Estonia and Lithuania has been more or less similar, the infection having a focal character mostly devoted to children's homes, kinder-gartens, and children up to 9 years of age (Taunené 1962; Kondratjeva *et al.* 1963; Jõgiste and Barotov 1993). Thanks to its focal character the prevalence of this infection has mostly been lower than 0.1%, and only single cases per year are reported nowadays. Historically, the highest number of cases has been reported for Estonia with more than 300 cases per year in 1958 and 1959. In Latvia, the numbers were 53 (in 1959 and 1960) and in Lithuania, there were between 70 and 90 cases per year between 1958 and 1960. Since 1973, less than 10 cases per year have been registered in Latvia except 1986, when 13 cases were reported. From 1986, five cases have been reported in Latvia: four cases in 1989 and one case in 1995. In Lithuania, 10 or less cases yearly have been reported since 1984. In Estonia, the number of reported cases decreased below 100 by 1964 and to 25 or less by 1972. In the end of the 1980s and the beginning of the 1990s, there was a permanent focus in one children's home in Estonia, which is now eliminated. In 1995, only one case was reported in Estonia.

Enterobius vermicularis *(Figure 5.4)*

The prophylactic measures against enterobiosis have been ineffective in all three Baltic countries, with no decrease in prevalence being achieved. Due to the capability of *E. vermicularis* to parasitize, mainly children, it would be incorrect to speak about prevalence of infection in the whole population. The data received from the health officials of the three countries seem to be incomparable, as in Lithuania the enterobiosis investigations have been kept separate from the control programme of other helminthosis since 1949. The data on Estonia taken from the official publication of the National Board of Health Protection (Jõgiste *et al.* 1995) give the prevalence calculated to the whole population. The same is the situation with the data from Latvia. Accordingly, the data given in Figure 5.4 are informative only concerning dynamics of prevalence of the infection, but the levels of prevalence are

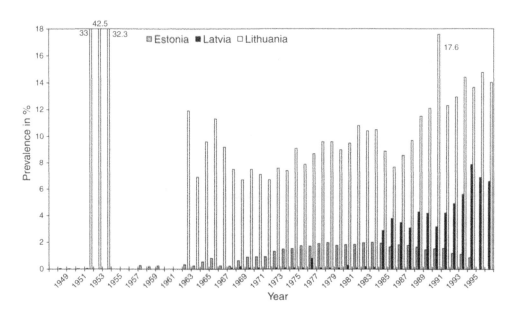

Figure 5.4 Prevalence of enterobiasis in three Baltic states (1948–1996).

incomparable. It appears that in the 1960s, in Estonia, only 3% of the population (mainly children) were studied for *E. vermicularis* eggs, and 16.2% of the people studied were positive. In the 1970s, 12.2% of the population was studied, with 14.5% prevalence. In the 1980s, 17.6% of the population was studied, and the prevalence was 10.3%. These results do not seem to indicate a decrease in prevalence, but it is more likely that by widening the range of people being studied the control measures were less and less directed to specific risk groups (Jõgiste and Barotov 1993).

Looking at the dynamics of prevalence of enterobiosis in Latvia (Figure 5.4) we see a sharp rise in prevalence beginning from 1984, which seems to be difficult to explain. It is most likely that changes were made in an overall helminthosis control strategy, more attention being paid to *E. vermicularis* infection, and the portion of the enterobiosis risk group (children) rose in the number of people tested. In conclusion, it can be said that the data available for Lithuania (12–15% positive for the people tested during the 1990s) seem to be characteristic for the risk groups of all the Baltic states.

Other helminthosis

Opisthorchis felinus

Historically, two foci of opisthorchiosis due to *O. felinus* in the region of the lagoon of Kursiu Marios and the Nemunas river in the Kaunas region (Biziulevicius and Krotas 1961; Biziulevicius 1969) have been described in Lithuania. During the post-war period no human infection has been registered. In Estonia, two imported cases of human opisthorchiosis of Siberian origin have been registered in the border city of Narva among people visiting relatives.

Echinococcus granulosus

From 1992, serological tests have been carried out for this infection in Lithuania. Enzyme linked immunosorbent assay (ELISA) test was used from 1992 to 1994 and indirect haemagglutination (IHA) test for 1995 and 1996. For this period, 1 350 persons have been tested (1 783 single tests carried out, 141 giving positive results). Some cases of echinococcosis with localization in liver were registered in Latvia during the 1960s. The prevalence in domestic animals in Latvia in 1965 according to the veterinary service was as follows: 2.8% in pigs, 0.8% in sheep, and 0.48% in cows. From the mid-1980s, 0.01% of cow carcasses and 0.001–0.002% of pig carcasses have been reported infected. In 1994–1995, 0.001% of sheep carcasses have been reported infected. The data of surgical intervention in humans in Latvia are as follows: 1984, one case (liver); 1994, six cases (liver, five; cerebral, one); 1995, four cases (liver). There are no reliable data available for Estonia.

Strongyloides stercoralis

For Lithuania and Estonia, few cases of *S. stercoralis* infections in humans in the early literature (Biziulevicius 1958; Kondratjeva *et al.* 1963) have been described. According to the data of the Communicable Disease Prevention and Control Centre of the Ministry of Health of Lithuania, two cases have been documented in humans since 1992.

Toxocara canis

Beginning from 1975, the patients with eosinophilia of unclear aetiology have been serologically tested for toxocariosis in Lithuania (Baiorinene and Ramanauskaite 1979). For the first

3 years, 1 524 patients were studied, of whom 86 were diagnosed for toxocariosis. The infection has been attributed to *T. canis*, since up to 13% of dogs are infected in Lithuania. Between 1992 and 1996, 7 720 patients have been tested for the infection by means of ELISA and 3 400 tests gave positive results. Concerning Estonia and Latvia, no toxocariosis studies in humans have been carried out to our best knowledge.

Fasciola hepatica

Single cases of human infection have been reported in Estonia (Kondratjeva *et al.* 1963) and Latvia.

Filariasis

In the end of the 1970s, and the beginning of the 1980s, about 100 cases of imported filariasis were registered in Tallinn (Estonia) (Märtin *et al.* 1979; Mihailova 1982; Mihailova *et al.* 1990). The people infected were the workers of the Estonian Fishing Company and the Estonian Shipping Company, mostly returning from Equatorial Guinea. In five cases, the infection was bound to *Lao lao*, three to *Wucheria* sp. and 80 to *Onchocerca* sp.

Protozoan parasites

Trichomonas vaginalis

Trichomoniosis data are available only for Estonia beginning from 1971. The prevalence of this infection has not changed much since. According to the registered and treated cases the prevalence was 0.38% in 1972 (5 222 cases). In the first half of the 1980s, there was an apparent decrease to 0.16%, but later this level returned to 0.373% (5 624 cases in 1994).

Intestinal protozoa

Data about intestinal protozoa are available only for Lithuania. In the end of the 1950s, pre-school children were tested for intestinal protozoan parasites in Vilnius (Pabrezaite 1958), 34.1% of whom were infected with *Giardia lamblia*, 10.6% had *Entamoeba coli*, 9.1% *Chilomastix mesnili*, 8.3% *Endolimax nana*, 4.2% *Jodamoeba bütschlii*, and 3.3% *Entamoeba hartmani*. *Entamoeba histolytica* was identified in faeces of one child. Between 1975 and 1977, 6 810 adult patients were tested for intestinal protozoa in the Kaunas Clinical Infectious Disease Hospital of whom 112 had intestinal protozoan infection (Stankaitite *et al.* 1979). In 40 cases, it was *G. lamblia*, and in 72 cases *E. coli* was identified. Since 1992, faecal examination for intestinal protozoan infection has been carried out in Lithuania. The data are given in Table 5.1.

Toxoplasma gondii

Toxoplasmosis studies have been carried out in Lithuania and Estonia, but the data are fragmentary and incomparable as different diagnostic tools have been used not to mention the array of the antigen preparations used. It is reported for Estonia that in the middle of the 1960s, 52.3% of the blood donors and 45.8% of the pregnant women tested in Tallinn were positive for *T. gondii* (Jõgiste and Kondratjeva 1965). There are data available from a study carried out at the end of the 1960s in Tartu indicating that among children between 7 and

Table 5.1 The results of faecal examination for intestinal protozoan infection in Lithuania

Year	Number of persons examined	Number of persons infected			
		Giardia lamblia	Entamoeba histolytica	Balantidium coli	Cryptosporidium
1992	164 639	918	3	2	24
1993	132 162	754	8	0	4
1994	137 972	786	2	0	0
1995	120 657	419	8	1	0
1996	133 775	566	2	0	0

12 years of age, 12.4% among ones living in the countryside and 4.8% living in towns were positive for *T. gondii* (Kääri 1970). In the same report, 11 cases of congenitally acquired toxoplasmosis are discussed, but it is not clear for how many years the study was carried out.

The results of the toxoplasmosis studies carried out in Lithuania indicate that 43.2% of the mothers, infants and children below 14 years studied were IgG positive for *T. gondii*, and only 6.2% IgM positive (Bajoriniene *et al.* 1991). In a more recent study of sero-prevalence in Lithuania, 50.9% of the people studied were positive for *T. gondii* (Rockiene 1997). The sero-prevalence in pregnant women was 40.3%. The lower prevalence in the latter case is explained by the fact that the average age in this group was about 10 years lower than in the whole study group. According to this investigation, the annual incidence rate of new infection in the whole study group was 0.76 and 0.55% for the study group of pregnant women. Among the slaughter-house workers, this indicator was 0.93%. Four cases of acute infection were identified in the study group of pregnant women (528) indicating the possible prevalence of 7.6 cases of congenital infection per 1 000 deliveries in Lithuania. The main source of infection underlined in this study has been the consumption of raw meat.

Lithuania is the only one of the three Baltic States where toxoplasmosis studies have been initiated, although so far having preliminary character. The current situation in Latvia and Estonia remains unclear, and there are no data available for any of the three countries, on the role of *T. gondii* in opportunistic infections of immuno-compromised patients.

Malaria

During the 1940s and the 1950s in the Soviet Union, a malaria control programme was carried out to get rid of this infection, covering the territory of the Baltic States as well. Malaria data available are given in Table 5.2. The high number of malaria cases in the second half of the 1940s and beginning of the 1950s can be explained by the fact that during the post-war period, large-scale immigration of people from other parts of Soviet Union began to the Baltic region. This is supported by data from Estonia, where the number of indigenous cases is known for these years, and it is insignificant compared to the amount of imported cases. It is only in 1948 when there seems to have been a local indigenous outbreak or imported cases were registered as local ones (people registering themselves as local inhabitants before being diagnosed). The high number of indigenous cases in Latvia in the end of the 1940s and the beginning of the 1950s might be explained with the possibility that people arriving from other parts of Soviet Union had registered themselves as local inhabitants before they were diagnosed for malaria. Lithuanian data between 1945 and 1957 include both imported and

Table 5.2 Malaria cases in Estonia, Latvia, and Lithuania

Year	Number of cases			Year	Number of cases			Year	Number of cases		
	Estonia	Latvia	Lithuania		Estonia	Latvia	Lithuania		Estonia	Latvia	Lithuania
1926	1	15		1950	168 (0)	288 (230)	259	1974	0	0	3
1927	4			1951	55 (0)	71 (44)	64	1975	12	2	3
1928	6			1952	35 (3)	43 (28)	19	1976	7	1	1
1929	0			1953	17	24 (9)	19	1977	13	2	15
1930	0			1954	3	20	20	1978	24	0	9
1931	0			1955	5	12	20	1979	16	1	11
1932	1	2		1956	1	4	10	1980	12	3	4
1933	0	3		1957	2	2	3	1981	9	0	6
1934	1			1958	2	9		1982	10	5	4
1935	1	2		1959	0	0		1983	16	2	8
1936	1	3		1960	0	2	0	1984	21	4	8
1937	4	3		1961	0	2	0	1985	9	1	9
1938	12	4		1962	0	1	1	1986	7	0	12
1939	15			1963	2	3	0	1987	7	3	19
1940				1964	5	1	0	1988	6	3	15
1941				1965	3	3	0	1989	14	4	12
1942	26			1966	2	0	0	1990	5	3	3
1943	(5)[a]			1967	1	2	2	1991	8	2	0
1944				1968	2	5	0	1992	3	6	2
1945	523 (15)		464	1969	5	0	0	1993	3	4	8
1946	522 (8)		1 632	1970	1	2	0	1994	11	2	7
1947	1 104 (11)	2 391 (2 391)	1 781	1971	3	4	1	1995		2	4
1948	946 (170)	2 963 (2 103)	2 946	1972	2	0	1	1996		3	12
1949	260 (7)	961 (721)	1 926	1973	4	4	2	1997			

Note
a Number in parentheses denotes indigenous cases when known.

indigenous cases. Since 1953 in Estonia, 1956 in Latvia and 1957 in Lithuania only imported cases have been reported. Beginning from the 1970s, the number of imported case have risen in Estonia and Lithuania. The disease was registered mostly among sailors and during the period of the Afghanistan war, among the returning soldiers in all three countries. Since 1992, Lithuania and Latvia have been facing a new problem of illegal immigrants, importing malaria from endemic countries. Most often, *Plasmodium vivax* is identified. Forty-five per cent of permanent water bodies are anophelogenic in Lithuania with *A. maculipennis* dominating. Due to the climatic conditions favourable for *P. vivax* to mature in mosquitoes, there is a potential possibility for indigenous malaria to reappear in the most southern one of the Baltic states.

In Estonia, most of the cases of imported malaria during the 1980s originated from Angola where the Estonian Fishing Company was running a shipyard in Luanda. In most cases, *Plasmodium falciparum* was identified (Andrejeva and Berseneva 1979; Mihailova *et al.* 1990).

Leishmaniosis

In 1989, one case of leishmaniosis was imported to Tallinn by a soldier returning from Afghanistan (Mihailova *et al.* 1990).

References

Andrejeva, I. and Berseneva, G. (1979). The potential epidemiological consequences of imported malaria in Tallinn. In: *Theoretical and practical questions in parasitology. Materials of the VIII Scientific-Coordination Conference on the Problems of Parasitology in the Baltics.* Tartu, pp. 47–49 (in Russian).

Bajoriniene, D., Arlauskene, A., and Balkkjawiczjus, B. (1991). Toxoplasmosis as anthropo-zoonosis and its immunodiagnostics in Lithuania. *Wiad. Parasitol.*, **37**(4), 457–460 (in Polish).

Baiorinene, D. and Ramanauskaite, M. (1979). Toxocariasis in children in the Lithuanian SSR. In: *Theoretical and practical questions in parasitology. Materials of the VIII Scientific-Coordination Conference on the Problems of Parasitology in the Baltics.* Tartu, pp. 65–66 (in Russian).

Biziulevicius, S. (1958). Helminthofauna of the inhabitants of the Lithuanian SSR. *Acta Parasitol. Lithuanica*, **1**, 52–68 (in Russian).

Biziulevicius, S. (1969). On the natural focus of opistorchosis on the sea-shore of Kursiu Marios. *Acta Parasitol. Lithuanica*, **9**, 5–10 (in Russian).

Biziulevicius, S. and Krotas, R. (1961). Diphyllobothriasis and opisthorchiasis in Lithuania. *Acta Parasitol. Lithuanica*, **3**, 93–98 (in Russian).

Erdmann, J. (1833). De virtute et viae medica extracti Filicis maris resinos ad taenias expellendas. *Dissertation*, Universitas Dorpatensis, Dorpat.

Jõgiste, A. and Barotov, O. (1993). Helminthosis in Estonia 1960–1989. *Eesti Arst*, **4**, 251–254 (in Estonian).

Jõgiste, A. K. and Kondratjeva, G. P. (1965). About the epidemiology of toxoplasmosis. In: *The Thesis of the Republican Congress of Epidemiologists, Microbiologists, Infectionists and Hygienists Devoted to the 25th Anniversary of the Estonian SSR (On the problems of reduction and elimination of infectious diseases)*, 24–26 June 1965, Tallinn. Tallinn, pp. 150–151 (in Russian).

Jõgiste, A., Pool, V., and Trei, T. (1995). Communicable diseases statistics in Estonia. *An Official Publication of the National Board for Health Protection.* 84pp., Tallinn.

Jõgiste, A., Pool, V., and Trei, T. (1996). Communicable diseases statistics in Estonia. Part 2. *An Official Publication of the National Board for Health Protection.* 100pp., Tallinn.

Karamarenkow, J. (1841). Nonnulla de Boothrocephalo lato ejusque expellendi quibusdam methodis. *Dissertation*, Universitas Dorpatensis, Dorpat.

Kondratjeva, G. P., Gorskaya, V. I., and Schamardin, M. B. (1963). Materials about the helminthofauna of inhabitants of the Estonian SSR. In: *The Thesis of the VI Scientific Conference of the Tallinn Scientific Research Institute for Epidemiology, Microbiology and Hygiene*, 29–30 November 1962, Ministry of Health of the Estonian SSR. Tallinn, pp. 112–120 (in Russian).

Kääri, H. (1970). The results of a study for toxoplasmosis in children and their mothers in the Estonian SSR. In: *The problems of parasitology in Baltics. Materials of the IV Scientific-Coordination Conference on the Problems of Parasitology in the Baltic Republics*. Riga, pp. 269–270 (in Russian).

Lieven, S. (1834). Nonnulla de taeniis, imprimis de Bothriocephalo lato, ejusque expellendi methodo peculiari. *Dissertation*, Universitas Dorpatensis, Dorpat.

Märtin, J. K., Mihailova, V. A., and Lepnjova, S. M. (1979). Diagnostics, clinics and treatment of the patients ill with filariasis. In: *Theoretical and practical questions in parasitology. Materials of the VIII Scientific-Coordination Conference on the Problems of Parasitology in the Baltics*. Tartu, pp. 52–54 (in Russian).

Mihailova, V. A. (1982). The treatment results of the filariasis patients. In: *The Thesis of the IV Republican Congress of the Epidemiologists, Microbiologists, Infectionists and Hygienists of the Estonian SSR*, 12–13 October 1982, Tallinn, pp. 84–85 (in Russian).

Mihailova, V. A., Subi, V. O., and Kapral, H. A. (1990). The problem of parasitological diseases imported from tropics by sailors and shipyard workers. In: *Actual questions of infectious deseases. The Thesis of a Conference*, 23–24 November, Tartu, pp. 106–107 (in Russian).

Pabrezaite, L. (1958). The distribution of intestinal protozoa in preschool children in Vilnius. *Acta Parasitologica Lithuanica*, **1**, 116–120 (in Russian).

Rockiene, A. (1994). The epidemiology of trichinellosis in Lithuania 1969–1992. In: *Trichinellosis. Proceedings of the Eight International Conference on Trichinellosis*, 7–10 September 1993, Ovieto, Italy. (Campbell, W. C. *et al.* eds). Instituto Superiore di Sanità Press, Rome, pp. 539–544.

Rockiene, L. (1997). The prognosis of congenital toxoplasmosis in Lithuania. *Hygiena Epidemiol.*, **58**(2), 39–45 (in Lithuanian).

Sangaila, I. and Federene, M. (1973). The results of helminthosis control in the Lithuanian SSR for the last 10 years (1962–1971). In: *The Materials of the VI Baltics Scientific-Coordination Conference on Parasitology*, 21–22 June, 1973, Academy Sciences of the Lithuanian SSR, Institute of Zoology and Parasitology, Vilnjus, pp. 155–156 (in Russian).

Stankaitite, S. Ju., Lazdauskene, D. K., and Vitene, T. I. (1979). Distribution of intestinal protozoa among the infectious diseases patients. In: *Materials of the VIII Scientific-Coordination Conference on the Problems of Parasitology in the Baltics*. Tartu, pp. 43–44 (in Russian).

Szydlowski, J. (1879). Beiträge zur mikroskopie der faeces. *Dissertation*, Universitas Dorpatensis, Dorpat.

Taunené, A. (1962). On hymenolepidosis in Lithuanian SSR. *Acta Parasitol. Lithuanica*, **4**, 115–123 (in Russian).

Vasiljeva, K. A. (1973). The influence of some sources of contamination of the Narva and Peipus watter basins of the Estonian SSR with the eggs of *Diphyllobothrium Latum*. In: *The Materials of the VI Baltics Scientific-Coordination Conference on Parasitology*, 21–22 June 1973, Academy of Sciences of the Lithuanian SSR, Institute of Zoology and Parasitology, Vilnjus, pp. 35–37 (in Russian).

6 Prevalence of human parasitic infections in Russia

Epidemiology past and present

V. P. Sergiev and N. N. Ozeretskovskaya

The number of inhabitants of the Russian North exceeds 12 million. Food-borne zoonotic helminthiases – *Opisthorchis felineus* infection, trichinellosis, *Diphyllobothrium dendriticum, D. luxi* (*D. klebanovskii*) infections, endemic *D. latum* infection, northern isolate of *Taenia saginatus* infection, and echinococcoses are the most widespread parasitic infections of the Russian North. One and a half million of human cases of opisthorchiasis exist in Russia. No direct correlation has been found between the intensity of egg output and the severity of infection. Acute opisthorchiasis (unknown till the 1960s of the past century) has been registered among non-immune newcomers to the endemic foci. The 'selective-radical' therapy with praziquantel in endemic foci of infection begun in the middle of the 1980s gave optimistic results. Outbreaks of trichinellosis in newcomers to Arctic and Subarctic regions have been registered in the late 1950s. *Thalarctos maritimus* and *Ursus arctos* meat accounted for these outbreaks. In 1976–1977, 47.6% of trichinellosis cases registered in Russia were from the North, only 5.5% were registered in 1996 and 22% in 2000. *Echinococcus granulosus* and *E. multilocularis* infections have been registered in the Russian North from the early 1930s with 58.1–64.2% predominance of the latter. Foci of *E. multilocularis* infection are developing in Kamchatka recently. *Diphyllobothrium dendriticum* accounts for five-sixths of human cases in mixed *D. latum* and *D. dendriticum* foci in the North. The mean prevalence of enterobiasis in children in 1995–1996 was 25.7%. Cases of ascariasis and giardiasis are registered. Malaria that in the 1920s of the past century was registered up to Arkhangelsk is now absent, but imported cases are registered regularly.

Introduction

The territory of the Russian Federation is up to $17\,000\,000\,km^2$ in size. Russia occupies the main portion of the northern part of the Eastern Hemisphere, with Finland, a tip of Norway in the west and a bit of Alaska in the east. Sixty-eight per cent of the territory of Russia belongs to northern regions according to geographical, climatic, and ethnographic criteria. The northern regions of Russia include the Arctic and Subarctic zones with Arctic deserts, tundra, forest-tundra, and taiga. The population of the northern territories of the country slightly exceeds 12 million subjects, 3.5 million of whom are children. Indigenous ethnic minorities (aboriginal) make up about 180 000 inhabitants with 62 200 children (Federal Programme 'Children of the North', 1995–2000). The Russian North is a great bank of natural resources – gas, oil, coal, diamonds, gold, nickel, uranium, etc. Many industrial centres were created in the European North, in Ural, and Siberia in the past decades. Oil and gas fields have been developing during the last three decades in the extreme northern territories of the eastern part of the country. Several big hydroelectric stations and water reservoirs were built in the Eastern Siberia in the 1960s and 1970s.

The transmission and the high prevalence of zoonotic parasitic infections are determined by the vast river basins, habitable for intermediate hosts of parasitic helminthes and by a continental climate with a high summer temperature in most of the northern regions. These conditions are sufficient for the development of larval forms of parasites. The social factor – the habit of consuming raw meat and raw fish is equally important. Therefore, the widespread zoonotic helminthic infections in man are food borne. Soil-transmitted intestinal helminthiases are of a rather low incidence on the outskirts of 60° N or are focused around heated agriculture complexes and greenhouses where eggs of geohelminthes are able to develop (Lebedev *et al.* 1996). Contagious helminthic infections (enterobiasis, hymenolepiasis) are spread in communities, especially in children. A survey was conducted by the Martsinovsky Institute of Medical Parasitology and Tropical Medicine (the Martsinovsky Institute) in the six administrative northern territories in 1995–1996, where 4 009 subjects (children and adults) were examined. It was demonstrated that up to 25.7% of persons had enterobiasis and 19.3% had other intestinal helminthiases (ascariasis, trichuriasis, or hymenolepiasis). Seroprevalence of toxocarosis in children and adults revealed in enzyme linked immunosorbent assay (ELISA) test was 2.6–5% in the Subarctic regions and 7.3–47.6% around 55–57° N altitude (Tomsk region) (Report for the Federal Programme 'Children of the North', 1996). There are no vector-borne parasitic infections in the Russian North. Malaria that was registered in the 1920s, in the European part of the country up to Archangelsk (Dobreitser 1924; Sergiev *et al.* 1968) is now absent even in its southern regions (imported cases only). Giardiasis and toxoplasmosis are the only protozoa infections officially registered in the population of the Russian North, but a wide range of intestinal protozoa was detected by special surveys (see below). Anyway, *Lamblia intestinalis* cysts were found in 4.6% tests of piped water in Tjumen and in 11.2% tests in Khabarovsk regions (Sergiev *et al.* 1998). The cases of cryptosporidiasis were reported from St Petersburg (Beyer *et al.* 1990). Data on the seroprevalence of toxoplasmosis are restricted mainly to the moderate climate zones of Russia.

Helminthiases are one of the serious problems of health care in Russia. The morbidity rate of enterobiasis in the USSR was above 1 000 per hundred thousand of the population and made up about 70% of registered cases of intestinal helminthic infections. Ascariasis made up about 20%, trichiurasis about 7% and hymenolepiasis about 3% (Sergiev 1991). The number of subjects infected with *O. felineus* has reached 1.5 million (Sergiev 1991). The State (USSR) Programme for the Control of Parasitic Infections including opisthorchiasis and echinococcosis was partly accomplished between 1985 and 1991. Also, the morbidity rate of opisthorchiasis was still 29.6 in 1999 (Siskova *et al.* 2001). The control of parasitic infections in the Russian Federal Programme 'Children of the North' 1995–2000 was carried out by the Martsinovsky Institute.

Helminthiases of the Russian North

Three hundred helminthological expeditions were planned and led by the creator of Russian helminthology – academician K. I. Scrijabin in 1919–1966. In 20 of these, helminthic infections of humans, synanthropic, and wild animals in the European and the Asiatic Russian North were studied. Opisthorchiasis (*O. felineus*) trichinellosis, diphyllobothriasis (*D. latum*) were revealed to be the principal zoonotic helminthic infections in the Russian North (Shikhobalova *et al.* 1969). *Diphyllobothrium dendriticum* was later registered in the population of the Taimir peninsula and of the North of Siberia (Klebanovskii 1980, 1985). The northern isolate of *Cysticercus bovis* – 'deer taeniasis' has been revealed in the indigenous population of the Yamal and the Taimir peninsulae. More than one case of *E. granulosus* and *E. multilocularis* for 100 000 of population per year are registered in the Kamchatka, Magadan, Chukotkan, Korjak regions, and in Yakutia (Bessonov 2001a,b). *Anisakis simplex* larvae have been found in 85.5% of 104 Siberian salmon examined, 56.7% of 231 of hump-backed salmon and in 91.2% of herring, in 89.7% in halibut, in 87.3% of Greenland halibut in the North Okchotsk Sea Region (Serdukov 1993; Vitomskova and Dovgalev 2001). The data on human anisakidosis in the northern regions of Russia are still anecdotal (Cibina and Siskova 2001).

Opisthorchiasis (cat liver fluke disease)

The first case of *O. felineus* (Jablokov 1979) infection in man was reported in 1892 by Professor K. V. Vinogradov from the University of Tomsk in West Siberia. The high prevalence of the infection in man, domestic and wild animals in the Ob-Irtish basin was discovered in the 1930s (Shikhobalova *et al.* 1969). *Opisthorchis felineus* infection exists as a zoonosis in the vast territory from the Baikal lake to the western borders of the continent of Europe. About two-thirds of nosoarea of opisthorchiasis is located in Russia (Zavoikin 1990).

Pathogen

Opisthorchis felineus a trematode (fluke) lives in intrahepatic bile and in pancreatic ducts. Adult flukes measure $4-13 \text{ mm} \times 1-3.5 \text{ mm}$. Oval eggs of yellow-brown colour measure $0.01-0.019 \text{ mm} \times 0.02-0.034 \text{ mm}$. The parasite delivers about 1 000 eggs per day. Eggs excreted with faeces and ingested by snails *Bithinia inflata*, develop in two months into cercaria to be released into water. Cercaria penetrate the skin and flesh of a second intermediate host – freshwater carp fishes. In six weeks, encysted larvae, metacercariae, become invasive and can be consumed by fish-eating wild and domestic mammals (cats, dogs, pigs) and by humans. Comparative study of *O. felineus* and *O. viverrini* somatic and metabolic antigens by electrophoresis and immunoblotting technique revealed that they share a number of proteins including p89 and p70. Protein p70 in both species is considered to be of high immunogenecity. Common epitopes in protein p89 and protein p70 have been found using monoclonal antibody technique. The possibility of the identity of these proteins was discussed (Kotelkin *et al.* 1996). The sequencing of *O. felineus* paramyosin cDNA showed only 66–70% of homologue with paramyosins of *Schistosoma mansoni* and *S. japonicum* (Shustov *et al.* 2001).

Epidemiology and geographical distribution

More than 80% of human opisthorchiasis cases in Russia belong to West Siberia. The highest prevalence and intensity of *O. felineus* infection in man are in the middle and middle-lower reaches of the Ob-Irtish river basin with its peak in Hanti-Manssi Autonomous District (HMAD) (Zavoikin 1990). The prevalence of *O. felineus* infection in the aboriginals (Hanti, Manssi, Komi-zirjane) of HMAD, appeared to be 100% in 10-year-olds. In spite of 15–20 years (1965–1980) of mass chloxyl (hexachlorparaxylol) treatment, the prevalence remains at this level until the end of a life. However, egg output, which was $21\,924 \pm 8\,502$ per gram in untreated groups, became two-fold lower. The percentage of patients with clinical manifestations of the infection among the treated and untreated individuals was the same: 34.3 and 34.9, respectively (Bronstein 1986a). Prevalence of the infection after mass treatment in HMAD settlers (mostly Russians living there for generations), was 91.7% in

15–19-year-olds. This went up to 100% in the 20–39-year-olds, with an egg output of 8 292 per gram, decreasing to 70% in older age groups. No correlation between the intensity of infection and the rate of morbidity have been found in either group (Bronstein and Ozeretskovskaya 1985).

In Jamalo-Nenezki Autonomous District (JNAD) (the western riverside of the Ob river's lower reaches) the prevalence of the disease after chloxyl mass treatment remained up to 63.6–70.6% with a low egg output (22–277 per gram). The mean prevalence of *O. felineus* infection in Komi-Permyatsky Autonomous District (KPAD) (the main hyperendemic foci in riverside parts of Kama basin) was up to 46%, with egg output of 392 ± 91 per gram. In the aboriginal of KPAD, the prevalence reached 73.9% by the age of 30–39 years, with a three-fold decrease at 60 years of age. In spite of a much lower intensity of infection, the percent of clinically manifesting cases of gastrocholepaty in KPAD population was two-fold higher than in HMAD (53.8 and 27.7%, respectively). The intensity of infection in the latter was almost 20-fold higher (Bronstein and Ozeretskovskaya 1985). The prevalence of infection in the upper-middle reaches of the Ob and the Irtish rivers is up to 40–60%. Opisthorchiasis is hypoendemic with 1–10% prevalence in the western riverside regions of the Enisei river and in some settlements of the Don and the Pripjat rivers basins. The incidence of human opisthorchiasis in the European part of the Russian North (Leningrad, Arkhangelsk regions) is low or sporadic (Zavoikin 1990).

Opisthorchiasis became an especially acute social and economic problem from the early 1960s with the boom of gas and oil resource exploration in West Siberia and a mass migration there of non-immune population. After having been known as a primarily chronic infection for 70 years (see above), the disease appeared as an acute and severe one with allergic manifestations in 14.5–26% of infected newcomers (Pavlov 1990). From 20 to 30% of these became infected in the first 2 years after arrival, about half got the infection in the next 5–10 years. Symptoms of gastrocholepathy in infected newcomers registered 5 years after arrival were three-fold more frequent than in non-infected ones: 48.4 and 18.6%, respectively (Ozeretskovskaya 1979, 1982). The special surveys conducted in HMAD and in JNAD in the late 1980s and the early 1990s revealed the absence of clinical signs of the acute phase of the disease in the aboriginal in the past (Ozeretskovskaya *et al.* 2001). However, 45% of settlers' group and 14.5% of newcomers (migrants) suffered from an acute phase. Sub-clinical course of the disease has been found in 85.1, 50.0, and 30.5% of the same groups, respectively. Trans-placental transmission of *O. felineus* antigens has been revealed in aboriginals and in some settlers, but not in newcomers. Minimal *in vitro* lymphocyte transformation responses to parasite antigen but normal responses to the mitogen PHA were found in aboriginals (Parfenov *et al.* 1989). Therefore, the long-term clinical and field study of *O. felineus* infection revealed principal differences in the pattern of the disease in the foci with the different endemicity and in different population groups. The earlier in life infection had been acquired and the heavier the infection (re-infection) was, appeared to associate with the less severe clinical manifestations. One presumes that the indigenous population may have developed some form of unresponsiveness to parasite antigens or capability to mount the effective level of immune responses that minimize parasite's development and damage (Ozeretskovskaya 1975, 1982; Ozeretskovskaya and Sergiev 1993). The appearance of genetic heterogeneity in response to infection [genetic adjustment (Capron 1995)] leads to optimal fitness of host–parasite relationships (Wakelin 1994) can probably play a role too.

The attempt to control opisthorchiasis by mass chloxyl treatment was a failure determined by the moderate activity of the drug itself and the high rate of re-infection due to the persistence of the habit of a raw fish consuming in the population (Bronstein *et al.* 1987; Sergiev 1989). The programme of 'Supressive-radical (or selective-radical) therapy of *O. felineus* infection in endemic foci' was proposed in the end of the 1980s (Sergiev 1989). That was based upon the principles of WHO programme of schistomiasis control and the programme of opisthorchiasis control in Thailand. According to these, azynox (praziquantel) treatment (40 mg per kg of body weight, if necessary repeated 6–12 months later) should be used in all clinically manifesting and/or intensively infected cases in foci with an endemicity of more than 40%. The idea was to diminish the unfavourable pressure of the high rate of infection in individuals rather than to eliminate infection in endemic foci.

Selective-radical biltricide (Bayer-Merck) therapy was used in KPAD, in an aboriginal settlement with a prevalence of the infection of 84.4% and a mean egg output of 1 588 per gram. The results showed an efficacy of 85.9% (cured or with a considerably diminished egg output) at six months and 81.5%, 12 months later. Side effects were registered in 89.7% of cases. Clinical efficacy of the treatment was 25.8% only (Bronstein *et al.* 1989). Praziquantel (India) therapy with the dose of 25–50 mg was applied in two settlements in INAD with a mixed population (aboriginals, settlers, migrants). The prevalence of *O. felineus* infection there was 40.5 and 45.7% and the mean egg output about 200 per gram. Mean cure-rate 6 and 12 months later was 92.5 and 90%, respectively. Side-effects of a pharmacological nature were registered in 45.3% of patients. Allergic symptoms appeared in 10% of settlers and in 30% of newcomers. In total 18% cases of side effects were revealed in aboriginal and 60% ones in settlers and migrants. Clinical efficacy of the treatment was revealed in 58.6% of patients, or two-fold higher than in KPAD (Bronstein *et al.* 1991). The difference of efficacy was connected possibly with a higher (average) doses of the drug, a lower intensity of infection in the patients of JNAD and/or with a higher potential to cure short-term opisthorchiasis. The lower percent of side-effects in aboriginal patients seems to be an additional marker of their tolerance to the effects of *O. felineus* antigens (Ozeretskovskaya and Sergiev 1994).

Clinical features

Chronic cholepathy and gastroduodenitis without marked changes of clinical and biochemical laboratory parameters are the most characteristic features of the infection (Ozeretskovskaya *et al.* 1985). There is no direct correlation between morbidity and intensity of infection. Morbidity in KPAD population was two-fold higher than in HMAD: 53.8 and 27.4%, respectively in spite of 20-fold higher egg output in the latter (Bronstein and Ozeretskovskaya 1985). In hyperendemic foci (HMAD, INAD), 85% of aboriginals and only 50% of settlers showed sub-clinical course of infection, but only 30% of migrants with 3.5 years history of infection were asymptomatic (Ozeretskovskaya and Sergiev 1993). The acute phase of the infection in migrants is manifested by high fever, myalgia, Loffler syndrome, and peripheral blood hypereosinophilia. In severe cases, hepatitis and myocarditis were noted while toxic shock has been registered in some severe cases (Ozeretskovskaya 1979; Pavlov 1990). Mass treatment of population of hyperendemic foci with chloxyl and later with praziquantel prevented the development of complicated forms of chronic opisthorchiasis with the ruptures of dilated bile ducts, bile peritonitis, and suppuration had been reported in the past (Jablokov 1979). However the development of chronic persistent hepatitis in 88.85% of examined in-door patients and 11.2% of chronic active hepatitis predominantly in young patients with super- and re-infections in the Novosibirsk region (mesoendemic area) was reported (Paltzev and Nepomnyashchikh 1998).

Chronic opisthorchiasis has been reported to aggravate the course of enteric bacterial infections, viral hepatitis, and to provoke chronic carrier of *Salmonella typhi* (Jablokov 1979). The mean titres of antibody to Vi-antigen in *S. tiphy* carriers with chronic *O. felineus* infection revealed to be significantly higher than in carriers without the latter 1 : 2 560 and 1 : 320 (Lepekhin *et al.* 1990). The titres of antibodies to Epstein–Barr virus antigens in chronic opisthorchiasis patients are significantly higher than in an adequate control's and directly correlate with the severity of parasitic disease and the amount of structural aberrations of lymphocyte's chromosomes (Ilyinskikh *et al.* 2001). Opisthorchiasis interferes with chemotherapy of tuberculosis increasing the percentage of treatment failures (Korablev 1990). Tendency to bronchospastic syndrome in newcomers to endemic foci with chronic opisthorchiasis promotes the chronic course of respiratory infections and the development of bronchial asthma and pneumosclerosis (Jablokov 1979; Ozeretskovskaya 1979). The incidence of primary cholangiocarcinoma in West Siberia was 16-fold higher than was its average incidence in the population of the USSR (Zhdanov 1990). It is believed that *O. felineus* infection promotes cholangiocarcinogenesis in indigenous populations of hyperendemic foci as do *O. viverrini* (Migasena 1990) and *Clonorchis sinensis*. This is thought to be through the lifespan process of proliferation of intrahepatic bile duct epithelial cells and by some kind of immunosuppression, characteristic to infection. The role of genetic predisposition of aboriginals to cancerogenesis can however, not be excluded (Ozeretskovskaya 1975; Bychkov and Jarotskii 1990). In golden hamsters with *O. felineus* and HSV-2 infections, the virus particles were found in hepatocites along with a massive proliferation of bile ducts epithelial cells. HSV-2 particles were found in the nuclei and cytoplasm of homogenized parasite cells by immunofluorescence method. The possible involvement of HSV-2 in the development of the primary liver carcinomatosis was discussed (Ivanskikh and Bliznjuk 1996). Disorders of reproductive function, of pregnancy as well as of physical undevelopment of children in hyperendemic areas of *O. felineus* infection have been reported (Jablokov 1979; Ozeretskovskaya *et al.* 1985).

Diagnosis of *O. felineus* infection is formed on clinical and on epidemiological data and verified by ovoscopy of feaces (Kato, Kato–Kaz, Allen–Ridley techniques) and/or duodenal juice. In the acute phase of the disease and in sero-epidemiological studies ELISA is predominantly used. ELISA gives 7.30.5 (log 2), 1 : 158 (mean geometric titre) in aboriginies, 7.70.4, 1 : 208 in settlers, and 8.70.5, 1 : 416 in migrants in the chronic phase of opisthorchiasis (Parfenov *et al.* 1989).

Treatment with praziquantel 40–60 mg per kilogram of body weight is effective up to 60–80% if re-infection is excluded. The side effects of the treatment of pharmacological nature were registered equally (90%) in all population groups. Side-effects of allergic nature appeared in 30% of newcomers only. The clinical efficacy of treatment in different hyperendemic foci in in-door patients was 33.3%, in out-door ones 25.8–58.6% (the latter – perhaps due to the lower intensity of infection in INAD) (Ozeretskovskaya and Sergiev 1993).

Prognosis of infection in aboriginal and settler groups of population that are under an administration of regular specific treatment is benign. Acute opisthorchiasis in non-immune migrants sometimes requires an urgent desensitizing therapy and later a proper follow-up for prevention of the development of erosive gastroduodenitis, duodenal ulcers, and chronic pulmonary syndrome (Ozeretskovskaya 1975, 1979, 1982; Ozeretskovskaya and Sergiev 1994).

Prevention

Infected fish is safe after boiling or hot smoking. Metacercariae are killed by freezing fish for 7 h at − 40°C, 14 h at − 35°C or 32 h at − 28°C. Salted fish is disinfected if the gravity of salt solution is up to 1.2 of water weight at 1–2°C for 10–40 days with respect to the mass of a fish. The prevention of river water pollution by human and animals feaces, the irrigation of riverside marshes, and the administration of molluscocides in the endemic foci of opisthorchiasis decrease the prevalence of parasites in fish (Prevention of parasitic infections on the territory of the Russian Federation 1997). From 99 to 100% of larvae mortality after a fast electron radiation of fresh fish is reported (Nazmov *et al.* 2001). Prazyquantel treatment of out-door patients at 40 mg per kilogram with 85.8% of efficacy in hyperendemic foci and 86.1% in mesoendemic ones, followed by 24.8–27% of side-effects, is proposed as one of the control measures for opisthorchiasis recently (Zavoikin *et al.* 2001).

Trichinellosis

High prevalence of *T. spiralis* infection with the vast range of animal hosts was revealed in the 1960s in the Russian Arctic from the Archangelsk region to Chukotka. The circulation of infection is considered to be supported by a high viability of parasite larvae in carcasses of wild animals (carnivores food) in winter. Also, by the migration of polar bears and polar foxes from North Atlantic in the west to the Bering sea in the east. *Thalarctos maritimus* can cover a distance up to 80 km and more in 24 h. It was believed that marine animals were involved in the transmission of *Trichinella* through carnivorous birds excrement contaminated by parasite larvae and washed into a sea or through bottom *Crustacea* consumed by seals. Six outbreaks of trichinellosis in man involving 97 cases were recorded from the late 1950s to the end of the 1960s in northern Europe and Asiatic part of the USSR (Shikhobalova *et al.* 1969). In 1957, a severe outbreak of trichinellosis involving more than 40 persons with one lethal case occurred in the north of the Leningrad region. The source of infection was undercooked and salted pork, also pigs had been fed by the carcasses of caged silver foxes (Ozeretskovskaya *et al.* 1958). The scheme of *Trichinella* circulation in the animals of the Arctic was suggested (Ozeretskovskaya and Uspensky 1957).

The clinical features and the course of disease from the bear *Trichinella* strain on the Bennet Island (*Th. maritimus*) and from KPAD (*Ursus arctos*) were characterized by a long incubation period (30–35 days). The disease is manifested by hypertermia, skin rashes, general oedema, delirium, hallucinations, and in some cases by acute abdominal pains with haemorrhage syndrome and a hypereosinophilia. The latter in the patient from the Bennet Island outbreak reached 88% (24 816 cells per millilitre) (Ozeretskovskaya and Uspensky 1957; Ozeretskovskaya 1968). The peculiarity of clinical features of trichinellosis from the northern strains of *Trichinella* (*T. spiralis var. nativa* today) is due to its poor compatibility with a human and a mice host that manifesting by a protracted intestinal development of the parasite, a violent cell infiltrations around the larvae in muscle tissue and the destruction of larvae, as a muscle biopsy of patients showed (Ozeretskovskaya 1968; Ozeretskovskaya and Pereverzeva 1976). The experimental study of *Trichinella* strain from *Th. maritimus* by Franz Josef Land revealed a stable low invasiveness for mice, the peculiarities of proteins content (disc electrophoresis in polyacrilamid gele) as compared with the laboratory (primarily synanthropic swine) strain in mice from the Martsinovsky Institute, a rather poor ability of the larvae incapsulation in mice muscle tissue; a higher sensitivity to thiabendazole. In contrast to the two-layered fibrous-hyaline capsules of larvae in animals from temperate areas capsules of arctic trichinella larva in polar bears and polar foxes have irregular form, multilayered, with unequal hyaline regeneration of encapsulated portion of the sarcoplasm (Ozeretskovskaya and Pereverzeva 1976). The study of 146 isolates of *Trichinella* from wild animals of various regions of Eurasia by cross-breeding methods revealed *T. spiralis var. nativa* in Arctic islands and in Galarctic zone of Russia (Shaikenov 1995). One presumes that there is a sympatric spread of *T. spiralis* and *T. pseudospiralis* in the nature and in synanthropic foci in Russia with infected birds as a link (Garkavi and Zverzhanovsky 1999). The high sera reactivity of patients with trichinellosis myocarditis and cardiomyodystrophia to α-myosin cardiac peptide (known to induce experimental allergic myocarditis, Wegmann *et al.* 1994) revealed recently (Ozeretskovskaya *et al.* 2000) corresponded with the clinical and morpho-logical pattern of allergic myocarditis developed in guinea pigs sensitized with somatic *Trichinella* antigens with lanoline and mineral oil (without BCG) (Ozeretskovskaya 1976).

The incidence of trichinellosis in the previously endemic foci of Ukraina, Belorussia, and the Central Europe regions of Russia was almost zero in the 1970s. The new economic development of the Russian North and the Northeast in the 1970s caused the mass influx of population from the regions of the country free of zoonotic helminthic infections. The inci-dence of trichinellosis in Russia became 16-fold higher during 1971–1975 and the source of infection in 96% of patients was bear or wild boar meat. The outbreaks were registered in the Magadan region, Kamchatka and Sakhalin Island, the Krasnoyarsk, and the Khabarovsk regions. Out of 377 cases of trichinellosis registered in Russia in 1976–1977, 178 (47.2%) belonged to the outbreaks in the North and the Northeast of the country. Bear meat accounted for 96.6% of these 178 cases. In the industrial region of Taimir peninsula 132 (62.3%) cases of trichinellosis were registered in 1984–1993, with 80 (37.7%) of them in rural areas. *Ursus arctos* meat accounted for 84 (39.6%) cases, pork accounted for the other 60.4% (Knjazev 1994). More than 200 cases of trichinellosis caused by bear meat were reg-istered in the Sakhalin and in Kamchatka in 1978. Trichinellosis caused by bear meat was up to 43% in 1979 and only about 30% in 1980. The decline of the portion of trichinellosis caused by bear meat was in some way the result of reviving of endemic trichinellosis in the Europe regions of Russia (Ozeretskovskaya 1979). The incidence of trichinellosis in Russia was about 700–1000 cases per year in 1990–1994. Forty-three outbreaks of infection with 594 cases including 54 from the bear and a wild boar meat were registered in 1995. Forty-three outbreaks with 789 cases of infection but with only 13 of them from bear meat were notified in 1996. In total, 1 383 cases (two lethal ones) of trichinellosis were registered in 1995–1997 in Russia, with pork accounting for 80% of them. Outbreaks connected with infested bear meat occurred in Krasnoyarsk region and in Kamchatka (Ozeretskovskaya 1997a). In 1998, the infection was contracted only in two cases through bear meat in the Krasnoyarsk region and in five cases through dog meat in the Irkutsk region and Kamchatka. In 2000, only 483 cases of trichinellosis were registered in Russia, 105 (21.9%) them in the Northeast part of the country, particularly in the Krasnoyarsk, Irkutsk, and Khabarovsk regions (81, or 16.8% cases). The general decline of the incidence of the infec-tion is due to the strict sanitary–veterinary checking of meat.

Sero-epidemiological study (indirect ELISA technique) including 713 persons (519 chil-dren) in the northern regions of Russia (the Federal Programme 'Children of the North', see above) in 1995–1996 revealed the highest seroprevalence of trichinellosis in HMAD and the lowest one in Chukotka: 11.1 and 1.5–5, respectively. The rather high percentage of diag-nostically significant titres in HMAD might be connected with the ethnically mixed group examined. Only children were examined in Chukotka (Table 6.1). In KPAD, seroprevalence was 5, but in 2.2 of the examined cases, the titres were of diagnostic significance (Table 6.1). In aboriginals of the Tomsk region and in the Taimir peninsula seroprevalence of trichinellosis was equal. However, diagnostic titres in the former were higher than that of epidemiological significance (11.1 and 7.1, respectively) and was the highest among the diagnostic titres in all surveyed regions (Table 6.1). In none of examined regions, outbreaks of trichinellosis was

Table 6.1 Seroprevalence[a] of helminthic infections of medical–social significance in
the North of Russia

Region	No. of children examined	No. of adults examined	Group of population	Infections			
				E. granulosus	E. multilocularis	T. spiralis	T. canis
KPAD	131	49	Mixed	2.1[b]	0[c]	5 (2.2)[d]	15.4 (7.7)
Tomsk	5	94	Aboriginals	6.4 (9.6)	12.8 (3.2)	7.1 (11.1)	14.1 (42.4)
Taimir peninsula	51		Aboriginals	(2)	(3.9)	7.8 (2)	5
HMAD	201	51	Mixed	5.9 (1.5)	3.9	11.1 (6.3)	14.2 (11.5)
Chukotka	131		Aboriginals (mostly)	15		1.5–5	4.6

Notes
a Epidemiologically significant titres, %.
b In children; in adults 6.1.
c In children; in adults 2.2.
d In brackets here and further – diagnostically significant titres.

registered during the survey. One presumes that in the Tomsk region some cases of trichinellosis might be diagnosed as acute opisthorchiasis. The lack of clinically manifesting trichinellosis in the indigenous population of the North of Russia in spite of a high seroprevalence of the infection even in children (see HMAD in Table 6.1), seems to be the result of an early and permanent infestation or of some kind of tolerance to parasite antigens (see section on 'Opisthorchiasis').

Clinically manifesting cases of trichinellosis in Russia are treated with albendazole at 5–10 mg per kg for 5–10 days. Glucocorticides are administered in the severe, complicated cases only for 3–7 days, simultaneously with albendazole. For moderately severe cases a combination of the latter with adenylate-cyclase cAMP-proteinkinase inhibitors (voltaren, brufen) and the replacement of glucocorticides by this combination in the severe cases as soon as possible are recommended. The preventive treatment with benzimidazole carbamates of persons who have consumed infected meat is widely used (Ozeretskovskaya and Sergiev 1994; Ozeretskovskaya *et al.* 1994).

Diphyllobothriases

Diphyllobothriasis had been registered from the 1930s through all territory of the Russian North from the Kolsky peninsula and Karelia in the West, to the Jamal peninsula, JNAD and HMAD in the West Siberia, the Krasnoyarsk region, Yakutia, Kamchatka, Sakhalin, and the Magadan region in the east. The higher prevalence of infection was revealed close to food-land served by water reservoirs than along coasts of large rivers. The prevalence of infection in the indigenous population of the littoral zone of the Enisei River in the 1930s had reached 52%. In the early 1960s it varied there from 1.4 to 19.2%. In the foci of the Kolsky peninsula, the prevalence of the infection by the same time was reduced from 23.7 to 1.53% (Shikhobalova *et al.* 1969). New foci of diphyllobothriases have been formed after the building of water reservoirs with hydro-energy complexes, in particular the one in Krasnoyarsk (Klebanovskii 1985). The morbidity rate per 100 000 population for diphillobothriasis in the USSR in 1987 was 13.1 (Sergiev 1993).

The main incidence of diphyllobothriases in man in Russia is due to *D. latum* infestation with synanthropic foci (Klebanovski 1980, 1985). The intermediate hosts are *Copepoda* (*Crustacea*) and freshwater fish – pike, perch, salmon, trouts, and eel. Even in sparsely populated regions of Siberia, such as Karim lakes system in the western riverside of the Konda (HMAD), plerocercoid larvae were found in 92% of pikes in lakes that were close to villages and in 7.1% of pikes in lakes practically not visited by man (Klebanovskii 1985). Synanthropic foci of *D. latum* infection are the most intensive in the middle and middle-lower reaches of the West and the East Siberia great rivers (but in the lower reaches of the Irtish river) (Klebanovskii 1985). The prevalence of *D. latum* infection in the population of middle reaches of the Ob river in the 1970s was 10%. In aboriginal populations (Komi, Hanti) of the Ob middle-lower reaches it was 43.6% and was only 25–7% in the lowest reaches of the river (Klebanovskii 1985). In the middle zone of the Krasnoyarsk water reservoir, the prevalence of *D. latum* infection in 1985 reached 11.7%; in 1–2-year-old children it was 4% in 1984 and 12.7% in 1985 (Plyuscheva *et al.* 1987).

Diphyllobothrium dendriticum is the causative agent of 'seagull's diphyllobothriasis'. The disease takes second place in incidence of diphyllobothriases in man (Klebanovskii 1980, 1985). The first intermediate host – *E. gracilis*, the second ones – *Coregonus peled*, *C. lavaretus pidshian*, *C. muksun*, and *Thimallus arcticus*. The definitive hosts are gulls, magpies, and crows. The high incidence of *D. dendriticum* infection in the aboriginals of the Taimir permits to consider that man served as the one of the (local) definitive hosts of the parasite. *Diphyllobothrium dendriticum* infection have been revealed in aboriginal population of Arctic zone of West and East Siberia, in Jakutia and Chykotka. In the indigenous population of the upper reaches of the Khatanga river (the Taimir peninsula) the prevalence of the infection was up to 11.8–23.5%, but in the lower reaches only 7–7.5% (Klebanovskii 1985). In the Sackha Republic (Yakutia) the prevalence of *D. dendriticum* infection in the middle-lower reaches of the Lena river was up to 6.6% in adults and 1% in children (Simonova 1995).

There are mixed – *D. dendriticum* and *D. latum* foci of diphyllobothriases in the Russian Arctic zone, where the former parasite makes about 5/6 of the incidence (Klebanovskii 1985). In 1991, the morbidity rate of diphyllobothriases in the Taimir peninsula was 2.36. The prevalence of diphyllobothriases in children of INAD was 19.7% (Khodakova *et al.* 1996). Diphillobothriasis (without species discrimination) was revealed in 12.9% of villagers and in 23.7% of reindeer-breeders in Chukotka (Lebedev *et al.* 1986). The zoonotic parasite *D. ditremum* in the Russian North is not socio-economically significant (an abortive course of infection) (Klebanovski 1985).

Diphyllobothrium klebanovskii revealed to be the main (if not the only) agent of the infection in humans in the northeast of Russia (Muratov and Posochov 1988). The main definitive host of *D. klebanovskii* is *U. arctos* (parasite was revealed in 16 out of 36 bears). *Canidae*, *Felidae*, and *Mustelidae* were infected for 1.7, 0.6, and 0.5%, respectively. Animals contracted the disease through consuming of *Oncorhynchus gorbusha* (humpbacked salmon), *O. keta*, and *Hucho perrui*. The prevalence of the infection in aboriginals in the North Pacific reaches 7.6% (Muratov 1993). Histological study of plerocercoids and adult worms of *D. klebanovskii* permitted to consider that it cannot be discriminated from *D. nikonhainense*, *D. luxi*, and *D. giljacicum* (Dovgalev *et al.* 1991). On the grounds of the priority of *D. luxi* description (Rutkevitch 1937), Dovgalev and Valovaja (1996) suggested to consider the latter as the main agent of diphyllobothriasis in the Pacific coast of the Russian Far East. The mean morbidity index of diphyllobothriasis in Russia was 15.7 (9.9 in children) in 1999, also reached 489.6

in EvenkyAD, 382 in Yakutia, 304.6 in NenezkyAD and 247.7 in TaimirAD (Siskova *et al.* 2001).

Taeniasis due to *Taenia saginata* infection

The Russian North is the zone of the widest distribution of *T. saginata* infection, in Russian literature – teniarinchosis. The prevalence of infection in the Vyatka region was up to 5% in 1943–1945, but decreased to 0.4% in 1957. The mean prevalence of infection was 0.34% in JNAD in 1965 and 0.53% in 1973 (Bessonov 1988). It was experimentally proved that *Cysticercus bovis* from the North reindeer (*Cervidae*) is not infective for cattle and vice versa – *C. bovis* from cattle (the southern isolates) are unable to reproduce in reindeers. Cysticercosis of the North isolate have been found located in the brain of reindeers (Kirichek *et al.* 1984). The raw brain of the North reindeer is the main source of infection for the aboriginals of the Yamal peninsula. The Mountain-Altai Autonomous District was the most intensive focus of taeniasis due to *T. saginata* in the West Siberia. The prevalence was up to 43.6% in 1929. It was 5.7% in the whole Altai region in 1929, but only 0.024% in 1974. The prevalence of infection in the Irkutsk region was 1.9% in 1959 and 1.5% in 1963. The disease is more prevalent in Buryatia (part of the Irkutsk region) where it reached up to 5.9% in 1959, but decreased to 0.1% in 1973. The prevalence of the infection in Yakutia was up to 40% in 1925 and it ranged from 1.8 to 15.2% in 1951–1970. The same figure for Sakhalin was 0.02% in 1980 (Bessonov 1988).

Echinococcoses

Echinococcus granulosus and *E. multilocularis* (genus *Alveococcus*, Abuladze, 1960, *Alveococcus multilocularis*) infections in man were recorded in the late 1950s and the early 1960s in the Archangelsk region, in the northern zone of Tjumen region including the Yamal peninsula, in the Novosibirsk and the Krasnoyarsk regions and in the Yakutia. Hospital and autopsy cases of *E. multilocularis* were up to 57% of all cases of disease in Yakutia, 58.1% in the north of the Krasnoyarsk and 64.2% in the Novosibirsk regions in 1937–1939. The highest morbidity per 100 000 of population was revealed in the valleys of the Lena, Amga, and Aldan rivers. Polar and red foxes had been considered as the principal definitive hosts of *E. multilocularis*. *Lemmus obensis*, *Microtus gregalis*, *M. oeconomus*, *Clethrionomus glareolus* and *Ondatra tibethica* were recorded as additional hosts. *Echinococcus granulosus* was found mainly in dogs and in some cases in wolves. Larval stages of the agent were present in sheep, swine, cattle in moderate climate areas, but in the Yakutia, Chukotka, and the Archangelsk region they were found in reindeers (Shikhobalova *et al.* 1969).

Echinococcus multilocularis infection is transmitted to man in natural foci through the shooting of game, by eating wild berries contaminated by the excrement of wild carnivores or by drinking polluted water from natural reservoirs. The survival time of onchospheres was up to 67 days in the excrement of foxes covered with snow, in a straw stack, at the bottom of a lake in the Siberian winter. Dogs are infected with *E. multilocularis* by hunting for small rodents. Transmission of *E. granulosus* infection is due to continuous contact of man with dogs infected by eating the internal organs of reindeer, domestic animals, and garbage containing the agent cysts. Man contracts hydatid disease through close contact with infected dogs (Shikhobalova *et al.* 1969).

Anthropogenic factors have changed the principally natural origin of *E. multilocularis* infection foci and led to the possibility of formation of synathropic (settlement's) ones

(Martinenko *et al.* 1984). Two hundred and seventeen cases of *E. multilocularis* infection were registered in the Korjak Autonomous District (Kamchatka) in 1955–1986, the morbidity rate was 6.1 in aboriginals and 0.3 in settlers and migrants (Stepchuk *et al.* 1990). The morbidity rates of *E. multilocularis* and *E. granulosus* were 9.5 and 15.6, respectively, in the same groups in the north of the Omsk region in 1989. But, clinical course of infection in the latter had been significantly more serious than in patients from two former groups. Migrants made 26% in the group with the severe course of disease and only 5% of moderately severe cases among 84 patients with *E. multilocularis* infection examined and followed up in the Clinic of the Martsinovsky Institute in 1961–1970. A 30-year-old man, a Moscow resident, died from *E. multilocularis* infection complicated by the hepato-renal insufficiency after 3 years of living in the Magadan region (Ozeretskovskaya 1979).

The morbidity rates of *E. multilocularis* and *E. granulosus* infections were 9.5 and 15.6, respectively, in the north of the Omsk region in 1989, at the same time sero-epidemiological survey (ELISA) gave 5.2 and 5% positive results, respectively (Klebanovski 1985), perhaps due to the quality of used diagnostic tools. In the mixed, synanthropic and natural foci of *E. granulosus* in the Khabarovsk region, 6.4% of the population were seropositive and contamination of soil by the parasite oncospheres was found in 6.6% of tests (Maslov, 1991). The morbidity rate (for 100 000 of population) of echinococcoses in Chukotka in 1992 was 2 for the total population and 23 for aboriginals; about a half of cases were *E. granulosus* infection. The lung localization of cysts was found in 40% of cases, which is supposed to be the specificity of the strain (Boytsov *et al.* 1992; Bessonov 2001a,b). It is of interest that among 10 patients with *E. granulosus* cysts in the lungs (ethnically mixed group) in HLA-B5 carriers the production of IFN-gamma was significantly higher than in HLA-A2 carriers (Ozeretskovskaya *et al.* 1992). One presumes that the polar fox/lemming strain of *E. granulosus* is less virulent than the red fox/voles, gerbils, muskrats strain spread in tajga-zone of Siberia.

Sero-epidemiological survey of the northern regions of Russia performed by the Martsinovski Institute in 1994–1996 (Report to the Ministry of Health of Russian Federation, 1996) revealed the predominance of *E. granulosus* infection in KPAD with its higher epidemiologically significant rate in adults (Table 6.1). In contrast, in aboriginals of the Tomsk region, the positive tests with *E. multilocularis* antigen were two-fold higher: 12.8 and 6.4. However, diagnostically significant rate for *E. granulosus* infection was almost three-fold higher than for *E. granulosus* (Table 6.1). In aboriginals of the Taimir peninsula, diagnostically positive tests were 2 for *E. granulosus* and 3.9 for *E. multilocularis*. In the mixed group of HMAD's population seroprevalence by *E. granulosus* prevailed. Perhaps that was due to the four-fold higher number of children in the examined group (Table 6.1). Seroprevalence of children (5–18-year-old) by *E. granulosus* was the highest in Chukotka: 15 (Table 6.1). It seems that the Tomsk region and Chukotka remain the active mixed foci of both infections. Seroprevalence in aboriginals of the Russian North by both infections was higher than of settlers.

Toxocarosis

Toxocarosis – *Toxocara canis* induced infection in humans, especially in children. As the paratenic hosts of the parasite, it is attracting a lot of attention of physicians of different specialities, of epidemiologists and of immunologists. *Toxocara canis* was revealed in 4.1% of dogs and caged wild animals in the European part of the Russian North (Avdyukhina and Lysenko 1994). Sero-epidemiological survey accomplished in Russia in 1984–1989 showed 6% of seroprevalence with Toxocara antigen (ELISA) in the healthy children of the Irkutsk region and 7.3% in Tjumen city. The mean seroprevalence of toxocarosis was 1.7% in Tjumen city and 2.4% in the Jakutia Republic (Avdyukhina and Lysenko 1994). Irkutsk city is considered to be an endemic foci of toxocarosis. The infection was revealed in 29.5% of dogs there. Seroprevalence of toxocarosis was 6% in children and 2.6% in adults. Seroprevalence of infection in the other towns and in villages ranged from 2 to 11.5%. There was no difference in the prevalence of infection between the Russian and indigenous population (Evenki): 7 and 7.9%, respectively (Kuprijanova 1989). The titres were usually 1 : 6400 and higher in the clinically manifesting cases (Avdyukhina and Lysenko 1994). Sero-epidemiological survey of 465 children and teenagers and 382 adults made by the Martsinovski Institute in 1995–1996 revealed seroprevalence of toxocarosis of up to 14.1–15.4 in KPAD, in the Tomsk region, and in HMAD. Diagnostically significant titres (1 : 400) were up to 11.5% in HMAD and 42.4% in the Tomsk region (Table 6.2). In the Taimir Peninsula and in Chukotka, where adult aboriginals mostly were examined, sero-prevalence was about three-fold lower: 5 and 4.6%, respectively (Table 6.2). An additional examination of 20 children and 118 adults in the Khabarovsk region showed 18.8% of epidemiologically significant positive tests and 42.6% of them of diagnostic value (Report for the Federal Programme 'Children of the North' 1996). The specific IgG titres reached 1 : 400–800, in some cases – 1 : 1280–1 : 2560 in 17 patients of the Clinic of the Martsinovski Institute with toxocarosis clinically manifesting by lympadenopathy, hepatomegaly or/and pulmonary infiltration, by the blood eosinophilia up to 3 914–18 360 cells per millilitre (Ozeretskovskaya 1997b). The incidence of infection in Russia rose from 15 cases in 1991, to 641 (69.7% children) cases in 1999 mostly due to the wide serological screening of patients with allergic manifestations. The mean morbidity index in Russia was 0.8 in 2000, also, it reached 8.7 in the Udmurtia Republik, 7.6 in the Perm region, 5.2 in the EvenskAD (the Krasnoyarsk region), and 5.5 in the Sakhalin Island (Siskova *et al.* 2001).

Intestinal helminthiases

Parasitological survey (Kato–Kaz, Allen–Riddley methods) of 30 388 persons, mainly children and few adults, in 15 out of 28 northern regions of Russia was performed in 1994–2000 (Table 6.2). The population of villages and day-care children institutions were included. While 34.6% of enterobiasis was revealed among examined subjects, 10.1% had other intestinal helminthiases. The highest percentage of *Enterobius vermicularis* infection was revealed in Khabarovsky kraj – 54.6%. From zero to 4.1% of ascariasis was found in the examined subjects in KPAD. Up to 3.7% of hymenolepiasis was reported in the Taimir population. Diphyllobothriasis was revealed only in 1% of those examined in both regions. Enterobiasis was found in 15% of examined persons (mostly children) in HMAD (where *O. felineus* infection was revealed up to 52.1%). The Tomsk region gave the most impressive results, where *O. felineus* infection was revealed in 54.3–67.1%. In addition, enterobiasis was found in 32.8%, ascariasis in 2.4%, and lambliasis in 3.4% (Report to the Federal Programme 'Children of the North' 1996). Enterobiasis in the north of Russia is distributed equally or even more intensively than in the other geographical zones of the country. This underlined the social significance of infection and the necessity of improving the living conditions of the population of the North. Indigenous ascariasis cases in the North were connected to focal greenhouses. Parasite eggs were able to develop under artificially warm conditions of greenhouses but not in the soil of natural environment (Lebedev *et al.* 1996). The mean morbidity rate for enterobiasis in Russia was 650.6 in 1999, also in the Arkhangelsk region it reached 6 038.1, in the Mari El and the Udmurtia Republic it was 6 489 and 6 138, respectively, and in NenezkyAD it was 5 537 (Siskova *et al.* 2001).

Table 6.2 Prevalence of helminth infections (data collected by the Martsinovsky Institute of Medical Parasitology and Tropical Medicine in the field during 1994–2000)

Territories	Number of settlements where investigation had been performed	Number of subjects examined	With enterobiasis among them	With other interstinal helminthiases among them
Altaiszkij kraj	7	1 805	851	14
Amurskaja oblast	17	2 416	893	188
Burjatskajia AO	4	1 042	501	4
Chukotskaja AO	19	1 883	350	11
Kamchatskaja oblast	3	805	355	2
Khabarovskij kraj	29	4 072	2 225	144
Khanti-Mansi AO	9	2 929	1 391	485
Komi-Permjatskij AO	9	1 963	320	147
Korjak AO	5	1 528	325	2
Primorskij kraj	6	1 105	138	56
Sakha-Yakutija Republic	6	2 474	371	118
Taimirskij AO	10	1 953	372	12
Tomskaja oblast	32	4 388	1 366	1 677
Tumenskaja oblast	9	481	245	133
Yamalo-Neneckij AO	6	1 542	810	75
Total	171	30 388	10 513	3 068

Protozoal infections in the Russian North

Malaria

Malaria, including *Plasmodium falciparum* infection was distributed far to the North of European and Asian parts of Russia. The extreme northern boundary of malaria reached 64° N (Sergiev *et al.* 1968). The incidence rate for malaria, predominantly *P. vivax* in the early 1920s fluctuated in Arkhangelsk region from 176 to 409 per 10 000 inhabitants. Local transmission was intensified by extremely hot summers in the early 1920s (Dobreitser 1924). Hundreds of *P. vivax* malaria cases with long incubation periods were registered in Yakutia even in the 1950s (Dukhanina 1962). Anti-malarial stations had been built in the country from the 1930s. The problem of malaria eradication was addressed by Martsinovsky and Sergiev at this time. The task of malaria liquidation as a mass infection (morbidity rate under 10 for 100 000 of the population) was established in 1949. The tempo of a decline of the infection prevalence was very high in 1934–1963. Contact insecticides as well as the system of mass chemo-prophilaxis of the population helped to achieve the eradication of malaria in the USSR in the 1960s (Sergiev *et al.* 1968). Nowadays, only sporadic imported cases of the disease are registered in the northern territories of Russia, but introduced cases are registered in the south of the country. Local transmission in the North has been absent for decades despite the presence of the vector.

Intestinal protozoa

A wide range of intestinal protozoa infections was detected among indigenous ethnic minorities and local Russians (settlers) living in the northern territories for generations. Prevalence of different intestinal protozoa revealed by examination of onetime collected faeces without any concentration techniques, fluctuated from 26 to 94% (Sorochenko 1968). The frequency of a detection of cysts and trophozoites of *Entamoeba histolytica* in healthy subjects in the Arkhangelsk region was not changed between the 1930s and the 1960s. Reported prevalence was from 6 to 10% (Epstein 1932; Sorochenko 1966). The spectrum of other intestinal protozoa detected among aboriginals and local Russians includes: *E. hart-manny* (prevalence fluctuated from 1 to 35%), *E. coli* (9–32%), *Endolimax nana* (5–34%), *Giardia lamblia* (9–28%), *Blastocystis hominis* (16–24%), *Iodamoeba butchlii* (4–15%), and *Chilomastix mesnili* (2–11%). Prevalence of *G. lamblia* in children was constantly higher than in adults. There was no gender difference in prevalence of intestinal protozoa. A higher prevalence sometimes among Russians or in aboriginals was detected in different surveys (Epstein and Zertchaninov 1932; Sorochenko 1966, 1968; Khudoshin 1968). Giardiasis contracted in Leningrad was reported in tourist groups from the USA (Centre for Disease Control 1986) and Finland (Jokipii *et al.* 1985). Three hundred and seventy three children under 7 years of age were examined for *Cryptosporidium oocysts* in the hospital of St Petersburg. The oocysts were revealed in 10 (2.7%) children. The infection manifested by fever, abdominal pains, and diarrhoea in two of them (Beyer *et al.* 1990).

Information on the prevalence of intestinal protozoa among inhabitants of the northern territories of Russia is extremely rare. Intestinal protozooses are not included in the list of notifiable diseases. Only giardiasis has been included recently. That is why general practitioners are not familiar with these infections. Fragmentary data presented above are results of occasional surveys that been carried out by solitary researchers.

References

Avdyukhina, T. I. and Lysenko, A. Ya. (1994). How many patients with visceral toxocarosis are there in Russia? *Med. Parasitol. Parasit. Bolezni*, **1**, 12–16 (in Russian).

Bessonov, A. S. (1988). *Teniarinchosis – Cysticercosis*. Nauka, Moscow, pp. 40–77 (in Russian).

Bessonov, A. S. (2001a). The types of echinococcosis–hydatidosis foci at the territory of Russia. *XXth International Congress of Hydatidology*, 4–8 June 2001, Kusadasi, Turkey, Abstract Book, p. 353.

Bessonov, A. S. (2001b) Echinococcoses in the Russian Federation. *Med. Parasitol. Parasit. Bolezni*, **4**, 3–8 (in Russian).

Beyer, T. V., Antikova, L. P., Gerbina, G. I., Sargaeva, V. G., and Sidorenko, N. V. (1990). Human cryptosporidiasis in Leningrad. *Med. Parasitol. Parasit. Bolezni*, **2**, 45–48 (in Russian).

Bronstein, A. M. (1986a). Prevalence of *Opisthorchis felineus* infection and diphyllobothriasis among the aboriginal population of Kishik settlement in Hanti-Manssi Autonomous District. *Med. Parasitol. Parasit. Bolezni*, **3**, 44–48 (in Russian).

Bronstein, A. M. (1986b). Prevalence of *Opisthorchis felineus* infection and diphyllobothriasis in population of Vanzetur settlement in Hanti-Manssi Autonomous District. *Med. Parasitol. Parasit. Bolezni*, **5**, 10–14 (in Russian).

Bronstein, A. M. (1987). Prevalence of *Opisthorchis felineus* infection in newcomers to one of the endemic foci of the middle Ob river region. *Med. Parasitol. Parasit. Bolezni*, **3**, 52–57 (in Russian).

Bronstein, A. M. and Ozeretskovskaya, N. N. (1985). Medical, stastistical and economical assesment some health parameters of settler population of endemic opisthorchiasis foci. *Med. Parasitol. Parasit. Bolezni*, **6**, 22–29 (in Russian).

Bronstein, A. M., Ozeretskovskaya, N. N., and Bitchkov, V. G. (1987). The analysis of the causes of unefficiency of chloxyl in the treatment of *Opisthorchis felineus* infection. *Med. Parasitol. Parasit. Bolezni*, **2**, 22–25 (in Russian).

Bronstein, A. M., Uchatkin, E. A., Romanenko, N. A., Kanzan, S. N., Veretennikova, N. L., and Sabgaida, T. P. (1989). Complex assesment of opisthorchiasis in Komi-Permjak Autonomous District *Med. Parasitol. Parasit. Bolezni*, **4**, 66–72 (in Russian).

Bronstein, A. M., Zolotuchin, B. A., Gizu, G. A., Sabgaida, T. P., and Parfenov, S. B. (1991). Clinical and epidemiological characterictics of opisthorchiasis foci and the results of prazyquantel treatment in Jamalo-Nenezkii Autonomous District. *Med. Parasitol. Parasit. Bolezni*, **5**, 12–15 (in Russian).

Boytsov, V. D., Telushkin, A. V., Tumolskaya, N. I., and Yarotskii, L. S. (1992). Clinical features of pulmonary hydatid disease. *Med. Parasitol. Parasit. Bolezni*, **2**, 12–14 (in Russian).

Bychkov, V. G. and Jarotskii, L. S. (1990). Probleme of parasites oncogenecity. *Med. Parasitol. Parasit. Bolezni*, **3**, 46–49 (in Russian).

Capron, A. (1995). Molecular language of parasites. In: *The year of Louis Pasteur international symposia*, Dakar, Senegal, Abstracts, p. 17.

Center for Disease Control (1986). Giardiasis in travellers returning from the Soviet Union. Advisory memorandum, p. 86.

Cibina, T. N. and Siskova, T. G. (2001). Diagnosis and prophylactic of anisakidosis at the territorry of the Russian Federation. *Med. Parasitol. Parasit. Bolezni*, **3**, 52–53.

Dobreitser, I. A. (1924). *Malaria in the USSR*. Narkomizdat, Moscow, 68 pp. (in Russian).

Dovgalev, A. S., Valovaja, M. A., Piskunova, U. A., Romanenko, N. A., Khodakova, V. I., and Artamoshin, A. S. (1991). Morphology of human diphyllobothriasis agent in the Far East. *Med. Parasitol. Parasit. Bolezni*, **6**, 42–46 (in Russian).

Dovgalev, A. S. and Valovaja, M. A. (1996). Species of the agent of diphyllobothriasis in the zone of the Pasific shore of Russia. *Med. Parasitol. Parasit. Bolezni*, **3**, 31–34 (in Russian).

Dukhanina, N. N. (1962). Tertian malaria with long incubation in the period of malaria eradication. In: *The Annals of the E.I. Martsinovsky Institute of Medical Parasitology and Tropical Medicine*, Medicina, Moscow, pp. 315–329 (in Russian).

Epstein, G. (1932). Some epidemiological data on amoebiasis. *Med. Parasitol. Parasit. Bolezni*, **1**, 200–234 (in Russian).

Epstein, G. and Zertchaninov, L. (1932). Data on distribution of intestinal protozoa among the population of the Ural region. *Med. Parasitol. Parasit. Bolezni*, **1**, 235–237 (in Russian).

Federal Programme 'Children of the North' for 1996–2001 (1996). In: *National plan of action in the interest of children*, 163 pp., Moscow (in Russian).

Garkavi, B. L. and Zverghanovsky, M. I. (1999). *Trichinella pseudospiralis* the possible agent of trichinellosis in Russia. *Med. Parasitol. Parasit. Bolezni*, **4**, 45–46.

Ilyinskikh, I. N., Ilyinskikh, E. N. and Ilyinskikh, N. N. (2001). Cardiological changes after tick encephalitis in man. In: *Actual Problems of Infectology and Parasitology. International Conference on the 110th Anniversary of Professor K. N. Vinogradov's discovery of the cat distomum in man*, Tomsk, Russia, p. 108.

Ivanskikh, V. I. and Bliznjuk, V. V. (1996). Impact of *Opisthorchis felineus* infection on the manifestation of *Herpes simplex* virus type 2 and possible involvement of the latter in the mechanism of the primary liver carcinogenesis. *Med. Parasitol. Parasit. Bolezni*, **2**, 23–26 (in Russian).

Jablokov, D. D. (1979). *Opisthorchiasis of a Man*. The University of Tomsk, Tomsk, p. 7.

Jokipii, L., Pohjola, S. and Jokipii, A. M. (1985). Cryptosporidiasis and giardiasis assotiated with travelling. *Gastroenterology*, **89**, 838–842.

Khodakova, V. I., Legonkov, Yu. A., Melnikova, L. I., Frolova, A. A., and Artamoshin, A. S. (1996). Prevalence of helminthiases in the children institutions in one of the regions of the extreme North. *Med. Parasitol. Parasit. Bolezni*, **4**, 31–33 (in Russian).

Khudoshin, V. A. (1968). Prevalence of lambliasis among children in an isolated community in the extreme North and the experience of control of this infection in children's institutions. *Med. Parasitol. Parasit. Bolezni*, **46**, 733–734 (in Russian).

Kirichek, V. S., Belousov, V. N., and Nikitin, A. S. (1984). New data on epidemiology of teniarinchosis in the extreme North regions. *Med. Parasitol. Parasit. Bolezni*, **6**, 27–33 (in Russian).

Klebanovskii, V. (1980). Probleme of diphyllobothriasis in the USSR today. In: *Zoonoses. Congressus cum participationi internationall*, Summa abstractorum. Streske Pleso, p. 157.

Klebanovskii, V. A. (1985). Diphyllobothriases. In: *Helminthiases of man (Epidemiology and control)*. Medicina, Moscow, pp. 164–178 (in Russian).

Knjazev, A. U. (1994). Ecological and epidemiological approach to control of biohelminthiases (echinococcoses, trichinellosis, diphyllobothriasis) in the conditions of anthropopressure in the Taimir peninsula. ScD Thesis, The Martsinovski Institute, Moscow, Russia.

Korablev, V. N. (1990). The pharmakokinetic of isoniozide in the blood of patients with tuberculosis and chronic trematodoses. *Antibiot Chemotherapy*, **6**, 45–46 (in Russian).

Kotelkin, A. T., Razumov, I. A., and Loktev, V. B. (1996). Comparative biochemical and immunological study of *Opisthorchis felineus* somatic and metabolic antigens. *Med. Parasitol. Parasit. Bolezni*, **2**, 18–23 (in Russian).

Kuprijanova, N. U. (1989). *Toxocara canis* areal in the USSR and epidemiology of toxocarosis in the West Siberia. ScD Thesis, The Central Institute of Postgraduation Specialisation, Moscow, Russia.

Lebedev, G. B., Sergiev, V. P., Romanenko, N. A., and Novosiltsev, G. I. (1996). Survival and development of *Ascaris lumbricoides* eggs in open soil and soil of greenhouses in Chukotka. In: *Children of the North: protection, survival and development under extreme natural conditions*, p. 80 (in Russian).

Lepekhin, A. V., Ratner, G. M., and Menjazeva, T. A. (1990). Cross-reacting *Opisthorchis felineus* and *Salmonella typhy* antigens and their possible impact on the typhus clinical course. *Med. Parasitol. Parasit. Bolezni*, **4**, 33–36 (in Russian).

Martinenko, V. B., Shubin, A. G., Mordosov, I. I., Isakov, S. I., Shakarov, A. G., and Suvorina, V. I. (1984). The possibility of formation the settlement's foci of *Echinococcus multilocularis*. *Med. Parasitol. Parasit. Bolezni*, **6**, 250–257 (in Russian).

Maslov, S. S. (1991). The peculiarities of epidemiology and control of echinococcoses in the foci of different types. ScD Thesis, The Martsinovsky Institute, Moscow, Russia.

Migasena, P. (1990). Opisthorchiasis an initiator and promoter in liver carcinogenesis. *Bull. Soc. Fr. Parasitol.* **8** (Suppl.), 360.

Muratov, I. V. (1993). Predatory terrestrial mammals as definitive hosts of *Diphyllobothrium klebanovskii*. *Med. Parasitol. Parasit. Bolezni*, **2**, 3–5 (in Russian).

Muratov, I. V. and Posochov, P. S. (1988). The agent of diphyllobothriasis in man – *Diphyllobothrium klebanovskii*. *Parasitologia*, **22**, 165–167 (in Russian).

Nazmov, V. P., Fedorov, K. P., Serbin, V. I., and Auslender, V. L. (2001). *Med. Parasitol. Parasit. Bolezni*, **2**, 26–27 (in Russian).

Opisthorchiasis and the control programme in Thailand (1990). In: *Symposium on medical parasitology*, Shanghai, 1990, Abstracts, pp. 40–41.

Ozeretskovskaya, N. N. (1968). Clinical and epidemiological peculiarities of trichinellosis in the different geographical regions of the USSR. *Med. Parasitol. Parasit. Bolezni*, **4**, 387–397 (in Russian).

Ozeretskovskaya, N. N. (1975). The differences of clinical features and the tolerance of treatment in opisthorchiasis patients due to some endogenous and exogenous factors. *Vestnik Akad. Medicin. Nauk*, **6**, 36–43 (in Russian).

Ozeretskovskaya, N. N. (1976). Immunological and immunopathological reactions in pathogenesis of helminthiases. In: *The basis of a general helminthology*, T. III, *Pathology and immunology in helminthiases*. (R. S. Schultz and E. V. Gvozdev, eds). Nauka, Moscow, pp. 169–214 (in Russian).

Ozeretskovskaya, N. N. (1979). Zoonotic helminthiases of the North-East and the Far-East of the USSR. Peculiarities of their clinical pattern in migrants and the prevention of the development of new anthropogenic foci. *Vestnik Akad. Medicin. Nauk*, **12**, 79–84 (in Russian).

Ozeretskovskaya, N. N. (1982). Intestinal parasitic infections. *Scand. J. Infect. Dis. Suppl.*, **36**, 46–51.

Ozeretskovskaya, N. N. (1997a). In: *ICT Country Status Reports 1995–1997 (June)*, p. 1. International Commission on Trichinellosis, Beltsville, MD, USA.

Ozeretskovskaya, N. N. (1997b). Blood eosinophilia and serum immunoglobulin E regulation in patients with helminthiases and allergic diseases. *Med. Parasitol. Parasit. Bolezni*, **2**, 3–9 (in Russian).

Ozeretskovskaya, N. N., Isaguliants, M. G., Carlenor, E., Poletaeva, O. G., Thors, C., and Linder, E. (2000). Organic visceral pathology in helminthiases coincides with antibody recognition of human α-cardiac myosin-derived peptide that induces experimental allergic myocarditis. In: *Trichinellosis, Xth International Conference on Trichinellosis*, France, Abstract Book, p. 104.

Ozeretskovskaya, N. N., Ivanova, M. G., and Mickhailiva, A. D. (1958). The efficacy of prednizolone in the therapy of trichinellosis. *Sovetskaya Medicina*, **9**, 11–14 (in Russian).

Ozeretskovskaya, N. N. and Pereverzeva, E. V. (1976). Clinical, epidemiological and parasitological features of the Arctic Trichinella strain (ATS). In: *Third international conference on bears – their biology and management*, Switzerland, Morges, pp. 391–402.

Ozeretskovskaya, N. N., Poletaeva, O. G., Sergiev, V. P., Isaguliants, M. G., Linder, E. and Out, T. A. Some factors determining the character of immune response and the clinical features of

Opisthorchis felineus infection in human. In: *Actual problems of infectology and parasitology. International Conference on the 110th Anniversary of Professor K. N. Vinogradov's discovery of the cat distomum in man*, Tomsk, Russia, pp. 25–26.

Ozeretskovskaya, N. N. and Sergiev, V. P. (1993). Mass chloxyl treatment of *Opisthorchis felineus* infection from the clinical and epidemiological positions. *Med. Parasitol. Parasit. Bolezni*, **5**, 6–13 (in Russian).

Ozeretskovskaya, N. N. and Sergiev, V. P. (1994). Specific and biological effects of chemotherapeutic drugs and their combinations with pathogenetic therapeutical remedies. *Med. Parasitol. Parasit. Bolezni*, **4**, 9–14 (in Russian).

Ozeretskovskaya, N. N., Tcherbakov, A. M., Sunzov, S. N., Sabgajda, T. P., Grigorjan, S. S., and Gervazieva, S. B. (1993). Blood eosinophils, serum immunoglobulines, circulation immunocomplexes content and interferones production in patients with *Echinococcus granulosus* infection regarding the parasite cysts location. *Med. Parasitol. Parasit. Bolezni*, **2**, 10–14 (in Russian).

Ozeretskovskaya, N. N. *et al.* (1994). In: *Trichinellosis* (W. C. Campbell, E. Pozio and F. Bruschi, eds). Instituto Superiori di Sanita Press, Rome, pp. 437–442.

Ozeretskovskaya, N. N. and Uspensky, S. M. (1957). Group's infestation by Trichinella by the meat of polar bear in the Soviet Arctic. *Sovetskaya Medicina*, **2**, 152–159.

Ozeretskovskaya, N. N., Zalnova, N. S., and Tumolskaya, N. I. (1985). In: *Clinical features and therapy of helminthiases*, pp. 95–112 (in Russian).

Paltzev, A. I., Nepomnyashchikh, D. I. (1998). Clinical, laboratory and pathomorphological study of the liver in patients with chronic opisthorchiasis. *Med. Parasitol. Parasit. Bolezni*, **4**, 28–32.

Parfenov, S. B., Ozeretskovskaya, N. N., Pomigalov, A. Yu., and Chistjakova, I. V. (1989). Clinical and immunological parameters in the different groups of patients with chronic opisthorchiasis in the endemic foci. *Med. Parasitol. Parasit. Bolezni*, **4**, 72–75 (in Russian).

Pavlov, B. A. (1990). In: *Acute opisthorchiasis*. The University of Tomsk, Tomsk, 145 pp. (in Russian).

Peshkov, M. (1932). Intestinal protozoa in the Russian and Burjat population in the Selenga Aimak. *Med. Parasitol. Parasit. Bolezni*, **1**, 237–239 (in Russian).

Plyuscheva, G. L. *et al.* (1987). Formation of diphyllobothriasis foci in the Krasnoyarsk waterbody. *Med. Parasitol. Parasit. Bolezni*, **1**, 64–67 (in Russian).

Prevention of parasitic infections on the territory of the Russian Federation (1997). Minzdrav of Russia, Moscow, pp. 65–69.

Report for the Federal Programme Children of the North (1996). Minzdrav of Russia, Moscow, 30 pp.

Rutkevitch, M. (1937). Loc. cit. Dovgalev, A. S., and Valovaja, M. A. (1996). Species of the agent of diphyllobothriasis in the zone of the Pacific shore of Russia. *Med. Parasitol. Parasit. Bolezni*, **3**, 31–34.

Serdukov, A. M. (1993). The problem of anizakidosis. *Med. Parasitol. Parasit. Bolezni*, **2**, 50–54 (in Russian).

Sergiev, P. G. *et al.* (1968). In: *Clinical features and epidemiology of infectious diseases*, vol. 9. Medicina, Moscow, pp. 37–115 (in Russian).

Sergiev, V. P. (1991). Registered and estimated cases of parasitic infections. *Med. Parasitol. Parasit. Bolezni*, **2**, 3–10 (in Russian).

Sergiev, V. P. (1993). Prevalence and distribution of parasitic infections in Russian Federation and in the former U.S.S.R. *Bull. Soc. Fr. Parasitol.*, **11**, 35–42.

Sergiev, V. P. and Be'er, S. A. (eds) (1988). In: *Opisthorchiasis. Theory and practice.* Moscow, pp. 155–176 (in Russian).

Sergiev, V. P., Drinov, I. D., and Malishev, N. A. (1998). In: *Problems of health care, approaches and perspectives.* Moscow, p. 26 (in Russian).

Shaikenov, B. (1995). Distribution of *Trichinella nativa*, *Trichinella nelsoni* and *Trichinella pseudospiralis* in Eurasia. *Med. Parasitol. Parasit. Bolezni*, **3**, 20–24 (in Russian).

Shikhobalova, N. P., Leikina, E. S., and Ozeretskovskaya, N. N. (1969). Principal helminthozoonoses of the population of the northern districts of the USSR (1969). *Arch. Environ. Health*, **19**, 365–380.

Shustov, A. I., Kotelkin, A. T., Sorokin, A. V., Ternovoy V. A., and Loktev, V. B. (2001). Paramyosin of *Opisthorchis felineus*: sequence of cDNA and the research of a recombinant fragment of paramyosin. In: *Actual problems of infectology and parasitology. International Conference on the 110th Anniversary of Professor K. N. Vinogradov's discovery of the cat distomum in man*, Tomsk, Russia, p. 62.

Simonova, N. F. (1995). Epidemiology and prevention of diphyllobothriases in the Lena river region. ScD Thesis, The Martsinovsky Institute, Moscow, Russia.

Siskova, T. G., Cibina, T. N., Sidorenko, A. G., and Yasinsky, A. A. (2001). *Med. Parasitol. Parasit. Bolezni*, **3**, 31–35.

Sorochenko, E. V. (1966). Helminthic diseases and intestinal protozoa in the population of the Nenetsk territory. *Med. Parasitol. Parasit. Bolezni*, **43**, 151–153 (in Russian).

Sorochenko, A. M. (1968). Intestinal protozoa in the population of the Nenetsk National District. *Med. Parasitol. Parasit. Bolezni*, **43**, 151–153 (in Russian).

Stepchuk, M. A., Blotsky, V. E., and Vapirov, V. M. (1990). *Echinococcus multilocularis* infection in the Korjak Autonomous District. *Med. Parasitol. Parasit. Bolezni*, **4**, 25–26.

Vitomskaya, Ye. A. and Dovgalev, A. S. (2001). The rate of Okhotsk sea fish infestation by Anisakis dangerous to man. *Med. Parasitol. Parasit. Bolezni*, **2**, 31–34.

Wakelin, D. (1994). Host populations: genetics and immunity. In: *Parasitic and infectious diseases*. Academic Press, New York, pp. 83–100.

Zavoikin, V. D. (1990). The structure of nosoareal of *Opisthorchis felineus* infection and the measures of its control. *Med. Parasitol. Parasit. Bolezni*, **3**, 26–30 (in Russian).

Zavoikin, V. D., Zelja, O. P., Mikhailov, M. M. and Bragin, V. V. (2001). Tactics of a wide prazyquantel administration in the complexes of measures of opisthorchiasis control. *Med. Parasitol. Parasit. Bolezni*, **2**, 13–17.

Zhdanov, V. M. (1990). Cancer in the epidemiologist's estimation. *Vestnik. Akad. Medicin. Nauk*, **3**, 26–30 (in Russian).

7 Prevalence of parasites in Canada and Alaska

Epidemiology past and present

Theresa W. Gyorkos, J. Dick MacLean,
Bouchra Serhir, and Brian Ward

The nature and distribution of parasite infections and diseases in Canada and Alaska are as diverse and widespread as their human host populations. Toxoplasmosis, trichinosis, and diphyllobothriasis, for example are more commonly found in Northern Inuit communities than in the multi-ethnic communities of the southern regions. Intestinal protozoal infections such as *Giardia* are cosmopolitan while trematode infections such as *Metorchis* are extremely focal. Travellers and immigrants who have acquired parasite infections abroad receive expert medical attention available in centres specializing in tropical medicine. Imported parasite infections which are transmitted via the oral–faecal route are usually self-limited or fail to survive locally because of the environmental and/or sanitary conditions.

Because of increased travel and trade, change in dietary habits, and frequency of immunocompromising diseases, newly emergent parasite diseases (e.g. toxoplasmosis, cryptosporidiasis, cyclosporiasis) have become important medical challenges despite the relatively inhospitable northern climate.

Introduction

Alaska and Canada, together, occupy a vast portion of the Northern Hemisphere, stretching from 41° to 83° latitude and from 179° to 52° longitude (Figure 7.1). The climate varies extensively by region. Northern Alaska and Canada have a polar climate with a brief growing period in the summer months while the sub-arctic regions have severe winters but somewhat longer, warmer summers. The southern part of Canada is temperate with a continental climate of four distinct seasons, where summer lasts from June to September. The occurrence and variety of parasites in both animal and human populations reflect this climatic and geographic diversity.

Canada's population numbers over 31 million and Alaska's, 600 000. Alaska's population is clustered around the larger cities of Juneau and Anchorage with sparse distribution elsewhere. Over 90% of Canada's population is distributed within 300 km of the United States border. The population is divided into 10 provinces and three territories.

Using a passive surveillance system, four parasite infections were required to be reported to Statistics Canada up until 2000 by the laboratory making the diagnosis: amoebiasis, giardiasis, malaria, and trichinosis. After 2000, amoebiasis and trichinosis were removed from the list of reportable diseases but cryptosporidiosis and cyclosporiasis were added. Provinces can add other parasites to their list of reportable diseases to reflect provincial interests and public health concerns. For a number of reasons, including a presumed high proportion of asymptomatic infections that never contact a health facility, the initiation of medical treatment based on a clinical diagnosis in the absence of laboratory confirmation, host–parasite dynamics related to

Figure 7.1 Map of Canada and Alaska (United States).

Table 7.1 Number of reported cases of notifiable parasite infections in Canada, by province and territory, and in Alaska, United States, in 1993

	Amoebiasis 006*	Giardiasis 007.1	Malaria 084	Trichinosis 124	Population (in thousands)
Alaska		7ᵃ	40		598.3
Canada	1 782	7 063	4 859		28 753.0
Newfoundland	1	71	00		581.1
Prince Edward Island	2	11	00		131.6
Nova Scotia	9	108	20		923.0
New Brunswick	4	138	00		750.9
Québec	201	748	360		7 208.8
Ontario	991	3 054	2 496		10 746.3
Manitoba	50	0	90		1 116.0
Saskatchewan	54	446	20		1 003.1
Alberta	126	941	340		2 662.3
British Columbia	3 411	4 681	530		3 535.1
Yukon	2	22	00		32.0
Northwest Territories	1	56	03		62.9

Note
a Giardiasis is not a reportable disease in the United States.

* means ICD 9 code number.

the probability of detection of parasites in stool specimens, and the variability in laboratory diagnostic capacity (in terms of personnel, equipment, techniques, etc.), the reporting of parasite infections is considered to be grossly underestimated in both Canada and Alaska. Table 7.1 is provided to show the number of cases of notifiable parasite diseases in 1993 for Alaska and for each Canadian province and territory. Table 7.2 presents the results of the 1987 intestinal parasite prevalence survey of provincial laboratories. Each province and state

Table 7.2 Prevalence (%) of *E. histolytica, G. lamblia*, non-pathogenic protozoa, and helminths identified in provincial laboratories, Canada, 1984

Province	E. histolytica	G. lamblia	Non-pathogenic protozoa[a]	Helminths
NFLD	0	1.25	NR	0.53
PEI	0.11	1.92	1.86	0.73
NS	1.55	3.33	18.76	2.23
NB	0.02	1.86	1.89	0.51
QUE[b]	4.43	13.02	42.38	16.34
ONT	3.75	3.84	29.52	4.87
MAN[c]	0.59	43.99	8.36	4.25
SASK	0.10	5.08	3.03	0.40
ALB	1.28	5.98	14.49	1.54
BC	1.61	4.21	14.11	4.51
CANADA	1.75	4.11	15.76	3.26

Source: Reprinted from Gyorkos *et al.* (1987).

Notes
a Includes all protozoa other than *E. histolytica* and *G. lamblia*.
b Based on mostly reference specimens.
c Excludes reports of *Blastocystis hominis* and presumed *B. hominis*.
NR – not recorded routinely.

has designated provincial and state laboratories which provide diagnostic, training, and advisory support to hospital-based laboratories and physicians and which maintain active communication with national health authorities (Population and Public Health Branch (Health Canada)) and in the United States, the Centers for Disease Control and Prevention (CDC). These two national agencies provide parasite reference services to the laboratories. Health Canada delegates reference activities in parasite serology to its National Centre for Parasitology–Serology (NCPS).

The parasite infections which will be described in this chapter have been selected to illustrate both the spectrum of endemic parasite infections and those parasites that are most important in terms of prevalence or morbidity, whether imported or endemic (Table 7.3). Where published evidence is lacking, expert opinion has been solicited to provide key points. Each parasite section highlights the prevalence of the infection in addition to selected epidemiological features pertinent in the general Canadian population or high risk groups.

Protozoal infections

Malaria

Although endemic from Québec to the prairie provinces in the early 1800s, malaria is now an imported disease in Canada. The chief risk groups are immigrants from endemic countries, travellers, missionaries, overseas workers, and foreign students. Between 1985 and 1995, imported malaria accounted for a total of five deaths in Canada, a greater number of deaths in Canadians abroad, and an average annual 436 cases reported to Statistics Canada. These mortality and morbidity figures reflect increasing travel trends to high risk malarious areas, the use of inadequate or lack of chemoprophylaxis, and delays in seeking medical attention on return to Canada (Svenson *et al.* 1995).

Table 7.3 Most important parasite infections and diseases of Canada

Parasite infection	Prevalence[a]	Distribution	High-risk groups	Research being undertaken by Canadians[b]
Malaria	Isolated cases, imported	General	Travellers, immigrants	Compliance with anti-malarials, drug resistance, vaccine development
Toxoplasmosis	Up to 53%, endemic	All regions	Pregnant women	Prenatal screening, diagnostic tools
Giardiasis	About 5%, endemic	All regions	Hikers, children in daycare centres	Prevalence, water management, detection in food
Amoebiasis	Limited foci, endemic	All regions	Native and Inuit populations, immigrants	Host defence mechanisms, diagnostic tools
Pinworm	Widespread, endemic	All regions	Children, families	
Intestinal worms (e.g. hookworm, whipworm, etc.)	Rare cases, imported	All regions	Immigrants	Epidemiology
Anisakiasis	Rare cases, endemic	Coastal regions	Raw fish eaters	Biogeography, taxonomy
Metorchiasis	Rare cases, endemic	Southern Canada	Raw fish eaters	Link with cancer, diagnostic tools
Diphyllobothriasis	Limited foci, endemic	Northern Canada	Raw fish eaters	Biogeography
Tapeworms	Rare cases, endemic	Southern Canada	Immigrants, rural inhabitants	Epidemiology, physiology
Pediculosis	1–10%, endemic	All regions	School children	Epidemiology, treatment
Scabies	Limited foci, endemic	All regions	Hospital population, care facilities	Epidemiology
Echinococcosis	<1%, endemic	Northern and western Canada	Immigrants, rural inhabitants	Immunity
Trichinellosis	<1%, endemic	Northern Canada	Native communities, immigrants	Molecular biology, immunology, ecology
Toxocariasis	Endemic	Southern Canada	Preschool children who eat dirt	Contamination in environment
Cryptosporidiosis	Endemic	Most regions	Immunocompromised persons	Detection techniques

Notes

a Imported = imported by immigrants and travellers into Canada, no local transmission; endemic = local transmission.

b Research interests are based on information provided in the *Directory of Parasitologists in Canada*, 2nd edn, 1996 published under the auspices of the Biological Survey of Canada, the Canadian Museum of Nature, and the Canadian Society of Zoologists.

Toxoplasmosis

There have been several studies of seroprevalence of toxoplasmosis in the general Canadian population, but most are dated from the 1960s and 1970s. Population-based seroprevalence estimates vary from as low as 2.2% in Cree and Ojibway Indians in Ontario and

Table 7.4 Seroprevalence studies of toxoplasmosis in pregnant women in Canada

First author	Year	Canadian location	Sample size	Test[a]	Cut-off titre	Seroprevalence (%)
Grimard	1996	Québec	2 141	Agg1	4 IU	11.2
Karim	1977	British Columbia	305	DT	1:8	25.0
Mackenzie	1974	Nova Scotia	65	IHA	—	13.7
Martineau	1974	Québec	1 162	IFA	4 IU	48.1
McDonald	1990	North Québec	30	IFA	4 IU	49.9
Proctor	1994	British Columbia	49	ELISA	—	20.4
Tanner	1987	North Québec	131 (Inuit)	IHA	1:32	72.0
			31 (Indian)			19.0
Viens	1977	Québec	4 136	IFA	4 IU	40.8

Notes

a ELISA: Enzyme-linked immunosorbent assay; IHA: indirect haemagglutination assay; DT: dye test; Agg1: agglutination test; IFA: indirect immunofluorescence assay.
— Not reported.

12% among Indian patients in northern Québec to as high as 48% among Inuit patients in northern Québec and 53% in Montréal, Québec. While the variability in the estimates of seroprevalence is great, the absence of large-scale or representative population-based studies using standardized serological tests makes it difficult to characterize the true magnitude of infection.

The primary focus of study on seroprevalence has been in women of reproductive age. Seroprevalence estimates reported in this population sub-group (Table 7.4) indicate not only an important variation in the magnitude of infection but also that congenital toxoplasmosis continues to be an important health concern. From available Canadian data, it has been estimated that between 140 and 1 400 cases of congenital toxoplasmosis occur annually, with severe manifestations at birth in 70–280 infants. Many of the remaining cases will experience sequelae over several years (Carter and Frank 1986).

Toxoplasma infection is generally acquired in one of two ways: from ingestion of raw or undercooked meat containing infective bradyzoïtes or from ingestion of oocysts excreted by felines, typically via oocyst contamination in cat litter or soil. In North America, there is evidence of both these types of transmission but the contribution of each has been difficult to assess. However, in northern native communities, the likelihood is that transmission is almost exclusively related to ingestion of raw and undercooked meat of various kinds. An epidemiologic study conducted in northern Québec in 1988 among 22 pregnant women revealed that seroconversion during pregnancy was related to skinning of animals and the frequent consumption of caribou meat while seropositivity was related to a history of ingestion of dried seal meat, fresh seal liver, and the frequent consumption of caribou meat (McDonald *et al.* 1990).

Oocyst contamination of water and other foods, particularly those which are water-washed, is possible and may play a role in transmission. It is suspected that the winter 1995 outbreak of toxoplasmosis in the greater Victoria region in British Columbia was due to contamination of the water supply, likely by free-roaming cougars (Stephen *et al.* 1996).

Giardiasis

Giardia lamblia is the most common cause of parasite-induced diarrhoea in immunocompetent hosts in Canada and Alaska. This organism has been found in a wide variety of wild and domestic animal species throughout this region; therefore human cases may be either indigenous (locally called 'beaver fever') or imported. The prevalence of *G. lamblia* carriage and symptomatic illness varies widely in Canada and Alaska with the highest rates reported in children (2–31%), adult staff of daycare facilities (8%), native communities, travellers, immigrants (4–67%), the immunocompromised, and the institutionalized. It has been shown that within 6 years of arrival in Canada, levels in immigrant populations decrease to that of the general Canadian population (estimated to be approximately 4%) (Gyorkos *et al.* 1987, 1992). The presence of *G. lamblia* cysts in virtually all of the recreational waterways of these regions also puts hikers, campers, boaters and hunters, and others using these waters at risk (Isaac-Renton and Philion 1992). In Canada, the number of giardiasis cases reported between 1985 and 1994 has varied from 6 500 to 9 500 per year, figures which probably reflect symptomatic cases and which would therefore underestimate the prevalence in the general population.

Amoebiasis

Entamoeba histolytica has been one of the most reported parasitic diseases of immunocompetent hosts in Canada and Alaska. Those at highest risk for symptomatic *E. histolytica* infection are travellers returning from the developing world and native North Americans. In the past, the overall prevalence of amoebic infection in Canada and Alaska was estimated at 1% (Burrows 1961). More recent data from Canada suggests a rate of 6.8 (reported) cases per 100 000 population, with a male to female ratio of almost 2:1, the majority in adults (CCDR 1996). Endemic amoebic disease has most often been observed in native populations. For example, communities in the Loon Lake region of Saskatchewan were reported to be at the centre of an area of endemicity which expanded to involve large parts of Alberta and the Northwest Territories from 1956 to 1973. The new technologies to distinguish between *E. histolytica* and *E. dispar* have not yet been applied in Canada and Alaska. It is very likely that once these tests are routinely available, a large majority of the asymptomatic individuals reported to have *E. histolytica* will prove to have *E. dispar* only. Published surveillance reports typically do not distinguish symtomatic disease (presumed *E. histolytica*) and asymptomatic cyst carriage (possibly *E. histolytica*, but much more likely *E. dispar*). As a result, the carriage prevalences cited earlier are certainly an overestimate of the true *E. histolytica* levels in the population.

Other intestinal protozoa

Estimates for the prevalence of other intestinal parasites in Canada and Alaska are based on limited data from studies in selected populations. As has been reported worldwide, symptomatic infection with microsporidial species and *Cryptosporidium parvum* is closely associated with HIV/AIDS. No overall prevalence data are available for microsporidia but prevalence rates for *Cryptosporidium* can be striking in domestic animals such as sheep (23%) and cattle (20%) (Olson *et al.* 1997). Estimates of human infection range from 0.2 to 8% (Ratnam *et al.* 1985; Kabani *et al.* 1995). The ageing sewage and water infrastructure of many urban centres in Canada and Alaska likely puts them at risk for major outbreaks of cryptosporidiosis. Cyclosporiasis is not endemic in Canada or Alaska. Up until the summer of 1996, the rare cases reported were always associated with travel to the developing world. In the spring months of 1996 and 1997 in both Canada and the United States epidemics of cyclosporiasis have been reported. The first outbreak was thought to have been caused by the importation of contaminated raspberries from Guatemala (Herwaldt and Ackers 1997). Little is known about the epidemiology of *Dientamoeba fragilis* in Canada or Alaska other than the fact that it is present in selected population sub-groups (e.g. daycare centre staff and children) (Keystone *et al.* 1984). The presence of non-pathogenic intestinal protozoa has not been systematically studied in Canada or Alaska. Data from large provincial laboratories show rates similar to those in other industrialized countries.

There has been no systematic effort to study the prevalence of *Trichomonas vaginalis* in Canada or Alaska. Prevalence rates have ranged from 0% in 57 Québec women with dysparunia (Bazin *et al.* 1994) to 7.3% for women attending an STD clinic in Nova Scotia between 1983 and 1985 (Pereira *et al.* 1990).

Helminth infections

The helminth fauna of Canada and Alaska is as diverse as the ethnic origins, habits, and geographic characteristics of their communities. Immigrants and travellers import helminths from around the world: the most clinically important being *Echinococcus granulosus*, cysticercosis, *Strongyloides stercoralis*, and lymphatic filariasis. The helminths presently indigenous to Canada and Alaska are zoonoses and occur in those populations that live in close proximity to wild animals and/or consume either raw fish or meat. These include the nematodes *Trichinella nativa*, *Anisakis simplex*, and *Pseudoterranova decipiens*, the cestodes *E. granulosus* and several species of *Diphyllobothrium*, and the trematode *Metorchis conjunctus*.

Cestode infections

Diphyllobothriasis

High prevalence rates have been recorded in small northern and native communities in which the consumption of raw and partially cooked fish is common (Freeman and Jamieson 1976). In recent years, the growing popularity of sushi and sashimi has resulted in increased reports of diphyllobothriasis at more southern latitudes (Ruttenber *et al.* 1984). At least six species of *Diphyllobothrium* capable of infecting humans are thought to exist in this region (*D. latum*, *D. dendriticum*, *D. ursi*, *D. lanceolatum*, *D. dalliae*, and *D. alascense*) although the precise taxonomic classification of some of these species is a matter of debate (Schantz 1996). Diphyllobothriasis (*D. dendriticum*) is particularly common in native populations across the Arctic, reaching prevalence rates of 80% in some communities. Unlike the historical reports from Scandinavia, diphyllobothriasis has not been associated with vitamin B_{12} deficiency in Canada or Alaska. However, the long-term health impact of human diphyllobothriasis is largely unstudied despite the high prevalence of this parasite in some populations.

Hydatid disease

Hydatid disease is an important zoonosis in populations engaged in hunting in Canada and Alaska. Although two species of *Echinococcus* are widely distributed in this region (*E. granulosus*, *E. multilocularis*), human disease is caused almost exclusively by *E. granulosus* (cystic echinococcosis) with rare cases caused by *E. multilocularis* (alveolar echinococcosis). Two biotypes of *E. granulosus* are found in Canada and Alaska based on the host specificity of the larval stage. The northern or cervid biotype is likely to be the ancestral form of this parasite and is widely distributed in the holoarctic and northern boreal forest (taiga) regions. This biotype cycles between northern herbivores (e.g. moose, caribou) and wild canines (principally wolves).

The European or sheep biotype likely evolved from the cervid biotype in association with the development of animal husbandry. *Echinococcus multilocularis* also has a wide circumpolar distribution cycling between rodent intermediate hosts (e.g. voles, lemmings) and canines (e.g. foxes, coyotes).

Cystic hydatid disease caused by the cervid biotype has long been recognized as a significant health problem in native and northern populations. Some data suggest that the clinical presentation of the northern biotype differs from the European form of *E. granulosus* in that cysts are more frequently found in the lungs and the course of infection may be more benign (Rausch 1986). Surveys in wild animals have demonstrated high prevalence rates in moose (19–59%), caribou (9.5%), and wolves (47%) in northern regions (Curtis *et al.* 1987; Messier *et al.* 1989). High rates of human infection across northern Canada and Alaska have also been demonstrated by autopsy studies (1% of Alaska natives; Arthaud 1970) and serosurveys in native communities (2–4% seropositive; Curtis *et al.* 1987).

Alveolar hydatid disease is one of the most fatal helminth infections of man with historical mortality rates as high as 63% (Wilson and Raush 1980). The prevalence of *E. multilocularis* in natural hosts is influenced by a variety of factors, including the relative densities of fox and rodent populations, periodicity of rodent numbers, and the diversity of prey species eaten by foxes. Nevertheless, very high infection rates have been well documented in arctic foxes (40–100%), voles (2–80%), and deer mice (15–22%). Despite the wide distribution of the parasite and high prevalence rates in the natural hosts, human cases of alveolar disease have rarely been reported and are localized in western Alaska (Stehr-Green *et al.* 1988). No cases have been recorded in humans living in the extensive sub-Arctic and Arctic regions of Canada where the parasite is enzootic. Although the likelihood of infection with *E. multilocularis* is also decreasing with lifestyle changes in native and northern communities, a recent survey in western Alaska has demonstrated seroprevalence rates ranging from 7–8/100 000 to as high as 98/100 000 (Schantz *et al.* 1995). Controlling *E. multilocularis* infection in domestic dogs as well as improved control of the dog population itself in these communities is probably the key to eliminating human alveolar hydatid infection (Stehr-Green *et al.* 1988).

Other cestode infections

Taenia saginata is regularly found in North American cattle herds, but reporting is limited and indigenous human cases are rare. *Taenia solium* is exceedingly rare or absent from Canadian and Alaskan swine herds (Schantz 1996). The number of reported imported cysticercosis cases is increasing however, probably as a result of immigration from endemic areas (e.g. Central and South America) and improved diagnostic capacity (e.g. Western blot, CT scan, MRI). *Hymenolepis nana* has been reported to be endemic in some institutionalized populations in southern Ontario (Tamblyn 1975).

Nematode infections

Trichinosis

Trichinosis in Canada and Alaska is caused by two distinct species, *Trichinella spiralis* and *T. nativa*. *Trichinella spiralis* has a life cycle that involves pigs and a variety of other carnivores in the southern regions of Canada. A common source of infection for humans is undercooked pork or bear meat. *Trichinella nativa* has a life cycle that involves a variety of carnivores, including northern bears and walrus. It differs biologically from *T. spiralis* in its behaviour in laboratory animal models and also, most remarkably, in its resistance to freezing. Poorly cooked bear has been the main source of human trichinosis in Alaska while the walrus has been the main source in the eastern Canadian Arctic region. While bear is eaten undercooked unintentionally, walrus is eaten raw by choice.

Trichinosis is a reportable disease in the United States and Canada with estimated incidence rates of 0.05–0.06 cases/100 000 per year. The regions with the the highest rates are Alaska and northern Canada (Northwest Territories and northern Québec) with rates of 1.8 cases/100 000 and 11 cases/100 000, respectively (MacLean *et al.* 1989). Focal serological studies have reported rates as high as 18% for *Trichinella* antibodies in native communities in northern Québec. Since *Trichinella* antibodies are thought to persist less than 5 years after a single exposure, these data suggest significant ongoing exposure in these communities. The clinical presentation of trichinosis in northern communities is often a prolonged diarrhoea (e.g. six weeks) rather than the classical myopathy and fever seen in *T. spiralis* infections around the world. Recent research suggests that the diarrhoeal presentation occurs upon re-infection of individuals who have pre-existing immunity to *Trichinella* (MacLean *et al.* 1989, 1992).

Toxocariasis

Toxocariasis is an endemic zoonosis both in urban and rural canines and felines. Outdoor playareas for children are often contaminated with *Toxocara* eggs with prevalence levels in sandbox and playarea soil from 2 to 38%. Dogs in Canada have stool egg prevalences for *Toxocara canis* of 2–57%, the highest rates occurring in puppies under 1 year of age. Human cases of both visceral larva migrans and *Toxocara* retinal lesions have been reported in small numbers. However, the true incidence is unknown as these infections are not reportable (Fanning *et al.* 1981). Several seroprevalence studies show the presence of *Toxocara* antibodies in 4–26% with higher levels in small children and in rural areas. In northern regions of Canada, *T. canis* is rare, being replaced by *T. leonina*, an ascarid of unknown human pathogenicity.

Anisakiasis

Although the anisakid worm *Pseudoterranova decipiens* is found in a high percentage of cod caught off the Atlantic coast of Canada and cod is eaten raw in some coastal communities, human anisakiasis has been only rarely reported as a human infection in this region. The most frequent presentation is a vomited or coughed up worm several days after raw fish consumption. The anisakid *Anisakis simplex* is more frequently found off the west coast of Canada and in Alaska and is a more invasive worm burrowing into the stomach or small intestinal wall with resulting pain.

Other nematode infections

Historically, the intestinal nematodes *Ascaris lumbricoides*, *Trichuris trichiura*, and hookworm have been endemic in Canada but they have now disappeared and are only seen as imported infections in travellers and immigrants. The pinworm, *Enterobius vermicularis*, remains endemic but its prevalence has not been systematically studied in Canada or Alaska. Chronic strongyloidiasis is seen in immigrants and WW II prisoners of war from Asia. Deaths are seen intermittently as a result of *Strongyloides stercoralis* hyperinfection in the immunocompromised. The chronic sequelae of filariasis (chyluria, lymphedema) are a problem in a small group of immigrants.

Trematode infections

There are two trematodes with an impact on human health that are endemic in Canada. *Metorchis conjunctus* infects fish, commonly the sucker, *Catostomus commersoni*. When an infected fish is eaten this fluke ascends the common bile duct and intra-hepatic bile ducts, which results in an acute illness of abdominal pain, anorexia, and low grade fever associated with a marked peripheral eosinophilia (MacLean *et al.* 1996). In native (aboriginal) communities where this fish is occasionally consumed raw, impressive levels of asymptomatic infection (up to 20%) have been noted in stool parasite surveys. A second endemic trematode problem is 'swimmers' itch', a short-lived itchy dermatitis caused by the cercaria of duck schistosomes. The distribution of the problem, which is found across Canada, is not well delineated but is focal, with certain lakes developing a reputation for the nuisance.

Ectoparasite infestations

The three most common types of human ectoparasitic infestations found in Canada are lice infestations caused by *Pediculus humanus capitis* and *Phthirus pubis*, scabies caused by *Sarcoptes scabiei*, and myiasis caused by different species of fly. Outbreaks of head lice occur most notably in the autumn because of the increased close contact among school children. Although there have been no surveys to document the prevalence of lice in the Canadian population, it has been estimated that between 1 and 10% of school children are infested with head lice at any one time (Chunge *et al.* 1991). Reports of body lice or pubic lice infestations are rare. Scabies has become relatively uncommon in Canada but continues to be of concern in certain high-risk groups, such as hospitalized populations and travellers (Belle *et al.* 1979; Jack 1993). Medical and public health authorities occasionally mention sporadic cases of scabies in immunosuppressed individuals and among preschool children attending daycare centres. Several different kinds of myiasis have been reported from each region in Canada (Gyorkos 1977). The most common fly species implicated in endemic myiasis has been the flesh-fly, *Wohlfahrtia vigil*, which causes furuncular myiasis. Imported cases also occur.

Conclusions

Parasite infections rarely cause death in Canada. Those primarily associated with mortality are toxoplasmosis, malaria, and the larval cestodes (*Echinococcus* and cysticerci). Significant morbidity does occur and many factors which exist today, and which are likely to increase in the future, will have an impact on the occurrence of parasite infections. Some of the more important factors include increasing travel (with increased 'at-risk' behaviours while travelling), increasing immigration to Canada from tropical and sub-tropical regions, increasing trade from tropical and sub-tropical regions (i.e. food products), and increasing diversity of food preparation and consumption patterns, among others. This situation requires that public health, medical, and laboratory personnel not only be aware of potential health concerns but also that they participate interactively in prevention and control activities to minimize adverse effects due to parasite infections. In addition, Canada has an active research community focusing on parasite-related research spanning the fields of molecular biology to epidemiology both in Canada and abroad. This provides a rich resource, in terms of parasite expertise, which supports local and international training activities and which enhances Canadian participation in international networks involved in global parasite prevention and control activities (i.e. Pan American Health Organization and World Health Organization programmes, Emerging Diseases Networks, Public Health Associations, and others).

References

Arthaud, J. B. (1970). Cause of death in 339 Alaskan natives as determined by autopsy. *Arch. Pathol.*, **90**, 433–438.

Bazin, S., Bouchard, C., Brisson, J., Morin, C., Meisels, A., and Fortier, M. (1994). Vulvar vestibulitis syndrome: an exploratory case-control study. *Obstet. Gynecol.*, **83**, 47–50.

Belle, E. A., D'Souza, T. J., Zarzour, J. Y., Lemieux, M., and Wong, C. C. (1979). Hospital epidemic of scabies: diagnosis and control. *Can. J. Public Health*, **70**, 133–135.

Burrows, R. B. (1961). Prevalence of amoebiasis in the United States and Canada. *Am. J. Trop. Med. Hyg.*, **10**, 172.

Carter, A. O. and Frank, J. W. (1986). Congenital toxoplasmosis: epidemiologic features and control. *Can. Med. Assoc. J.*, **135**, 618–623.

CCDR (Canada Communicable Disease Report) (1996). Health Canada.

Chunge, R. N., Scott, F. E., Underwood, J. E., and Zavarella, K. J. (1991). A review of the epidemiology, public health importance, treatment and control of head lice. *Can. J. Public Health*, **82**, 196–200.

Curtis, M. A., Rau, M. E., Tanner, C. E., Prichard, R. K., Faubert, G. M., Olpinski, S., and Trudeau, C. (1987). Parasitic zoonoses in relation to fish and wildlife harvesting by Inuit communities in Northern Québec, Canada. In: *Proceedings of the 7th International Congress on Circumpolar Health*, Umea, Sweden.

Fanning, M., Hill, A., Langer, H. M., and Keystone, J. S. (1981). Visceral larva migrans (toxocariasis) in Toronto. *Can. Med. Assoc. J.*, **124**, 21–26.

Freeman, R. S. and Jamieson, J. (1976). Parasites of Eskimos at Igloolik and Hall Beach, Northwest Territories. In: *Proceedings of the 3rd International Congress on Circumpolar Health*, Toronto, Canada.

Gyorkos, T. W. (1977). A review of human myiasis in Canada. *Can. Dis. Weekly Rep.*, **3–26**, 101–104.

Gyorkos, T. W., MacLean, J. D., Viens, P., Chheang, C., and Kokoskin-Nelson, E. (1992). Intestinal parasite infection in the Kampuchean refugee population 6 years after resettlement in Canada. *J. Infect. Dis.*, **166**, 413–417.

Gyorkos, T. W., Meerovitch, E., and Prichard, R. (1987). Estimates of intestinal parasite prevalence in 1984: report of a 5-year follow-up survey of provincial laboratories. *Can. J. Public Health*, **78**, 185–187.

Herwaldt, B. L. and Ackers, M. L. (1997). An outbreak in 1996 of cyclosporiasis associated with imported raspberries. The *Cyclospora* Working Group. *N. Engl. J. Med.*, **336**, 1548–1556.

Isaac-Renton, J. L. and Philion, J. J. (1992). Factors associated with acquiring giardiasis in British Columbia residents. *Can. J. Public Health*, **83**, 155–158.

Jack, M. (1993). Scabies outbreak in an extended care unit – a positive outcome. *Can. J. Infect. Control*, **8**, 11–13.

Kabani, A., Cadrain, G., Trevenen, C., Jadavji, T., and Church, D. L. (1995). Practice guidelines for ordering stool ova and parasite testing in a pediatric population. The Alberta Children's Hospital. *Am. J. Clin. Pathol.*, **104**, 272–278.

Keystone, J. S., Yang, J., Grisdale, D., Harrington, M., Pillon, L., and Andreychuk, R. (1984). Intestinal parasites in metropolitan Toronto day-care centres. *Can. Med. Assoc. J.*, **131**, 733–735.

MacLean, J. D., Arthur, J. R., Ward, B. J., Gyorkos, T. W., Curtis, M. A., and Kokoskin, E. (1996). Common-source outbreak of infection due to the North American liver fluke *Metorchis conjunctus*. *Lancet*, **347**, 154–158.

MacLean, J. D., Poirier, L., Gyorkos, T. W., Proulx, J.-F., Bourgeault, J., Corriveau, A., Illisituk, and Staudt, M. (1992). Epidemiology and serologic definition of primary and secondary trichinosis in the Arctic. *J. Infect. Dis.*, **165**, 908–912.

MacLean, J. D., Viallet, J., Law, C., and Staudt, M. (1989). Trichinosis in the Canadian Arctic: report of five outbreaks and a new clinical syndrome. *J. Infect. Dis.*, **160**, 513–520.

McDonald, J. C., Gyorkos, T. W., Alberton, B., MacLean, J. D., Richer, G., and Juranek, D. (1990). An outbreak of toxoplasmosis in pregnant women in northern Québec. *Can. J. Infect. Dis.*, **161**, 769–774.

Messier, F., Rau, M. E., and McNeill, M. A. (1989). *Echinococcus granulosus* infections and moose-wolf population dynamics in southwestern Québec. *Can. J. Zool.*, **676**, 216–219.

Olson, M. E., Thorlakson, C. L., Deselliers, L., Morck, D. W., and McAllister, T. A. (1997). *Giardia* and *Cryptosporidium* in Canadian farm animals. *Vet. Parasitol.*, **68**, 375–381.

Pereira, L. H., Embil, J. A., Haase, D. A., and Manley, K. M. (1990). Cytomegalovirus infection among women attending a sexually transmitted disease clinic: association with clinical symptoms and other sexually transmitted diseases. *Am. J. Epidemiol.*, **131**, 683–692.

Ratnam, S., Paddock, J., McDonald, E., Whitty, D., Jong, M., and Cooper, R. (1985). Occurrence of *Cryptosporidium* oocysts in fecal samples submitted for routine microbiological examination. *J. Clin. Microbiol.*, **22**, 402–404.

Rausch, R. L. (1986). Life cycle patterns and geographic distribution of *Echinococcus* species. In: *The biology of Echinococcus and hydatid disease* (R. C. A. Thompson and A. J. Lumbery, eds). George Allen and Unwin, London, UK, pp. 44–80.

Ruttenber, A. J., Weniger, B. G., Sorvillo, F., Murray, R. A., and Ford, S. L. (1984). Diphyllobothriasis associated with salmon consumption in Pacific Coast states. *Am. J. Trop. Med. Hyg.*, **33**, 455–459.

Schantz, P. M. (1996). Tapeworms (cestodiasis). *Gastroenterol. Clin N. Am.*, **25**, 637–653.

Schantz, P. M., Chai, J., Craig, P. S., Eckert, J., Jenkins, D. J., Macpherson. C. N. L., and Thakur, A. (1995). Epidemiology and control of hydatid disease. In: *Echinococcus and hydatid disease* (R. C. A. Thompson and A. J. Lumbery, eds). George Allen and Unwin, London, UK, pp. 233–331.

Stehr-Green, J. A., Stehr-Green, P. A., Schantz, P. M., Wilson, J. F., and Lanier, A. (1988). Risk factors for infection with *Echinococcus multilocularis* in Alaska. *Am. J. Trop. Med. Hyg.*, **38**, 380–385.

Stephen, C., Haines, D., and Bollinger, T. *et al.* (1996). Serological evidence of *Toxoplasma* infection in cougars on Vancouver Island, BC. *Can. Vet. J.*, **37**, 241.

Svenson, J. E., MacLean, J. D., Gyorkos, T. W., and Keystone, J. (1995). Imported malaria. Clinical presentation and examination of symptomatic travelers. *Arch. Intern. Med.*, **155**, 861–868.

Tamblyn, S. E. (1975). Intestinal parasites in an institution – Ontario: preliminary report. *Epidemiol. Bull.*, **19**, 53–54.

Wilson, J. F. and Rausch, R. L. (1980). Alveolar hydatid disease. A review of clinical features of 33 indigeneous cases of *Echinococcus multilocularis* infection in Alaskan eskimos. *Am. J. Trop. Med. Hyg.*, **29**, 1340–1355.

Part 3

Important parasites in the North

Biology, pathogenesis and epidemiology

8 *Acanthamoeba* spp.

Jadwiga Winiecka-Krusnell

Opportunistic and facultative parasites of genus *Acanthamoeba* belong to the group of primary free-living amoebae widely distributed in nature – in water, air, and soil. The organisms are well adapted to adverse environmental conditions and are resistant to disinfectants. They play a part in the cycling of nutrients by grazing on bacteria on biofilms. *Acanthamoeba* species may also act as reservoir hosts for human pathogens like *Legionella pneumophila*, being responsible for maintaining, protecting, and transmitting pathogenic bacteria in the environment. Some of the *Acanthamoeba* species are pathogenic to humans. The disease may present as granulomatous amoebic encephalitis (GAE) or disseminated invasion localized in different tissues in immunocompromised individuals or as *Acanthamoeba* keratitis (AK) in immunocompetent, soft contact lens wearers.

Taxonomy

The taxonomic position of *Acanthamoeba* has been subjected to various changes over the years. Due to nomenclatural difficulties, the isolated organisms were earlier classified as *Hartmannella*. In the recently updated systematics both genera are separated at the level of suborders (Page 1988). The genus *Acanthamoeba* today contains approximately 20 species. Classification of new isolates at the species level requires morphological and molecular measures (Pussard and Pons 1977; Gast *et al.* 1996).

Morphology

Acanthamoeba exists in two forms. The trophozoite (25–40 μm) is a proliferative stage, characterized by spine-like pseudopodia (acanthopodia) and slow locomotion. The cell contains a single nucleus with a large, central karyosome, fine granulated cytoplasm with multiple digestive vacuoles, and a prominent pulsating vacuole. The cyst (12–30 μm) a dormant stage, is spherical or star-shaped and has double walls with pores (ostioles) (Martinez 1985) (Figure 8.1). Cysts are highly viable and resistant to a variety of environmental conditions, including ambivalent temperatures, desiccation, amoebicidal drugs, and disinfectants. In a recent report (Mazur *et al.* 1995), cysts have been demonstrated to retain their viability and virulence even after 24 years of storage.

Distribution

Acanthamoeba spp. have been isolated from the environment in both warm and cold climates and from animals and humans in all parts of the world. They are present in different habitats, including natural and artificial water bodies, coastal sediments, soil, and air (Rodriguez-Zagaroza 1994). Tolerance to high chlorine concentrations promotes their presence in tap

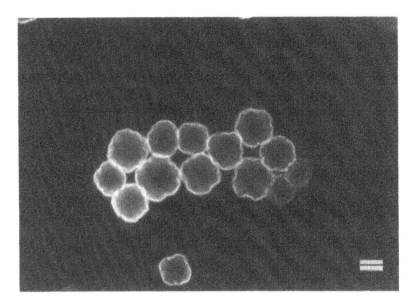

Figure 8.1 Acanthamoeba spp. cysts stained with fluorescein conjugated anti-cyst wall monoclonal
antibody. Bar = 10 μm.

water, swimming pools, dental units, and medicinal pools. In Scandinavian countries,
the presence of *Acanthamoeba* species has been reported in indoor swimming pools and in
periodically freezing, outdoor swimming areas (Cerva and Huldt 1974; Vesaluoma *et al.*
1995; Brown and Cursons 1977).

Acanthamoeba can be naturally infected by pathogenic bacteria which multiply and survive
intracellulary. Despite the well-documented relationship with *Legionella pneumophila*
(Rowbotham 1980), an increasing number of gram-negative endosymbionts in *Acanthamoeba*
and other members of free-living amoebae group have been reported (Winiecka-Krusnell
and Linder 2001). The significance of *Acanthamoeba* in maintaining and transmitting of
human pathogens is not well recognized and the clinical and epidemiological relevance of
bacterial infections in amoebae should be reconsidered.

Pathogenicity and epidemiology

The pathogenic features of *Acanthamoeba* spp. have been discovered relatively recently.
Environmental strains pathogenic to laboratory animals represent potential human
pathogens. *Acanthamoeba* has been isolated from stools, throats, and nasal cavities of asymptomatic individuals. The discrepancy observed between the low number of identified cases
and the common presence of *Acanthamoeba* in the environment can be a result of either the
low virulence of the pathogen or the resistance developed by humans under frequent contacts with parasites. Commonly, the genus *Acanthamoeba* represents facultative, opportunistic
parasites. Both cysts and trophozoites are considered invasive. The skin and upper respiratory
tract are recognised as a portal of entry in immunodeficient and chronically ill patients.
Infection may present as a skin ulceration, respiratory tract, ear or nose infection, or gastrointestinal symptoms. The parasites which spread by a haematogenous route may invade the

central nervous system producing GAE, a chronic and fatal disease (Martinez 1985). GAE is a rare condition – 103 cases have been reported around the world, the majority of them in the United States. In recent years, an increased number of *Acanthamoeba* invasions have been reported in HIV and AIDS patients (Martinez and Visvesvara 1997).

Ocular acanthamoebiasis – AK is a disease occurring among immunocompetent and otherwise healthy people. Several species have been involved in human ocular infections. Parasites enter their hosts through damaged cornea. The disease is painful and devastating. Cellular infiltration in the cornea may result in its perforation and loss of visual acuity (Auran *et al.* 1987). Although antibody response during AK is low, a successful immunization against *Acanthamoeba* infection has been obtained in an animal model by combined subconjunctival and intramuscular administration of the parasite antigen (Alizadeh *et al.* 1995). The incidence of AK is low. Approximately 1 in 250 000 contact lens wearers develop infection (Seal 1994). More than 700 cases have been reported since the first AK case was detected in a Texas rancher (Martinez and Visvesvara 1997). Few cases have been reported from Sweden, Norway, and Denmark, which may suggest that the condition is very rare in Scandinavian countries (Brincker *et al.* 1988; Aasly and Bergh 1992; Stenevi *et al.* 1992; Nilsson and Montan 1994; Skarin *et al.* 1996). The exact prevalence, however, is unknown since reporting of identified cases is often not obligatory.

Initially, corneal trauma and contact with contaminated water were considered to be the main risk factors. The recent epidemiological data however, indicate that the disease is strongly correlated with soft contact lens wearer (Moore *et al.* 1987). Inappropriate hygiene and disinfection of lenses and lens cases using tap water or home-made washing solution may result in contamination with *Acanthamoeba*. High water content and gel-like structure of modern soft lenses promotes adherence of amoebae which are then easily transported to the eye. Cysts of *Acanthamoeba* are resistant to several commercial lens disinfectants (Ahearn and Gabriel 1997). Moist heat disinfection is effective if the recommended procedure is respected (Seal 1994). The increasing use of contact lenses results in a growing population at risk. Compliance with lens care instructions and education and information about risk factors are essential for preventing AK.

References

Aasly, K. and Bergh, K. (1992). *Acanthamoba* keratitis: report of the first Norwegian cases. *Acta Ophthalmol.*, **7**, 698–701.

Ahearn, D. G. and Gabriel, M. M. (1997). Contact lenses, disinfectants and *Acanthamoeba* keratitis. *Adv. Appl. Microbiol.*, **43**, 35–56.

Alizadeh, H., He, Y., McCulley, J. P., Ma, D., Stewart, G. L., Via, M., Haehling, E., and Niederkorn, J. Y. (1995). Successful immunization against *Acanthamoeba* keratitis in a pig model. *Cornea*, **14**, 180–186.

Auran, J. D., Starr, M. B., and Jakobiec, F. A. (1987). *Acanthamoeba* keratitis. A review of the literature. *Cornea*, **6**, 2–26.

Brincker, P., Gregersen, E., and Prause, J. U. (1988). *Acanthamoeba* keratitis, clinico-pathological report of 2 cases. *Acta Ophthalmol.*, **66**, 210–213.

Brown, T. J. and Cursons, R. T. M. (1977). Pathogenic free-living amoebae (PFLA) from frozen swimming areas in Oslo, Norway. *Scand. J. Infect. Dis.*, **9**, 237–240.

Cerva, L. and Huldt, G. (1974). *Limax* amoebae in five swimming pools in Stockholm. *Folia Parasitol.*, **21**, 71–75.

Gast, R. J., Ledee, D. R., Fuerst, P. A., and Byers, T. J. (1996). Subgenus systematics of *Acanthamoeba*: four nuclear 18S rDNA sequence types. *J. Eukaryot. Microbiol.*, **43**, 498–504.

Martinez, A. J. (1985). *Free-living amebas: natural history, prevention, diagnosis, pathology and treatment of disease.* CRC Press, Boca Raton, FL.

Martinez, A. J. and Visvesvara, G. S. (1997). Free-living amphizoic and opportunistic amebas. *Brain Pathol.*, **7**, 583–598.

Mazur, T., Hadas, E., and Iwanicka I. (1995). The duration of the cyst stage and the viability and virulence of *Acanthamoeba* isolates. *Trop. Med. Parasitol.*, **46**, 106–108.

Moore, M. B., McCulley, J. P., Newton, C., Cobo, L. M., Foulks, G. N., O'Day, D. M., Johns, K. J., Driebe, W. T., Wilson, L. A., Epstein, R. J., and Dougham, D. J. (1987). *Acanthamoeba* keratitis: a growing problem in soft and hard contact lens wearers. *Ophthalmology*, **94**, 1654–1661.

Nilsson, S. E. and Montan, P. G. (1994). The hospitalised cases of contact lens induced keratitis in Sweden and their relation to lens type and wear schedule: results of a three-years retrospective study. *Contact Lens Assoc. Ophthalmol. J.*, **20**, 97–101.

Page, F. C. (1988). *A new key to fresh water and soil gymnamoebae.* Fresh Water Biological Association, Cumbria, UK.

Pussard, M. and Pons, R. (1977). Morphologie de la pario kystique et taxonomie du genre *Acanthamoeba* (Protozoa: Amoebidae). *Protistologica*, **13**, 557–598.

Rodriguez-Zaragoza, S. (1994). Ecology of free-living amoebae. *Crit. Rev. Microbiol.*, **20**, 225–241.

Rowbotham, T. J. (1980). Preliminary report on the pathgenicity of *Legionella pneumophila* for freshwater and soil amoebae. *J. Clin. Pathol.*, **33**, 1179–1183.

Seal, D. V. (1994). *Acanthamoeba* keratitis. *Br. Med. J.*, **308**, 1116–1117.

Skarin, A., Floren, I., Miörner, H., and Stenevi, U. (1996). *Acanthamoeba* keratitis in the south of Sweden. *Acta Ophthalmol. Scand.*, **74**, 593–597.

Stenevi, U., Floren, I., and Miörner, H. (1992). *Acanthamoeba* keratit-en ny diagnos i Sverige. *Läkartidningen*, **89**, 127–128.

Vesaluoma, M., Kalso, S., Jokipii, L., Warhurst, D., Pönkä, A., and Tervo, T. (1995). Microbiological quality in Finnish public swimming pools and whirlpools with special reference to free living amoebae: a risk factor for contact lens wearers? *Br. J. Ophthalmol.*, **79**, 178–181.

Winiecka-Krusnell, J. and Linder, E. (2001). Bacterial infections of free-living amoebae. *Res. Microbiol.*, **7**, 613–619.

9 Entamoeba histolytica/dispar

Ewert Linder

Entamoeba histolytica infects humans and some higher non-human primates. It is thought to cause an estimated 40 000–100 000 deaths annually. Intestinal infection with *E. histolytica* – and the associated severe manifestations of invasive amoebiasis – is a global problem, but the true prevalence of the intestinal infection in the northern countries – as indeed globally – is not known. The reason for this is that available prevalence figures, to a large extent, have been based on laboratory findings of '*E. histolytica* cysts' in faeces. This finding needs to be related to the recent observation that the majority of such cysts can be identified as non-pathogenic *Entamoeba dispar. Entamoeba histolytica* and *E. dispar* are morphologically indistinguishable (Diamond and Clark 1993). The two species have previously been called pathogenic and non-pathogenic *E. histolytica.*

The laboratory techniques for distinguishing *E. histolytica* and *E. dispar* are available but not yet in general use. Thus, even if the available information is sufficient to estimate the health impact of *E. histolytica* infection, we will only slowly, with the development of more widely available laboratory diagnostics, become aware of the prevalence of the infection in areas with few cases of clinical disease, as in the northern areas we are dealing with in this chapter.

The parasite

Entamoeba histolytica lacks many organelles seen by electron microscopy in eukaryotes, such as mitochondria, Golgi apparatus, rough endoplasmic reticulum, and microtubules. This was earlier interpreted as evidence for *E. histolytica* being an ancestral eukaryote. However, ribosomal RNA-based molecular phylogeny shows that *E. histolytica* diverged from other eukaryotes much later than lineages containing all the missing organelles. Direct evidence for secondary loss of organelles has recently been shown by the identification, in the *E. histolytica* nuclear genome, of genes that in other eukaryotes reside in the mitochondrion. Interestingly a putative mitochondrial remnant has been identified recently in the cytoplasm of *E. histolytica* (Mai *et al.* 1999; Tovar *et al.* 1999).

Life cycle

Protozoa of the genus *Entamoeba* have two stages: the infectious cyst and the disease-producing trophozoite. Infection occurs directly from person to person, by the faecal–oral route; transmission occurs by insects and food. Cysts survive in contaminated food, water, and vegetables. The cysts pass the stomach and excyst in the small intestine. The emerging trophozoites migrate to the colon where interactions with the mucosa causes functional and morphological damage. Encystation occurs in the large intestine, apparently as a result of interactions with the altered environment, including the presence of the bacterial flora. Infectious cysts are excreted with the stools. In acute amoebic diarrhoea, trophozoites are also excreted.

The basis for pathogenicity of *E. histolytica*

Adherence and chemotaxis

Entamoeba histolytica extracellular killing of host cells is contact dependent. Adherence to human colonic epithelial cells and mucins is mediated by a galactose-specific Gal/GalNAc lectin (Saffer and Petri 1991). Galactose-specific binding of the soluble adherence lectin has been extensively studied, and anti-lectin antibodies have been used to develop an assay for the identification of *E. histolytica* in stools (see subsequently). Monoclonal antibodies to the extracellular cysteine-rich region of the 170-kDa subunit have been shown to interfere with adhesion to target cells (Mann and Lockhart 1998).

Fibronectin fragments also appear to stimulate directed trophozoite migration and chemokinesis (Franco *et al.* 1997). Adherence of *E. histolytica* to target cells by means of the described surface lectin induces a cytolytic response involving secretion of proteolytic substances and pore-forming peptides.

Penetration, invasion, and cytolysis

Entamoeba histolytica virulence is related to a number of amoebic components (lectins, cysteine proteinases, and lytic peptides), the intestinal bacterial flora, as well as other host factors. Invasive amoebas resist lysis by serum complement in contrast to non-invasive ones. Trophozoites are selective in their interactions with bacteria, and the parasite recognition of glycoconjugates plays an important role in amoebic virulence (Padilla-Vaca *et al.* 1999).

Active migration of *E. histolytica* trophozoites through extracellular matrices might play a role in host tissue destruction. Trophozoites degrade fibronectin bound to their surface and adhere to substrate-bound fibronectin, producing local matrix degradation. Tissue invasion has been linked to amoeba collagenase production, and especially degradation of collagen type I (see Ravdin 1988). Cytopathic and cytolytic activities of viable trophozoites against mammalian nucleated cells and red blood cells have been extensively studied. Cytolytic peptides of *E. histolytica* capable of formation of ion channels in target cell membranes exist in amoebic cytoplasmic granules. They have been described as amoebapores lysing several types of mammalian cells and as hemolysins based on their lytic effect on erythrocytes (Jansson *et al.* 1994). Amoebapores, 77-residue peptides, are structural and functional analogues of NK-lysin and granulysin of porcine and human cytotoxic lymphocytes. These peptides may be related to members of the gene family of saposin-like proteins having antibacterial activity (Banyai and Patthy 1998). They have been proposed to be major pathogenicity factors of *E. histolytica* (Bracha *et al.* 1999). That differences in abundance and activity of the lytic polypeptides may have an impact on the pathogenicity of amoebae is suggested by the comparison between amebapores of *E. histolytica* and *E. dispar* (Nickel *et al.* 1999).

Clinical manifestations, diagnostics, and treatment of *E. histolytica* infection

Colitis and severe diarrhoea

In Nordic countries amoebiasis cases are of two categories, immigrants or residents of tropical and sub-tropical regions and travellers. Short-term travellers as opposed to long-term travellers and residents of endemic areas seem to have a significantly higher rate of *E. histolytica* to *E. dispar* infection (Walderich *et al.* 1997).

The incubation time is dependent on the infectious dose and may vary from a few days to months. Usually symptoms appear after 2–4 weeks. Acute amoebic colitis is manifested as abdominal pain, tenderness, watery diarrhoea, and frequent bloody stools. In proctoscopy, ulcerations with a 'punced out' appearance are typically associated with *E. histolytica* colitis. Severe disease is more commonly seen in association with malnutrition, pregnancy, immunosuppressive therapy, and in children. The enterotoxic activity of *E. histolytica* appears to be due to several factors, including secretagogues, like neurohormones and prostaglandins, which may originate both from the parasite and the host (see Ravdin 1988).

The diagnostics is based on finding haematophagous *E. histolytica* trophozoites in microscopy. However, this is a method of low sensitivity (see, e.g. Walderich *et al.* 1997). As pointed out in the paragraph on diagnostic methods, the finding of cysts of *E. histolytica/dispar* is largely misinterpreted as diagnostic for *E. histolytica* infection.

Extraintestinal infection and amoebic liver abscess

Colonic perforations are commonly seen in fulminant colitis. This may lead to peritonitis which is reported to occur at a frequency of 3–5%. In highly endemic areas, such as Mexico, amoebic liver abscess has been found in 5.8% of autopsies. Amoebic liver abscess is due to haematogeneous dissemination of amoebas. Tissue destruction is caused by proteolytic enzymes in part released from host polymorphonuclear leucocytes which are lysed by amoebas. Trophozoites may be found in a minority of cases in the abscess fluid by microscopy. The parasites are located at the edge of the abscess in contact with hepatocytes (Healy 1988). The diagnosis is based on radiology and serology. Parasite nucleic acids can be demonstrated by polymerase chain reaction (PCR) (see subsequently).

In most cases of invasive amoebiasis a specific humoral antibody response is seen. Several methods are in routine use in diagnostic laboratories measuring antibodies against *E. histolytica*. The sensitivity of diagnostic serology is in the order of 80%. False negative results are seen especially during the early phase of amoebic liver abscess formation, possibly due to excess amoebic antigen being released during the invasive process. Positive serology in a non-endemic area may be a more reliable diagnostic tool than the same assay in an endemic region.

Identification of E. histolytica/dispar

The common procedure for stool examination involves formol-ether (ethylacetate) concentration of faecal cysts and ova (Ridley and Hawgood 1956). The identification of *Entamoeba* species in human stool is based on the size of the trophozoite and cyst, nuclear morphology, and number of nuclei in mature cysts as well as on the morphology of the chromatoid bodies that appear during encystation.

Cysts of *E. histolytica* are 10–16 in diameter with four. Glycogen in a vacuole and chromatoid bodies are distinct in the immature cysts, becoming more diffuse upon maturation. Morphologically *E. histolytica* and *E. dispar* cysts are indistinguishable. This is, however, possible using novel techniques (see subsequently).

Prevalence of E. histolytica/dispar and of E. histolytica

As discussed earlier, prevalence numbers in the literature on '*E. histolytica*' are based on routine stool examination, and in fact may be either *E. histolytica* or *E. dispar*. Thus, despite the estimated high rates of *E. histolytica/dispar*-infected individuals (50 million cases globally is a figure in the literature, see Ravdin 1988; the reported prevalence in Sweden is about 500 yearly cases), amoebiasis apparently occurs only in a few per cent of these individuals.

The reason for this low figure is twofold. First, the relative prevalence of *E. histolytica* is about 1 in 10 (Anonymous 1997). Second, even in *E. histolytica*-infected individuals, symptoms related to amoebiasis seem to be lacking in at least one-third of the cases. The figure may be even lower as it may be impossible to rule out other causes of enteric symptoms, such as enterotoxic *Escherichia coli* and enteric viruses (Walderich *et al.* 1997), and in fact there is data suggesting that in most cases colonization with *E. histolytica* is asymptomatic (Braga *et al.* 1996; Gilchrist and Petri 1999).

The reported high number of clinically diagnosed cases of amoebic infection is a problem as it has consequences with respect to treatment and handling of both diarrhoeal patients and asymptomatic individuals. This risk of coincidental presence of *E. dispar* cysts in patients with diarrhoea is of course higher in situations where *E. histolytica/dispar* is prevalent. The reported *E. histolytica/dispar* prevalence seems to be around 10% in tropical and sub-tropical countries, but may be higher: *E. histolytica/dispar* was isolated in 345 out of 3 536 individuals (9.7%) of subjects living in rural communities around Delhi but there was no increase in the prevalence rate of bowel symptoms in the culture-positive compared to the culture-negative subjects (Anand *et al.* 1993).

There is some variation in the reported relative proportion of *E. histolytica* among *E. histolytica/dispar* cyst carriers and so far the estimated 10% (Anonymous 1997) is based on rather limited materials. Of the 313 individuals studied 15% had *E. histolytica/dispar* cysts in the stools at the Seychelles and 8/40 (20%) had *E. histolytica* (Sargeaunt 1992). In north-eastern Brazil 14 out of 155 individuals (9.0%) carried *E. histolytica/dispar* and 4 out of 10 stools (29%) had *E. histolytica* by detection of the Gal/GalNAc lectin in stool infection (Braga *et al.* 1996).

The presence of *E. histolytica/dispar* cysts in stools apparently correlates to sexual activities that allow for oral–faecal contamination and thus not to homosexuality *per se* (see Ravdin 1988). *Entamoeba histolytica/dispar* is frequently found among male homosexuals, with a prevalence of 25–35% and in some communities over 50%. Sargeaunt *et al.* found that among 52 isolates of *E. histolytica/dispar* from 470 stools no pathogenic zymodemes were found (Sargeaunt *et al.* 1983) and in Recife only 1 of 77 stools had cysts of *E. histolytica/dispar*, and this was of non-pathogenic zymodeme (*E. dispar*) (de Alencar *et al.* 1996). The conclusion from a careful evaluation is that there are no data to suggest that finding cysts of *E. histolytica/dispar* implies that *E. histolytica* is a pathogen in homosexual men. (Goldmeier *et al.* 1986). However, of 28 symptomatic amoebic patients studied retrospectively in Tokyo, almost half had positive serology for *Treponema pallidum* or HIV, and indicated that they engaged in homosexual practices (Ohnishi and Murata 1997).

This finding of *E. histolytica/dispar* cysts has been associated with intestinal symptoms, 'the gay bowel syndrome'. It is important, however, to critically evaluate the significance of finding *E. histolytica/dispar* cysts in stools of patients as chronic abdominal pain and frequent bowel disturbance are common symptoms experienced by more than 15% of apparently healthy people. In areas endemic for *E. histolytica* infection, these symptoms are often diagnosed as 'non-dysenteric intestinal amoebiasis'. In a study addressing this problem more than 60% of cyst-positive as well as cyst-negative patients with symptoms showed either complete or partial response to treatment strategy for irritable bowel syndrome. Thus it was concluded that chronic bowel symptoms, such as pain in the abdomen and frequent bowel disturbance, have no association with either past or present infection with *E. histolytica* and that '... the clinical entity of non-dysenteric intestinal amoebiasis, if it exists, must be extremely rare'. Most patients with such symptoms are likely to have 'irritable bowel syndrome' (Anand *et al.* 1997).

Serology, as discussed later, is a potent diagnostic tool for invasive amoebiasis, but serology has also been used for seroepidemiology. While only 3% of older children, in the Brazilian study refered to earlier, carried *E. histolytica*, 40% developed serologic evidence of having experienced pathogenic *E. histolytica* infection (Braga *et al.* 1996). Thus positive serology reflects the presence of *E. histolytica* in a community – which needs to be considered in the evaluation of diagnostic significance of antibodies against *E. histolytica*.

Treatment

Two classes of anti-amoebic drugs are in use: luminal amoebicides (such as diloxanide furoate and paromycin) and tissue amoebocides (such as 5-nitroimidazoles). To treat invasive infection, tissue amoebocide treatment is followed by treatment with luminal amoebicides. If *E. histolytica/dispar* is found in symptomatic patients it should not be assumed that *E. histolytica* is the cause of the symptoms and other explanations for the symptoms should also be considered. Chemoprophylaxis is never appropriate (Anonymous 1997).

Laboratory methods to distinguish between E. histolytica and E. dispar

Isoezyme analysis

By starch gel or polyacrylamide gel electrophoresis of cultures of *E. histolytica/dispar* from stools it is possible to distinguish between pathogenic and non-pathogenic strains using electrophoretic banding patterns of four enzymes: hexokinase, malic enzyme, glucosephosphate isomerase, and phosphoglucomutase. These methods have been used to distinguish *Entamoeba* isolates into 18 zymodemes, of which zymodeme II seems to be a marker for invasive *E. histolytica* 'in the Western world' (Healy 1988).

PCR to detect E. histolytica

DNA hybridization technology is capable of distinguishing between *E. histolytica* and *E. dispar* (Clark and Diamond 1991; Clark 1993) and the methods have evolved to become important tools in the demonstration of *E. histolytica* in clinical samples. A number of PCR primers designed to detect *E. histolytica* have been reported. Sequence analysis of homologous cDNA clones derived from pathogenic and non-pathogenic isolates of *E. histolytica* revealed 10% nucleic acid substitutions (Tannic and Burchard, 1991). Oligonucleotide primers specific for the gene encoding the 30-kDa molecule of pathogenic *E. histolytica* detected the parasite in 19 liver abscess fluids from 14 patients with a presumptive amoebic liver abscess. Only 2 of the 19 samples were positive microscopically (Tachibana *et al.* 1992). Primers detecting *E. histolytica* actin gene (Huber *et al.* 1988) detected the parasite in 22 out of 23 suspected liver abscess cases from Nicaragua (Linder *et al.* 1997).

Detection of E. histolytica-specific antigens

Immunofluorescence can detect *E. histolytica/dispar* trophozoites and cysts with high sensitivity, but antibody reagents specific for *E. histolytica* cysts have been difficult to achieve (see, e.g.

Perez *et al.* 1987), and despite encouraging reports in the literature on the detection of cysts in stool samples using monoclonal antibodies as markers, no such reagents are currently available commercially.

An ELISA that detects the Gal/GalNAc lectin antigen and can distinguish between *E. histolytica* and *E. dispar* has been introduced commercially (Petri 1996). It appears to be useful, but its use is restricted to fresh stool samples. The target antigen appears to be rapidly destroyed even in frozen samples.

Prospects for vaccine development

The application of molecular biologic techniques over the past decade has seen a tremendous growth in our knowledge of the biology of *E. histolytica*, and has led to the identification and structural characterization of three potential amoebic vaccine antigens: the serine-rich *E. histolytica* protein (SREHP), the 170-kDa subunit of the Gal/GalNAc binding lectin, and the 29-kDa cysteine-rich protein. Such a putative vaccine may be an oral combination 'enteric pathogen' vaccine, capable of inducing protective mucosal immune responses to several clinically important enteric pathogens, including *E. histolytica* (Ryan *et al.* 1997; Stanley 1997). An understanding of the amoeba's obligatory encystment pathway should provide an approach for interrupting the transmission (Eichinger 1997).

References

Anand, A. C., Reddy, P. S., Saiprasad, G. S., and Kher, S. K. (1997). *Lancet*, **349**, 89–92.

Anand, B. S., Tuteja, A. K., Kaur, M., Alam, S. M., Aggarwal, D. S., Mehta, S. P., and Baveja, U. K. (1993). *Dig. Dis. Sci.*, **38**, 1825–1830.

Anonymous (1997). *Epidemiol. Bull.*, **18**, 13–14.

Banyai, L. and Patthy, L. (1998). *Biochim. Biophys. Acta – Prot. Struc. & Mol. Enzymol.*, **1429**, 259–264.

Bracha, R., Nuchamowitz, Y., Leippe, M., and Mirelman, D. (1999). *Mol. Microbiol.*, **34**, 463–472.

Braga, L. L., Lima, A. A., Sears, C. L., Newman, R. D., Wuhib, T., Paiva, C. A., Guerrant, R. L., and Mann, B. J. (1996). *Am. J. Trop. Med. Hyg.*, **55**, 693–697.

Clark, C. G. (1993). In: *Diagnostic molecular mircobiology* (D. H. Persing, T. F. Smith, F. C. Tenover, and T. J. White, eds). *American Society for Microbiology, Washington*, pp. 468–474.

Clark, C. G. and Diamond, L. S. (1991). *Mol. Biochem. Parasitol.*, **49**, 297–302.

de Alencar, L. C., Magalhaes, V., de Melo, V. M., Aka, I., Magalhaes, M., and Kobayashi, S. (1996). *Rev. Soc. Bras. Med. Trop.*, **29**, 319–322.

Diamond, L. S. and Clark, C. G. (1993). *J. Eukaryot. Microbiol.*, **40**, 340–344.

Eichinger, D. (1997). *Bioessays*, **19**, 633–639.

Franco, E., Vazquez-Prado, J., and Meza, I. (1997). *J. Infect. Dis.*, **176**, 1597–1602.

Gilchrist, C. A. and Petri, W. A. (1999). *Curr. Opin. Microbiol.*, **2**, 433–437.

Goldmeier, D. *et al.* (1986). *Lancet*, **1**, 641–644.

Healy, G. R. (1988). John Wiley & Sons Inc., New York.

Huber, M., Garfinkel, L., Gitler, C., Mirelman, D., Revel, M., and Rozenblatt, S. (1988). *Mol. Biochem. Parasitol.*, **31**, 27–33.

Jansson, Å., Gillin, F., Kagardt, U., and Hagblom, P. (1994). Coding of hemolysins within the ribosomal RNA repeat on a plasmid in *Entamoeba histolytica*. *Science*, **263**, 1440–1443.

Linder, E., Isaguliants, M., Olsson, M., Lundin, L., Zindrou, S., Corrales, E., Tellez, A., Leiva, B., Morales, W., Rivera, T., and Cabrera, J. M. (1997). *Arch. Med. Res.*, **28**, 314–316.

Mai, Z., Ghosh, S., Frisardi, M., Rosenthal, B., Rogers, R., and Samuelson, J. (1999). *Mol. Cell. Biol.*, **19**, 2198–2205.

Mann, B. J. and Lockhart, L. A. (1998). *J. Eukaryot. Microbiol.*, **45**, 13S–16S.

Nickel, R., Ott, C., Dandekar, T., and Leippe, M. (1999). *Eur. J. Biochem.*, **265**, 1002–1007.

Ohnishi, K. and Murata, M. (1997). *Epidemiol. Infect.*, **119**, 363–367.

Padilla-Vaca, F., Ankri, S., Bracha, R., Koole, L. A., and Mirelman, D. (1999). *Infect. Immun.*, **67**, 2096–2102.

Perez, d. S. E., Perez, S. I., Perozo, R. G., de, D. D., Romer, H., and Tapia, F. J. (1987). *Trans. R. Soc. Trop. Med. Hyg.*, **81**, 624–626.

Petri, W., Jr. (1996). *J. Invest. Med.*, **44**, 24–36.

Ravdin, J. I. e. (1988). John Wiley & Sons Inc., New York.

Ridley, D. S. and Hawgood, B. C. (1956). *J. Clin. Pathol.*, **9**, 74–76.

Ryan, E. T., Butterton, J. R., Zhang, T., Baker, M. A., Stanley, S. L., Jr., and Calderwood, S. B. (1997). *Infect. Immun.*, **65**, 3118–3125.

Saffer, L. D. and Petri, W. (1991). *Infect. Immun.*, **59**, 4681–4683.

Sargeaunt, P. G. (1992). *Arch. Med. Res.*, **23**, 265–267.

Sargeaunt, P. G., Oates, J. K., MacLennan, I., Oriel, J. D., and Goldmeier, D. (1983). *Br. J. Ven. Dis.*, **59**, 193–195.

Stanley, S. L., Jr. (1997). *Clin. Microbiol. Rev.*, **10**, 637–649.

Tachibana, H., Kobayashi, S., Okuzawa, E., and Masuda, G. (1992). *Int. J. Parasitol.*, **22**, 1193–1196.

Tannich, E. and Burchard, G. D. (1991). *J. Clin. Microbiol.*, **29**, 250–255.

Tovar, J., Fischer, A., and Clark, C. G. (1999). *Mol. Microbiol.*, **32**, 1013–1021.

Walderich, B., Weber, A., and Knobloch, J. (1997). *Am. J. Trop. Med. Hyg.*, **57**, 70–74.

10 *Giardia intestinalis*

Jadwiga Winiecka-Krusnell and Agneta Aust Kettis

The parasitic flagellate *Giardia intestinalis* is one of the most frequently identified intestinal protozoan in humans. The primitive organization of the cell and some molecular similarity to bacteria suggests that the organism branched very early in the evolution. The parasite is distributed worldwide. Giardiasis, endemic in developing countries, appears repeatedly as a water-borne epidemic in the industrialized world. Relatively harmless and self-limited disease in immunocompetent adults may present as persistent and devastating diarrhoea in malnourished children. Morphologically identical parasites exist in several wild and domestic animals. Their role in the transmission of human disease is not sufficiently proven.

Despite the over 300-year-old medical history, the progress in the research and understanding of the biology of the parasite and the pathogenesis of the disease has been relatively slow. *Giardia* and giardiasis are still subjected to controversial opinions concerning the zoonotic nature, interspecies transmission, host specificity, and taxonomy of the parasite.

Taxonomy

Giardia is the first protozoan parasite of man discovered as early as 1681 by Antony van Leeuwenhoek and described later by Lambl in 1859. It is a member of order Diplomonadida belonging to class Zoomastigophorea. The taxonomic name of the organism has been subjected, during the past years, to many changes. As many as seven different binomial labels of the parasite can be traced in the past. *Giardia intestinalis* and *G. lamblia* used currently are synonymous names for the human parasite, which, together with parasites of wild and domestic mammals, belong to *G. duodenalis* morphological group (Filice 1952).

The recent research work provides data on *Giardia* being in the position of a 'missing link' between prokaryotes and eukaryotes and data indicating genetic heterogeneity of isolates suggesting that a human parasite is not a single species but may represent a species complex (Andrews *et al.* 1989; Kabnik and Peattie 1991).

Morphology

Two stages, trophozoite and cyst (Figure 10.1A and B), undergoing encystation and excystation complete the simple life cycle of *Giardia*. The motile and dividing trophozoite has a primitive organization of a cell that lacks several intracellular organelles, including mitochondria. It is pear-shaped and has an approximate size of 10–20 by 5–15 μm. The cell contains two nuclei, four pairs of flagella, and a rod-shaped median body. Its flattened ventral side contains an adhesive disc, which promotes attachment to epithelial cells in the intestine. The dormant cyst, an infective form, is oval, 8–12 by 7–10 μm in size and contains

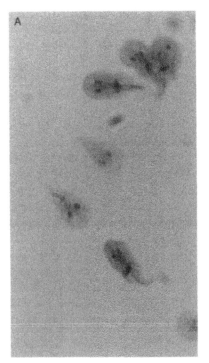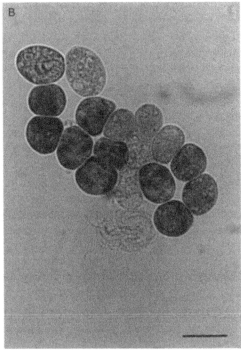

Figure 10.1 Giardia intestinalis: (A) trophozoites stained with Giemsa stain and (B) iodine-stained cysts. Bar = 10 μm. (See Colour Plate I.)

four nuclei. Fibrillar structures and crescent-shaped chromidial bodies present in the cyst are residual material from the flagella and the adhesive disc of the trophozoite stage. Ingested cysts, stimulated to excystation by the acidic pH in the stomach, develop into trophozoites, which colonize the upper part of the intestinal tract adhering to the surface of the entero-cytes. There is no evidence for the intracellular penetration by *Giardia* trophozoites in humans. Stimulated by bile salts and elevated pH in the intestine, parasites later undergo encystation and are excreted with faeces.

Distribution and transmission

Giardia intestinalis has a worldwide distribution and represents the most frequently detected intestinal parasite in humans. The reported prevalence varies between 0.5 and 30% depend-ing on the area. In Sweden, during recent years *Giardia* has been identified and reported in approximately 1 500 cases annually (Smittsamma sjukdomar 2001).

Seasonal trends in giardiasis and dependence on sex and age have also been noticed (Addis *et al.* 1992). Higher prevalence is usually observed in sub-tropical and tropical coun-tries and in regions with poor sanitation where giardiasis may be endemic (Meloni *et al.* 1993). The disease is more frequent in children than in adults. Parasites are transmitted directly from person to person by faecal–oral contamination with cysts or by ingestion of cyst-contaminated water or food. Ingestion of 10–100 cysts is enough to result in infection (Rendtorff 1954). Infection occurs frequently in day-care centres and within families of infected children (Pickering *et al.* 1984). Transmission among male homosexuals has been described as a result of anal–oral sexual behaviour (Schmerin *et al.* 1978).

Despite the endemic form of transmission in certain regions (Fraser and Cooke 1991; Isaac-Renton *et al.* 1996), giardiasis appears repeatedly as a water-borne epidemic (Neringer *et al.* 1987; Moore *et al.* 1993). Travellers, campers, and hikers may become infected after drinking contaminated, untreated surface or ground water. Community-wide outbreaks occur as a result of faecally contaminated municipal water supplies. Man is the main reservoir for human infections. An infected individual can produce 10^5–10^7 cysts/gram of stool (Feachem *et al.* 1983). One human stool deposited in a medium-sized water reservoir (9.5×10^6 l) can lead to a density of 6–7 cysts/l (DeRegnier *et al.* 1989). Cysts are well adapted to survival and transmission in the environment; in the cold and humid conditions they retain their viability for several weeks. Although the zoonotic nature of the parasite and its host specificity are still controversial, there are some evidences for inter-species transmission (Davies and Hibler 1979; Thompson *et al.* 1988; Majewska 1994). Wild and domestic animals infected by morphologically identical parasites may represent reservoirs for human infections and thus contribute to the environmental transmission of the parasite.

Pathophysiology and clinical features

The parasite was first discovered by Anthony van Leeuwenhook in the diarrhoeic stool but he did not consider *Giardia* as associated with the symptomatic disease (Dobell 1931). Today, there is no question that it causes symptomatic diarrhoeal illness.

The infective dose for giardiasis is low. The incubation period for symptomatic infection is 1–2 weeks but could vary from 1 to 45 days or more. The majority of infected individuals remain asymptomatic, sometimes amounting to 60% (Ortega and Adam 1997). Giardiasis has a range of clinical expression from asymptomatic cyst passage to a syndrome of chronic diarrhoea, malabsorption, and weight loss. This spectrum is likely to be related to virulence of the *Giardia* strain and the host's immune response to the parasite. Classification of strains relies on antigen, isoenzymes, chromosome, and DNA pattern. However, no tissue invasion is observed and the presence of a toxin has not been demonstrated to date; nor have any specific virulence factors been identified.

Giardia is able to perturb the structure and the function of the intestinal mucosa. At the same time additional factors contribute to diarrhoea and malabsorption (Katelaris and Farthing 1992; Farthing 1993). Changes of the small intestinal architecture from normal to sub-total villous atrophy will be seen in the human infection. The morphological abnormalities are associated with disaccharide activities and contribute to osmotic diarrhoea (Welsh *et al.* 1984). In animal models impaired solute and electrolyte absorption, together with impaired glucose absorption can be demonstrated. Several explanations for these abnormalities have been proposed. The *Giardia* trophozoites attach to the enterocytes and a recently identified mannose-binding lectin, which interacts with the mannose residues on the cells, contributes to disruption of microvilli.

It is suggested that the parasite might produce cytopathic substances such as proteinase, which could cleave the proteins in the microvillous membranes (Lev *et al.* 1986).

There is also increasing evidence that T-cell activation within the intestinal mucosa can produce villous atrophy (Mac Donald and Spencer 1988).

Bacterial overgrowth could contribute to malabsorption by the conjugation of bile salts (Tandon *et al.* 1977). It has been shown that concentration of pancreatic enzymes are reduced, which may relate to a direct effect of the pancreatic proteins (Katelaris *et al.* 1991).

Acute giardiasis has characteristic symptoms with watery diarrhoea often developing into steatorrhea, nausea, abdominal discomfort, bloating, and weight loss. Although giardiasis, to a great extent, is self-limiting in healthy, immunocompetent individuals a proportion of these continue having diarrhoea and steatorrhea. In symptomatic patients with diarrhoea around 50% will have evidence of malabsorption of fat and protein as well as D-xylose, vitamin A, and vitamin B_{12}. Secondary lactase deficiency is well recognized and may take many weeks to recover even after clearance of parasite. Recurrent diarrhoea after treatment for *Giardia* would rather be lactose intolerance than relapse of infection. Malabsorption in giardiasis is well documented and is also responsible for weight loss as well as failure to thrive. Even when the infection is asymptomatic, malabsorption may occur. Extraintestinal manifestations in unusual cases have been described (Shaw and Stevens 1987; Clyne and Eliopoulus 1989).

Immune response

Humoral, cellular, and macrophage-mediated mechanisms are each important components of the host response (den Hollander *et al.* 1988). The critical site for this response is the gut. Therefore the secretory antibodies are the most important (Taylor and Wenman 1987; Ljungström and Castor 1992). IgA is the predominant antibody class detected, although these antibodies do not appear to be cytotoxic (Heyworth 1992). Cellular mechanisms, particularly T-helper cells, are necessary for the development of secretory antibody response (Heyworth *et al.* 1987). Macrophages may also act as effector cells, their phagocytic activity for *Giardia* trophozoites being increased in the presence of specific antibodies.

Current evidence suggests that anti-*Giardia* sIgA acts in clearing *Giardia* from the gut lumen by trophozoite agglutination and/or inhibition of the flagellar motility (Char *et al.* 1992). Individuals with hypo- or agammaglobulinemia are at risk of chronic giardiasis although individuals with HIV and AIDS do not seem to develop symptomatic disease at a higher number. The secretory immunity in the intestinal lumen is more important for clearance than the cell-mediated responses within the mucosa. Despite the limits of the humoral immune response there is some evidence that antibodies provide protection against newly acquired infection or reinfection. Human milk appears to be protective by non-immune mechanisms. It is cytotoxic against *Giardia* trophozoites by the activity of free fatty acids (Reiner *et al.* 1986) although it also may contain anti-*Giardia* antibodies, which may be protective to breast-fed infants (Nayak *et al.* 1987).

Diagnosis

The diagnosis is mostly established by a typical history, together with identification of the parasite in faecal specimens by microscopy. Sensitive and specific ELISA tests for *Giardia* antigens have been developed. Humoral IgG antibody titres are not helpful in diagnosis, nor do IgM antibodies always reflect acute infection. Specific DNA probes and amplification techniques such as PCR are available.

Treatment

All patients with symptomatic giardiasis should be treated. Several effective treatment alternatives exist. Most patients respond to a single course of treatment but in refractory cases multiple or combination courses are required. Metronidazol given for five days or a single dose of tinidazol are the treatments of choice. Furazolidone for 7–10 days is commonly used for children in many parts of the world. Other drugs such as mepacrin, paromomycin, and albendazol will equally be recommended against giardiasis.

Many patients develop a post-*Giardia* lactose intolerance and present persistent intestinal symptoms but actually have a disaccharidase deficiency (McIntyre *et al.* 1986) and usually improve with time and lactose-free diet. The need for treatment of asymptomatic carriers is currently being discussed. Treatment is indicated if reinfection is unlikely. Different opinions prevail about treating asymptomatic individuals when faecal–oral spread cannot be prevented and rapid reinfection may occur, as in day-care centres.

Prevention

It seems highly unlikely that *Giardia* spp. will ever be eliminated from the environment since they can survive for weeks or months outside the host. Prevention includes proper handling and treatment of community water supplies, appropriate disposal of human and animal waste, and good personal hygiene.

References

Addiss, D. G., Davis, J. P., Roberts, J. M., and Mast, E. E. (1992). Epidemiology of giardiasis in Wisconsin: increasing incidence of reported cases and unexplained seasonal trends. *Am. J. Trop. Med. Hyg.*, **47**, 13.

Andrews, H. R., Adams, M., Boreham, P. F. L., Mayrhofer, G., and Meloni, B. P. (1989). *Giardia intestinalis*: electrophoretic evidence for a species complex. *Int. J. Parasitol.*, **19**, 183–190.

Char, S., Cevallos, A. M., Yamson, P. *et al.* (1992). Impaired IgA response to *Giardia* heat shock antigen in children with persistent diarrhoea and giardiasis. *Gut*, **34**, 38–40.

Clyne, C. A. and Eliopoulos, G. M. (1989). Fever and urticaria in acute giardiasis. *Arch. Intern. Med.*, **149**, 939–940.

Davies, R. B. and Hibler, C. P. (1979). Animal reservoirs and cross-species transmission of *Giardia*. In: *Proceedings of Symposium on Waterborne Transmission of Giardiasis, US EPA*, 18–20 September 1987, Cincinnati, OH, 104.

DeRegnier, D. P., Cole, L., Schupp, D. G., and Erlandsen, S. L. (1989). Viability of *Giardia* cysts suspended in lake, river and tap water. *Appl. Environ. Microbiol.*, **55**, 1223–1229.

Dobell, C. (1931). *Antony van Leeuwenhoek and his "Little Animals"*. John Bale, Sons and Danielsson, London, UK.

Farthing, M. J. (1993). Diarrhoeal disease: current concepts and future challenges. Pathogenesis of giardiasis. *Trans. R. Soc. Trop. Med. Hyg.*, **87**(Suppl. 3), 17–21.

Feachem, R. G., Bradley, D. J., Garelick, H., and Mara, D. D. (1983). *Sanitation and disease. Health aspects of excreta and wastewater management*. Wiley, New York.

Filice, F. P. (1952). Studies on the cytology and life history of *Giardia* from the laboratory rat. *Univ. Cal. Publ. Zool.*, **57**, 53–146.

Fraser, G. G. and Cooke, K. R. (1991). Endemic giardiasis and municipal water supply. *Am. J. Public Health*, **81**, 760.

Heyworth, M. F. (1992). Relative susceptibility of *Giardia muris* trophozoites to killing by mouse antibodies of different isotypes. *J. Parasitol.*, **78**, 73–76.

Heyworth, M. F., Carlson, J. R., and Ermak, T. H. (1987). Clearance of *Giardia muris* infection requires helper/inducer T lymphocytes. *J. Exp. Med.*, **165**, 1743–1748.

den Hollander, R., Riley, D., and Befus, D. (1988). Immunology of giardiasis. *Parasitol. Today*, **4**, 124–131.

Isaac-Renton, J., Morehead, W., and Ross, A. (1996). Longitudinal studies of *Giardia* contamination in two community drinking water supplies: cyst levels, parasite viability and health impact. *Appl. Environ. Microbiol.*, **62**, 47.

Kabnik, K. S. and Peattie, D. A. (1991). *Giardia*: a missing link between prokaryotes and eukaryotes. *Am. Scientist*, **79**, 34–43.

Katelaris, P. H. and Farthing, M. J. G. (1992). Diarrhoea and malabsorption in giardiasis: a multifactorial process. *Gut*, **33**, 295–297.

Katelaris, P. H., Seow, F., and Ngu, M. C. (1991). The effect of *Giardia lamblia* trophozoites on lipolysis *in vitro*. *Parasitology*, **103**, 35–39.

Lev, B., Ward, M., Keusch, G. T., *et al.* (1986). Lectin activation in *Giardia lamblia* by host protease: a novel host–parasite interaction. *Science*, **232**, 71–73.

Ljungström, I. and Castor, B. (1992). Immune response to *Giardia lamblia* in a waterborne outbreak of giardiasis in Sweden. *J. Med. Microbiol.*, **36**, 347–352.

Mac Donald, T. T. and Spencer, J. (1988). Evidence that activated mucosal T cells play a role in the pathogenesis of enteropathy in human small intestine. *J. Exp. Med.*, **167**, 1341–1349.

Majewska, A. C. (1994). Successful experimental infection of human volunteer and mongolian gerbils with *Giardia* of animal origin. *Trans. R. Soc. Trop. Med. Hyg.*, **88**, 360.

McIntyre, P., Boneham, P. F. L., Phillips, R. E., *et al.* (1986). Chemotherapy in giardiasis: clinical responses and in vitro drug sensitivity of human isolates in axenic culture. *J. Pediatr.*, **108**, 1008–1010.

Meloni, B. P., Thompson, R. C. A., Hopkins, R. M., Reynoldson, J. A., and Gracey, M. (1993). The prevalence of *Giardia* and other intestinal parasites in children, dogs and cats from Aboriginal communities in the Kimberley. *Med. J. Aust.*, **158**, 157.

Moore, A. C., Herwaldt, B. L., Craun, G. F., Calderon, R. L., Highsmith, A. K., and Juranek, D. D. (1993). Surveillance for waterborne disease outbreaks – United States, 1991–1992, *Morb. Mort. Weekly Rep.*, **42**(ss-5), 1.

Nayak, N., *et al.* (1987). Specific secretory IgA in the milk of Giardia *lamblia* – infected and uninfected women. *J. Infect. Dis.*, **155**, 724–727.

Neringer, R., Andersson, Y., and Eitrem, R. (1987). A water-borne outbreak of giardiasis in Sweden. *Scand. J. Infect. Dis.*, **19**, 85.

Ortega, Y. R. and Adam, R. D. (1997). *Giardia*: overview and update. *Clin. Infect. Dis.*, **25**, 545–550.

Pickering, L. K., Woodward, W. E., Dupont, H. L., and Sullivan, P. (1984). Occurence of *Giardia lamblia* in children in day care centers. *J. Pediatr.*, **104**, 522–526.

Reiner, D. S., Wang, C. S., and Gillin, F. D. (1986). Human milk kills *Giardia lamblia* by generating toxic lipolytic products. *J. Infect. Dis.*, **154**, 825–832.

Rendtorff, R. C. (1954). The experimental transmission of human intestinal protozoan parasites. I. *Giardia lamblia* cysts given in capsules. *Am. J. Hyg.*, **59**, 209–220.

Schmerin, M. J., Jones, T. C., and Klein, H. (1978). Giardiasis: association with homosexuality. *Ann. Intern. Med.*, **88**, 801–803.

Shaw, R. A. and Stevens, M. B. (1987). The reactive arthritis of giardiasis. *J. Am. Med. Assoc.*, **25**, 2734–2735.

Smittsamma sjukdomar (2001) Årsrapport från avdelningen för epidemiologi, SMI, Stockholm, Sweden.

Tandon, B. N. Tandon, R. K., Satpathy, B. K. *et al.* (1977). Mechanism of malabsorption in giardiasis: a study of bacterial flora and bile salt deconjugation in upper jejunum. *Gut*, **18**, 176–181.

Taylor, G. D. and Wenman, W. M. (1987). Human response to *Giardia lamblia* infection. *J. Infect. Dis.*, **155**, 137–140.

Thompson, R. C. A., Meloni, B. P., and Lymbery, A. J. (1988). Humans and cats have genetically identical forms of *Giardia*: evidence of a zoonotic relationship. *Med. J. Aust.*, **148**, 207.

Welsh, J. D., Poley, J. R., Hensley, J. *et al.* (1984). Intestinal disaccharidase and alkaline phosphatase activity in giardiasis. *J. Pediatr. Gastroenterol. Nutr.*, **3**, 37–40.

11 Microsporidia spp.

J. I. Ronny Larsson and Marianne Lebbad

The microsporidia constitute a phylum of small, spore-forming unicellular parasites. New species are being described in rapid succession, and it is obvious that about 1 000 species known today are only a minor fraction. The real number probably exceeds one million. Their history started in 1857 with *Nosema bombycis*, the famous agent of the 'pébrine' disease of the silk moth. However, before long it was apparent that microsporidia were not restricted to insect hosts. At present microsporidia have been found in all major groups of animals. They are especially common and important parasites of insects, crustaceans, and fish. Approximately 10% of the species are parasites of vertebrates, with about 100 species from fish. Thus, far a little more than 10 species have been described from homoiothermous vertebrates (Canning and Lom 1986). Many of the species, at least of invertebrates, appear to be host specific.

Even if *N. bombycis* was originally described as a fungus, microsporidia have usually been considered to be protozoa. Cavalier-Smith (1983), apparently erroneously, included them among the primitive Archaezoa. Recent information suggests that the microsporidia have fungal affinities (discussed by Weiss and Vossbrinck 1999).

In the light microscopic era the minute size was a limitation. The first revolution came with the electron microscope, which revealed a strange and unique cytological world. The explosive appearance of new species in the last two decades is a product of electron microscopy, and ultrastructural documentation is nowadays necessary when describing new microsporidia. Recently, molecular biology has started a second revolution. So far this technique has not given us a huge number of new species, but it has facilitated identification and revealed relationships, and will undoubtedly contribute profoundly to our understanding of the evolution of microsporidia. Recently, classification and identification have been surveyed by Sprague (1977), Weiser (1977), Issi (1986), Larsson (1986, 1988, 1999), Canning (1989), and Sprague *et al.* (1992).

Cytology

Microsporidia are eukaryotic organisms, producing spores of unique construction. All life cycle stages are intracellular. The cytology is of normal eukaryotic type with a few important specializations (Vávra 1976; Larsson 1986; Vávra and Larsson 1999). All life cycle stages lack mitochondria and the nuclei are either isolated or coupled as diplokarya. Further, the RNA is of prokaryotic type (Curgy *et al.* 1980; Vossbrinck *et al.* 1987). While presporal stages, even to the specialist, are difficult to identify as microsporidia, the spores, which have a thick and complex spore wall and a unique infection apparatus, are so characteristic that with basic knowledge of the group it is normally no problem to recognize their microsporidian nature.

The infectious stage is the spore. Living spores of most microsporidia range in size between 1 and 5 μm, but a small number of species produce larger spores (up to 40 μm). The spore shape is, for most species, oval or pyriform, less commonly rod-like or spherical (Figure 11.1A–D). A small number of genera, like *Cougourdella* and *Caudospora*, have spores of unique shape for the genus. It is often possible to identify microsporidia to a genus using the spore shape in combination with the way the spores are produced (Larsson 1983b, 1988, 1999).

The spore wall is normally composed of three layers: an external more or less complex exospore, a median wide, seemingly structureless endospore containing chitin (Vávra 1976), and the internal plasma membrane (Figure 11.1C).

The infection apparatus is composed of three components. The anterior half of the spore is occupied by a system of lamellae or sacs delimited by unit membranes (the polaroplast) (Figure 11.1C). The polaroplast is normally divided into two, or sometimes three, structurally different parts. It surrounds the anterior part of a long thread-like organelle (the polar filament) which is attached to an anchoring apparatus at the anterior pole of the spore. The filament proceeds straight backwards through the centre of the spore, approximately to the middle of the spore, while the posterior part usually is coiled up in one, or sometimes more, layers of coils in the posterior half of the spore (Figure 11.1C). The filament is composed of concentric layers of different electron density, and it is externally covered by a unit membrane which is continuous with the membrane component of the polaroplast. The width of the filament is either uniform from the anterior to the posterior end (isofilar filament) or the anterior part of the filament is wider (anisofilar filament, Figure 11.1C). All species belonging to the same genus have either isofilar or anisofilar filament. The third component of the extrusion apparatus is the posterior vacuole, a membrane-lined cavity at the posterior end of the spore.

Infection

Most microsporidia enter the new host through the gut, even if other routes of infection are known to occur. In the gut of the correct host, the conditions stimulate the spore to eject the polar filament (Figure 11.1D), which is everted like a finger of a glove. During this process the protein of the filament coils is rearranged to form an infection tube (Weidner 1976). In the laboratory some microsporidia discharge their filament easily when stimulated by hydrogen peroxide, certain salt solutions, or change of pH (Vávra and Maddox 1976). In the light microscopic era the artificial ejection of the polar filament was the proof of the microsporidian nature of the organism in study.

There are different theories for the ejection process (Dall 1983; Undeen 1990). According to one of them, which has so far not been proven, ionophore molecules in the plasma membrane participate in an exchange of cations or protons, causing an osmotic imbalance (Dall 1983). This leads to a rapid inflow of water into the spore. The posterior vacuole and the chambers of the polaroplast swell, and the increased pressure causes the explosive discharge of the polar filament. A second theory is based upon the observed decrease of the trehalose level during ejection. The disaccharid trehalose is degraded into smaller molecules causing an increased osmotic pressure (Undeen 1990).

The spore functions as a syringe. The tip of the newly formed tube penetrates a host cell and the parasite is injected safely into this (Figure 11.1D). The infectious stage (the sporoplasm) consists of the nucleus (or diplokaryon) of the spore and the cytoplasm. The plasma membrane of the sporoplasm is not the plasma membrane of the spore – this remains as the

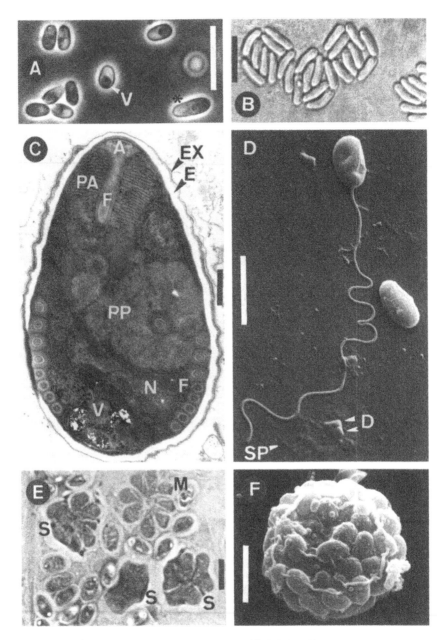

Figure 11.1 Light and electron microscopic aspects of microsporidia. (A) Living, lightly pyriform spores of two size classes (*Glugea anomala*, * = macrospore; phase contrast, bar = 10 μm). (B) Living rod-shaped spores in eight-sporous sporophorous vesicles (*Resiomeria odonatae*; interference phase contrast, bar = 10 μm). (C) Longitudinally sectioned mature spore exhibiting the characteristic organelles: layered spore wall, polaroplast divided into two regions, anisofilar polar filament (*Trichoctosporea pygopellita*; transmission electron microscopy, bar = 0.5 μm). (D) Mature spores; one spore has extruded the polar filament and transformed it into an infection tube through which the infectious cell (the sporoplasm with nuclei coupled as diplokarya) has left the spore (*Nosema tractabile*; scanning electron microscopy, bar = 10 μm). (E) Reproduction by multiple budding both in the vegetative phase (merogony) and the sporogony (*Systenostrema corethrae*; light microscopy, Giemsa stain, bar = 10 μm). (F) Polysporous sporophorous vesicle (*Vavraia holocentropi*; scanning electron microscopy, bar = 5 μm). Abbreviations: A = anchoring apparatus, D = diplokaryon, E = endospore, EX = exospore, F = polar filament, M = merogony, N = nucleus, PA = anterior part of polaroplast, PP = posterior part of polaroplast, S = sporogony, SP = sporoplasm, V = posterior vacuole.

internal component of the spore wall. The new plasma membrane is formed from the unit membranes of the polaroplast during the ejection process.

Many microsporidia infecting insects combine horizontal transmission by spores, released into nature from the infected host, with vertical transmission from the female to the offspring (Becnel 1994; Becnel and Andreadis 1999). The oocytes are infected in the ovary and the larvae are already infected when they hatch.

Reproduction in the host cell

The life cycle of most microsporidia comprises two sequences of proliferation. The first, which is called merogony, starts directly upon infection (Figure 11.1E). It yields daughter cells (merozoites) with the potential either to repeat the merogony or to enter the second phase of reproduction, the sporogony. The daughter cells of the sporogony (sporoblasts) mature to spores without further division. Both cycles of reproduction proceed as binary fission, plasmotomy (splitting of a multinucleate plasmodium), or schizogony (multiple budding), and for each species of microsporidia the modes of reproduction are constant (Figure 11.1E).

Spores are either produced free in the host cell or are enclosed in an envelope, formed by the microsporidium, called a sporophorous vesicle (Figure 11.1B,F). This is a secretory material, most commonly produced by the sporont at the beginning of the sporogony (sporontogenetic sporophorous vesicle). In a few genera of microsporidia, for example *Pleistophora* and *Trachipleistophora*, the sporophorous vesicle is initiated already in the merogonial part of the life cycle (merontogenetic sporophorous vesicle). This means that the merontogenetic vesicle not only encloses the daughter cells of one sporont, but also collects the daughter cells of all sporonts produced by one merogonial mother cell. The envelope of a sporophorous vesicle is normally not visible in a light microscopic preparation, but the occurrence of groups of spores in regular numbers indicates the presence of vesicles (Figure 11.1B).

The characteristic organelles of the spore are formed in the sporoblast stage, or in some species, earlier in the sporogony. The sporulating microsporidium covers itself with a more or less complex layer of material outside the plasma membrane. This layer survives as the exospore layer of the spore. With very few exceptions, the polar filament is generated in the sporoblast stage from a vesicular area (Golgi apparatus) close to the nucleus (Vávra 1976). The length of the filament increases until the spore is mature. In *Enterocytozoon*, and a few more genera, the polar filament is initiated already in the sporogonial plasmodium (Desportes *et al.* 1983). The endospore layer is the last structure of the spore to form. Consequently, immature spores could be identified as such from the shorter polar filament and the thinner endospore layer.

Sexual processes have been observed in a small number of microsporidia. Zygotes are normally formed at the beginning of the sporogony, and clear cases have been observed in species shifting from diplokaryotic condition in the merogonial part of the life cycle to isolated nuclei in the sporogonial stages and spores. There are varying opinions about the interpretation of the sexual processes (Canning 1988; Flegel and Pasharawipas 1995). Most reports of sexuality in microsporidia are not based on observed zygote production but on the presence of synaptonemal complexes (coupled chromosomes in the reductional division) visible in ultrathin sections of sporonts (Larsson 1986).

Life cycles

Complete life cycles have been elucidated for only few microsporidian species. At present, life cycles comprising one or two host species are known, and up to three morphologically different types of spores are produced in different parts of the cycle. It seems that most microsporidia complete their life cycle in one host and only one spore type is produced. Life cycles involving alternate hosts are so far restricted to mosquitoes and copepods (Becnel 1994). Life cycles of vertebrate microsporidia appear comparatively simple, and no involvement of alternate hosts is known.

In an increasing number of cases of microsporidian infections of insects it has been revealed that the life cycle involves two sequences of sporogony in the same host, yielding two kinds of slightly different spores (Iwano and Ishihara 1991). The first sporogony occurs in the early phase of infection, yielding spores that differ from the second type of spores in that the endospore layer is weakly developed and the polar filament is shorter, which are typical signs of the spores being immature. These spores germinate in the cell where they have been produced, and their function is to facilitate the dispersal of the microsporidia inside the host where they were produced. The second type of spore, which is completely mature, is produced in a different tissue. This is the infectious stage for new hosts.

Interactions with the host cell

Normally microsporidia develop in the cytoplasm of the host cell, but occasionally intranuclear development, like in *Nucleospora salmonis*, has been observed (Hedrick *et al.* 1990).

In the most simple cases the host cells are used up by the proliferating microsporidia, but no particular modifications of the host cells are visible, except that the cell, and often also the nucleus of the host cell, become hypertrophic. The microsporidium reproduces freely in the cytoplasm. In other cases the microsporidia are separated from the cytoplasm by an envelope formed by the host (parasitophorous vacuole). There are also well-known cases where the cytology of the host cell is reorganized under the influence of the microsporidia. *Chytridiopsis* species establish such close connection to the host nucleus that the developing microsporidium finally lies in a shallow invagination of the nucleus (Sprague *et al.* 1972). Another example is the close association between *Buxtehudea scaniae* and the mitochondria of the host cell which aggregate around the microsporidium (Larsson 1980).

In the infected tissue cell walls might be destroyed, transforming the tissue into a syncytium where the developmental stages of the microsporidium are floating among host nuclei. Microsporidia of fish sometimes induce the production of tumours. The best known example is the xenoma caused by *Glugea anomala*, a common skin parasite of sticklebacks. The microsporidium completely rules the infected epidermis cell, inducing nuclear fission, cell hypertrophy, and encapsulation processes (Canning *et al.* 1982). The final xenoma is a rounded body, a few millimetres in diameter, which contains hundreds, or even thousands, of host nuclei and millions of microsporidia. It is encapsulated by numerous layers of fibroblasts and material from the host cell surface coat.

The clinical aspects of microsporidiosis have been recently reviewed by Kotler and Orenstein (1999).

Human microsporidiosis

At least 10 species of microsporidia are known from humans, most of them being found only once. Six of them belong to genera exclusive to man: *Brachiola, Enterocytozoon, Septata, Vittaforma,* and *Trachipleistophora.* Before the advent of HIV infection and AIDS, microsporidia had only occasionally been observed in man. Later, microsporidia were commonly observed in AIDS patients, and they were often considered to be opportunistic pathogens. The first human case of sufficiently substantiated microsporidial infection was reported in 1959 (Matsubayashi *et al.* 1959). Nearly 25 years later the first human case was observed in Scandinavia (Bergquist *et al.* 1984). Up to 1994 totally 10 sufficiently documented cases of human microsporidial infection in persons not infected with HIV have been reported (Weber *et al.* 1994). The first Scandinavian case connected with AIDS was observed in 1992 (Højlyng *et al.* 1993). As a result of antiretroviral therapy no cases of microsporidiosis have been observed in HIV-positive persons in the northern countries in the last few years.

The sources of microsporidia infecting humans and the modes of transmission are still uncertain. Whether other animals are involved in the transmission to man is unknown, but one of the species, *Enterocytozoon bieneusi*, has recently been found in faecal material from pigs (Deplazes *et al.* 1996). The following species are the most important and best known microsporidia pathogenic to man.

Encephalitozoon cuniculi, described by Levaditi *et al.* (1923), was, in 1922, the first microsporidium to be recognized as a parasite of mammals. Contrary to the microsporidia of invertebrates, this species has a wide host range, including hares, rabbits, mice, various carnivores, elephants, and primates, including man. Brain and kidneys are the primary sites of infection, and it has been proven that the parasite is transmitted across the placenta. Few cases of *E. cuniculi* infection have been observed in connection with HIV infection. In AIDS patients the parasite causes widely disseminated infection involving nearly all organ systems. A second *Encephalitozoon* species, *E. hellem*, described by Didier *et al.* (1991), was isolated from the eyes of AIDS patients with keratoconjunctivitis. *Encephalitozoon hellem* is known to provoke systemic infection, also comprising the bronchial epithelium and the kidney tubules.

Brachiola vesicularum, described by Cali *et al.* (1998), was isolated from an AIDS patient. It causes myositis with muscular weakness and pain. Infection has not been found in tissues other than muscle cell cytoplasm and fibres.

Enterocytozoon bieneusi, described by Desportes *et al.* (1983), is the smallest microsporidium reported to infect humans (spore length 1–1.6 μm). This is the most commonly diagnosed of the microsporidian species infecting man. It is apparently persisting in populations in light infections which are below the threshold of detection. It cannot be excluded that *E. bieneusi* is a natural parasite of man with worldwide distribution. The infection is restricted to the gut where enterocytes of the small intestine are destroyed, causing severe diarrhoea. This species has also been identified from immunocompetent and HIV-negative patients with self-limited traveller's diarrhoea (Sobottka *et al.* 1995).

Septata intestinalis, described by Cali *et al.* (1993) and recently transferred to the genus *Encephalitozoon* by Hartskeerl *et al.* (1995), has mainly been identified in HIV-infected patients. It is mainly a parasite of the gut, where it invades enterocytes throughout the intestine, but disseminates to other epithelial cells, for example in the kidneys, liver, or respiratory tract.

The new genus *Vittaforma* was established by Silveira and Canning (1995) for a species previously known as *Nosema corneum*. It was originally isolated from the cornea of an immunocompetent patient.

Trachipleistophora hominis, described by Hollister *et al.* (1996), was originally found in skeletal muscles of AIDS patients, but has later also been observed in cornea scrapings and washings from the nasopharyngeal tract. A second species, *T. anthropophthera*, was recently found in brain, kidneys, liver, and a variety of other organs in an AIDS patient (Vávra *et al.* 1998). This is a dimorphic species, yielding two kinds of morphologically different spores occurring together.

Diagnosis of human microsporidiosis

Diagnosis of microsporidia depends upon detecting the organism in patient specimens by microscopic or, more recently, by molecular biology methods (Weber *et al.* 1999; Weiss and Vossbrinck 1999).

Histochemical methods such as modified trichrome staining (Weber *et al.* 1992) and chitin-staining fluorochromes (Van Gool *et al.* 1993) are simple and useful light microscopic techniques for detection of microsporidia. However, they do not allow for species differentiation which is becoming increasingly important for the therapy. For instance *E. intestinalis* can be cured by albendazole therapy, but to patients infected with *E. bieneusi* that will be of little benefit (Van Gool and Dunkert 1995).

Transmission electron microscopy is the standard method for identification of microsporidia at species level. However, this is a laborious procedure which needs specialized equipment as well as experience, and it is of limited use for routine diagnostic purposes. Polymerase chain reaction (PCR) with specific primers is more convenient for detection and identification of microsporidia at species level in patient specimens. Screening for microsporidia by histochemical methods, followed by confirmation and identification using PCR may become the model for the laboratory diagnosis of microsporidiosis (Franzen and Müller 1999).

Microsporidia in the North

Microsporidia are common parasites in all parts of the world where they have been looked for, but no country has a well-known fauna. They are also common in the northern countries, with a high proportion of species new to science (Larsson 1983a, 1988, 1999; Widtmann 1990; Bylén 1994). Recent investigations have treated *E. cuniculi*, a wide-spread parasite of various mammals (Petri 1969; Nordstoga 1972; Waller 1979), microsporidia associated with AIDS (Højlyng *et al.* 1993), *Nosema apis*, the important cosmopolitan parasite of honey bees (Fries 1988), *Thelohania contejeani*, destructive to crayfish (Skurdal *et al.* 1990; Gydemo and Westin 1988), *Pleistophora mirandellae*, and other fish parasites (Pekkarinen 1995).

References

Becnel, J. (1994). Life cycles and host–parasite relationships of microsporidia in culicine mosquitoes. *Folia Parasitol.*, **41**, 91–96.

Becnel, J. and Andreadis, T. G. (1999). Microsporidia in insects. In: *The microsporidia and microsporidiosis* (M. Wittner and L. M. Weiss, eds). ASM Press, Washington, pp. 447–501.

Bergquist, N. R., Stintzing, G., Smedman, L., Waller, T., and Andesson, T. (1984). Diagnosis of encephalitozoonosis in man by serological tests. *Br. Med. J.*, **288**, 902.

Bylén, E. (1994). Ultrastructure and taxonomy of the microsporidia (Protozoa: Microspora) with emphasis on the microsporidia infecting midges (Diptera: Chironomidae). Thesis, Lund.

Cali, A., Kotler, D. P., and Orenstein, J. M. (1993). *Septata intestinalis*, n. g., n. sp., an intestinal microsporidian associated with chronic diarrhea and dissemination in AIDS patients. *J. Eukaryot. Microbiol.*, **40**, 101–102.

Cali, A., Takvorian, P. M., Lewin, S., Rendel, M., Sian, C. S., Wittner, M., Tanowitz, H. B., Keohane, E., and Weiss, L. M. (1998). *Brachiola vesicularum*, n. g., n. sp., a new microsporidian associated with AIDS and myositis. *J. Eukaryot. Microbiol.*, **45**, 240–251.

Canning, E. U. (1988). Nuclear division and chromosome cycle in microsporidia. *BioSystems*, **21**, 333–340.

Canning, E. U. (1989). Phylum Microspora. In: *Handbook of the Protoctista* (L. Margulis, J. O. Corliss, M. Melkonian, and D. J. Chapman, eds). Jones and Bartlett Publishers, Boston, MA, pp. 53–72.

Canning. E. U. and Lom, J. (1986). *The microsporidia of vertebrates.* Academic Press, London, UK.

Canning, E. U., Lom, J., and Nicholas, J. P. (1982). Genus *Glugea* Thélohan, 1891 (Phylum Microspora): redescription of the type species *Glugea anomala* (Moniez, 1887) and recognition of its sporogonic development within sporophorous vesicles (pansporoblastic membranes). *Protistologica,* **18,** 193–210.

Cavalier-Smith, T. (1983). A 6-kingdom classification and a unified phylogeny. In: *Endocytobiology II* (W. Schwemmler and H. E. A. Schenn, eds). De Gruyter, Berlin, pp. 265–279.

Curgy, J.-J., Vávra, J., and Vivarès, C. (1980). Presence of ribosomal RNA with prokaryotic properties in microsporidia, eukaryotic organisms. *Biol. Cell.,* **38,** 49–52.

Dall, D. J. (1983). A theory for the mechanism of polar filament extrusion in microsporidia. *J. Theor. Biol.,* **105,** 647–659.

Deplazes, P., Mathis, A., Müller, C., and Weber, R. (1996). Molecular epidemiology of *Encephalitozoon cuniculi* and first detection of *Enterocytozoon bieneusi* in faecal samples of pigs. *J. Eukaryot. Microbiol.,* **43,** 93S.

Desportes, I., Le Charpentier, Y., Galian, A., Bernard, F., Cochand-Priollet, B., Lavergne, A., Ravisse, P., and Modigliani, R. (1983). Occurrence of a new microsporidian: *Enterocytozoon bieneusi* n. g., n. sp., in the enterocytes of a human patient with AIDS. *J. Protozool.,* **32,** 250–254.

Didier, E. S., Didier, P. J., Friedberg, D. N., Stenson, S. M., Orenstein, J. M., Yee, R. W., Tio, F. O., Davis, R. M., Vossbrinck, C., Millichamp, N., and Shadduck, J. A. (1991). Isolation and charaterization of a new human microsporidian *Encephalitozoon hellem* n. sp. from three AIDS patients with keratoconjunctivitis. *J. Infect. Dis.,* **163,** 617–621.

Flegel, T. W. and Pasharawipas, T. (1995). A proposal for typical eukaryotic meiosis in microsporidians. *Can. J. Microbiol.,* **41,** 1–11.

Franzen, C. and Müller, A. (1999). Molecular techniques for detection, species differentiation, and phylogenetic analysis of microsporida. *Clin. Microbiol. Rev.,* **12,** 243–285.

Fries, I. (1988). Contribution to the study of *Nosema* diseases *(Nosema apis Z.)* in honey bee *(Apis mellifera)* colonies. Thesis, Uppsala.

Gydemo, R. and Westin, L. (1988). Observations on *Thelohania contejeani* infestation in an *Astacus astacus* pond population. *J. Aquat. Prod.,* **2,** 125–137.

Hartskeerl, R. A., van Gool, T., Schuitema, A. R. J., Didier, E. E., and Terpstra, W. J. (1995). Genetic and immunological characterization of the microsporidian *Septata intestinalis* Cali, Kotler and Orenstein, 1993: reclassification to *Encephalitozoon intestinalis. Parasitology,* **110,** 277–285.

Hedrick, R. P., Groff, J. M., McDowell, T. S., Willis, M., and Cox, W. T. (1990). Hematopoietic intranuclear microsporidian infections with features of leukemia in chinook salmon *Onchorhynchus tshawytscha. Dis. Aquat. Org.,* **8,** 189–197.

Højlyng, N., Nielsen, A., Wandall, J., Blom, J., Mølbak, K., Chauhan, D., and Petersen, E. (1993). First cases of microsporidiosis in Scandinavian patients with AIDS. *Scand. J. Infect. Dis.,* **25,** 667–669.

Hollister, W. S., Canning, E. U., Weidner, E., Field, A. S., Kench, J., and Marriott, D. J. (1996). Development and ultrastructure of *Trachipleistophora hominis* n. g., n. sp., after *in vitro* isolation from an AIDS patient and inoculation into athymic mice. *Parasitology,* **112,** 143–154.

Issi, I. V. (1986). Microsporidia as a phylum of parasitic protozoa. *Protozoology (Leningrad),* **10,** 6–136 (in Russian).

Iwano, H. and Ishihara, R. (1991). Dimorphism of spores of *Nosema* sp. in cultured cell. *J. Invert. Pathol.,* **57,** 211–219.

Kotler, D. P. and Orenstein, J. M. (1999). Clinical syndromes associated with microsporidiosis. In: *The microsporidia and microsporidisis* (M. Wittner and L. M. Weiss, eds). ASM Press, Washington, pp. 258–292.

Larsson, R. (1980). Insect pathological investigations on Swedish Thysanura. II. A new microsporidian parasite of *Petrobius brevistylis* (Microcoryphia, Machilidae); description of the species and creation of two new genera and a new family. *Protistologica,* **16,** 85–101.

Larsson, R. (1983a). Studies on the cytology and taxonomy of the microsporidia (Protozoa, Microspora). Thesis, Lund.

Larsson, R. (1983b). Identifikation av mikrosporidier (Protozoa, Microspora). *Mem. Soc. Fauna Flora Fenn.*, **59**, 33–51.

Larsson, R. (1986). Ultrastructure, function and classification of microsporidia. *Prog. Protist.*, **1**, 325–390.

Larsson, J. I. R. (1988). Identification of microsporidian genera (Protozoa, Microspora) – a guide with comments on the taxonomy. *Arch. Protist.*, **136**, 1–37.

Larsson, J. I. R. (1999). Identification of microsporidia. *Acta Protozool.*, **38**, 161–197.

Levaditi, C., Nicolau, S., and Schoen, R. (1923). L'étiologie de l'éncéphalite. *C. R. Acad. Sci.*, **177**, 985–988.

Matsubayashi, H., Koike, T., Mikata, T., and Hagiwara, S. (1959). A case of *Encephalitozoon*-like body infection in man. *Arch. Pathol.*, **67**, 181–187.

Nordstoga, K. (1972). Nosematosis in blue foxes. *Nord. Vet. Med.*, **24**, 21–24.

Pekkarinen, M. (1995). *Pleistophora mirandellae* Vaney & Conte, 1901 (Protozoa: Microspora) infection in the ovary of the roach, *Rutilus rutilus* (L.), from Finnish coastal waters. *Mem. Soc. Fauna Flora Fenn.*, **71**, 19–32.

Petri, M. (1969). Studies on *Nosema cuniculi*. *Acta Pathol. Microbiol. Scand.*, **204** (Suppl.), 1–91.

Silveira, H. and Canning, E. U. (1995). *Vittaforma corneae* n. comb. for the human microsporidium *Nosema corneum* Shadduck, Meccoli, Davis & Font, 1990, based on its ultrastructure in the liver of experimentally infected athymic mice. *J. Eukaryot. Microbiol.*, **42**, 158–165.

Skurdal, J., Qvenild, T., Taugbøl, T., and Fjeld, E. (1990). A 6-year study of *Thelohania contejeani* parasitism of the noble crayfish, *Astacus astacus* L., in lake Steinsfjorden, S. E. Norway. *J. Fish Dis.*, **13**, 411–415.

Sobottka, I., Albrecht, H., Schottelius, J., Schmetz, C., Bentfeld, M., Laufs, R., and Schwartz, D. A. (1995). Self-limited traveller's diarrhea due to a dual infection with *Enterocytozoon bieneusi* and *Cryptosporidium parvum* in an immunocompetent HIV-negative child. *Eur. J. Clin. Microbiol. Infect. Dis.*, **14**, 919–920.

Sprague, V. (1977). Systematics of the microsporidia. In: *Comparative pathobiology* (L. A. Bulla, Jr. and T. C. Cheng, eds), Vol. 2. Plenum Press, New York, pp. 1–510.

Sprague, V., Becnel, J. J., and Hazard, E. I. (1992). Taxonomy of phylum Microspora. *Crit. Rev. Microbiol.*, **18**, 285–395.

Sprague, V., Ormières, R., and Manier, J.-F. (1972). Creation of a new genus and a new family in the microsporidia. *J. Invert. Pathol.*, **20**, 228–231.

Undeen, A. H. (1990). A proposed mechanism for the germination of microsporidian (Protozoa: Microspora) spores. *J. Theoret. Biol.*, **142**, 223–235.

Van Gool, T. and Dunkert, J. (1995). Human microsporidiosis: clinical, diagnostic and therapeutic aspects of an increasing infection. *J. Clin. Microbiol. Infect.*, **1**, 75–85.

Van Gool, T., Snijders, F., Reiss, P., Eeftinck Schattenkerk, J. K. M., van den Bergh Weerman, M. A., Bartelsman, J. F. W. M., Bruins, J. J. M., Canning, E. U., and Dankert, J. (1993). Diagnosis of intestinal and disseminated microsporidial infections in patients with HIV by a new rapid fluorescence technique. *J. Clin. Pathol.*, **46**, 694–699.

Vávra, J. (1976). Structure of the microsporidia. In: *Comparative pathobiology* (L. A. Bulla, Jr. and T. C. Cheng, eds), Vol. 1. Plenum Press, New York, pp. 1–85.

Vávra, J. and Larsson, J. I. R. (1999). Structure of the microsporidia. In: *The microsporidia and microsporidiosis* (M. Wittner and L. M. Weiss, eds). ASM Press, Washington, pp. 7–84.

Vávra, J. and Maddox, J. V. (1976). Methods in microsporidiology. In: *Comparative pathobiology* (L. A. Bulla, Jr. and T. C. Cheng, eds), Vol. 1. Plenum Press, New York, pp. 281–319.

Vávra, J., Yachnis, A. T., Shadduck, J. A., and Orenstein, J. M. (1998). Microsporidia of the genus *Trachipleistophora* – causative agents of human microsporidiosis: Description of *Trachipleistophora anthropophthera* n. sp. (Protozoa: Microsporidia). *J. Eukaryot. Microbiol.*, **45**, 273–283.

Vossbrinck, C. R., Maddox, J. V., Friedman, S., Debrunner-Vossbrinck, B. A., and Woese, C. R. (1987). Ribosomal RNA sequence suggests microsporidia are extremely ancient eukaryotes. *Nature*, **326**, 411–414.

Waller, T. (1979). Serology and sensitivity of *Encephalitozoon cuniculi* in relation to diagnosis and control of encephalitozoonosis in domestic rabbits. Thesis, Uppsala.

Weber, R., Bryan, R. T., Owen, R. L., Wilcox, C. M., Gorelkin, L., and Visvesvara, G. S. (1992). Improved light-microscopical detection of microsporidia in stool and duodenal aspirates. *N. Engl. J. Med.*, **326**, 161–166.

Weber, R., Bryan, R. T., Schwartz, D. A., and Owen, R. L. (1994). Human microsporidial infections. *Clin. Microbiol. Rev.*, **7**, 426–461.

Weber, R., Schwartz, D. A., and Deplazes, P. (1999). Laboratory diagnosis of microsporidiosis. In: *The microsporidia and microsporidiosis* (M. Wittner and L. M. Weiss, eds). ASM Press, Washington, pp. 315–362.

Weidner, E. (1976). The microsporidian spore invasion tube. *J. Cell Biol.*, **71**, 23–34.

Weiser, J. (1977). Contribution to the classification of microsporidia. *Věstn.Čs. Spol. Zool.*, **41**, 308–321.

Weiss, L. M. and Vossbrinck, C. R. (1999). Molecular biology, molecular phylogeny, and molecular diagnostic approaches to the microsporidia. In: *The microsporidia and microsporidiosis* (M. Wittner and L. M. Weiss, eds). ASM Press, Washington, pp. 129–171.

Widtmann, S. (1990). On the microsporidians (Protozoa, Microsporidia) fauna of cladocerans (Crustacea, Cladocera) and copepods (Copepoda) in the water bodies of Lithuania with description of three new species. *Ekologija*, **2**, 104–111 (in Russian with English summary).

12 *Pneumocystis carinii*

Ewert Linder, Kerstin Elvin, and Mats Olsson

Some 90 years ago, Carlos Chagas the discoverer of Chagas' disease, American trypanosomiasis, made an observation of pulmonary cysts in *Schizotrypanum cruzi*-infected individuals (Chagas 1909). This finding was followed up by Carinii working on rats infected with *Trypanosoma lewisi*. In these rats suffering from trypanosomiasis, lungs contained cysts similar to those observed by Chagas, which were considered as the schizogonic stage of the trypanosome. In 1912 the Delanoë couple in Paris redescribed and reached a different conclusion on the origin of the 'pulmonary cysts of Carinii' in the lungs of *T. lewisi*-infected rats. The 'pneumocysts of Carinii' were identified as a new parasite of rats by Delanoë and Delanoë and were called by them as *Pneumocystis carinii* in their 1912 publication (Delanoë and Delanoë 1912). This significant contribution is acknowledged in the current classification (Haase 1997). It took another 10 years before Vavra in Prague associated *P. carinii* with the severe plasma cell pneumonia affecting undernourished children in postwar Europe (Vavra and Kucera 1970; see also Gajdusek 1957). The original publication was followed by an article by Vavra and Jirovec a year later. However, for more than a decade the etiological link between pneumocystis and plasma cell pneumonia was disputed. Not only in the United States where only a few cases were seen, but also in European countries like Finland there was a reluctance to accept *Pneumocystis* as the etiological agent of plasma cell pneumonia (Stenbäck *et al.* 1968). Despite the fact that the high standards of neonatology brought down neonatal mortality to a record low, a premature baby only had a 50–50 chance of avoiding a deadly *Pneumocystis* infection at the Children's hospital in Helsinki at that time. The seriousness of 'plasma cell pneumonia' in prematurely born babies in the 1950s (Vanek and Jirovec 1952; Gajdusek, 1957) can be compared to pneumocystis pneumonia (PCP) in AIDS 30 years later (Figure 12.1).

Nomenclature, host specificity, and taxonomy

Delanoë and Delanoë thought that *P. carinii* could be a *coccidian* parasite, a concept surviving for about 80 years. However, there was always a degree of uncertainty, and the similarities to fungi was stressed by several investigators during that time (Vavra and Kucera 1970). The recognition of human *pneumocystis* as a separate species by Frenkel should be noted. He redescribed *P. jiroveci* n. spp. from humans, including the trophozoite precyst, cyst, intracystic bodies, and empty cyst and showed the forms from man to be distinct with respect to morphology, biology, and physiology compared with those of *P. carinii* from rats. He concluded that 'These two forms should be regarded as separate species, and forms from other hosts should tentatively be regarded as distinct.' (Frenkel 1976).

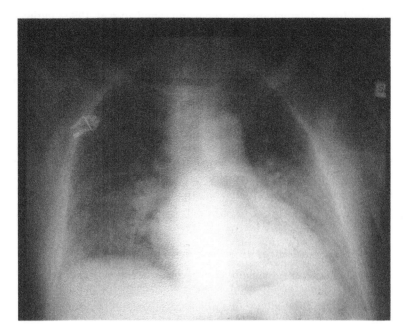

Figure 12.1 Appearance of PCP in a case of AIDS as seen in a chest X-ray a few days before exitus.

Today it is generally agreed that sufficient data is available to regard *P. carinii* as a fungus (Cushion *et al.* 1990; Eriksson 1994; Wakefield and Miller 1996), for example, in contrast to the situation in protozoa, the *P. carinii* genes for the enzymes thymidylate synthase (TS) and dihydrofolate-reductase (DHFR) encode separate proteins (Edman *et al.* 1989). On the basis of TS, DHFR, actin, and beta-tubulin gene sequences, *P. carinii* can be placed taxonomically near the ascomycetes and it has been classified as a member of a new family and order (Pneumocystidaceae, Pneumocystidales) in the class Archiascomycetes. Members of this class of saphrophytic and parasitic plant pathogens, for example, the causative agent of peach leaf curl disease and witch brooms, *Taphrina*, may represent the earliest diverging group of ascomycetes (Eriksson 1994; Miller and Wakefield 1996; Eriksson and Winka 1997).

Recent data indicate that *P. carinii* represents a diverse group of organisms. Thus a revised trinomial nomenclature for *P. carinii* has been suggested based on the host of origin: *P. carinii* f. sp. *hominis*, f. sp. *oryctolagi* (rabbit), etc. (Bartlett *et al.* 1994).

Genetic diversity of especially two loci has been studied. These are the mitochondrial large sub-unit rRNA (mt LSU) and the internal transcribed spacer (ITS) of the ribosomal RNA genes (Lee *et al.* 1993; Latouche *et al.* 1994, 1996, 1997b; Lu *et al.* 1995; Jiang *et al.* 1996; Keely *et al.* 1996; Wakefield 1996; Wakefield *et al.* 1997). Organisms isolated from one mammalian host species do not cause infection in another mammalian host species; i.e. *P. carinii* in man is not a zoonosis. Host specificity of *P. carinii* has been suggested by unsuccessful transmission studies using several mammalian-derived isolates of *P. carinii*, including human *P. carinii* (Furuta and Katsamoto 1987; Gigliotti *et al.* 1993; Aliouat *et al.* 1994; Frenkel *et al.* 1996). Human-derived *P. carinii* is genetically distinct from parasites isolated from other species of mammalian hosts. Rats and humans can harbour distinct types of *P. carinii* that are sufficiently different to suggest that *P. carinii* from the two hosts could be different species

(Stringer 1993). Antigenic differences have been observed between *P. carinii* isolates from human isolates obtained from different areas of the world (Smulian *et al.* 1993), in different hosts from the same region (Bauer *et al.* 1993), and even between *P. carinii* organisms obtained from the same host. Genetic diversity between *P. carinii* isolates from different hosts is seen both at the chromosomal and at sequence levels. Genetic diversity has also been demonstrated in *P. carinii* isolates from the same host, but the extent of diversity is, as expected, lower than that between isolates from different species.

Nucleotide sequence variations of *P. carinii* can be used for typing and studying the epidemiology of *P. carinii* infections (Lu *et al.* 1995; Miller and Wakefield 1996; Hauser *et al.* 1997). Two types of nucleotide sequences (designated types A and B) have been found in the internal transcribed spacer region 1 (ITS1), and three types of nucleotide sequences (designated types a, b, and c) have been found in the ITS2 region. Of the six possible combination types, four have been detected: types Ac, Bb, Ba, and Bc.

Cell biology, life cycle, and interactions with the host

Several aspects of the life cycle, transmission, pathogenesis, and host response in *P. carinii* infection remain unknown. However, in recent years we have gained much information on the complex interplay of *P. carinii* with host inflammatory cells, release of cytokines, generation of toxic metabolites, and involvement of both cellular and humoral immunity. Although much remains unknown about the pathogenesis and host response, studies of *P. carinii* in recent years have provided us with an increasing base of knowledge about this organism and its relationship to the host. These studies have led to a better understanding of mechanisms of *P. carinii* attachment and injury to host cells, new information about the interaction of *P. carinii* with pulmonary epithelial cells and the interplay of the organism with host inflammatory cells (Su and Martin 1994).

Two dominant stages of *P. carinii* have been identified in the mammalian lung, the mature thick walled cysts or asci containing eight intracystic bodies or ascospores and the polymorphic 'trophozoites' (Figure 12.2). The latter arise from ascospores liberated from ruptured

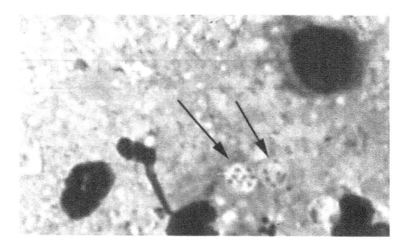

Figure 12. 2 Pneumocystis carinii organisms with eight 'intracystic bodies' in broncho-alveolar lavage sample from a patient with PCP seen in a Giemsa-stained preparation.

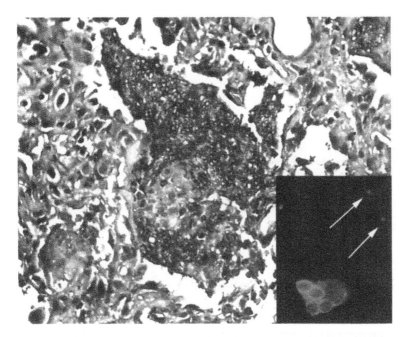

Figure 12.3 Pneumocystis carinii organisms identified in histological section of infected lung tissue and in broncho-alveolar lavage fluid obtained from a patient with fulminant PCP. Immunoperoxidase staining of intra-alveolar masses of *Pneumocystis carinii* is seen. Using fluorescence microscopy of lavage fluid both asci (cysts) and trophozoites are seen using monoclonal anti-pneumocystis antibody 3F6 as the marker (100). (See Colour Plate II.)

asci which can be seen as empty, banana-shaped shells in electron micrographs of bronchial lavage fluid sediments from infected lungs. Alveolar spaces filled with organisms can be seen in pathology specimens and both stages identified in lavage fluids by immunocytochemistry, (Figure 12.3).

The composition of the cyst wall is distinct from the surface of the trophozoites. The thick middle electron-lucent layer of the cyst wall seen in transmission electron microscopy (Figure 12.4) contains important immunogens. The cyst wall contains carbohydrates (De Stefano *et al.* 1989) and some appear to contain beta-1-3-glucan (Goheen *et al.* 1994). Degradation by glucanase and chitinase confirms that this layer contains branched glucan and chitin. The susceptibility of the polysaccharide-rich electron-lucent layer to proteolysis reveals that proteins are also relevant in building up the cyst-wall glucan skeleton (Roth *et al.* 1997). Interestingly beta-D-glucan can be detected in sera obtained from patients with PCP (Yasuoka *et al.* 1996).

The trophozoites have a plasma membrane surrounded by a glycocalyx. Upon development of trophozoites to thin walled asci and subsequently thick walled asci, a typical trilaminated wall is seen (Figure 12.1). Numerous tubular extensions of the ascus wall are apparently involved in the attachment of the organism to type-I lung epithelial cells. Binding seems to involve penetration, but not invasion, into the host cell cytoplasm without damaging the plasma membrane.

All *P. carinii*, irrespective of their host of origin, express an abundant mannosylated surface glycoprotein, commonly referred to as gp120, or because its size varies in different species, as surface glycoprotein A (Gigliotti 1992). The protein moiety of the major high molecular

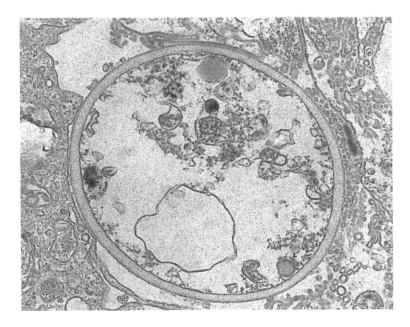

Figure 12.4 Ultrastructural appearance of *Pneumocystis* ascus (cyst) in lung tissue from a patient with fatal PCP.

weight surface antigen, represented by numerous isoforms, is encoded by different genes. These proteins are post-transcriptionally modified by carbohydrates and lipids. *Pneumocystis* surface glycoproteins, especially the mannose-rich major surface glycoproteins, appear to mediate adherence to type-I epithelial cells. Extracellular matrix proteins are necessary for the attachment of *Pneumocystis* to host cells, a phenomenon well known in several other host–parasite interactions. Integrin receptors on the epithelial cells take up extracellular matrix proteins fibronectin and vitronectin produced by alveolar macrophages in the inflammatory reaction in the alveolar space (Limper *et al.* 1996, 1997; Limper 1997).

The more rapid turnover of plasma membrane constituents of type-II cells may inhibit attachment of *Pneumocystis*. Also, type-II cells may have a direct toxic effect on *Pneumocystis* (Pesanti and Shanley 1988). There is evidence for *Pneumocystis*-induced reduced alveolar type-II epithelial cell function upon secretagogue stimulation (Rice *et al.* 1993).

Pulmonary surfactants appear to play a role in the disease process, containing mainly phospholipids and four surfactant-specific proteins (SP-A, SP-B, SP-C, and SP-D). Pulmonary surfactant protein A (SP-A), an alveolar glycoprotein containing collagen-like and carbohydrate-recognition domains, binds *P. carinii* and enhances adherence to alveolar macrophages (McCormack *et al.* 1997). Also, SP-D seems to facilitate binding of the parasite to alveolar macrophages. SP-A and SP-D may also mediate attachment to the type-I pneumocyte. In rodent models of PCP a reduction in phospholipid levels as well as a decrease in SP-B have been suggested to play an important role in the hypoxemic respiratory insufficiency associated with PCP (Beers *et al.* 1997).

Several aspects of *Pneumocystis* metabolism have been clarified recently, which have led to the recognition of novel therapeutic possibilities. Lipid transfer from human alveolar epithelial cells to *P. carinii* has been demonstrated (Furlong *et al.* 1997), and uncommon lipids have been identified in *P. carinii*. (Kaneshiro *et al.* 1989). The organism has the shikimic acid pathway that leads to the formation of compounds which mammals cannot synthesize (e.g. folic acid); hence drugs that inhibit these pathways are effective against the pathogen. The high concentration of free fatty acids and the relatively low level of triglycerides in *P. carinii* suggest that fatty acids may represent major carbon sources for ATP production by the organism (Ellis *et al.* 1996). *Pneumocystis carinii* also possesses the biochemical pathway for *de novo* synthesis of the CoQ benzoquinone ring (Sul and Kaneshiro 1997). Instead of ergosterol (the major sterol of higher fungi), *P. carinii* synthesizes distinct delta(7), C-24-alkylated sterols. An unusual C-32 sterol, pneumocysterol, has been identified in human-derived *P. carinii*. Another signature lipid discovered is *cis*-9,10-epoxy stearic acid. CoQ(10), identified as the major ubiquinone homologue, is synthesized *de novo* by *P. carinii* (Kaneshiro 1998).

Pneumocystis carinii glycoprotein A stimulates interleukin-8 production and inflammatory cell activation in alveolar macrophages and cultured monocytes (Lipschik *et al.* 1996). The role of cytokines in PCP has been discussed by Perenboom *et al.* (1996). Vitronectin, fibronectin, and gp120 antibody enhance macrophage release of TNF-alpha in response to *P. carinii* (Neese *et al.* 1994) and the intercellular adhesion molecule-1 which is important in leukocyte accumulation is enhanced in lung epithelium during *P. carinii* infection, in part, through TNF-alpha-mediated mechanisms (Yu and Limper 1997).

During recovery from PCP in a dexamethasone rat model, the trophozoite-to-cyst ratio significantly changed. The number of cysts decreased from week 0 to week 4 whereas there was a much lesser decrease in the number of trophozoites suggesting that in particular the cyst form of *P. carinii* is sensitive to the host response mounted during recovery. The results support the possibility that trophozoites may multiply by extracystic asexual fission (Sukura 1995).

Immune response to clear the infection is manifested by phagocytosis by alveolar macrophages added by CD4+ lymphocytes and serum antibodies.

Pneumocystis carinii adheres to alveolar macrophages and is engulfed and digested in the presence of an opsonizing antibody. The adherence of *P. carinii* occurs via multiple pathways and stimulates oxidative burst, cytokine production, phagocytosis, and killing of the organism. A number of cytokines are involved in pneumonia. TNF-α and IL-1b are considered to be important mediators of host resistence against *P. carinii* (Chen *et al.* 1992a,b) while IL-6 is involved in the inflammatory and antibody responses during resolution of the infection (Chen *et al.* 1993). *Pneumocystis carinii* induces production of TNF-α by alveolar macrophages *in vivo* and *in vitro*. (Krishnan *et al.* 1990; Pesanti *et al.* 1991; Kandil *et al.* 1994). TNF-α release is also augmented by opsonization of *P. carinii* with not only antibodies but also adhesive glycoproteins (Neese *et al.* 1994)

Epidemiology and transmission

Pneumocystis carinii is ubiquitous and apparently encountered by the population in many geographical areas without disease manifestation (Pifer *et al.* 1978; Smulian *et al.* 1993). Most humans appear to be exposed to *Pneumocystis* within 2 years after birth (Hong 1991). An increased rate of seropositivity is seen with age (Meuwissen and Leeuwenberg 1972; Meuwissen *et al.* 1973) in children in Europe, the US, and Gambia.

Despite the global presence of the organism, geographical differences are seen in prevalence of PCP. PCP is more frequently detected in northern Europe and the US as compared with southern Europe, Africa, and Asia (Lucas *et al.* 1988; Elvin *et al.* 1989). Several factors may account for this difference. It has been attributed to the fact that under conditions of crowding and poor hygiene, other pathogens, such as *Mycobacterium tuberculosis* in the environment of immunosuppressed individuals, will cause severe and even fatal disease within a short time, so that *Pneumocystis* would be masked or not have sufficient time to establish itself. The clinical profile of Indian patients with HIV bears much resemblance to those seen in African countries owing perhaps to the similar background of poverty, malnutrition, and spectrum of endemic infection (Giri *et al.* 1995). However, differences in virulence or degree of exposure cannot be ruled out as contributing factors explaining the observed geographical differences (Chaisson and Moore 1997). The fact that PCP is less common in African HIV patients living in Europe (Biggar 1986) does not exclude the possibility that there are genetic differences in susceptibility between hosts.

In part, the geographic differences may be explained by the difference in diagnostic capacity between these regions. Indirect evidence for *P. carinii* being prevalent in Gambia is provided by serological data which indicate a high prevalence of *P. carinii* (Wakefield *et al.* 1990).

More than 80% of the reported PCP cases are homo- and bisexual men and injection drug users. The remaining cases are hemophiliacs, persons who get infected by heterosexual contacts, blood transfusion recipients, and children infected pre- or perinatally (Bunikowski *et al.* 1992).

The pathophysiology of HIV infection is incompletely understood, but is in large part related to the destruction of helper CD4 lymphocytes. This results in immune dysfunction and the development of a variety of opportunistic infections.

Significant differences in CD4 lymphocyte numbers were observed among 12 AIDS-defining illnesses, oral candidiasis, and asymptomatic infection, allowing them to be grouped into five general catgories based on mean CD4 count: (a) asymptomatic infection (ASX) – CD4 greater than $500/mm^3$; (b) oral candidiasis (O-C) and tuberculosis (TB) – range $250–500/mm^3$; (c) Kaposi's sarcoma (KS), lymphoma (LYM), and cryptosporidiosis (CRS) – range $150–200/mm^3$; (d) PCP, disseminated *Mycobacterium avium* complex (MAC), herpes simplex ulceration (HSV), toxoplasmosis (TOX), cryptococcosis (CRC), oesophageal candidiasis (E-C) – range $75–125/mm^3$; (e) cytomegalovirus retinitis – less than $50/mm^3$ (Crowe *et al.* 1991). This correlation depicted in Graph 12.1, can explain the relative difference in occurrence of TB and PCP in Africa as compared with Europe.

The risk of development of PCP in immunodeficient patients with CD4 counts below 200 motivates prophylactic treatment (Guss 1994). This has led not only to a decrease in PCP

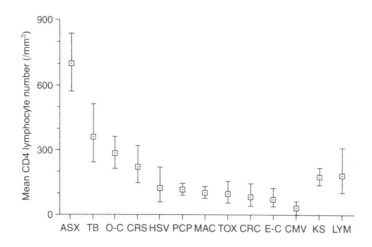

prevalence but also to extrapulmonary *P. carinii* infections. Prior to the advent of the HIV-1 epidemic, only 16 cases of extrapulmonary *Pneumocystosis* had been reported in individuals who were immunocompromised by a variety of underlying diseases. Since the beginning of the HIV-1 and related PCP epidemic, at least 90 cases of extrapulmonary pneumocystosis have been reported (Ng *et al.* 1997).

The central role played by CD4 T-lymphocytes is illustrated by the observed relationship between CD4 T-lymphocyte counts and the time of appearance of opportunistic infections. There is a definite clinical association of PCP with low CD4 lymphocyte counts as described above. The major surface glycoprotein of *Pneumocystis* can induce a protective CD4 T-cellular immunity mediated by IFN-γ, which has several identified protective effects. IFN-γ has an effect on macrophages, activating them to produce toxic superoxide intermediates and NO which are toxic to *Pneumocystis*. IFN-γ also has an effect on the expression of integrins on alveolar cells, thereby decreasing receptor-mediated binding of the parasites. Thus lack of IFN-γ in experimental *Pneumocystosis* is a typical interstitial cell pneumonia. Antibody-mediated killing of *Pneumocystis* is another protective mechanism, even if its relative role appears to be varied under different conditions, depending on the type of immunodeficiency.

In experimentally induced immunodeficiency caused by simian immunodeficiency virus, SIV, the monkey equivalent of HIV infection, the CD4 cell counts of <50 cells/μl was associated with PCP, cytomegalovirus meningoencephalitis, lymphoid depletion, and thymic atrophy (Shibata *et al.* 1997).

Two hypotheses, which are not mutually exclusive, have been advanced to explain the development of PCP in immunosuppressed patients: activation of a latent infection and *de novo* infection. Reactivation, in analogy to the relationship between immunosuppression and toxoplasmosis, is assumed to be related to the demonstrated seropositivity developing in children (see above) and to some experimental studies demonstrating latent infections in immunosuppressed rats (Frenkel *et al.* 1966). Experimental studies suggest that the activation of a latent infection of *P. carinii* occurs in SIV-infected rhesus monkeys: PCR-based methods and *P. carinii*-specific bands of DNA amplification, but not histopathologic examination, detected *P. carinii* in the liver, kidney, spleen, adrenal gland, testis, brain, and other organs examined (Furuta *et al.* 1993). However, no *P. carinii*. could be detected with monoclonal antibodies and PCR in broncho-alveolar lavage fluid (BAL) and autopsy lung tissue from immunocompetent subjects (Millard and Heyret 1988; Peters *et al.* 1992).

Several observations support the hypothesis of *de novo* infection. The host immune response to *P. carinii* can completely eliminate the organism from the host (Chen *et al.* 1993), and in animal experiments the normal host is capable of clearing the organism within a year (Vargus *et al.* 1995) It has been suggested that *Pneumocystis* infection might be acquired, as deep mycoses, from environmental sources through the respiratory tract (Dei Cas *et al.* 1992).

In the literature there are several reports consistent with nosocomial transmission and outbreaks. An outbreak of PCP in three patients within a 6-week period was reported. Two patients had acute lymphoblastic leukaemia and one had brain-stem glioma. They shared common features of immunosuppression and absence of cotrimoxazole prophylaxis and had been nursed in the same room (Cheung *et al.* 1994).

There are reports of outbreaks of PCP in hospitals and the genetically distinct genotypes associated with both separate and recurrent episodes of PCP (Keely *et al.* 1995, 1996; Tsolaki *et al.* 1996; Keely and Stringer 1997; Latouche *et al.* 1997d) favour the concept of *de novo* infection. In some of these reports, person-to-person transmission of *P. carinii* was considered likely (Hirschl *et al.* 1992; Hennequin *et al.* 1995).

We, like others, have encountered difficulties in inducing experimental PCP using the classical immunosuppression regimens published by several workers. Our initial failure to induce PCP in immunosuppressed rats was followed by success when *P. carinii*-infected rats obtained from Costa Rica were housed together with non-infected immunosuppressed rats in Helsinki (Sukura *et al.* 1991b). Hughes has already demonstrated in 1982 that rats acquire *P. carinii* infection by an air-borne route and eliminated food, water, and soil as sources of infection (Hughes 1982). The air-borne mode of transmission is further supported by recent experiments with SIV-infected rhesus macaques (Vogel *et al.* 1993) and with 'sentinel' rats housed near immunosuppressed *P. carinii*-infected rats (Sepkowitz 1993). Transient carriage of *P. carinii* may occur in immunocompetent rats housed near immunosuppressed *P. carinii*-infected rats. Within 6 weeks, *P. carinii* DNA became detectable in the lungs and by 8 weeks, in the blood. *Pneumocystis carinii* DNA disappeared rapidly from the lungs and sera after sentinel rats were isolated away from corticosteroid-treated rats. Sporadic findings of *P. carinii* in human immunocompetent hosts are also reported (Jacobs *et al.* 1991; Calderón *et al.* 1996; Heresi *et al.* 1997). High levels of anti-*P. carinii* antibody titres were seen in health care staff working in AIDS units (Leigh *et al.* 1993), but this was not confirmed in a Scandinavian study, where attempts to demostrate *P. carinii* DNA in sputa also failed (Lundgren *et al.* 1997).

The hypothesis that *P. carinii* infection is acquired through the inhalation of air containing the infectious stage of the parasite, fits with the reclassification of *P. carinii* as a fungus. There is accumulating direct evidence for an air-borne route of transmission. The DNA of *P. carinii* has been demonstrated not only in the air surrounding both *P. carinii*-infected rats and patients with PCP but also in air samples from rural locations (Bartlett *et al.* 1996b; Olsson *et al.* 1996; Wakefield 1996). Even stronger evidence for air-borne transmission is the demonstated presence of the same species-specific genotypes of *P. carinii* in the air and in infected individuals. This has been shown both in experimentally infected rabbits and rats (Latouche *et al.* 1997a) and in patients with (Bartlett *et al.* 1997; Olsson *et al* 1998).

An obvoius question is, what is the transmissive stage of *P. carinii*? So far there is no conclusive answer. It is remarkable that infected lungs containing the two identified stages of the organism, cysts or asci and the trophic form, are not very infectious, for example, Hughes (1982) showed that rats were infected through air, but failed to induce the infection by exposing rats to infected lung tissue.

Pathogenesis of PCP

Today the term 'plasma cell pneumonia' is replaced by *Pneumocystis* pneumonia (PCP), which reflects not only the recognition of the etiological agent, but also the histopathologically different appearance of the lung tissue in AIDS-associated PCP and the severe interstitial reaction seen in infected small children.

Infection with *P. carinii* typically results in a pneumonia which is histologically seen to consist of an eosinophilic foamy alveolar exudate associated with a mild plasma cell interstitial infiltrate. Special stains show that cysts of *P. carinii* lie within the alveolar exudate. Atypical histological appearances may occasionally be seen, including a granulomatous pneumonia and diffuse alveolar damage (Robillard *et al.* 1965; Foley *et al.* 1993).

Cells of the macrophage lineage mediate extremely important normal functions of the immune system. The CD4 glycoprotein is the major cellular receptor for HIV. Macrophages mediate the elimination of *P. carinii* from the living host. The uptake and degradation of ^{35}S-labelled *P. carinii* by cultured macrophages has been quantified, demonstrating significant degradation of *P. carinii* over 6 h (Limper *et al.* 1997). Quantitative differences have been observed in phagocytosis and degradation of *P. carinii* by alveolar macrophages in AIDS and non-HIV patients *in vivo* (Wehle *et al.* 1993). CD4, although expressed at low levels on the surface of alveolar macrophages, appears to be critical in the HIV-1 infection of these cells. (Lewin *et al.* 1996).

Extrapulmonary infection with *P. carinii* is an uncommon event involving the skin of the external auditory canal, the mastoid area, and the glabrous skin (Hennessey *et al.* 1991; Litwin and Williams 1992).

The classical model for the study of *P. carinii* exposed to PC, is corticosteroid-immunosuppressed rats developing PCP when exposed to the parasite (Frenkel *et al.* 1966; Sukura *et al.* 1991a). PCP also occurs spontaneously in other rodents such as SCID mice, athymic mice, and voles. Furthermore a sub-clinical PCP occurs in young rabbits at the end of their first month of life (Dei Cas *et al.* 1998). Impaired cellular immunity has been shown to be more important than impaired humoral immunity (Frenkel *et al.* 1966; Walzer *et al.* 1989). In mice the disease can be cleared by adoptive transfer of CD4 cells but not of CD8 cells (Harmsen and Stankiewicz, 1990). In HIV-infected individuals a low CD4+ cell count is the single most important risk factor for PCP (Masur *et al.* 1989; Lidman *et al.* 1992; Lundgren *et al.* 1994). Treatment with immunosuppressive drugs involves a risk for developing PCP, which is time and dose dependent. The combined effect of immunosuppression and drug-dependant differences in the effects on different host cell types apparently accounts for the differences seen among immunosuppressive drugs with respect to their capacity to induce PCP. In addition to corticosteroids, cyclosporin A, a highly specific inhibitor of T-cell activation, used in treatment of transplant recipients, is associated with a risk for developing PCP. The pathogenic effect of corticosteroids is not fully known, but apart from affecting the immune system, for example, depressing the production of TNF-α and IL-6 (Perenboom *et al.* 1996), they induce a decrease in the alveolar cell surface glycocalyx, an effect which might enhance organism attachment.

In severe cases of PCP, interstitial fibrosis, granulomatous inflammations, pulmonary cavitations, and pneumothorax can develop (Travis *et al.* 1991). Using sensitive immunocytochemical techniques, *P. carinii* has been found in the interstitial compartment, pleura, and vessel walls during autopsy and in open lung biopsy specimens (Murry and Schmidt 1992). Extrapulmonary *P. carinii* was previously rarely seen. With the use of local treatment with Penthamidine aerosol a number of cases have been reported. Various authors have estimated the risk of extrapulmonary *P. carinii* as 0.5–3% in AIDS patients (Raviglione 1990; Telzak *et al.* 1990). These are probably underestimations since the clinical picture is very variable and adequate sensitive staining is not routinely performed on material outside the lung during an autopsy.

Perspectives

Several fundamental questions regarding the biology of *P. carinii*, such as identification of the transmissive stage and the presence of reservoirs in nature, remain to be answered, and methods need to be developed for the long-term culture and *in vitro* differentiation of the parasite. The use of genetic markers are likely to increase our understanding of the transmission of the organism and the mechanisms leading to severe infection.

With changes in the epidemiology and the application of newer treatments and prophylactic regimens, the types of pulmonary diseases that occur in HIV-infected persons are changing. The important advance is treatment with Zidovudine (AZT) which remains the most important agent in slowing the progression of the disease and has resulted in prolonging survival. New ways to assess the progression of HIV disease and new anti-retroviral treatments are available and we see an increased survival despite worsening immunosuppression (Schneider and Rosen 1997). However, the number of individuals at risk for developing PCP appears to be relatively constant as immunosuppression due to therapeutics in an increasing group of organ transplant recipients and in patients with malignancies compensates for the decrease seen in AIDS patients. The incidence of PCP and the overall mortality from PCP is declining (Montaner *et al.* 1994). Interestingly, compounds active against fungal cell walls are selectively active against the cyst forms of *P. carinii* (Bartlett *et al.* 1996a). With the reclassification of *Pneumocystis* as a fungus and with an increasing knowledge about *Pneumocystis* metabolism, especially its lipid metabolism, one may expect further innovative therapeutic strategies in the future (Kaneshiro 1998). Studies aiming at the characterization of metabolic pathways of the lipid metabolism of *Pneumocystis* are potentially of importance for the design of new therapeutic agents with anti-*P. carinii* activity (Kaneshiro *et al.* 1989).

References

Aliouat, E., Mazars, E., Dei-Cas, E., Delcourt, P., Billaut, P., and Camus, D. (1994). *J. Eukaryot. Microbiol.*, **41**, 71S.

Bartlett, M., Cushion, M. T., Fishman, J. A., Kaneshiro, E. S., Lee, C.-H., Leibowitz, M. J., Lu, J.-J., Lundgren, B., Peters, S. E., Smith, J. W., Smulian, A. G., Staben, C., and Stringer, J. R. (1994). *J. Eucaryot. Microbiol.*, **41**, 121S–122S.

Bartlett, M. S., Current, W. L., Goheen, M. P., Boylan, C. J., Lee, C. H., Shaw, M. M., Queener, S. F., and Smith, J. W. (1996a). *Antimicrob. Agents Chemother.*, **40**, 1811–1816.

Bartlett, M. S., Lu, J. J., Lee, C. H., Durant, P. J., Queener, S. F., and Smith, J. W. (1996b). *J. Eukaryot. Microbiol.*, **43**, 44S.

Bartlett, M. S., Vermund, S. H., Jacobs, R., Durant, P. J., Shaw, M. M., Smith, J. W., Tang, X., Lu, J. J., Li, B., Jin, S., and Lee, C.-H. (1997). *J. Clin. Microbiol.*, **35**, 2511–2513.

Bauer, N., Paulsrud, J., Bartlett, M., Smith, J., and Wilde, C. (1993). *J. Clin. Microbiol.*, **61**, 1315–1319.

Beers, M., Atochina, E., Lipschik, G., and Beck, J. (1997). *Fifth International Workshops on Opportunistic Protists & Fifth General Meeting of the European Concerted Action on Pneumocystis Research* (Abstract O15).

Biggar, R. (1986). *Br. Med. J.*, **293**, 1453–1454.

Bunikowski, R., Estermann, J., and Koch, M. A. (1992). *Med. Klin.*, **87**, 1–7.

Calderón, E., Regordán, C., Medrano, F., Ollero, M., and Varela, J. (1996). *Lancet*, **347**, 977.

Chagas, C. (1909). *Memorias do Instituto Osvaldo Cruz*, **1**, 159–161.

Chaisson, R. E. and Moore, R. D. (1997). *J. Acq. Immune Def. Synd. Hum. Retrovirol.*, **16**, S14–S22.

Chen, W., Gigliotti, F., and Harmsen, A. (1993). *J. Clin. Microbiol.*, **61**, 5406–5409.

Chen, W., Havell, E., and Harmsen, A. (1992a) *Infect. Immun.*, **60**, 1279–1284.

Chen, W., Havell, E., Moldawer, L., McIntyre, K., Chizzonite, R., and Harmsen, A. (1992b). *J. Exp. Med.*, **176**, 713–718.

Cheung, Y. F., Chan, C. F., Lee, C. W., and Lau, Y. L. (1994). *J. Paediatr. Child Health*, **30**, 173–175.

Crowe, S. M., Carlin, J. B., Stewart, K. I., Lucas, C. R., and Hoy, J. F. (1991). *J. Acq. Immune Def. Synd.*, **4**, 770–776.

Cushion, M., Hong, S., Steele, P., Stringer, S., Walzer, P., and Stringer, J. (1990). *Ann. NY Acad. Sci.*, **616**, 415–420.

De Stefano, J. A., Cushion, M. T., Trinkle, L. S., and Walzer, P. D. (1989). *J. Protozool.*, **36**, 65S–66S.

Dei Cas, E., Brun-Pascaud, M., Bille-Hansen, V., Allaert, A., and Aliouat, E. (1998). *FEMS Immunol. Med. Microbiol.*, **22**, 163–168.

Dei Cas, E., Cailliez, J. C., Palluault, F., Aliouat, E. M., Mazars, E., Soulez, B., Suppin, J., and Camus, D. (1992). *Eur. J. Epidemiol.*, **8**, 460–470.

Delanoë, P. and Delanoë, M. (1912). *Bull. Soc. Path. Exot.*, **5**, 599–602.

Edman, U., Edman, J. C., Lundgren, B., and Santi, D. (1989). *Proc. Natl. Acad. Sci. USA*, **86**, 6503–6507.

Ellis, J. E., Wyder, M. A., Zhou, L., Gupta, A., Rudney, H., and Kaneshiro, E. S. (1996). *J. Eukaryot. Microbiol.*, **43**, 165–170.

Elvin, K. M., Lumbwe, C. M., Luo, N. P., Bjorkman, A., Kallenius, G., and Linder, E. (1989). *Trans. R. Soc. Trop. Med. Hyg.*, **83**, 553–555.

Eriksson, O. (1994). *Syst. Ascomycetum*, **13**, 165–180.

Eriksson, O. E. and Winka, K. (1997). *Myconet*, **1**, 1–16.

Foley, N. M., Griffiths, M. H., and Miller, R. F. (1993). *Thorax*, **48**, 996–1001.

Frenkel, J. K. (1976). *Nat. Cancer Inst. Monogr.*, **43**, 13–30.

Frenkel, J. K., Good, J. T., and Shultz, J. A. (1966). *Lab. Investig.*, **15**, 1559–1577.

Furlong, S. T., Koziel, H., Bartlett, M. S., McLaughlin, G. L., Shaw, M. M., and Jack, R. M. (1997). *J. Infect. Dis.*, **175**, 661–668.

Furuta, T., Fujita, M., Mukai, R., Sakakibara, I., Sata, T., Miki, K., Hayami, M., Kojima, S., and Yoshikawa, Y. (1993). *Parasitol. Res.*, **79**, 624–628.

Furuta, T. and Katsamoto, U. (1987). *Jpn. J. Exp. Med.*, **57**, 11–17.

Gajdusek, D. (1957). *Pediatrics*, **19**, 543–565.

Gigliotti, F. (1992). *J. Infect. Dis.*, **165**, 329–336.

Gigliotti, F., Harmsen, A., Haidaris, C., and Haidaris, P. (1993). *J. Clin. Microbiol.*, **61**, 2886–2890.

Giri, T. K., Pande, I., Mishra, N. M., Kailash, S., Uppal, S. S., and Kumar, A. (1995). *J. Commun. Dis.*, **27**, 131–141.

Goheen, M. P., Bartlett, M. S., Current, W. L., Shaw, M. M., and Smith, J. W. (1994). *J. Eukaryot. Microbiol.*, **41**, 89S.

Guss, D. A. (1994). *J. Emergency Med.*, **12**, 375–384.

Haase, G. (1997). *Infect. Immun.*, **65**, 4365–4366.

Harmsen, A. and Stankiewicz, M. (1990). *J. Exp. Med.*, **172**, 937–945.

Hauser, P. M., Francioli, P., Bille, J., Telenti, A., and Blanc, D. S. (1997). *J. Eukaryot. Microbiol.*, **44**, 16S.

Hennequin, C., Page, B., Roux, P., Legendre, C., and Kreis, H. (1995). *Eur. J. Clin. Microbiol. Inf. Dis.*, **14**, 122–126.

Hennessey, N. P., Parro, E. L., and Cockerell, C. J. (1991). *Arch. Dermatol.*, **127**, 1699–1701.

Heresi, G., Caceres, E., Atkins, J., Rueben, J., and Doyle, M. (1997). *Clin. Infect. Dis.*, **25**, 739–740.

Hirschl, M., Derfler, K., Janata, O., Heinz, G., Sertl, K., and Balcke, P. (1992). *Clin. Nephrol.*, **37**, 105.

Hong, S. T. (1991). *Kisaengchunghak Chapchi*, **29**, 355–361.

Hughes, W. T. (1982). *J. Infect. Dis.*, **145**, 842–848.

Jacobs, J., Libby, D., Winters, R., Gelmont, D., Fried, E., Hartman, B., and Lawrence, J. (1991). *N. Engl. J. Med.*, **109**, 246–250.

Jiang, B., Lu, J. J., Li, B., Tang, X., Bartlett, M. S., Smith, J. W., and Lee, C. H. (1996). *J. Clin. Microbiol.*, **34**, 3245–3248.

Kandil, O., Fishman, J., Koziel, H., Pinkston, P., Rose, R., and Remold, H. (1994). *Infect. Immun.*, **62**, 644–650.

Kaneshiro, E. S. (1998). *Int. J. Parasitol.*, **28**, 65–84.

Kaneshiro, E. S., Cushion, M. T., Walzer, P. D., and Jayasimhulu, K. (1989). *J. Protozool.*, **36**, 69S–72S.

Keely, S. C. and Stringer, J. R. (1997). *J. Clin. Microbiol.*, **35**, 2745–2747.

Keely, S. P., Baughman, R. P., Smulian, A. G., Dohn, M. N., and Stringer, J. R. (1996). *AIDS*, **10**, 881–888.

Keely, S. P., Stringer, J. R., Baughman, R. P., Linke, M. J., Walzer, P. D., and Smulian, A. G. (1995). *J. Infect. Dis.*, **172**, 595–598.

Krishnan, V., Meager, A., Mitchell, D., and Pinching, A. (1990). *Clin. Exp. Immunol.*, **80**, 156–160.

Latouche, S., Olsson, M., Polack, B., Brun-Pascaud, M., Bernard, C., and Roux, P. (1997a). *J. Eukaryot. Microbiol.*, **44**, 46S–47S.

Latouche, S., Ortona, E., Mazars, E., Margutti, P., Tamburrini, E., Siracusano, A., Guyot, K., Nigou, M., and Roux, P. (1997b). *J. Clin. Microbiol.*, **35**, 383–387.

Latouche, S., Poirot, J.-L., Bernard, C., and Roux, P. (1997c). *J. Clin. Microbiol.*, **35**, 1687–1690.

Latouche, S., Poirot, J. L., Lavrard, I., Miltgen, M., Nigou, M., and Roux, P. (1996). *J. Eukaryot. Microbiol.*, **43**, 56S–57S.

Latouche, S., Poirot, J. L., Maury, E., Bertrand, V., and Roux, P. (1997d). *AIDS*, **11**, 549.

Latouche, S., Roux, P., Poirot, J. L., Lavrard, I., Hermelin, B., and Bertrand, V. (1994). *J. Clin. Microbiol.*, **32**, 3052–3053.

Lee, C.-H., Lu, J.-J., Bartlett, M. S., Durkin, M., Liu, T.-H., Wang, J., Jiang, B., and Smith, J. (1993). *J. Clin. Microbiol.*, **31**, 754–757.

Leigh, T., Millet, M., Jameson, B., and Collins, J. (1993). *Thorax*, **48**, 619–621.

Lewin, S., Sonza, S., Irving, L., McDonald, C., Mills, J., and Crowe, S. (1996). *IDS Res. Hum. Retrovir.*, **12**, 877–883.

Lidman, C., Berglund, O., Tynell, E., and Lindback, S. (1992). *Scand. J. Infect. Dis.*, **24**, 157–160.

Limper, A. H. (1997). *Am. J. Resp. Cell Mol. Biol.*, **16**, 110–111.

Limper, A. H., Hoyte, J. S., and Standing, J. E. (1997). *J. Clin. Investig.*, **99**, 2110–2117.

Limper, A. H., Standing, J. E., and Hoyte, J. S. (1996). *J. Eukaryot. Microbiol.*, **43**, 12S.

Lipschik, G. Y., Treml, J. F., and Moore, S. D. (1996). *J. Eukaryot. Microbiol.*, **43**, 16S–17S.

Litwin, M. A. and Williams, C. M. (1992). *Ann. Intern. Med.*, **117**, 48–49.

Lu, J. J., Bartlett, M. S., Smith, J. W., and Lee, C. H. (1995). *J. Clin. Microbiol.*, **33**, 2973–2977.

Lucas, S., Goodgame, R., Kocjan, G., and Serwadda, D. (1988). *AIDS*, **3**, 47–48.

Lundgren, B., Elvin, K., Rothman, L., Ljungström, I., Lidman, C., and Lundgren, J. (1997). *Thorax*, **52**, 422–424.

Lundgren, J., Pedersen, C., Glumeck, N., Gatell, J., Johnson, A., Ledergerber, B., Vella, S., and Nielsen, J. (1994). *Br. Med. J.*, **308**, 1068–1073.

Masur, H., Ognibene, F., Yarchoan, R., Shelhamer, J., Baird, B., Travis, W., Suffredini, A., Deyton, L., Kovacs, J., Falloon, J., and al, e. (1989). *Ann. Intern. Med.*, **111**, 223–231.

McCormack, F. X., Festa, A. L., Andrews, R. P., Linke, M., and Walzer, P. D. (1997). *Biochemistry*, **36**, 8092–8099.

Meuwissen, J. H. and Leeuwenberg, A. D. (1972). *Trop. Geogr. Med.*, **24**, 282–291.

Meuwissen, J. H., Leeuwenberg, A. D., Heeren, J., and Stumpel, A. (1973). *J. Infect. Dis.*, **127**, 209–210.

Millard, P. and Heyret, A. (1988). *J. Pathol.*, **154**, 365–370.

Miller, R. F. and Wakefield, A. E. (1996). *J. Med. Microbiol.*, **45**, 233–235.

Montaner, J. S., Le, T., Hogg, R., Ricketts, M., Sutherland, D., Strathdee, S. A., O'Shaughnessy, M., and Schechter, M. T. (1994). *AIDS*, **8**, 693–696.

Murry, C. and Schmidt, R. (1992). *Hum. Pathol.*, **23**, 1380–1387.

Neese, L. W., Standing, J. E., Olson, E. J., Castro, M., and Limper, A. H. (1994). *J. Immunol.*, **152**, 4549–4556.

Ng, V. L., Yajko, D. M., and Hadley, W. K. (1997). *Clin. Microbiol. Rev.*, **10**, 401–418.

Olsson, M., Lidman, C., Latouche, S., Björkman, A., Roux, P., Linder, E., and Wahlgren, M. (1998). *J. Clin. Microbiol.*, **36**, 1737–1740.

Olsson, M., Sukura, A., Lindberg, L. A., and Linder, E. (1996). *Scand. J. Infect. Dis.*, **28**, 279–282.

Perenboom, R. M., Beckers, P., Van Der Meer, J. W., Van Schijndel, A. C., Oyen, W. J., Corstens, F. H., and Sauerwein, R. W. (1996). *J. Leukoc. Biol.*, **60**, 710–715.

Pesanti, E., Tomicic, T., and Donta, S. (1991). *J. Protozool.*, **38**, 28S–29S.

Pesanti, E. L. and Shanley, J. D. (1988). *J. Infect. Dis.*, **158**, 1353–1359.

Peters, S., Wakefield, A., Sinclair, K., Millard, P., and Hopkin, J. (1992). *J. Pathol.*, **166**, 195–198.

Pifer, L., Hughes, W., Stagno, S., and Woods, D. (1978). *Pediatrics*, **61**, 35–41.

Raviglione, M. (1990). *Rev. Infect. Dis.*, **12**, 1127–1138.

Rice, W. R., Singleton, F. M., Linke, M. J., and Walzer, P. D. (1993). *J. Clin. Invest.*, **92**, 2278–2782.

Robillard, G., Bertrand, R., Gregoire, H., Berdnikoff, G., and Favreau-Ethier, M. (1965). *J. Can. Assoc. Radiol.*, **16**, 161–168.

Roth, A., Wecke, J., Karsten, V., and Janitschke, K. (1997). *Parasitol. Res.*, **83**, 177–184.

Schneider, R. F. and Rosen, M. J. (1997). *Curr. Opin. Pulm. Med.*, **3**, 151–158.

Sepkowitz, K. (1993). *Clin. Infect. Dis.*, **17**, S416–S422.

Shibata, R., Maldarelli, F., Siemon, C., Matano, T., Parta, M., Miller, G., Fredrickson, T., and Martin, M. A. (1997). *J. Infect. Dis.*, **176**, 362–373.

Smulian, A., Sullivan, D., Linke, M., Halsey, N. A., Quinn, T. C., MacPhail, A. P., Hernandez-Avila, M. A., Hong, S. T., and Walzer, P. D. (1993). *J. Infect. Dis.*, **167**, 1243–1247.

Stenbäck, F., Dammert, K., and Räsänen, O. (1968). *Ann. Paediat. Fenn.*, **74**, 61–65.

Stringer, J. R. (1993). *Infect. Agents Dis.*, **2**, 109–117.

Su, T. H. and Martin, W. n. (1994). *Annu. Rev. Med.*, **45**, 261–272.

Sukura, A. (1995). *Apmis*, **103**, 300–306.

Sukura, A., Lindberg, L., and Soveri, T. (1991a). *Acta Vet. Scand.*, **32**, 135–137.

Sukura, A., Lindberg, L. A., Soveri, T., Guerrero, O., Chinchilla, M., Elvin, K., and Linder, E. (1991b). *Acta Vet. Scand.*, **32**, 135–137.

Sul, D. and Kaneshiro, E. S. (1997). *J. Eukaryot. Microbiol.*, **44**, 60S.

Telzak, E., Cote, R., Gold, J., Campbell, S., and Armstrong, D. (1990). *Rev. Infect. Dis.*, **12**, 380–386.

Travis, W., Hoffman, G., Leavitt, R., Pass, H., and Fauci, A. (1991). *Am. J. Surg. Pathol.*, **15**, 315–333.

Tsolaki, A., Miller, R., Underwood, A., Banerji, S., and Wakefield, A. (1996). *J. Infect. Dis.*, **174**, 141–156.

Vanek, J. and Jirovec, O. (1952). *Zentralbl. Bacteriol.*, **158**, 120–127.

Vargas, S. L., Hughes, W. T., Wakefield, A. E., and Oz, H. E. (1995). *J. Infect. Dis.*, **172**, 506–510.

Vavra, J. and Kucera, K. (1970). *J. Protozool.*, **17**, 463–483.

Vogel, P., Miller, C. J., Lowenstine, L. L., and Lackner, A. A. (1993). *J. Infect. Dis.*, **168**, 836–843.

Wakefield, A. and Miller, R. (1996). *J. Med. Microbiol.*, **45**, 233–235.

Wakefield, A. E. (1996). *J. Clin. Microbiol.*, **34**, 1754–1759.

Wakefield, A. E., Keely, S. P., Stringer, J. R., Christensen, C. B., Ahrens, P., Peters, S. E., Bille-Hansen, V., Henriksen, S. A., Jorsal, S. E., and Settnes, O. P. (1997). *Apmis*, **105**, 317–321.

Wakefield, A. E., Pixley, F., Banerji, S., Sinclair, K., Miller, R. F., Moxon, E. R., and Hopkin, J. M. (1990). *Mol. Biochem. Parasitol.*, **43**, 69–76.

Walzer, P., Kim, C., Linke, M., Pogue, C. L., Huerkamp, M. J., Chrisp, C. E., Lerro, A. V., Wixson, S. K., Hall, E., and Schultz, L. D. (1989). *Infect. Immun.*, **57**, 62–70.

Wehle, K., Schirmer, M., Dunnebacke, H. J., Kupper, T., and Pfitzer, P. (1993). *Cytopathology*, **4**, 231–236.

Yasuoka, A., Tachikawa, N., Shimada, K., Kimura, S., and Oka, S. (1996). *Clin. Diagnostic Lab. Immunol.*, **3**, 197–199.

Yu, M. L. and Limper A. H. (1997). *Am. J. Physiol.*, **273**, L1103–L1111.

13 *Toxoplasma gondii – biology and pathogenesis*

Eskild Petersen

History

Toxoplasma gondii is a coccidium with the domestic cat and other felids as its definitive host and a wide range of birds and mammals as intermediate hosts. It was first described by Nicolle and Manceaux (1908) from a rodent *Ctenodactylus gundi* and by Splendore (1908) in a rabbit. The name *Toxoplasma* is derived from the crescent shape of the tachyzoite (in Greek: toxo = arc, plasma = form). The knowledge of the full life cycle of *T. gondii* was not completed until 1970, when the sexual phase of the life cycle was identified in the intestine of the cat, by demonstrating oocysts in cat faeces and characterizing them biologically and morphologically (Dubey *et al.* 1970a,b).

Taxonomy

Toxoplasma gondii is placed in the phylum Apicomplexa, Levine 1970; class: Sporozoasida, Leukart 1879; subclass Coccidiasina, Leukart 1879. Traditionally, all coccidia until 1970 were classified into family Eimeriidae. After the discovery of the coccidian cycle of *T. gondii* in 1970, *T. gondii* has been placed in family Eimeriidae or Sarcocystidae or Toxoplasmatidae by different authorities.

Life cycle

The definitive host is the domestic cat and other felidae (Frenkel *et al.* 1970), where the sexual cycle takes place in the intestinal epithelial cells. Infected cats excrete oocysts which are infectious to virtually all warm blooded animals. There are three infectious stages of the parasite: the tachyzoite (the rapidly dividing form) in tissues, the bradyzoite (the slowly dividing form) inside cysts in tissues, and the sporozoites in the oocyst in cat faeces.

The enteroepithelial cycle in the definitive host – the cat

Five morphologically distinct asexual stages of *T. gondii* develop in entrocytes before gameto-gony begins (Dubey and Frenkel 1972). It is believed that merozoites develop into gamonts which occur throughout the small intestine, but are most prevalent in the ileum, where they are found 3–15 days after infection.

The microgamete (the male gamete) is biflagellate and fertilizes the macrogamete (the female gamete) within the enterocyte. Fertilization initiates oocyst wall formation. The oocyst is the developed zygote which is the product of the sexual reproduction through the fertilization of the macrogamete by the microgamete. The oocysts are discharged into the intestinal lumen by rupture of the epithelial cells, and thereafter excreted in cat faeces. The oocyst sporulates within 1–5 days after excretion, depending on aeration, humidity, and temperature, by dividing into two sporocysts. Each sporocyst contains four sporozoites. Thus, there are eight sporozoites in one oocyst. The sporulated oocyst can remain infectious in the environment for months, even years, in cold and dry climates (reviewed by Dubey 1977).

The prepatent period (interval between ingestion and shedding of oocysts) after the ingestion of tissue cysts is 3–10 days, with the peak oocyst production between five and eight days after a patent period varying from 7 to 20 days (Dubey and Frenkel 1972, 1976). Cats not previously infected with *T. gondii* shed oocysts after ingesting each of the infective stage of the parasite: the tachyzoite, the bradyzoite, and the sporozoite (Frenkel *et al.* 1970; Dubey and Frenkel 1976). The prepatent period varies according to which stage of *T. gondii* the cat is infected with, with a short (3–10 days) prepatent period when the oral inoculum contained bradyzoites and a long prepatent period (>21 days) when the inoculum contained tachyzoites or sporozoites (Freyre *et al.* 1989). Cats previously infected with *T. gondii*, and which produced oocysts during the previous infection, are generally immune against renewed oocyst shedding, but immunity is not life long (Frenkel and Smith 1982; Dubey 1995).

The asexual cycle in the definitive host – the cat

Bradyzoites released from the enteroepithelial cycle penetrate the lamina propria below the epithelial cell in the intestine of the cat and multiply as tachyzoites. The tachyzoites are disseminated throughout the body within a few days, eventually encysting in tissues. The extraintestinal cycle in the cat differs from the similar cycle in non-feline intermediate host in two aspects: (1) tachyzoites have not been demonstrated in feline intestinal epithelial cells, whereas they do occur in non-feline intermediate hosts (Dubey and Frenkel 1973) and (2) the enteroepithelial types of *T. gondii* are non-infectious to mice by any route (Dubey and Frenkel 1976), which suggests that the feline enteroepithelial forms do not give rise to tachyzoites.

Intermediate host

Toxoplasma gondii tachyzoites are disseminated throughout the body of the intermediate host in macrophages, lymphocytes, and are free in the plasma. Tachyzoites continue to divide within the host cell by endodyogeny (internal division into two) until the host cell is filled with parasites. At a given time when the dividing tachyzoites cannot be contained with the host cell which burst, the tachyzoites are released and seek new host cells to repeat the process. Depending on the strain of *T. gondii* and the host resistance, tachyzoites may be found for days, or even months, after acute infection. For example, tachyzoites persist in foetal membranes for weeks after infection of the mother or the dam, and are nearly always present in placentas of mothers at the time of parturition, if the foetus was infected *in utero*. Some time after infection the tachyzoites transform to bradyzoites in tissue cysts. The signals responsible for the transformation are not known, and it is still debated whether signals from the host immune system are needed. Bradyzoites also divide by endodyogeny.

Bradyzoites are enclosed in a thin cyst wall. Tissue cysts may be found as early as three days after infection but are usually not numerous until seven weeks after infection (Derouin and Garin 1991; Dubey and Frenkel 1976). Intact tissue cysts do not probably cause any inflammation and may persist for life. It has been suggested that tissue cysts may switch from the bradyzoite stage to the tachyzoite stage throughout the life of the tissue cysts, producing new tachyzoites which may give rise to new tissue cysts thus ensuring a prolonged infective stage (Hérion and Saavedra 1993). If the intermediate host is eaten by another warm-blooded animal, tissue cysts are able to infect a new host.

Less than 50% of cats shed oocysts after ingesting tachyzoites or oocysts, whereas almost all cats shed oocysts after ingesting tissue cysts (Dubey and Frenkel 1976). Cats infected with oocysts and tachyzoites probably give rise to bradyzoites, which after a variable time may disseminate to the intestinal mucosa and start the enteroepithelial cycle with the resulting production of oocysts (Freyre *et al.* 1989).

Morphology and ultrastructure

The tachyzoite

The tachyzoite is crescent shaped and is approximately $2 \times 6 \, \mu m$ in size. The tachyzoite has a pellicle, sub-pellicular microtubules, a polar ring, a conoid, rhoptries, micronemes, mitochondria, endoplasmatic reticulum, Golgi apparatus, ribosomes, rough surface endoplasmatic reticulum, micropores, apicoplast, and a well defined nucleus.

The nucleus is situated in the central or posterior part of the cell. The pellicle consists of three membranes. The polar ring encircles the conoid, a cylindrical cone which consists of six to eight fibrillar elements arranged like a compressed spring. The 22 sub-pellicular microtubules originate from the polar ring and run longitudinally for almost the entire length of the cell (Sulzer *et al.* 1974) and probably provide a frame for the parasite.

The rhoptries are 4–10 club-shaped, gland-like structures with an anterior narrow neck and posterior sac-like end reaching as far as the nucleus. The rhoptries have a secretory function associated with host cell penetration. When the parasite has attached to the host cell, the contents of the rhoptries are discharged through the conoid (Nichols *et al.* 1983). The micronemes are rice-grain-like structures, usually less than 100 in number, situated at the conoidal end of *T. gondii* without any defined function but may participate in the invasion of the host cell (Joiner and Dubremetz 1993). In addition to the rhoptries and the micronemes, the parasite contains dense granules which also appears to have a secretory function (Charif *et al.* 1990).

The functions of the conoid, rhoptries, and micronemes are not fully known. The conoid can rotate, extend, and retract and is important when the parasite searches for an attachment site at the host cell, as the parasite can rotate, glide, and twist. Myosin has been found in the apical end of the parasite (Schwartzman and Pfefferkorn 1983), and actin has been found both at the apical end and distributed throughout the cytoplasm (Endo *et al.* 1988). The motion observed during parasite entry corresponds to the orientation of the sub-pellicular microtubules, and it is likely that the microtubules are the basis of the motility system.

After entry into the host cell, the parasite is surrounded by a parasitophorous vacuole membrane (PVM). The PVM contains numerous intra-vacuolar tubules (Sibley *et al.* 1985; Sibley and Krahenbuhl 1988). *Toxoplasma gondii* enters the host cell by active invasion (Werk 1985).

Endodyogeny is a process in which two progenies form within *T. gondii* and consume it from within. The Golgi apparatus divides first, and the anterior cell membranes of the progenies are formed at the anterior end. The nuclear membranes remains intact and the chromosomes do not condense at metaphase. The progenies move towards the cell membrane of the parent parasite as they continue to grow, and at last the inner membrane of the parent parasite disappears and the outer membrane fuses with the inner membrane of the progenies, and two new tachyzoites are formed.

The bradyzoite and tissue cysts

The bradyzoite (brady = slow) is the organism dividing slowly within a tissue cyst (Frenkel 1973) and is a synonym of cystozoite. A tissue cyst is a collection of bradyzoites surrounded by a well-defined host cell membrane. The bradyzoites are slender and measure approximately $7 \times 1.5 \ \mu m$ (Mehlhorn and Frenkel 1980). The bradyzoites also multiply by endodyogeny. Tissue cysts range from 5 to 60 μm in the brain and 100 μm in other tissues (Dubey 1993) and contain four to several hundred bradyzoites.

Tissue cysts may develop in any tissue but are most prevalent in neural and muscular organs, like eye and brain, skeletal and cardiac muscles. The tissue cyst develops in the host cell cytoplasm and its wall is partly of host origin (Ferguson and Hutchison 1987; Sims *et al.* 1988). In older cysts, degenerating bradyzoites may occasionally be found (Pavesio *et al.* 1992).

Bradyzoites differ only slightly from the tachyzoites. They are more slender than tachyzoites and their nucleus is located more to the posterior end compared to that of the tachyzoites. The contents of the rhoptries of bradyzoites are electron dense in older cysts (Ferguson and Hutchison 1987). The prepatent period in cats following infection by bradyzoites is shorter (3–10 days) than following infection with tachyzoites (21 days or more) (Dubey and Frenkel 1976). The transition from tachyzoites to bradyzoites can be observed *in vitro*, and occur through an intermediate stage which expresses both usually exclusive tachyzoite and bradyzoite antigens (Tomavo *et al.* 1991; Bohne *et al.* 1992; Soete *et al.* 1993). The external mechanisms behind stage conversion from tachyzoites to bradyzoites are not known, but studies have shown that nitric oxide and IFNγ may be among other factors involved (Bohne *et al.* 1994).

Cyst formation and rupture

The factors that influence tissue cyst formation and induce differentiation of different stages are not known in detail. The transition from tachyzoite to bradyzoite is followed by a shift in stage-specific antigens (Bohne *et al.* 1993). Nitric oxide production by *T. gondii*-infected macrophages seem an important mediator in reducing parasite multiplication and initiation of bradyzoite formation (Bohne *et al.* 1994).

Tissue cysts are more numerous in animals in the chronic stage of infection with developed immunity than in animals in the acute stage of infection, but tissue cysts have been found in mice infected for only three days (Dubey and Frenkel 1976) and in tissue cell culture systems not influenced by any immune mechanism (Lindsay *et al.* 1991, 1993). The role of the developing immune response in the formation of tissue cysts is therefore still undecided. The cyst contains both host-derived and *T. gondii* antigens (Sims *et al.* 1988). The factors that determine the rupture of the cyst are unknown. Parasites can be released by the calcium ionophore, monensin (Endo *et al.* 1982), which implicates a role of calcium. One theory is that cysts rupture from time to time and the released bradyzoites are destroyed by host immune response or give rise to new cysts or tachyzoites, depending on the immune status of the host. Chronic infection can be relapsed reliably in rodents by corticosteroids or by relapsing, depleting, or interfering with cytokine production (Gazzinelli *et al.* 1993).

Enteroepithelial stages

The gametes contain the microgamete (the male gamete) and the macrogamete (the female gamete). The oocyst is the developed zygote which is a product of the sexual reproduction through the fertilization of the macrogamete by the microgamete. When excreted in cat faeces, the oocyst contains a sporont, which sporulates within 24–72 h after excretion by dividing into two round masses called sporoblasts. The sporoblasts elongate and mature into sporocysts, and within each sporocyst four sporozoites develop. Thus the sporulated oocysts contain eight sporozoites.

Five morphologically distinct stages (types) of *T. gondii* develop in the intestinal epithelial cells before gametogony begins (Dubey and Frenkel 1972). With repeating nuclear division, merozoite formation starts with the development of an anterior membrane complex near the nucleus. Immature merozoites are formed when the new nuclei are included in the newly formed anterior membrane complex. The immature merozoites lie near the outer membrane of the mature meront. The outer membrane invaginates around each merozoite, and the now mature merozoites are released from the meront resulting in host cell death. Released mature merozoites initiate gametogony. The gamonts are found 3–15 days after infection in the ileum.

The macrogamete is sub-spherical and contains a single centrally located nucleus and several PAS-positive granules (Ferguson *et al.* 1975).

During microgametogenesis the nucleus of the male gamonts divides to produce 10–21 nuclei. Each microgamete is biflagellate, appear laterally compressed, and has a large mitochodrium. Microgametes are fewer in number than macrogametes. The microgamete swims to the mature macrogamete, attach by the perforatorium, and penetrate the cell membrane, and we now have the fertilized gamete, the oocyst (Dubey and Frenkel 1972).

Oocyst wall formation is initiated by fertilization, and five layers are formed around the pellicle of the gamete (Ferguson *et al.* 1975). The oocysts are now discharged into the intestinal lumen by rupture of the epithelial cells. The mature, unsporulated oocyst, the sporont, is spherical and 10–12 µm in diameter. The wall consists of two colourless layers, and the sporont almost fills the oocyst.

Sporulation

Sporulated oocysts are 11 × 13 µm in diameter and each sporulated oocyst contains two ellipsoidal sporocysts, which measure 6 × 8 µm each. Each sporocyst contains four sporozoites, which are 2 × 6–8 µm in size. Ultrastructurally the sporozoite is similar to the tachyzoite, except that there is an abundance of micronemes and rhoptries in the former. There is no crystalloid body, nor are there any refractile bodies in the sporozoites, which are found in conventional coccidian sporozoites (Dubey 1993).

Genome and antigenic structure

Genome

Toxoplasma gondii is haploid except during the sexual division in the intestine of the cat. Sporozoites are the results of meiosis and seem to follow classical Mendelian laws (Pfefferkorn *et al.* 1977). The total haploid genome contains approximately 8×10^7 bp (Cornelissen *et al.* 1984), and a 36 kb circular mitocondrial DNA (Borst *et al.* 1984), which has been partly sequenced (Ossorio *et al.* 1991). Nine chromosomes have been identified by pulse-field gel electrophoresis, and a molecular karyotype constructed by using probes from low-copy number genes (Sibley and Boothroyd 1992a). The tubulin genes have been described, and both contain introns (Boothroyd *et al.* 1987; Nagel and Boothroyd 1988). Only the B1 gene has been found to be tandemly repeated (Burg *et al.* 1989), but other sequences repeated many times have been identified (McLeod *et al.* 1991; Blanco *et al.* 1992). The *T. gondii* DNA has been characterized and a genetic nomenclature for *T. gondii* has been proposed (Sibley *et al.* 1991).

Toxoplasma gondii rRNA has the usual large and small sub-units (Gagnon *et al.* 1993). Sequence analysis of the small sub-unit rRNA suggests that *T. gondii* is phylogenetically related to *Sarcocystis*, but separate from *Plasmodium* (Guay *et al.* 1993).

In 1997 the third DNA-containing organelle, the apicoplast, was described in apicomplexan parasites including *T. gondii* and *Plasmodia* spp. (Fichera and Roos 1997). The apicoplast contains several novel enzymatic pathways, which may be potential new drug targets (Roos 1999; Soldati 1999).

Antigens

Two-dimensional gel electrophoresis has identified more then 1 000 spots after ^{35}S-methionine labelling of tachyzoites cultured *in vitro* (Handman *et al.* 1980), which probably reflects at least the number of different proteins present. The different stages of the parasite share common antigens and express stage-specific antigens, both between oocysts and other stages (Kasper and Ware 1985) and between bradyzoites and tachyzoites (Woodison and Smith 1990).

The antigens involved in the attachment of *T. gondii* to the host cell, and antigens involved in penetration of the host cell and formation of the parasitophorous vacuole have received special attention (Cesbron-Delauw *et al.* 1989; Charif *et al.* 1990; Achbarou *et al.* 1991; Leriche and Dubremetz 1991; Saavedra *et al.* 1991; Ossorio *et al.* 1992).

The most abundant surface antigen is a 30-kDa protein (Burg *et al.* 1988). The P30-antigen (surface antigen 1, SAG1) appears to be conserved between different *T. gondii* isolates (Bülow and Boothroyd 1991; Sibley and Boothroyd 1992b). The SAG1 is found only in tachyzoites, where it constitutes up to about 5% of the total tachyzoite protein. It is distributed evenly on the surface of the tachyzoite and in the tubular network of the parasitophorous vacuole and is shed from the surface of the parasite at the moving junction between the parasite and the host cell during the invasion process (Dubremetz *et al.* 1985). A *T. gondii* mutant of the RH strain lacking the SAG1 surface antigen has been described (Kasper 1987), which is able to grow in continuous *in vitro* culture. Surface neoglycoproteins have been identified (Robert *et al.* 1991), but their role in antigenicity and attachment/invasion is still unknown.

A 22-kDa SAG2 has been sequenced (Prince *et al.* 1990), and a 43-kDa SAG3 has recently been cloned and sequenced and has been found to have structural similarities to SAG1 (Cesbron-Delauw *et al.* 1994). SAG4, is a 18-kDa bradyzoite-specific antigen (Ödberg-Ferragut *et al.* 1996). The first-sequenced bradyzoite-specific antigen was a 30-kDa antigen belonging to the heat shock protein family (Bohne *et al.* 1995).

The major surface proteins of 43-, 35-, 30-, and 22 kDa are anchored by a glycosylphosphatidyl-inositol anchor (Nagel and Boothroyd 1988; Tomavo *et al.* 1989). Early work suggested the presence of a 'penetration-enhancing factor' secreted from the rhoptries (Lycke and Norrby 1966); several rhoptry proteins (ROP) have been identified later: ROP1, which is a 61-kDa rhoptry protein (Schwartzman and Krug 1989; Ossorio *et al.* 1992),

ROP2, 3, and 4 (Gelder *et al.* 1993), ROP5 (Leriche and Dubremetz 1991), ROP6 (Dubremetz *et al.* 1987), and ROP7 (Leriche and Dubremetz 1991). ROP2 has recently been cloned and sequenced as a 54-kDa protein (Saavedra *et al.* 1991).

Five dense granule proteins have been identified, and have been shown to associate with the intravacuolar network: GRA1, 2, and 4 (Prince *et al.* 1989) and the PVM's, GRA1 and 3 (Achbarou *et al.* 1991).

Excreted–secreted antigens (ESA) (Decoster *et al.* 1988; Cazabonne *et al.* 1994), are secreted from the rhoptries and the dense granules. At least three ESA's are located in the dense granules, GRA1(P23), GRA2(P28.5) and GRA3(P21), and are released inside the parasitophorous vacuole (Leriche and Dubremetz 1990, 1991). The GRA3 antigen is inserted into the PVM after secretion (Ossorio *et al.* 1994), and GRA2 and GRA5 are associated with the cyst wall (Torpier *et al.* 1993). The 23-kDa calcium-binding protein is also found in the parasitophorous vacuole of the host cell (Cesbron-Delauw *et al.* 1989). The ESA's were reviewed by Cesbron-Delauw and Capron (1993). Different alleles have been demonstrated for the SAG1, GRA2, and GRA4 antigens thus far (Meisel *et al.* 1996).

Bradyzoite-specific antigens are just beginning to be described. One of the first was bradyzoite antigen 1, BAG1, belonging to the heat shock protein family (Bohne *et al.* 1995); the first surface antigen expressed exclusively in brazyzoites was named SAG4 (Ödberg-Ferragut *et al.* 1996).

Biochemistry

Early work on the biochemistry of coccidia has been reviewed by Wang (1982). None of the coccidia, including *T. gondii*, have been grown in cell-free medium. *Toxoplasma gondii* is incapable of purine synthesis and depends entirely on the host cell for preformed purines (Pfefferkorn 1990). Several purine and pyrimidine salvage enzymes have been identified (Pfefferkorn 1990; Iltzsch 1993; Manafi *et al.* 1993). *Toxoplasma gondii* appears only to be able to salvage uracil, and although it can convert thymine to thymidine, there is no salvage pathway for thymidine (Iltzsch 1993; Pfefferkorn and Pfefferkorn 1977).

Toxoplasma gondii cannot use preformed folates as mammalian cells can; the dihydrofolate reductase (DHFR) enzyme is therefore a major target for antibiotic agents against *T. gondii* (Derouin and Chastang 1989; Roos 1993). Little is known about the lipids and carbohydrates of *T. gondii*. The parasite has a low cholesterol/phospholipid ratio, many unsaturated fatty acid chains, and large amounts of phosphatidylcholine (Gallois *et al.* 1988). A 6-kDa carbohydrate antigen seems to be responsible for early *T. gondii*-specific IgM antibody production, and it appears that *T. gondii* is capable of both N- and O-glycosylation (Schwartz and Tomavo 1993).

Pathogenesis

Host cell invasion

Early work on invasion of the host cell has been reviewed by Werk (1985), and the invasion process has recently been reviewed by Schwartzman and Saffer (1992), Dubremetz and Schwartzman (1993), and Kasper and Mineo (1994). The invasion of the host cell is an active process involving attachment, host cell membrane penetration, formation of a moving junction with the host cell membrane, formation of the PVM, and subsequent closure of the host cell membrane after entry. The invasion process is calcium dependent (Bonhomme *et al.* 1993).

The major *T. gondii* surface protein, SAG1, has been implicated in attachment and penetration (Grimwood and Smith 1992; Mineo *et al.* 1993), and host cell laminin is also involved in attachment (Joiner 1991b; Furtado *et al.* 1992a,b). There are higher concentrations of laminin-binding proteins (Joiner *et al.* 1989) and Fc-binding sites at one end of the tachyzoite (Budzko *et al.* 1989). Considering the wide range of animal hosts, it is likely that *T. gondii* has several molecules which can be important in cell adhesion (Dubremetz and Schwartzman 1993; Mineo *et al.* 1993).

After entry into a host cell, *T. gondii* is surrounded by the PVM, partly derived from the host cell membrane and partly containing parasite material. The PVM lacks host cell plasma membrane markers, and freeze fracture analysis indicates that the PVM may completely lack intramembranous particles and may consist of only a phospholipid bilayer. The dense granules probably contribute to intravacuolar tubules (Achbarou *et al.* 1991; Dubremetz *et al.* 1993). The main phospholipid of the rhoptries is phosphatidylcholine (Foussard *et al.* 1991), which may be identical to the phospholipids found in the PVM (Joiner 1991a). *Toxoplasma gondii* phospholipase has been suggested to play a role in invasion, possibly by softening of the host cell membrane after attachment (Saffer and Schwartzman 1991). The micropore is formed by an invagination of the outer membrane of the pellicle (Nichols and Chiappino 1987).

Toxoplasma gondii may infect almost any host cell, but it seems to have a preference for monocytes, macrophages, and muscle cells. In the central nervous system *T. gondii* can infect and grow in neurons, astrocytes, microglia, and oligodendrocytes (Fischer *et al.* 1997).

Host pathogenicity

When dealing with pathogenicity and virulence, only natural infection by the oral route is considered, where infection is acquired from oocysts or tissue cysts, and not artificial infection routes like intraperitoneal, subcutaneous, or intracerebellar inoculation, unless specifically mentioned. Pathogenicity is determined by the virulence of the strain and the susceptibility of the host species, usually the mice (Ferguson and Hutchison 1981; Suzuki *et al.* 1989).

Toxoplasma gondii usually infect the host without producing any clinical signs. After ingestion, bradyzoites, or sporozoites penetrate the intestinal wall, often multiplying in cells lining the lamina propria and epithelium. However, infection may spread to other tissues within a short time after ingestion; *T. gondii* was isolated from mesenteric lymph nodes of cats 4 h after feeding on tissue cysts (Dubey and Frenkel, 1972). Infection is disseminated to distant organs through the blood and lymphatics.

An infected host may die from necrosis of the intestine and mesenteric lymph nodes before other organs are severely damaged (Dubey and Frenkel 1973), or focal necrosis may develop in many organs. The clinical picture is determined by damage to different organs, especially organs such as the eye, the heart, and the adrenals. Necrosis is caused by the intracellular growth of tachyzoites. *Toxoplasma gondii* does not produce a toxin.

Usually by about the third week after infection, the tachyzoites begin to disappear from the visceral tissues (in mice), and tissue cysts are found in increasing numbers in neural and muscular tissues. Tachyzoites may persist longer in the spinal cord and brain because immunity is less effective in neural organs than in visceral tissues.

The intracellular tachyzoite, and later the bradyzoite, are found within the parasitophorous vacuole within the host cell endocytic system. The host cell is unable to fuse the lysosomes and the parasitophorous vacuole, which explains the ability of the parasite to survive intracellularly (Joiner *et al.* 1990). *Toxoplasma gondii* resist the phagosomes by creating a membranous network between the parasite and the host cell (Sibley and Krahenbuhl 1988).

How *T. gondii* survives intracellularly is not completely known. Immunity to *T. gondii* is mainly cell mediated and this subject was recently reviewed by Gazzinelli *et al.* (1993). The fate of tissue cysts is not fully known. It has been proposed that tissue cysts may at times rupture during life of the host. The released bradyzoites may be destroyed by the host's immune response (Frenkel 1990), which may cause focal necrosis and inflammation, but hypersensitivity may also play a role in such reactions. The rupture of tissue cysts is rarely observed histologically (Ferguson *et al.* 1989).

Virulence and strain differences

Strain differentiation and population biology

Restriction fragment length polymorphism (RFLP) provides a method for identifying genomic DNA differences between parasites of different isolates or between or among organisms of phylogenic proximity, and has also been applied to *T. gondii* (Cristina *et al.* 1991a,b; Sibley and Bootroyd 1992b)

Using RFLP on polymerase chain reaction-amplified specific single copy genes obtained from different RH lines showed unique patterns, except in three isolates cloned from the same line (Howe and Sibley 1994).

Sequencing of the *sag*1 gene after amplification has shown that mouse-virulent and -avirulent isolates can be distinguished by mutations in the 3′ region (Rinder *et al.* 1995)

Strain differences between different isolates of *T. gondii* have been established by isoenzyme electrophoresis (Barnert *et al.* 1988; Dardé *et al.* 1988) and by immunological methods (Ware and Kasper 1987), and the relationship between isoenzyme pattern (zymodemes) and pathogenicity (virulence) has been described (Dardé *et al.* 1988, 1992). Furthermore, differences in chromosome size between different isolates have been demonstrated by pulse-field gel electrophoresis (Candolfi *et al.* 1989).

Strain-specific differences between different isolates of *T. gondii* have been demonstrated by immunoblot, immune-precipitation with antisera, isoenzyme analysis, and DNA typing techniques (Ware and Kasper 1987; Barnert *et al.* 1988; Weiss *et al.* 1988; Bülow and Boothroyd 1991). Polymorphism seems to be limited and for the few single loci examined in detail, only two alleles have been identified (Boothroyd and Sibley 1993). Virulent stains can now be differentiated from avirulent strains by their reactivity with certain monoclonal antibodies (Gross *et al.* 1991; Bohne *et al.* 1993), and random amplified polymorphic DNA (RAPD) polymerase chain reaction can distinguish between mouse-virulent and -avirulent strains (Guo *et al.* 1997).

Recent evidence suggests that the host may harbour different strains of *T. gondii*, and it has been hypothesized that only those hosts harbouring strains with the ability to induce *Toxoplasma* encepalitis are able to develop this manifestation during immunosuppression (Araujo *et al.* 1997).

Virulence

Different strains of *T. gondii* cause different degrees of pathogenecity in different hosts (Suzuki *et al.* 1989, 1993). Generally, humans, cattle, horses, rats, and old-world monkeys belong to the resistant species, whereas mice, guinea pigs, hamsters, and new-world monkeys are sensitive (Darcy and Zenner 1993).

Virulence of *T. gondii* has been traditionally measured in a susceptible host (e.g. mice). Based on studies in mice, some *T. gondii* strains are considered more virulent than others. Virulence is influenced by the stage of the parasite (tachyzoite, bradyzoite, or sporozoite), the route of inoculation (oral, intraperitoneal, and sub-cutaneous), and the susceptibility of the host. In mice, oocysts from the M-7741 isolate needed an inoculum size approximately 10–100 times lesser than tissue cysts, producing both earlier symptoms and more deaths than tissue cysts with inoculums of the same size (Dubey and Frenkel 1973). However, according to Dubey and Beattie (1988) there are no truly avirulent strains of *T. gondii*: 100 000 oocysts of all strains of *T. gondii* tested were lethal to mice by the oral route.

More severe infections are found in pregnant or lactating mice than in non-lactating mice. Concomitant infection may make the host more susceptible or resistant to *T. gondii* infection (Remington 1970). New-born kittens were more prone to severe toxoplasmosis and death compared to adult, non-immune cats (Dubey and Frenkel 1972).

The most well-known virulent RH strain of *T. gondii* was isolated in 1939 from a 6-year-old boy (with initials R. H.) in mice (Sabin 1941). Five of the eight mice inoculated with the brain of this boy died within 21 days and three mice were not infected. Thus, the RH strain was virulent for mice on its first isolation. Although the RH strain of *T. gondii* can be virulent in many hosts, including humans, it is avirulent for adult rats and adult dogs (Dubey and Beattie 1988). The virulence of the RH strain has been changed after passages in mice (Yano and Nakabayashi 1986). Virulence in mice has been linked to expression of the SAG1, where it was found that virulent stain expressed higher amounts of SAG1 compared to avirulent isolates, and this was related to the number of a 27-bp repeat in the SAG1-promoter region (Windeck and Gross 1996). Others have found that the virulent RH isolate has increased levels of DNA-polymerase activity compared to mice -avirulent isolates (Makioka and Ohtomo 1995). High expression of a heat shock protein, HSP, has been linked to virulence (Lyons and Johnson 1995).

Mice with severe combined immuno deficiency, SCID, or nude mice die from acute infection with *T. gondii*, but survive if reconstituted with spleen cells from immune mice or if kept covered with sulfadiazine (Johnson 1992).

Natural resistance and host specificity

Toxoplasma gondii strains may vary in their pathogenicity in a given host. Certain strains of mice are more susceptible than others (Suzuki *et al.* 1993; McLeod *et al.* 1984), and appear to be regulated, at least in part, by H-2- and H-13-linked genes (Jones and Erb 1985). However, more recently, genes of the H-2 and D/L locus have been implicated in the regulation of the development of brain cysts and *T. gondii* encephalitis (Brown and McLeod 1990; Suzuki *et al.* 1991). At least five genes are involved in the regulation of *T. gondii* infection in mice (McLeod *et al.* 1989; Blackwell *et al.* 1993). In mice *T. gondii* can be transferred during pregnancy from the mother to the litter (Beverley 1959); *Toxoplasma gondii* infection in rhesus monkeys has been described as a model for congenital toxoplasmosis (Schoondermark *et al.* 1993).

Most mammals and birds can be infected with *T. gondii*. The natural resistance to infection varies between species. In general, mice are more susceptible than rats are (Jacobs 1956). The severity of infection in individual mice within the same strain may vary (Suzuki *et al.* 1989), and certain host species are genetically resistant to clinical toxoplasmosis.

For example, adult rats do not become ill while young rats can die because of toxoplasmosis. Mice of any age are susceptible to clinical *T. gondii* infection. Adult dogs are resistant whereas puppies are fully susceptible. Cattle and horses are amongst the most-resistant intermediate hosts for clinical toxoplasmosis, in contrast to certain marsupials and new-world monkeys, which are the most susceptible to infection (Dubey and Beattie 1988).

References

Achbarou, A., Mercereau-Puijalon, O., Sadak, A., Fortier, B., Leriche, M. A., Camus, D., and Dubremetz, J. F. (1991). Differential targetting of dense granule proteins in the parasitophorous vacuole of *Toxoplasma gondii*. *Parasitology*, **103**, 321–329.

Araujo, F., Slifer, T., and Kim, S. (1997). Chronic infection with *Toxoplasma gondii* does not prevent acute disease or colonization of the brain with tissue cysts following reinfection with different strains of the parasite. *J. Parasitol.*, **83**, 521–522.

Barnert, G., Hassl, A., and Aspöck, H. (1988). Isoenzyme studies on *Toxoplasma gondii* isolates using isoelectric focusing. *Z. Bakteriol. Hyg. A*, **268**, 476–481.

Beverley, J. K. A. (1959). Congenital transmission of toxoplasmosis through successive generations of mice. *Nature*, **183**, 1348–1349.

Blackwell, J. M., Roberts, C. W., and Alexander, J. (1993). Influence of genes within the MHC on mortality and brain cyst development in mice infected with *Toxoplasma gondii*: kinetics of immune regulation in BALB H-2 congenic mice. *Parasite Immunol.*, **15**, 317–324.

Blanco, J. C., Angel, S. O., Maero, E., Pszenny, V., Serpente, P., and Garberi, J. C. (1992). Cloning of repetitive DNA sequences from *Toxoplasma gondii* and their usefulness for parasite detection. *Am. J. Trop. Med. Hyg.*, **46**, 350–357.

Bohne, W., Gross, U., and Heesemann, J. (1993). Differentiation between mouse-virulent and -avirulent strains of *Toxoplasma gondii* by a monoclonal antibody recognizing a 27-kilodalton antigen. *J. Clin. Microbiol.*, **31**, 1641–1643.

Bohne, W., Heesemann, J., and Gross, U. (1992). Coexistence of heterogeneous populations of *Toxoplasma gondii* parasites within parasitophorous vacuoles of murine macrophages as revealed by a bradyzoite-specific monoclonal antibody. *Parasitol. Res.*, **79**, 485–487.

Bohne, W., Gross, U., Ferguson, J. P., and Heesemann, J. (1995). Cloning and characterization of a bradyzoite-specifically expressed gene (hsp30/bag1) of *Toxopalsma gondii*, related to genes encoding small heat-shock proteins of plants. *Mol. Microbiol.*, **16**, 1221–1230.

Bohne, W., Heesemann, J., and Gross, U. (1994). Reduced replication of *Toxoplasma gondii* is necessary for induction of bradyzoite-specific antigens: a possible role for nitric oxide in triggering stage conversion. *Infect. Immun.*, **62**, 1761–1767.

Bonhomme, A., Pingret, L., Bonhomme, P., Michel, J., Balossier, G., Lhotel, M., Pluot, M., and Pinon, J. M. (1993). Subcellular calcium localization in *Toxoplasma gondii* by electron microscopy and by X-ray and electron energy loss spectroscopies. *Microsc. Res. Tech.*, **25**, 276–285.

Boothroyd, J. C., Burg, J. L., Nagel, S. D., Perelman, D., Kasper, L. H., Ware, P. L., Prince, J. B., Sharma, S. D., and Remington, J. S. (1987). Antigen and tubulin genes of *Toxoplasma gondii*. In: *Molecular strategies of parasitic invasion* (N. Agabian, H. Goodman, and N. Nogueira, eds). A. R. Liss, New York, pp. 237–250.

Boothroyd, J. C. and Sibley, L. D. (1993). Population biology of *Toxoplasma gondii*. *Res. Immunol.*, **144**, 14–16.

Borst, P., Overdulve, J. P., Weijers, P. J., Fase-Fowler, F., and van den Burg, M. (1984). DNA circles with cruciforms from *Isospora (Toxoplasma) gondii*. *Biochem. Biophys. Acta*, **781**, 100–111.

Brown, C. R. and McLeod, R. (1990). Class I MRC genes and CD8+ T cells determine cyst number in *Toxoplasma gondii* infection. *J. Immunol.*, **145**, 3438–3441.

Budzko, D. B., Tyler, L., and Armstrong, D. (1989). Fc receptors on the surface of *Toxoplasma gondii* trophozoites: a confounding factor in testing for anti-Toxoplasma antibodies by indirecimmunofluorescence. *J. Clin. Microbiol.*, **27**, 959–961.

Bülow, R. and Boothroyd, J. C. (1991). Protection of mice from fatal *Toxoplasma gondii* infection by immunization with p30 antigen in liposomes. *J. Immunol.*, **147**, 3496–3500.

Burg, J. L., Grover, C. M., Pouletty, P., and Boothroyd, J. C. (1989). Direct and sensitive detection of a pathogenic protozoan, *Toxoplasma gondii*, by polymerase chain reaction. *J. Clin. Microbiol.*, **27**, 1787–1792.

Burg, J. L., Perelman, D., Kasper, L. H., Ware, P. L., and Bootroyd, J. C. (1988). Molecular analysis of the gene encoding the major surface antigen of *Toxoplasma gondii*. *J. Immunol.*, **141**, 3584–3591.

Candolfi, E., Arveiler, B., Mandel, J. L., and Kien, T. (1989). Structure du genome de *Toxoplasma gondii*. Premiers résultats. *Bull. Soc. Française Parasitol.*, **7**, 27–32.

Cazabonne, P., Bessieres, M. H., and Seguela, J. P. (1994). Kinetics study and characterisation of target excreted/secreted antigens of immunoglobulin G, M, A and E antibodies from mice infected with different strains of *Toxoplasma gondii*. *Parasitol. Res.*, **80**, 58–63.

Cesbron-Delauw, M.-F. and Capron, A. (1993). Excreted/secreted antigens of *Toxoplasma gondii* – their origin and role in the host–parasite interaction. *Res. Immunol.*, **144**, 41–44.

Cesbron-Delauw, M.-F., Guy, B., Torpier, G., Pierce, R. J., Lenzen, G., Cesbron, J. Y., Charif, H., Lepage, P., Darcy, F., Lecocq, J. P., and Capron, A. (1989). Molecular characterisation of a 23-kilodalton major antigen secreted by *Toxoplasma gondii*. *Proc. Natl. Acad. Sci. USA*, **86**, 7537–7541.

Cesbron-Delauw, M.-F., Tomavo, S., Beauchamps, P., Fourmaux, M.-P., Camus, D., Capron, A., and Dubremetz, J.-F. (1994). Similarities between the primary structures of two distinct major surface proteins of *Toxoplasma gondii*. *J. Biol. Chem.*, **269**, 16217–16222.

Charif, H., Darcy, F., Torpier, G., Cesbron-Delauw, M.-F., and Capron, A. (1990). *Toxoplasma gondii*: characterization and localization of antigens secreted from tachyzoites. *Exp. Parasitol.*, **71**, 114–124.

Cornelissen, A. W. C. A., Overdulve, J. P., and van den Ploeg, M. (1984). Determination of nuclear DNA of five Eucoccidian parasites, *Isospora (Toxoplasma) gondii*, *Sarcocystis cruzi*, *Eimeria tenella*, *E. acervulina* and *Plasmodium berghei* with special reference to gametogenesis and meiosis in *I. (T.) gondii*. *Parasitology*, **88**, 531–553.

Cristina, N., Liaud, M.-F., Santoro, F., Oury, B., and Ambroise-Thomas, P. (1991b). A family of repeated DNA sequences in *Toxoplasma gondii*: cloning, sequence analysis, and use in strain characterization. *Exp. Parasitol.*, **73**, 73–81.

Cristina, N., Oury, B., Ambroise-Thomas, P., and Santoro, F. (1991a). Restriction-fragment-length polymorphisms among *Toxoplasma gondii* strains. *Parasitol. Res.*, **77**, 266–268.

Darcy, F. and Zenner, L. (1993). Experimental models of toxoplasmosis. *Res. Immunol.*, **144**, 16–23.

Dardé, M. L., Bouteille, B., and Pestre-Alexandre, M. (1988). Isoenzymic characterization of seven strains of *Toxoplasma gondii* by isoelectrofocusing in polyacrylamide gels. *Am. J. Trop. Med. Hyg.*, **39**, 551–558.

Dardé, M. L., Bouteille, B., and Pestre-Alexandre, M. (1992). Isoenzyme analysis of 35 *Toxoplasma gondii* isolates and the biological epidemiological implications. *J. Parasitol.*, **78**, 786–794.

Decoster, A., Darcy, F., and Capron, A. (1988). Recognition of *Toxoplasma gondii* excreted and secreted antigens by human sera from acquired and congenital toxoplasmosis: identification of markers of acute and chronic infection. *Clin. Exp. Immunol.*, **73**, 376–382.

Derouin, F. and Chastang, C. (1989). In vitro effects of folate inhibitors on *Toxoplsma gondii*. *Antimicrob. Agenta Chemother.*, **33**, 1753–1759.

Derouin, F. and Garin, Y. J. F. (1991). *Toxoplasma gondii*: blood and tissue kinetics during acute and chronic infections in mice. *Exp. Parasitol.*, **73**, 460–468.

Dubey, J. P. (1977). *Toxoplasma, Hammondia, Besnoitia, Sarcocystis*, and other tissue cyst-forming coccidia of man and animals. In: *Parasitic protozoa* (J. P. Kreier, ed.), vol. 3. Academic Press, New York, pp. 101–237.

Dubey, J. P. (1993). *Toxoplasma, Neospora, Sarcocystis*, and other tissue cyst-forming coccidia of humans and animals. In: *Parasitic protozoa* (J. P. Kreier, ed.), vol. 6. Academic Press, New York, pp. 1–158.

Dubey, J. P. (1995). Duration of immunity to shedding of *Toxoplasma gondii* oocysts by cats. *J. Parasitol.*, **81**, 410–415.

Dubey, J. P. and Beattie, C. P. (1988). *Toxoplasmosis of animals and man*. CRC Press, Boca Raton, FL, 220 pp.

Dubey, J. P. and Frenkel, J. K. (1972). Cyst-induced toxoplasmosis in cats. *J. Protozool.*, **19**, 155–177.

Dubey, J. P. and Frenkel, J. K. (1973). Experimental *Toxoplasma* infection in mice with strains producing oocysts. *J. Parasitol.*, **59**, 505–512.

Dubey, J. P. and Frenkel, J. K. (1976). Feline toxoplasmosis from acutely infected mice and the development of *Toxoplasma* cysts. *J. Protozool.*, **23**, 537–546.

Dubey, J. P., Miller, N. L., and Frenkel, J. K. (1970a). The *Toxoplasma gondii* oocyst from cat feces. *J. Exp. Med.*, **132**, 636–662.

Dubey, J. P., Miller, N. L., and Frenkel, J. K. (1970b). Characterization of the new fecal form of *Toxoplasma gondii*. *J. Parasitol.*, **56**, 47–56.

Dubremetz, J. F., Achbarou, A., Bermudes, D., and Joiner, K. A. (1993). Kinetics and pattern of organelle exocytosis during *Toxoplasma gondii* host-cell interaction. *Parasitol. Res.*, **79**, 402–408.

Dubremetz, J. F., Rodriguez, C., and Ferreira, E. (1985). *Toxoplasma gondii*: redistribution of monoclonal antibodies on tachyzoites during host cell invasion. *Exp. Parasitol.*, **59**, 24–32.

Dubremetz, J. F., Sadak, A., Taghy, Z., and Fortier, B. (1987). Characterisation of a 43-kDa rhoptry antigen of *Toxoplasma gondii*. In: *Host–parasite cellular and molecular interactions in protozoal infection* (K. P. Chang and D. Snary, eds). Springer Verlag, Berlin, pp. 365–369.

Dubremetz, J. F. and Schwartzman, J. D. (1993). Subcellular organelles of *Toxoplasma gondii* and host cell invasion. *Res. Immunol.*, **144**, 31–33.

Endo, T., Sethi, K. K., and Piekarski, G. (1982). *Toxoplasma gondii*: calcium ionophore A23187-mediated exit of trophozoites from infected murine macrophages. *Exp. Parasitol.*, **53**, 179–188.

Endo, T., Yagita, K., Yasuda, T., and Nakamura, T. (1988). Detection and localisation of actin in *Toxoplasma gondii*. *Parasitol. Res.*, **75**, 102–106.

Ferguson, D. J. P. and Hutchison, W. M. (1981). Comparison of the development of avirulent and virulent strains of *Toxoplasma gondii* in the peritoneal exudate of mice. *Ann. Trop. Med. Parasitol.*, **75**, 539–546.

Ferguson, D. J. P. and Hutchison, W. M. (1987). An ultrastructural study of the early development and tissue cyst formation of *Toxoplasma gondii* in the brains of mice. *Parasitol. Res.*, **73**, 483–491.

Ferguson, D. J. P., Hutchison, W. M., and Pettersen, E. (1989). Tissue cyst rupture in mice chronically infected with *Toxoplasma gondii*: an immunocytochemical and ultrastructural study. *Parasitol. Res.*, **75**, 599–603.

Ferguson, D. J. P., Hutchison, W. M., and Siim, J. C. (1975). The ultrastructural development of the microgamete and formation of the oocyst wall of *Toxoplasma gondii*. *Acta Pathol. Microbiol. Scand. Sect. B*, **83**, 491–505.

Fichera, M. E. and Roos, D. S. (1997). A plastid organelle as a drug target in apicomplexan parasites. *Nature*, **390**, 407–409.

Fischer, H.-G., Nitzgen, B., Reichmann, G., Gross, U., and Hadding, U. (1997). Host cells of *Toxoplasma gondii* encystation in infected primary culture from mouse brain. *Parasitol. Res.*, **83**, 637–641.

Foussard, F., Leriche, M. A., and Dubremetz, J. F. (1991). Characterization of the lipid content of *Toxoplasma gondii* rhoptries. *Parasitology*, **102**, 367–370.

Frenkel, J. K. (1973). Toxoplasmosis: parasite life cycle, pathology and immunology. In: *The coccidia* (D. M. Hammond and P. L. Long, eds). University Park Press, Baltimore, MD, pp. 344–410.

Frenkel, J. K. (1990). Transmission of toxoplasmosis and the role of immunity in limiting transmission and illness. *J. Am. Vet. Med. Assoc.*, **196**, 233–239.

Frenkel, J. K., Dubey, J. P., and Miller, N. L. (1970). *Toxoplasma gondii*: fecal stages identified as coccidian oocysts. *Science*, **167**, 893–896.

Frenkel, J. K. and Smith, D. D. (1982). Immunization of cats against shedding of *Toxoplasma* oocysts. *J. Parasitol.*, **68**, 744–748.

Freyre, A., Dubey, J. P., Smith, D. D., and Frenkel, J. K. (1989). Oocyst-induced *Toxoplasma gondii* infection in cats. *J. Parasitol.*, **75**, 750–755.

Furtado, G. C., Cao, T., and Joiner, K. A. (1992a). Laminin on tachyzoites of *Toxoplasma gondii* mediates parasite binding to the b1 integrin receptor ab6b1 on human foreskin fibroblasts and Chinese hamster ovary cells. *Infect. Immun.*, **60**, 4925–4931.

Furtado, G. C., Slowik, M., Kleinman, H. K., and Joiner, K. A. (1992b). Laminin enhances binding of *Toxoplasma gondii* tachyzoites to J774 murine macrophage cells. *Infect. Immun.*, **60**, 2337–2342.

Gagnon, S., Sogin, M. L., Levesque, R. C., and Gajadhar, A. A. (1993). Molecular cloning, complete sequence of the small subunit ribosomal RNA coding region and phylogeny of *Toxoplasma gondii*. *Mol. Biochem. Parasitol.*, **60**, 145–148.

Gallois, Y., Foussard, F., Girault, A., Hodbert, J., Tricaud, A., Mauras, G., and Motta, C. (1988). Membrane fluidity of *Toxoplasma gondii*: a fluorescence polarization study. *Biol. Cell*, **62**, 11–15.

Gazzinelli, R. T., Denkers, E. Y., and Sher, A. (1993). Host resistance to *Toxoplasma gondii*: model for studying the selective induction of cell-mediated immunity by intracellular parasites. *Infect. Agents Dis.*, **2**, 139–149.

Gelder, P. van, Bosman, F., Meuter, F. van de Heuverswyn, H., and Hérion, P. (1993). Serodiagnosis of toxoplasmosis by using a recombinant form of the 54-kilodalton rhoptry antigen expressed in *Escherichia coli*. *J. Clin. Microbiol.*, **31**, 9–15.

Grimwood, J. and Smith, J. E. (1992). *Toxoplasma gondii*: the role of a 30-kDa surface protein in host cell invasion. *Exp. Parasitol.*, **74**, 106–111.

Gross, U., Müller, W. A., Knapp, S., and Heesemann, J. (1991). Identification of a virulence-associated antigen of *Toxoplasma gondii* by use of a mouse monoclonal antibody. *Infect. Immun.*, **59**, 4511–4516.

Guay, J.-M., Dubois, D., Morency, M.-J., Gognon, S., Mercier, J., and Levesque, R. C. (1993). Detection of the pathogenic parasite *Toxoplasma gondii* by specific amplification of ribosomal sequences using comultiplex polymerase chain reaction. *J. Clin. Microbiol.*, **31**, 203–207.

Guo, Z. G., Gross, U., and Johnson, A. M. (1997). *Toxoplasma gondii* virulence markers identified by random amplified polymorphic DNA polymerase chain reaction. *Parasitol. Res.*, **83**, 458–463.

Handman, E., Goding, J. W., and Remington, J. S. (1980). Detection and characterization of membrane antigens of *Toxoplasma gondii*. *J. Immunol.*, **124**, 2578–2583.

Hérion, P. and Saavedra, R. (1993). The immunobiology of toxoplasmosis. *Res. Immunol.*, **144**, 7–79.

Howe, D. K. and Sibley, L. D. (1994). *Toxoplasma gondii*: analysis of different laboratory stocks of the RH strain reveals genetic heterogeneity. *Exp. Parasitol.*, **78**, 242–245.

Iltzsch, M. H. (1993). Pyrimidine salvage pathways in *Toxoplasma gondii*. *J. Eukaryot. Microbiol.*, **40**, 24–28.

Jacobs, L. (1956). Propagation, morphology, and biology of *Toxoplasma*. *Ann. NY Acad. Sc.*, **64**, 154–179.

Johnson, L. L. (1992). SCID mouse models of acute and relapsing chronic *Toxoplasma gondii* infections. *Infect. Immun.*, **60**, 3719–3724.

Joiner, K. A. (1991a). Rhoptry lipids and parasitophorous vacuole formation: a slippery issue. *Parasitol. Today*, **7**, 226–227.

Joiner, K. A. (1991b). Cell attachment and entry by *Toxoplasma gondii*. *Behring Inst. Mitteilungen*, **88**, 20–28.

Joiner, K. A. and Dubremetz, F. (1993). *Toxoplasma gondii*: a protozoan for the nineties. *Infect. Immun.*, **61**, 1169–1172.

Joiner, K. A., Fuhrman, S. A., Mietinnen, H., Kasper, L. L., and Mellman, I. (1990). *Toxoplasma gondii*: fusion competence of parasitophorous vacuoles in Fc receptor transfected fibroblasts. *Science*, **249**, 641–646.

Joiner, K. A., Furtado, G., Mellman, I., Kleinman, H., Mietinnen, H., Kasper, L. H., Hall, L., and Fuhrman, S. A. (1989). Cell attachment and invasion by tachyzoites of *Toxoplasma gondii*. *J. Cell. Biochem.*, **13E**, 64.

Jones, T. C. and Erb, P. (1985). H-2 complex-linked resistance in murine toxoplasmosis. *J. Infect. Dis.*, **151**, 739–740.

Kasper, L. H. (1987). Isolation and characterisation of a monoclonal anti-P30 antibody resistant mutant of *Toxoplasma gondii*. *Parasite Immunol.*, **9**, 433–445.

Kasper, L. H. and Mineo, J. R. (1994). Attachment and invasion of host cells by *Toxoplasma gondii*. *Parasitol. Today*, **10**, 184–188.

Kasper, L. H. and Ware, P. L. (1985). Recognition and characterization of stage specific oocyst sporozoite antigens of *Toxoplasma gondii* by human antisera. *J. Clin. Investig.*, **75**, 1570–1577.

Leriche, M. A. and Dubremetz, J. F. (1990). Exocytosis of *Toxoplasma gondii* dense granules into the parasitophorous vacuole after host-cell invasion. *Parasitol. Res.*, **76**, 559–562.

Leriche, M. A. and Dubremetz, J. F. (1991). Characterisation of the protein contents of rhoptries and dense granules of *Toxoplasma gondii* tachyzoites by subcellular fractionation and monoclonal antibodies. *Mol. Biochem. Parasitol.*, **45**, 249–260.

Lindsay, D. S., Dubey, J. P., Blagburn, B. L., and Toivio-Kinnucan, M. A. (1991). Examination of tissue cyst formation by *Toxoplasma gondii* in cell cultures using bradyzoites, tachyzoites and sporozoites. *J. Parasitol.*, **77**, 126–132.

Lindsay, D. S., Toivio-Kinnucan, M. A., and Blagburn, B. L. (1993). Ultrastructural determination of cytogenesis by various *Toxoplasma gondii* isolates in cell culture. *J. Parasitol.*, **79**, 289–292.

Lycke, N. and Norrby, R. (1966). Demonstration of a factor of *Toxoplasma gondii* enhancing the penetration of *Toxoplasma* parasites into cultured host cells. *Br. J. Exp. Pathol.*, **47**, 248–256.

Lyons, R. E. and Johnson, A. M. (1995). Heat shock proteins of *Toxoplasma gondii*. *Parasite Immunol.*, **17**, 353–359.

Makioka, A. and Ohtomo, H. (1995). An increased DNA polymerase activity associated with virulence of *Toxoplasma gondii*. *J. Parasitol.*, **81**, 1021–1022.

Manafi, M., Hassl, A., Sommer, R. and Aspöck, H. (1993). Enzymatic profile of *Toxoplasma gondii*. *Lett. Appl. Microbiol.*, **16**, 66–68.

McLeod, R., Estes, R. G., Mack, D. G., and Cohen, H. (1984). Immune response to mice ingested *Toxoplasma gondii*. *J. Infect. Dis.*, **149**, 234–244.

McLeod, R., Mack, D., and Brown, C. (1991). *Toxoplasma gondii* – new advances in cellular and molecular biology. *Exp. Parasitol.*, **72**, 109–121.

McLeod, R., Skamene, E., Brown, C. R., Eisenhauer, P. B., and Mack, D. G. (1989). Genetic regulation of early survival and cyst number after peroral *Toxoplasma gondii* infection of AXB/BXA recombinant inbred and B10 congenic mice. *J. Immunol.*, **143**, 3031–3034.

Mehlhorn, H. and Frenkel, J. K. (1980). Ultrastructural comparison of cysts and zoites of *Toxoplasma gondii*, *Sarcocystis muris* and *Hammondia hammondi* in skeletal muscle of mice. *J. Parasitol.*, **66**, 59–67.

Meisel, R., Stachelhaus, S., Mevelec, M. N., Reichmann, G., Dubremetz, J. F., and Fischer, H. G. (1996). Identification of two alleles in the GRA4 locus of *Toxoplasma gondii* determining a differential epitope which allows discrimination of type I versus type II and III strains. *Mol. Biochem. Parasitol.*, **81**, 259–263.

Mineo, J. R., McLeod, R., Mack, D., Smith, J., Khan, I. A., Ely, K. H., and Kasper, L. H. (1993). Antibodies to *Toxoplasma gondii* major surface protein (SAG-1, P30) inhibit infection of host cells and are produced in murine intestine after peroral infection. *J. Immun.*, **150**, 3951–3964.

Nagel, S. D. and Boothroyd, J. C. (1988). The a- and b-tubulins of *Toxoplasma gondii* are encoded by single copy genes containing multiple introns. *Mol. Biochem. Parasitol.*, **29**, 261–273.

Nichols, B. A. and Chiappino, M. L. (1987). Cytoskeleton of *Toxoplasma gondii*. *J. Protozool.*, **34**, 217–226.

Nichols, B. A., Chiappino, M. L., and O'Connor, G. R. (1983). Secretion from the rhoptries of *Toxoplasma gondii* during host-cell invasion. *J. Ultrastruct. Res.*, **83**, 85–98.

Nicolle, C. and Manceaux, L. (1908). Sur un infection a corps de *Leishman* (ou organismes voisins) du *gondi*. *Cahiers Recherche de la Herbdomaire Séances Academie Science*, **147**, 763–766.

Ödberg-Ferragut, C., Soête, M., Engels, A., Samyn, B., Loyens, A., van Beeumen, J., Camus, D., and Dubremetz, J.-F. (1996). Molecular cloning of the *Toxoplasma gondii* sag4 gene encoding an 18 kDa bradyzoite specific surface protein. *Mol. Biochem. Parasitol.*, **82**, 237–244.

Ossorio, P. N., Dubremetz, J.-F., and Joiner, K. A. (1994). A soluble secretory protein of the intracellular parasite *Toxoplasma gondii* associates with the parasotophorous vacuole membrane through hydrophobic interactions. *J. Biol. Chem.*, **269**, 15350–15357.

Ossorio, P. N., Schwartzman, J. D., and Boothroyd, J. C. (1992). A *Toxoplasma gondii* rhoptry protein associated with host cell penetration has unusual charge asymmetry. *Mol. Biochem. Parasitol.*, **50**, 1–16.

Ossorio, P. N., Sibley, L. D., and Boothroyd, J. C. (1991). Mitochondrial-like DNA sequences flanked by direct and inverted repeats in the nuclear genome of *Toxoplasma gondii*. *J. Mol. Biol.*, **222**, 525–536.

Pavesio, C. E. N., Chiappino, M. L., Setzer, P. Y., and Nichols, B. A. (1992). *Toxoplasma gondii*: differentiation and death of bradyzoites. *Parasitol. Res.*, **78**, 1–9.

Pfefferkorn, E. R. (1990). The cell biology of *Toxoplasma gondii*. In: *Modern parasite biology: cellular immunological and molecular aspects* (D. J. Wyler, ed.). Freeman, New York, pp. 26–50.

Pfefferkorn, E. R. and Pfefferkorn, L. C. (1977). *Toxoplasma gondii*: characterization of a mutant resistant to 5-fluorodeoxyuridine. *Exp. Parasitol.*, **42**, 44–55.

Pfefferkorn, E. R., Pfefferkorn, L. C., and Colby, E. D. (1977). Development of gametes and oocysts in cats fed cysts derived from clones trophozoites of *Toxoplasma gondii*. *J. Parasitol.*, **63**, 158–159.

Prince, J. B., Araujo, F. G., Remington, J. S., Burg, J. L., Boothroyd, J. C., and Sharma, S. D. (1989). Cloning of cDNA encoding a 28 kilodalton antigen of *Toxoplasma gondii*. *Mol. and Biochem. Parasitol.*, **34**, 3–14.

Prince, J. B., Auer, K. L., Huskinson, J., Parmley, S. F., Araujo, F. G., and Remington, J. S. (1990). Cloning, expression, and cDNA sequence of surface antigen p22 from *Toxoplasma gondii*. *Mol. Biochem. Parasitol.*, **43**, 97–106.

Remington, J. S. (1970). Toxoplasmosis: recent developments. *Annu. Rev. Med.*, **21**, 201–218.

Rinder, H., Thomschke, A., Dardé, M. L., and Löscher, T. (1995). Specific DNA polymorphisms discriminate between virulence and non-virulence to mice in nine *Toxoplasma gondii* strains. *Mol. Biochem. Parasitol.*, **69**, 123–126.

Robert, R., Leynia de la Jarrige, P., Mahaza, C., Cottin, J., Marot-Leblond, A., and Senet, J.-M. (1991). Specific binding of neoglycoproteins to *Toxoplasma gondii* tachyzoites. *Infect. Immun.*, **59**, 4670–4673.

Roos, D. S. (1993). Primary structure of the dihydrofolate reductase-thymidylate synthase gene from *Toxoplasma gondii*. *J. Bio. Chem.*, **268**, 6269–6280.

Roos, D. S. (1999). The apicoplast as a potential therapeutic target in Toxoplasma and other apicomplexan parasites: some additional thoughts. *Parasitol. Today*, **15**, 41.

Saavedra, R., deMeuter, F., Decourt, J. L., and Hérion, P. (1991)., Human T-cell clone identifies a potentially protective 54-kDa protein antigen of *Toxoplasma gondii* cloned and expressed in *Escherichia coli*. *J. Immunol.*, **147**, 1975–1982.

Sabin, A. B. (1941). Toxoplasmic encephalitis in children. *J. Am. Med. Assoc.*, **116**, 801–807.

Saffer, L. D. and Schwartzman, J. D. (1991). A soluble phospholipase of *Toxoplasma gondii* associated with host cell penetration. *J. Protozool.*, **38**, 454–460.

Schoondermark, E. van de Ven, Melchers, W., Galama, J., Camps, W., Eskes, T., and Meuwissen, J. (1993). Congenital toxoplasmosis: an experimental study in rhesus monkeys for transmission and prenatal diagnosis. *Exp. Parasitol.*, **77**, 200–211.

Schwartz, R. T. and Tomavo, S. (1993). The current status of the glycobiology of *Toxoplasma gondii*: glycosylphosphatidylinositols, N- and O-linked glycans. *Res. Immunol.*, **144**, 24–31.

Schwartzman, J. D. and Krug, E. C. (1989). *Toxoplasma gondii*: characterization of monoclonal antibodies that recognize rhoptries. *Exp. Parasitol.*, **68**, 74–82.

Schwartzman, J. D. and Pfefferkorn, E. R. (1983). Immunofluorescent localization of myosin at the anterior pole of the coccidian *Toxoplasma gondii*. *J. Protozool.*, **30**, 657–661.

Schwartzman, J. D. and Saffer, L. D. (1992). How *Toxoplasma gondii* gets into and out of host cells. In: *Subcellular biochemistry, intracellular parasites*, (J. L. Avila and J. R. Harris, eds), vol. 18. Plenum Press, New York: pp. 333–364.

Sibley, L. D. and Boothroyd, J. C. (1992a). Construction of a molecular karyotype for *Toxoplasma gondii*. *Mol. Biochem. Parasitol.*, **51**, 291–300.

Sibley, L. D. and Boothroyd, J. C. (1992b). Virulent strains of *Toxoplasma gondii* comprise a single clonal lineage. *Nature*, **359**, 82–85.

Sibley, L. D. and Krahenbuhl, J. L. (1988). Modification of host cell phagosomes by *Toxoplasma gondii* involves redistribution of surface proteins and secretion of a 32 kDa protein. *Eur. J. Cell Biol.*, **47**, 81–87.

Sibley, L. D., Pfefferkorn, E. R., and Boothroyd, J. C. (1991). Proposal for a uniform genetic nomenclature in *Toxoplasma gondii*. *Parasitol. Today*, **7**, 327–328.

Sibley, L. D., Weidner, E., and Krahenbuhl, J. L. (1985). Phagosome acidification blocked by intracellular *Toxoplasma gondii*. *Nature*, **315**, 416–419.

Sims, T. A., Hay, T., and Talbot, I. C. (1988). Host–parasite relationship in the brains of mice with congenital toxoplasmosis. *J. Pathol.*, **156**, 255–261.

Soete, M., Fortier, D., Camus, D., and Dubremetz, J. F. (1993). *Toxoplasma gondii*: kinetics of bradyzoite tachyzoite interconversion in vitro. *Exp. Parasitol.*, **76**, 259–264.

Soldati, D. (1999). The apicoplast as a potential therapeutic target in *Toxoplasma* and other apicomplexan parasites. *Parasitol. Today*, **15**, 5–7.

Splendore, A. (1908). Un nuovo protoaoz parassita de' conigli incontrato nelle lesioni anatomiche d'une malattia che ricorda in molti punti il Kala-azar dell'uomo. Nota prelininaire pel. *Rev. Soc. Sci. Sao Paulo*, **3**, 109–112.

Sulzer, A. J., Strobel, P. L., Springer, E. L., Roth, I. L., and Callaway, C. S. (1974). A comparative electron microscopic study of the morphology of *Toxoplasma gondii* by freeze-etch replication and thin sectioning technique. *J. Protozool.*, **21**, 710–714.

Suzuki, Y., Conley, F. K., and Remington, J. S. (1989). Differences in virulence and development of encephalitis during chronic infection vary with strain of *Toxoplasma gondii*. *J. Infect. Dis.*, **159**, 790–794.

Suzuki, Y., Joh, K., Orellana, M. A., Conley, F. K., and Remington, J. S. (1991). Gene(s) within the H-2D region determines the development of toxoplasmic encephalitis in mice. *Immunology*, **74**, 732–739.

Suzuki, Y., Orellana, M. A., Wong, S.-Y., Conley, F. K., and Remington, J. S. (1993). Susceptibility to chronic infection with *Toxoplasma gondii* does not correlate with susceptibility to acute infection in mice. *Infect. Immun.*, **61**, 2284–2288.

Tomavo, S., Fortier, B., Soete, M., Ansel, C., Camus, D., and Dubremetz, J. F. (1991). Characterization of bradyzoite-specific antigens of *Toxoplasma gondii*. *Infect. Immun.*, **59**, 3750–3753.

Tomavo, S., Schwartz, R. T., and Dubremetz, J. F. (1989). Evidence for glycosyl-phosphatidylinositol anchoring of *Toxoplasma gondii* major surface antigens. *Mol. Cell. Biol.*, **9**, 4576–4580.

Torpier, G., Charif, H., Darcy, F., Liu, J., Darde, M.-L., and Capron, A. (1993). *Toxoplasma gondii*: differential location of antigens secreted from encysted bradyzoites. *Exp. Parasitol.*, **77**, 13–22.

Wang, C. C. (1982). Biochemistry and physiology of coccidia. In: *Biology of the coccidia* (P. L. Long, ed.). University Park Press, Baltimore, PID, pp. 167–228.

Ware, P. L. and Kasper, L. H. (1987). Strain-specific antigens of *Toxoplasma gondii*. *Infect. Immun.*, **55**, 778–783.

Weiss, L. M., Udem, S. A., Tanowitz, H., and Wittner, M. (1988). Western blot analysis of the antibody response of patients with AIDS and toxoplasma encephalitis: antigenic diversity among *Toxoplasma* strains. *J. Infect. Dis.*, **157**, 7–13.

Werk, R. (1985). How does *Toxoplasma gondii* enter host cells? *Rev. Infect. Dis.*, **7**, 449–457.

Windeck, T. and Gross, U. (1996). *Toxoplasma gondii* strain-specific transcript levels of SAG1 and their association with virulence. *Parasitol Res.*, **82**, 715–719.

Woodison, G. and Smith, J. E. (1990). Identification of the dominant cyst antigens of *Toxoplasma gondii*. *Parasitology*, **100**, 389–392.

Yano, K. and Nakabayashi, T. (1986). Attenuation of the virulent RH strain of *Toxoplasma gondii* by passages in mice immunized with *Toxoplasma* lysate. *Biken J.*, **29**, 31–37.

14 *Toxoplasma* gondii – epidemiology

Babill Stray-Pedersen

Toxoplasma gondii infection is widespread among humans and warm-blooded animals and birds throughout the world. Approximately one-third of the entire world's population have serological evidence of past infection of *Toxoplasma*. However, there is considerable variation from one geographic area to another, with seropositivity varying from 2 to 90% in the adult population. Such differences may be explained by great variance in exposure to the different sources of infection.

Major routes of infection

Humans usually acquire the infection incidentally by ingestion. Consumption of under-cooked or raw or minced meat containing tissue cysts represents important sources. *Toxoplasma* cysts have been recovered from pork (5–30%), mutton (12–25%), and more rarely from beef (0–9%) and chicken (Frenkel and Dubey 1972). Handling of raw or undercooked meat and poor kitchen hygiene have been associated with an increased risk of infection (Kapperud *et al.* 1996). The tissue cysts are destroyed by freezing ($-10°C$ overnight), thawing, cooking until change of colour ($>66°C$), desiccation, or exposure to water for more than 30 min (Kotula *et al.* 1991; Dubey 2000). Cooking in a microwave oven, however, is not sufficient to kill the cysts (Lundén and Uggla 1992).

Another source of infection is contact with the oocyst either by consumption of unwashed fruits, berries, and vegetables, which may have oocysts on their surface, or by direct contact with cat faeces. In children, direct ingestion of oocysts from dirt in sandboxes and yards where cats have defaecated, is supposed to be the major route of infection.

Admittedly, the importance of these two main routes of contamination: both consumption of tissue cysts and contact with cat faeces, has been questioned (Klapper and Morris 1990), since the prevalence *of Toxoplasma* antibodies appears to be the same among vegetarians and meat-eaters and *Toxoplasma* seropositivity is rather rare in young children.

Occasionally, infection can occur after self-inoculation in the laboratory, after inhalation of sprouted oocysts, and after blood transfusion or organ transplantation. Parasites have been isolated from sputum, saliva, urine, and semen, but transmission through such body fluids has never been proved. In fact, the only proven route of human-to-human transmission is the transmission of tachyzoites across the placenta to the foetus during the primary stage of infection in pregnant women.

Variations in prevalence rates

World wide the prevalence of *Toxoplasma* immunity varies according to age, and it also differs from one country to another and even within the same country. Seropositivity is highest in areas with warm, moist climate where chances of survival of the oocysts are good, and lower in cold regions and high altitude. In central and southern Europe, many countries in Africa, and in South America, more than half of the population is infected with *T. gondii*, while in some countries with tropical climates, such as Guatemala, Costa Rica, and Tahiti, between 85 and 95% of the adult population have antibodies (Asburn 1992). In contrast only 11% of adults living in Iceland and 6% living in north Norway are infected (Jennum *et al.* 1998a; Jonsdottir 1989). Dry climate may also render the oocysts uninfective. Therefore, in environments with low rainfall or in which soil does not retain moisture, there is less risk of infection from oocysts. In Norway, the dry, cold inland has significantly fewer infections than the coastal areas (Stray-Pedersen *et al.* 1979). Altitude has also been shown to be inversely related to antibody prevalence. In Columbia, South America, 53% of the population living below 2 500 m elevation had *Toxoplasma* antibodies, while only 43% of those living between 2 500 and 4 500 m had the antibodies. This may be due either to the climate in mountaineous areas being unfavourable for oocyst development or to the presence of fewer cats in the area.

The differences in prevalence may also be ascribed to variations in hygiene and eating habits. In France, with the 'French cuisine' and a preference for undercooked and raw meat, the prevalence is especially high, with antibody prevalence exceeding 50% before the age of 10 years. The same is observed in different Asian groups that have a habit of eating raw meat. Otherwise, the Japanese and Chinese custom of cooking meat in small pieces minimizes the risk of acquiring the infection from meat (Wallace 1976).

In Europe and in the United States, a decrease in *Toxoplasma* infection has been observed over the last 20 years. In France, a reduction in antibody prevalence from 80 to 69% during the last decades has been reported in the age group of 20–24 years (Jeanell *et al.* 1988). In Sweden, the same tendency has been observed with a decrease from 28 to 10% in the same age group (Forsgren *et al.* 1991). In other countries the impression today is that an actual reduction in infection is taking place (Gilbert 2000). These changes might perhaps be ascribed to the increased consumption of instant or ready cooked food, to the more common use of home freezers for food storage, or to a general decrease in the infection rate of animals.

Age and sex

The prevalence of *Toxoplasma* antibodies increases with age and the acquisition is directly dependent on the prevalence of *Toxoplasma* in the environment. All babies born to infected mothers are seropositive. However, after 6–12 months the passively transferred maternal antibodies have disappeared, and the baby is seronegative, except in a very few cases with congenital infection (Stray-Pedersen 1993). Once the child starts to walk and become mobile, contact with the parasite in the form of oocysts in soil, sandboxes, and playgrounds, are possible. The local food habits may also influence the infection during childhood.

In the later years, however, most seroconversions occur in the age group of 15–35 years. In Norway, a steady increase in seroprevalence was observed in the higher age groups (Stray-Pedersen and Jennum 1992). Only 23% of patients of the age of 45, admitted to a district hospital, had antibodies, while 46% of those aged 84 years or higher had evidence of past *Toxoplasma* infection, indicating an annual incidence of approximately 0.6%.

Generally, there seems to be no sex differences in the incidence of infection (Beverly *et al.* 1976). In Norway, the same prevalence of *Toxoplasma* antibodies was found in military recruits and pregnant women of the same age (Stray-Pedersen *et al.* 1979). In one study, however, boys up to the age of 15 years had a higher incidence of infections than girls (Beverly *et al.* 1976). It is possible that boys, having more outdoor activities than girls, have a higher risk of infection by oocysts. On the other hand, more women than men above 25 years are infected, and this has been ascribed to women having more contact with cats, with raw meat during cooking, and doing more gardening than men. The increased prevalence is thus probably due to higher level of exposure to the parasites.

Risk factors for infection

In Norway, a case control study has been performed for identification of preventable risk factors for *Toxoplasma* infection during pregnancy (Kapperud *et al.* 1996). The case patients were identified through a serologic prenatal screening program. Pregnant women with serologic evidence of recent primary *Toxoplasma* infection (63 cases) were compared with 128 seronegative controls matched by age, stage of pregnancy, expected date of delivery, and geographic area. Multivariate analyses indicated that the following factors were independently associated with an increased risk of *Toxoplasma* infection:

a   eating raw or undercooked minced meat products (OR = 4.1);
b   eating unwashed, raw vegetables, or fruits (OR = 2.4);
c   eating raw or undercooked mutton (OR = 11.4);
d   eating raw or undercooked pork (OR = 3.4);
e   cleaning the cat litter box (OR = 5.5); and
f   washing the kitchen knife infrequently after preparation of raw meat prior to handling of other food items (OR = 7.3).

In an univariate analysis travelling to countries outside Scandinavia was identified as a significant risk factor (Kapperud *et al.* 1996). However, this was not due to the travel itself, but related to the different modes of infection. Even from Sweden travelling to southern Europe or areas with a significantly higher prevalence is regarded as a risk factor (Evengard *et al.* 1997). Thus, in the Nordic countries seronegative pregnant women should be advised to stay within the region during pregnancy, especially since preventive measure to avoid *Toxoplasma* infection may be difficult to apply abroad. Although attempts to avoid undercooked meat and unwashed or raw vegetables, and fruits would reduce the risk of infection, most meals taken while travelling are usually eaten in hotels or restaurants where the traveller has no influence on kitchen hygiene and food handling practices (Kapperud *et al.* 1996).

The risk factors connected to consumption of raw or undercooked meat are directly related to the seroprevalence of *Toxoplasma* infection in meat-producing animals. In Norway, the highest prevalence of *Toxoplasma* antibodies has been found in sheep (18%), followed by cattle (5%) and pigs (2.6%) (Skjerve *et al.* 1996). In Sweden, similar seroprevalence (19%) was found in different flocks of sheep (Lundén *et al.* 1992).

Some people may have occupations that increase their risk of infection. These include slaughterhouse workers and greenhouse workers. Actually, in Denmark, these workers were found to have remarkably high *Toxoplasma* antibody prevalence (Lings *et al.* 1994). In other studies veterinarians and other humans with intimate contact with animals have high antibody prevalence (Asburn 1992). Daily contact with cats or living in a neighbourhood with cats does not represent a risk factor according to a Norwegian study; however, daily contact with kittens less than 1 year of age or living in a household with a cat that uses a cat litter box and especially cleaning the cat litter box were associated with increased risk (Kapperud *et al.* 1996). Contamination of hands after contact with cat faeces or infected soil, when gardening, may also be a potential hazard (Frenkel 2000).

Seroprevalence in pregnant women

The frequency of *Toxoplasma* infection during pregnancy depends upon the number of seronegative women of child-bearing age and on the prevailing infection risk. It is thus of importance to know the seroprevalence of pregnant women. In some parts of Asia and in northern areas only a small percentage of pregnant women are infected, while in Africa, South America, and southern Europe, more than 50% have antibodies. The prevalence of antibodies in Europe is shown in Figure 14.1. Even within our continent great variations exist. Different surveys of pregnant women in the Nordic countries have been performed in the last decade.

Finland

Lappalainen (1995) studied nearly 17 000 pregnant women in the Helsinki area. She found an overall prevalence of *Toxoplasma* seropositivity of 20.3%. The incidence of primary infection was 2.4 per 1 000 pregnancies at risk (Lappalainen *et al.* 1992).

Denmark

In Denmark, Lebech *et al.* (1993) found that the prevalence of IgG antibodies was 27.4% in 5 000 pregnant women, while IgM antibodies were found in 0.5%. In the period from 1992

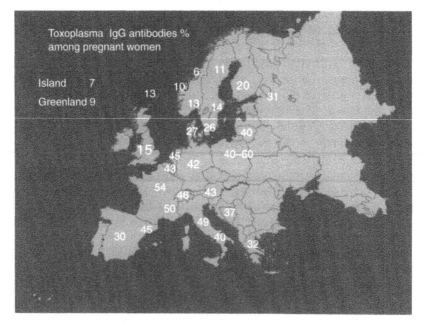

Figure 14.1 The prevalence of *Toxoplasma* IgG antibodies in pregnant populations in different countries in Europe. (See Colour Plate III.)

to 1996 nearly 100 000 blood samples collected on filter paper cards from new borns have been investigated. Whenever the new born's filter paper was positive (27.8%) the frozen blood sample from the mother collected during the first trimester was analyzed for *Toxoplasma* IgG antibodies. Out of 64 888 seronegative women, 141 mothers had acquired *Toxoplasma* infection during pregnancy. These women gave birth to 143 children, of which 27 were diagnosed with congenital *Toxoplasma* infection. Thus the estimated incidence of primary *Toxoplasma* infection in seronegative pregnant women was 2.7 per 1 000 and the birth prevalence of congenital infection in live-born children was 4.2 per 100 000 new borns. The mother to child transmission rate was 19.1% (Lebech *et al.* 1998). Denmark introduced a national neonatal screening program for toxoplasmosis in 1999.

Norway

In the 1990s a prenatal screening program of *Toxoplasma* infection was performed for 35 940 pregnant women comprising 11 of Norway's 19 counties (Jennum *et al.* 1998b). Of the women screened, 10.9% were *Toxoplasma* IgG positive. Evidence of primary infection during pregnancy was shown in 47 women, giving an incidence of 1.7 per 1 000 seronegative women. A significantly higher incidence was detected in the city of Oslo (4.6 per 1 000) than in the rest of the country (0.9 per 1 000) and among foreign women compared to Norwegian women (6 per 1 000 versus 1.5 per 1 000). Congenital infection was detected in 11 infants, giving an overall transmission rate of 23%, 13% in the first trimester, 39% in the second, and 50% in the third trimester (Jennum *et al.* 1998b). During the 1-year follow-up only one infant born to an untreated mother was found to be clinically affected (unilateral chorioretinitis and loss of vision). In Norway, the seroprevalence varied throughout the country. The lowest prevalences were detected in the North (6.7%) and in the inland counties (8.2%). A significantly higher prevalence was detected in the southern counties (13.4%), where mild coastal climate prevails (Jennum *et al.* 1998a).

Iceland

Jónsdóttir and Arnadóttir (1998) tested 279 pregnant women. The overall prevalence was 7.2%; 6.1% in those below 35 years of age, and 19.2% in women (26 women) above 35 years. One hundred and five pregnant women from Faroe islands and 129 women from Greenland have also been tested; the *Toxoplasma* IgG antibody prevalence was 13.3 and 9.3%, respectively (Lebech 1993).

Sweden

Seroepidemiological studies of pregnant women from four different regions in Sweden in the late 1980s revealed a significant trend in *Toxoplasma* seroprevalence from Gotland island (26%) through Örebro (18%) and Stockholm area (18%) to north Sweden (12%) (Figure 14.1) (Ljungström *et al.* 1995). A neonatal screening program was performed during 1997–1998 in the Stockholm and Malmö areas. Nearly 41 000 children were studied. The prevalence of *Toxoplasma* IgG antibodies in pregnant women was 14.0% in the Stockholm area and 25.7% in Malmö (Evengard *et al.* 2001). Twelve seroconversions were detected and three children were congenitally infected giving a birth prevalence of congenital toxoplasmosis of 0.7 per 10 000 children.

In the Nordic countries, the prevalence of *Toxoplasma* antibodies is rather low compared to the rest of Europe, varying from 6 to 26% among pregnant women. Immigrant women seem to have a higher prevalence. The incidence of infection in pregnancy varies from 0.5 to 5 per 1 000 seronegative women. This estimate may be less than the annual infection incidence in non-pregnant population because some of the pregnant women may take precautions to avoid acquiring *Toxoplasma* infection during pregnancy.

References

Asburn, D. (1992). General epidemiology. In: *Human toxoplasmosis* (D. O. Ho-Yen and A. W. L. Joss, eds). Oxford University Press, New York, pp. 15–19.

Beverly, J. K. A., Fleck, D. G., Kwantyes, W., and Ludlamd, G. B. (1976). Age–sex distribution of various diseases with particular reference to toxoplasmic lymphadenopathy. *J. Hyg.*, **76**, 215–228.

Dubey, J. P. (2000). The scientific basis for prevention of *Toxoplasma gondii* infection studies on tissue cysts survival, risk factors and hygienic measures. *Congenital Toxaplasmosis*, 271–275.

Evengard, B., Forsgren, M., and Uggla, A. (1997). Toxoplasmos. *Lækartidningen*, **36**, 3249–3253.

Evengard, B., Petersson, K., Engman, M. L. *et al.* (2001). Low incidence of *Toxoplasma* during pregnancy in new-borns in Sweden. *Epidemiol. Infect.*, **172**, 121–127.

Frenkel, J. K. (2000). Biology of *Toxoplasma gondii*. In: *Congenital toxoplasmosis* (P. Ambroise-Thomas and E. Petersen, eds). Springer-Verlag, France, pp. 237–249.

Frenkel, J. K. and Dubey, J. P. (1972). Toxoplasmosis and its prevention in cats and man. *J. Infect. Dis.*, **126**, 664–673.

Forsgren, M., Gille, E., Ljungstrom I., *et al.* (1991). *Toxoplasma* antibodies in pregnant women in Stockholm in 1969, 1979 and 1987. *Lancet*, **337**, 1431–14.

Gilbert, R. (2000). Epidemiology of infection in pregnant women. In: *Congenital toxoplasmosis* (P. Ambroise-Thomas and E. Petersen, eds). Springer-Verlag, France, pp. 237–249.

Jeanell, D., Costagliola, D., Niel, G. *et al.* (1990). What is known about prevention of congenital toxoplasmosis. *Lancet*, **336**, 359–361.

Jennum, P. A., Kapperud, G., Stray-Pedersen, B. *et al.* (1998a). Prevalence of *Toxoplasma gondii* specific immunoglobulin G antibodies among pregnant women in Norway. *Epidemiol. Infect.*, **120**, 87–92.

Jennum, P. A., Stray-Pedersen, B., Melby, K. K. *et al.* (1998b). Incidence of *Toxoplasma gondii* infection in 35,940 pregnant women in Norway and pregnancy outcome for infected women. *J. Clin. Microbiol.*, 2900–2906.

Jónsdóttir, K. E. and Arnadóttir, T. (1998). Mælingar á mótefnum gegn bogfrymlum i nokkrum hópum Íslendinga. *Icelandic Med. J.*, **74**, 255–263.

Kapperud, G., Jennum, P. A., Stray-Pedersen, B. *et al.* (1996). Risk factors for *Toxoplasma gondii* infection in pregnancy. Results of a prospective case-control study in Norway. *Am. J. Epidemiol.*, **144**, 405–412.

Klapper, P. E. and Morris, D. J. (1990). Screening for viral and protozoal infections in pregnancy. A review. *Br. J. Obst. Gynecol.*, **97**, 974–983.

Kotula, A. W., Dubey, J. P., Sharar, A. K. *et al.* (1991). Effect of freezing on infectivity of *Toxoplasma gondii* tissue cysts in pork. *J. Food Protection*, **54**, 687–690.

Lappalainen, M. (1995). *Toxoplasma infection during pregnancy. A prospective cohort study and a cost-benefit analysis*. Thesis, Department of Virology, University of Helsinki, Hakapaino Oy, Helsinki.

Lappalainen, M., Koskela, P., Hedman, K. *et al.* (1992). Incidence of primary toxoplasma infections during pregnancy in southern Finland: a prospective cohort study. *Scand. J. Infect. Dis.*, **24**, 97–104.

Lebech, M., Andersen, O., Christensen, N. C. *et al.* and the Danish congenital toxoplasmosis study group. (1998). Feasibility of neonatal screening for Toxoplasma infection in the absence of prenatal treatment. *Lancet*, **353**, 1834–1837.

Lebech, M., Larsen, S. O., and Petersen, E. (1993). Prevalence, incidence and geographical distribution of Toxoplasma gondii antibodies in pregnant women in Denmark. *Scand. J. Infect. Dis.*, **25**(6), 751–756.

Ljungstrøm I., Gille, E., Nokes, J. *et al.* (1995). Seroepidemilology of *Toxoplasma gondii* among pregnant women in different parts of Sweden. *Eur. J. Epidemiol.*, **11**, 149–156.

Lings, S., Lander, F., and Lebech, M. (1994). Antimicrobial antibodies in Danish slaughterhouse workers and greenhouse workers. *Int. Arch. Occup. Environ. Health*, **65**(6), 405–409.

Lundén, A., Carlsson, U., and Näslund, K. (1992). Toxoplasmosis and border disease in 54 Swedish sheep flocks: seroprevalence and incidence during one gestation period. *Acta Vet. Scand.*, **33**, 175–184.

Lundén, A. and Uggla, A. (1992). Infectivity of *Toxoplasma gondii* in mutton following curing, smoking, freezing, or microwave cooking. *Int. J. Food Microbiol.*, **15**, 357–363.

Skjerve, E., Tharaldsen, J., Waldeland, H. *et al.* (1996). Antibodies to *Toxoplasma gondii* in Norwegian slaughtered sheep, pigs and cattle. *Bull. Scand. Soc. Parasitol.*, **6**(1), 11–17.

Stray-Pedersen, B. (1993). Toxoplasmosis in pregnancy. In: *Infectious diseases. Challenges for the 1990s.* (G. L.Gilbert, ed.). Balliere's Clinical Obstetrics and Gynecology. Bailliere Tindall, London, vol. 17, pp. 107–137.

Stray-Pedersen, B. and Jennum, P. A. (1992). Current status of toxoplasmosis in pregnancy in Norway. *Scand. J. Infect. Dis.* (Suppl. 84), 80–83.

Stray-Pedersen, B., Pedersen, J. O., and Omland, A. (1979). Estimation of the incidences of *Toxoplasma* infections among pregnant women from different areas in Norway. *Scand. J. Infect. Dis.*, **11**, 247–252.

Wallace, G. D. (1976). The prevalence of toxoplasmosis on the Pacific islands, and the influence of ethnic group. *Am. J. Trop. Med. Hyg.*, **25**, 48–53.

15 *Trichomonas vaginalis*

Pål Wölner-Hanssen

Epidemiology

Trichomoniasis is the most common non-viral sexually transmitted disease world-wide. It has been estimated that more than 180 million people are infected with *Trichomonas vaginalis* (Brown 1972). However, according to the National Disease and Therapeutic Index survey, numbers of physician visits for trichomonal vaginitis in the US declined from a high of 1.3 million in 1974 to less than 600 000 in 1987 (Kent 1991). According to a study from Denmark, trichomoniasis has become a rare infection in that country. That is, in 1967, 19% and in 1997, 2% of specimens analysed in the Statens Serum Institut, Copenhagen, were positive for *T. vaginalis* (Dragsted *et al.* 2001). Also in Sweden and Norway, trichomoniasis is an unusual diagnosis (A. Hallén and B. Stray-Pedersen, personal communications). By contrast, in Estonia (the only Baltic country with available data) trichomoniasis was reported among approximately 6 000 individuals out of 1 500 000 inhabitants (1 : 250) in 1994 (Lazdane and Bukovskis 1997).

Microbiology

Anatomy

Trichomonas vaginalis is a protozoon with variable size and shape. The size varies from 10 to 20 μm. The shape tends to be ellipsoidal in axenic cultures, and amoeboid *in vivo*. The microorganism has four free anterior flagella and one recurrent flagellum attached to an undulating membrane. The longest flagella is 6–18 μm long. The undulating membrane extends for about one-half to two-thirds of the length of the cell. The organism is very mobile, and moves with jerky movements by the flagella and the undulating membrane. At the posterior end, a slender spike projects for a distance corresponding to less than one-third to over one-half the cell length. This spike is the prolongation of an organelle that runs axially through the body – the axostyle. Mitochondria are lacking in trichomonads. Relatively large, dense granules called hydrogenosomes are aligned in three rows along the axostyle. These paraxostylar granules are characteristic for trichomonads. Any cell with the shape and size of *T. vaginalis*, containing the characteristic paraxostylar granules in typical arrangement, can be regarded as belonging to this species (Honigberg and Brugerolle 1989).

   Trichomonas vaginalis is often accompanied by other pathogens in the vagina, many of which produce symptoms that are similar to those produced by the protozoae. For example, among pregnant women, *T. vaginalis* tends to be accompanied by *Neisseria gonorrhoeae*, *Chlamydia trachomatis*, *Ureaplasma urelyticum*, *Mycoplasma hominis*, and *Candida* species (Pastorek *et al.* 1996).

Culture

The *T. vaginalis* culture is 100% specific and therefore considered a 'gold standard' for trichomoniasis diagnosis. Trichomonas culture is the most sensitive diagnostic procedure, but will miss up to 15% of infected cases (Lossick 1988). Detailed aspects of *T. vaginalis* cultivation were reviewed by Lindstead (1989). A number of different media have been studied. Two media commonly used for *T. vaginalis* cultures are the modified Diamond's medium (tryptose-yeast extract maltose medium, TYM)[42] and the Feinberg–Whittington medium (Feinberg and Whittington 1957). The two media seem to be equally effective for the isolation of *T. vaginalis* (Krieger *et al.* 1988). Diamond's medium and modified Diamond's medium can detect 97 and 90% of isolates from vaginal secretions, respectively. A modified thioglycolate medium supplemented with yeast extract, horse serum, and antimicrobial agents was as reliable as the more expensive Diamond's medium according to one recent study (Poch *et al.* 1996). Disadvantages with trichomonas culturing include: sensitivity to modes of transport to the laboratory, incubation-time in the laboratory of 2–7 days, requiring at least three wet mount evaluations by technical staff, and relatively high costs.

Another culture method for *T. vaginalis* is the InPouch™ TV culture method (BioMed Diagnostics Inc., Santa Clara, USA). The pouch, which can be used for specimen transport and culture, can maintain trichomonad viability for periods from 41 to 131 days (Borchardt and Smith 1991). Specimens may be sent by regular mail to the laboratory and examined under the microscope after 24 h incubation. The test has a sensitivity of four organisms per millilitre (Borchardt *et al.* 1992), but requires up to three days of incubation and microscopic evaluation.

Antigen detection

Monoclonal antibodies to *T. vaginalis* have been used to identify vaginal trichomoniasis. Broadly reactive, fluorescein isothiocyanate-conjugated monoclonal antibodies to *T. vaginalis* are used for this test. In one study, the direct fluorescent antibody test was 86% sensitive, 99% specific, and had a positive predictive value of 96% (prevalence 15%) (Yule *et al.* 1987). The sensitivity of this test is also related to the concentration of protozoa in the vaginal fluid. In women with positive wet mounts, more than 90% of cases are detected with the antibody test. However, among women with negative wet mounts, the antibody test has a sensitivity of only 69–77%.

An immunoassay developed for the detection of *T. vaginalis* in vaginal swabs, showed a sensitivity of 93%, a specificity of 97.5%, and a positive predictive value of 82% (prevalence: 9%) when culture based on modified Diamond's medium, was used as reference (Krieger *et al.* 1988).

DNA hybridization

Recently, DNA probing and gene amplification have been used also for the diagnosis of *T. vaginalis*. A commercial test, based on synthetic oligonucleotide probes (Affirm VP Microbial Identification Test, developed by MicroProbe Corporation, Bothell, WA), was evaluated by Briselden and Hillier (1994). The test system, designed for use in the physician's office, had a sensitivity of 83%, a positive predictive value of 100% and a negative predictive value of 98% in a population with a 9% prevalence of *T. vaginalis*. DeMeo *et al.* (1996) also studied the Affirm VP test comparing it with microscopy of vaginal 'wet mounts' and with cultures in Diamond's medium. In that study of 615 women with signs or symptoms of vaginitis, the DNA test had a sensitivity of 90% and a specificity of 99.8%. In fact, polymerase chain reaction (PCR) analysis is sensitive enough to be used on introital specimens. Witkin *et al.* (1996) compared introital specimens with specimens from the endocervix and from the posterior vaginal vault in 219 pregnant women. Introital testing had a sensitivity of 95.5% and a specificity of 100%. Screening for *T. vaginalis* based on PCR of urine specimens only, might have a sensitivity problem. That is, among 51 women positive by vaginal wet prep or culture, 65% of urine specimens were PCR-positive (Lawing *et al.* 2000). By contrast, van Der Schee *et al.* (1999), identified trichomoniasis by PCR on vaginal swab of 10 and on urine specimen of 11 of 200 Dutch women.

Pathogenesis

Trichomonas vaginalis infects, almost exclusively, the epithelium of the lower genital tract. In women, the vagina, urethra, and Skene's glands are the main sites of infection.

Some infected individuals are asymptomatic, while others may have severe local symptoms. Symptomatology may be related to trichomonas virulence factors or to host factors. Antigenic heterogeneity, phenotypic variation, cytotoxicity, proteinases, and secretion of immunogens into the culture medium are possible virulence factors of *T. vaginalis* (Lehker and Alderete 1990). Attachment of *T. vaginalis* to epithelial cells seems to be dependent upon a 30-kDa cysteine proteinase (CP30) on the surface of the protozoa (Mendoza-Lopez *et al.* 2000). Contact-dependent cytotoxicity relies on specific surface proteins (adhesins) which mediates the interaction of *T. vaginalis* with epithelial cells (Arroyo *et al.* 1992). Studies of the interaction between trichomonads and the epithelial layer of human amniotic membrane have demonstrated damaged and desquamated cells in areas where parasites were in direct contact with the target cells (Mirhaghani and Warton 1996). A cysteine proteinase with a molecular mass of 65 kDa on the plasma membrane of *T. vaginalis* has recently been shown to be involved in *T. vaginalis* cytotoxicity (Alvarez-Sanchez *et al.* 2000). Cytotoxicity may also be caused by soluble factors, including free lactic or acetic acids (Pindak and Gardner 1993), and the so-called 'cell-detatching factor' (Garber *et al.* 1989). Cell-detatching factor is a 200-kDa glycoprotein that causes detachment of monolayer cells.

Table 15.1 Significant relationships between clinical findings and numbers of protozoa per 400 × field in 67 wet mounts from women with trichomoniasis

Manifestations	Mean number of protozoa when manifestation		P
	Present	Absent	
Symptoms			
Yellow discharge	16.5 ± 19.8	9.4 ± 13.6	0.04
Signs			
Strawberry cervix	16.2 ± 16.9	8.2 ± 15.0	<0.001
Increased discharge	13.8 ± 17.0	5.4 ± 11.1	<0.001
Purulent discharge	16.5 ± 17.8	3.8 ± 8.8	<0.001
Genital erythema	14.7 ± 17.7	8.4 ± 13.8	0.004
Clue cells	14.1 ± 17.5	4.7 ± 8.4	0.004
KOH-test pos.	13.6 ± 15.4	10.0 ± 17.3	0.03

Source: Wölner-Hanssen *et al.*, unpublished data.

Production of cell detaching factor does correlate with clinical presentation of trichomoniasis (Garber *et al.* 1989). The capability to produce subcutaneous abscesses in mice, haemolytic activity, and adherence to HeLa cells have also been regarded as markers of *T. vaginalis* pathogenicity. However, the clinical manifestations most strongly associated with trichomoniasis (colpitis macularis and yellow discharge), do not correlate with these features (Krieger *et al.* 1990b). Moreover, wet mounts are more often positive in women with symptomatic than in those with asymptomatic trichomoniasis (Krieger *et al.* 1990a). As shown in Table 15.1, the concentration of *T. vaginalis* in vaginal fluid is related to severity of signs and symptoms of inflammation (P. Wölner-Hanssen, unpublished data).

References

Alper MM, Barwin BN, McLean WM, McGilveray IJ, and Sved S (1985). Systemic absorption of metronidazole by the vaginal route. *Obstet. Gynecol.*, **65**: 781–784.

Alvarez-Sanchez ME, Avila-Gonzales L, Becerril-Garcia C, Fattel-Facenda LV, Ortega-Lopez J, and Arroyo R. (2000). A novel cysteine proteinase (CP65) of *Trichomonas vaginalis* involved in cytotoxicity. *Microb. Pathog.*, **28**: 193–202.

Arroyo R, Engbring J, and Alderete JF (1992). Molecular basis of host cell recognition by *Trichomonas vaginalis*. *Mol. Microbiol.*, **6**: 853–862.

Borchardt KA and Smith RF (1991). An evaluation of an InPouch TV culture method for diagnosing *Trichomonas vaginalis* infection. *Genitourin. Med.*, **67**: 149–152.

Borchardt KA *et al.* (1992). A clinical evaluation of trichomoniasis in San Jose, Costa Rica using the InPouch TV test. *Genitourin. Med.*, **68**: 328–330.

Briselden AM and Hillier SL (1994). Evaluation of Affirm VP microbial identification test for *Gardnerella vaginalis* and *Trichomonas vaginalis*. *J. Clin. Microbiol.*, **32**: 148–152.

Brown MT (1972). Trichomoniasis. *Practitioner*, **209**: 639.

DeMeo LR, Draper DL, McGregor JA, Moore DF, Peter CR, Kapernick PS, and McCormack WM (1996). Evaluation of a deoxyribonucleic acid probe for the detection of *Trichomonas vaginalis* in vaginal secretions. *Am. J. Obstet. Gynecol.*, **174**: 1339–1342.

Dragsted DM, Farholt S, and Lind I (2001). Occurrence of trichomoniasis in women in Denmark, 1967–1997. *Sex. Transm. Dis.*, **28**: 326–329.

Feinberg JG and Whittington MJA. (1957). A culture medium for *Trichomonas vaginalis* Donne and species of *Candida*. *J. Clin. Pathol.*, **10**: 327–329.

Fouts AC, and Kraus SJ (1980). *Trichomonas vaginalis*: reevaluation of its clinical presentation and laboratory diagnosis. *J. Infect. Dis.*, **141**: 137–143.

Garber GE, Lemchuk-Favel LT, and Bowie WR (1989). Isolation of a cell-detaching factor of *Trichomonas vaginalis*. *J. Clin. Microbiol.*, **27**: 1548–1553.

Honigberg BM and Brugerolle G (1989). Structure. In: Honigberg BM, ed. *Trichomonads parasitic in humans*. New York: Springer-Verlag, pp. 5–35.

Kent HL (1991). Epidemiology of vaginitis. *Am. J. Obstet. Gynecol.*, **165**: 1168–1176.

Krieger JN *et al.* (1988). Diagnosis of trichomoniasis: comparison of conventional wet-mount examination with cytologic studies, cultures, and monoclonal antibody staining of direct specimen. *JAMA*, **259**: 1223–1227.

Krieger JN, Torian BE, Hom J, and Tam MR. (1990a). Inhibition of *Trichomonas vaginalis* motility by monoclonal antibodies is associated with reduced adherence to HeLa cell monolayers. *Infect. Immun.*, **58**: 1634–1639.

Krieger JN, Wölner-Hanssen P, Stevens C, and Holmes KK. (1990b). Characteristics of *Trichomonas vaginalis* isolates from women with and without colpitis macularis. *J. Infect. Dis.*, **161**: 307–311.

Lawing LF, Hedges SR, and Schwebke JR (2000). Detection of trichomoniasis in vaginal and urine specimens from women by culture and PCR. *J. Clin. Microbiol.*, **38**: 3585–3588.

Lazdane G and Bukovskis M (1997). Epidemiology of sexually transmitted diseases in the Baltic countries. *Acta. Obstet. Gynecol. Scand.*, **76**(suppl. 164): 128–131.

Legator MS, Connor TH, and Stoekel M (1975). Detection of mutagenic activity of metronidazole and niridazole in body fluid of humans and mice. *Science*, **188**: 1118–1119.

Lehker MW and Alderete JF (1990). Properties of *Trichomonas vaginalis* grown under chemostat controlled growth conditions. *Genitourin. Med.*, **66**: 193–199.

Lindstead D (1989). Cultivation. In: Honigberg BM, ed. *Trichomonads parasitic in humans*. New York: Springer-Verlag, pp. 91–111.

Lossick JG (1988). The diagnosis of vaginal trichomoniasis. *JAMA*, **259**: 1230.

Lossick JG (1989). Therapy of urogenital trichomoniasis. In: Honigberg BM, ed. *Trichomonads parasitic in humans*. New York: Springer-Verlag, pp. 324–341.

Mason PR and Forman L (1982). Polymorphonuclear cell chemotaxis to secretions of pathogenic and nonpathogenic *Trichomonas vaginalis*. *J. Parasitol.*, **68**: 457–462.

McLellan R, Spence MR, Brockman M, Raffel L, and Smith JL (1982). The clinical diagnosis of trichomoniasis. *Obstet. Gynecol.*, **60**: 30–34.

Mendoza-Lopez MR *et al.* (2000). CP30, a cysteine proteinase involved in *Trichomonas vaginalis* cytoadherence. *Infect. Immun.*, **68**: 4907–4912.

Mirhagani A and Warton A (1996). An electron microscope study of the interaction between *Trichomonas vaginalis* and epithelial cells of the human amnion membrane. *Parasitol. Res.*, **82**: 43–47.

Narcisi EM and Secor WE (1996). In vitro effect of tinidazole and furazolidone on metronidazole-resistant *Trichomonas vaginalis*. *Antimicrob. Agents. Chemother.*, **40**: 1121–1125.

Pastorek II JG, Cotch MF, Martin DH, and Eschenbach DA (1996). Clinical and microbiological correlates of vaginal trichomoniasis during pregnancy: for the vaginal infection and prematurity study group. *Clin. Infect. Dis.*, **23**: 1075–1080.

Pindak FF and Gardner WA (1993). Contact-independent cytotoxocity of *Trichomonas vaginalis*. *Genitourin. Med.*, **69**: 35–40.

Poch F, Levin D, Levin S, and Dan M (1996). Modified thioglycolate medium: a simple and reliable means for detection of *Trichomonas vaginalis*. *J. Clin. Microbiol.*, **34**: 2630–2631.

Rust JH (1976). An assessment of metronidazole tumorigenicity: studies in mouse and rat. In: Finegold S, ed. *Metronidazole. Proceedings of the International Metronidazole Conference*. Princeton, NJ: Excerpta Medica, pp. 138–144.

van Der Schee C *et al.* (1999). Improved diagnosis of *Trichomonas vaginalis* infection by PCR using vaginal swabs and urine specimens compared to diagnosis by wet mount microscopy, culture, and fluorescent staining. *J. Clin. Microbiol.*, **37**: 4127–4130.

Witkin SS, Inglis SR, and Polaneczky M (1996). Detection of *Chlamydia trachomatis* and *Trichomonas vaginalis* by polymerase chain reaction in introital specimens from pregnant women. *Am. J. Obstet. Gynecol.*, **175**: 165–167.

Wölner-Hanssen *et al.* (1989). Clinical manifestations of vaginal trichomoniasis. *JAMA*, **261**: 571–576.

Yule A, Gelan MCA, Oriel JD, and Ackers JP (1987). Detection of *Trichomonas vaginalis* antigen in women by enzyme immunoassay. *J. Clin. Pathol.*, **40**: 566–568.

16 *Anisakis* spp.

Bjørn Berland

The nematodes *Anisakis simplex*, 'herring-worm' or 'whale-worm', and *Pseudoterranova decipiens*, 'cod-worm' or 'seal-worm', require marine mammals, whales and seals, as definitive hosts, while many marine fishes serve as intermediate or transport hosts for their third stage larvae. In abnormal hosts, such as man, live larvae ingested with raw fish bore into the gut wall, causing nausea and gastric or abdominal pains. This human affliction, anisakidosis/ anisakiasis, is common in countries, such as Japan, where raw fish is consumed. *Anisakis simplex* larvae may cause serious allergy. The biology, morphology, and nomenclature of these worms, and clinical features, diagnosis and prevention of the affliction are reviewed.

Introduction

In coastal areas in northern countries, people have always used marine fish for food, and the presence on and in their viscera and flesh of 'worms' was well known. These worms, being encapsulated nematode larvae, are, in Norway, known collectively as 'kveis' and have been regarded as being 'normal' and of no consequence for the value of fish as food, but admittedly may reduce its aesthetic value. In northern Europe, there is no tradition for eating raw fish; fresh fish were boiled or fried, or were preserved for later consumption by drying (stockfish), salting (e.g. herring) and salting + drying (bacalao); deep freezing was only a temporary option in cold winters. In the Far East, mainly in Japan, consumption of raw fish is very common.

History

In the 1950s, a small number of people in Holland were admitted to hospital with acute abdominal pains. In sections of resected oedematous pieces of digestive tract, a worm, unknown to the medical profession, was found. All affected persons had consumed a Dutch speciality – a very lightly salted herring – shortly before becoming ill. In Japan, cases of atypical intestinal diseases were reported in the late 1950s. When the first Dutch publication appeared in 1960 (Thiel *et al.* 1960), the Japanese recognized the cause of an old problem, and the disease was acknowledged as a public health problem in Japan.

The worm seen in histological sections is a nematode larva, genus *Anisakis*, found not only in herring, but also in many marine fishes and squids. Based on its generic name, this new human disease became known as *anisakiasis* in English, but *anisakiosis* is recommended (see *Veterinary Parasitology*, **29**: 229–326). The initial publication of the Dutch work led to great and varied research efforts in several countries, mainly in Holland and Japan. The basic biological knowledge – its identity, biology, life cycle, and distribution – were needed, but also new cases, infectivity and behaviour in abnormal hosts, experimental infection and the clinical picture, and host reaction were described. As consumption of live larvae is a prerequisite, it became important to study their tolerance to temperature, salts, acids, and other factors, in order to render them harmless. In the early years the nematode was named *Anisakis* sp. larva, but gradually it was accepted as the larva of *A. simplex*. The accumulated knowledge on *Anisakis* and anisakiosis was reviewed by Oshima (1972) and Smith and Wootten (1978). A number of Japanese publications on biological data, diagnosis, and clinical aspects were edited by Ishikura and Namiki (1989) and Ishikura and Kikuchi (1990). The latter two books provide many references to Japanese literature on the subject.

Biology and nomenclature

Anisakis simplex, known as 'herring-worm' and 'whale-worm', is only one larval type among several, occurring in marine fishes (Figure 16.1). The other important one is *Pseudoterranova*

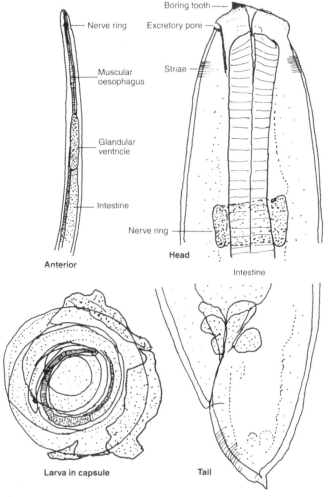

Figure 16.1 Anisakis simplex larva.

decipiens – 'cod-worm' or 'seal-worm' – formerly known as *Porrocaecum decipiens*, *Terranova decipiens*, and *Phocanema decipiens* (Figure 16.2). A third larval type is *Contracaecum/ Phocascaris* (Figure 16.3). Species in these genera live as sexually mature adult males and females in the stomach of their definitive warm-blooded hosts. *Anisakis simplex* occurs mainly in porpoises and whales (cetaceans). It is also found in seals (pinnipeds), *P. decipiens* is found in various seals, and *Contracaecum/Phocascaris* also in seals (*Contracaecum* species also occur in birds, the taxonomy of these genera is not resolved). Having similar morphology and biology, they are all placed in the family Anisakidae, and their larvae are potentially able to cause anisakiosis in abnormal hosts. For this reason the less specific anisakidosis is now gaining acceptance in the literature, although *A. simplex* larva is responsible for the majority of human cases.

Eggs are voided into the sea with host faeces, and in the egg the tiny larva develops and moults. It has been generally assumed that the second stage larva hatches from the egg, but Køie *et al.* (1995) found a second very thin cuticle surrounding the larva; thus, the larva hatching from the egg is a third stage one. This view is contested by Measures and Hong (1995), who by electron microscopy did not see this second sheath.

Suitable small invertebrates, mainly small crustaceans, which transfer the larvae to higher levels in the food chain, must ingest the sinking eggs/larvae. For *A. simplex*, krill (euphausids) are the main hosts, transferring them to various fishes, such as herring, blue whiting, mackerel

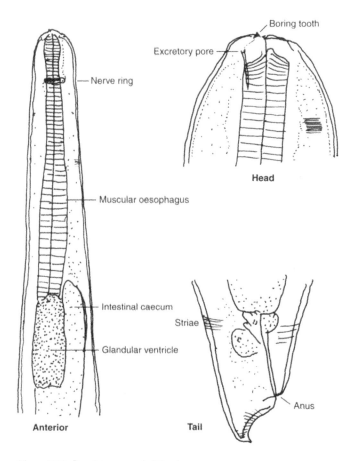

Figure 16.2 Pseudoterranova decipiens larva.

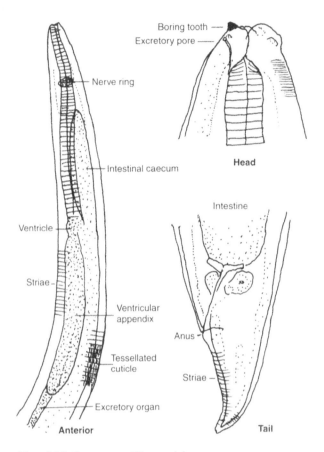

Boring tooth ―
Excretory pore ―――
Nerve ring
Intestinal caecum
Head
Intestine
Ventricle
Striae
Ventricular appendix
Anus ―
Tessellated cuticle
Striae ―
Excretory organ
Anterior
Tail

Figure 16.3 Contracaecum/Phocascaris larva.

and many others. *Pseudoterranova decipiens* eggs/larvae are transferred to fish via small bottom-dwelling crustaceans. Arriving in the fish gut, the larvae bore through into its viscera, body cavity, or muscles. The fish reacts by producing a connective tissue capsule around each larva, which remain dormant until it is damaged or digested. Being regularly challenged by new larvae, the infection in any fish tends to build up slowly; consequently large and old fish usually have a larger larval burden than small and young fish. If a fish, or fish viscera thrown overboard, is eaten by another fish, the larvae will repeat this process of boring into viscera and muscles. In this way old and large carnivorous fish, such as cod, hake, and ling, may harbour very high numbers of encapsulated third stage *A. simplex* larvae.

Anisakis simplex larvae not only lodge on and in the viscera of their fish hosts, but also in the somatic muscles, mainly in their belly flaps. When gutted, most larvae will be removed.

Pseudoterranova decipiens larvae are mainly found in the thick muscles, and where seals are common; this 'seal-worm' is also common in many commercial fishes, causing problems for the fishing industry. Due to such serious problems in Atlantic Canada, much work on its population biology has been done in that country (see Bowen 1990).

When the 'correct' final host – seal or whale – eats the fish host, the larvae become free, bore partly into the stomach wall, grow and moult twice before reaching the final fifth adult stage. Often, many specimens form tight clusters in craters in the stomach wall.

The anisakid larvae found in fish are all third stage. Lips are absent, but a small boring tooth is present on the head, and there are no visible reproductive organs. The shape of the head, tail, and anterior digestive tract is used for distinguishing each larva. *Anisakis simplex* (Figure 16.1) forms a 3–4 mm tight flat spiral, while *P. decipiens* (Figure 16.2), being larger and yellowish-brown, is shaped more like a corkscrew. Both have a glandular ventricle, but *P. decipiens* also has an intestinal caecum. *Contracaecum/Phocascaris* larvae have opposed intestinal caecum and ventricular appendix, and near the head, the cuticle has characteristic transverse striae and longitudinal ribs, giving it a tessellated appearance (Figure 16.3). Similar drawings are found in the literature, but to actually see these characters in whole worms through a microscope, fixed/dead specimens should be made transparent by clearing them in lactic acid, glycerol, phenol–ethanol, or lactophenol.

Clinical features, diagnosis and treatment

In warm-blooded hosts, anisakid larvae will attempt to develop to maturity; in their normal hosts they will succeed, but in abnormal ones – man being one – they may fail. In man, clinical symptoms of anisakidosis include nausea, vomiting, abdominal and gastric pains, diarrhoea, and sometimes urticaria. Some ingested larvae pass through the digestive tract and are voided via anus, but the majority attempt to attach to, and bore into, the mucosa of the digestive tract, challenging the host's defences, resulting in oedema, eosinophilia, and often eliciting acute gastric pains. In Japan, suspected cases of gastric anisakidosis are now routinely diagnosed by gastric endoscopy, and forceps attached to the instrument can easily remove any visible larvae. Intestinal anisakidosis is difficult to diagnose by endoscopy; in such cases X-rays, ultrasound, immunodiagnosis, and serodiagnosis (see Ishikura and Kikuchi 1990) are options. *Anisakis simplex* larvae may even migrate into other organs, such as liver and pancreas, deep into the wall of the digestive tract and even into the abdominal cavity. Larva embedded in gut wall elicit oedema with eosinophilia, immune reactions, inflammatory lesions; ulcers and vanishing tumours; in histological sections degenerate larvae surrounded by necrotic tissue may be seen (see Ishikura and Kikuchi 1990). Anisakidosis is not lethal, but if not diagnosed and treated it may linger on with diffuse symptoms.

According to Ishikura *et al.* (1992) 14 000 reported cases of anisakidosis were known, of these about 90% were gastric, and 10% were intestinal. In countries other than Japan, these values may be reversed; one reason being that the Japanese doctors are very adept at diagnosing the gastric variety. Reviewing the known cases of anisakidosis, Ishikura *et al.* (1992) stated that of the more than 12 000 cases in Japan, *P. decipiens* was involved in only 335. Outside Japan, anisakidosis was reported from 19 countries; a table gives a total of 392 *A. simplex* cases known for the Netherlands. New cases are still reported from various countries, in France 55 are known (Bouree *et al.* 1995). Ishikura *et al.* (1998) revised the accumulated anisakidosis cases worldwide by end of 1997 to 35 000, about 32 300 of them from Japan. Searching *Anisakis* on the Internet yields many more recent reports.

Public health and prevention

Anisakidosis can easily be avoided by abstaining from raw or improperly prepared marine fish, but cultural food preferences are hard to change. *Anisakis simplex* larvae can live for weeks in seawater, but die quickly in strong brine. The Dutch experience with anisakidosis forced the authorities to impose regulations regarding salt, temperature, and storage time to kill larvae in herring. Larvae are killed by proper heating, 60–70°C, and deep-freezing at −20°C or below for some days. In continental Europe, herring are commonly marinated in brine and acetic acid; Karl *et al.* (1995) studied the effects of salt and acid on the survival time of *Anisakis* larvae. It is worth noting that dead larvae (i.e. frozen) fluoresce under UV light, and they can be removed by hand.

Anisakis simplex larvae occur in wild Atlantic marine salmon, but not in farmed salmon fed food pellets. Angot and Brasseur (1993) found no anisakid larvae in farmed salmon, and I have searched several thousand farmed salmon for parasites, without finding any *A. simplex* larvae. If fish are to be consumed raw, farmed marine Atlantic salmon should be the species of choice.

Some people are allergic to various foods, including fish. But do anisakid larvae also play a role in allergy? Recently, a number of papers indicate that. Pozo *et al.* (1996), reporting specific and intense immune response to *A. simplex* extract, concluded that *A. simplex* is able to induce anaphylactic reactions, and that allergy to it should be suspected in patients with allergic symptoms after ingestion of fish. Corres *et al.* (1996) reported on 28 cases of allergy caused by this nematode. Searching the web for *Anisakis* yields many more reports on allergic reactions, caused mainly by consuming raw or under-prepared fish and squids. Several websites give general advice on nematodes in fish, sea-food safety, and fish recipes.

New developments

Anisakis simplex and *P. decipiens* were, until some years ago, regarded as monotypic species; they occur in whales and seals in all cool seas. However, by enzyme electrophoresis, research carried out mainly in Rome, Italy and reviewed by Bullini *et al.* (1997), the assumed simple *A. simplex* is shown to be composed of five sibling species, that is, genetically good species that cannot be distinguished on morphology. The same applies to *P. decipiens* also, with five siblings, three of them in the North Atlantic; their specific identities were resolved by Paggi *et al.* (2000). The species *Contracaecum osculatum* in seals also consists of five siblings. Some of the larvae in these genera can by electrophoresis even be identified to sibling species. In an ecological context, the species and their hosts have co-evolved.

Live *A. simplex* larvae, found in marinated herring, were shown on German TV in 1987, leading to instant reduced demand for fish. The public became aware of possible anisakid larvae in food fishes, and the German authorities established rules for permissible statistical numbers of larvae in imported fish. Other countries tightened control with fish products crossing their borders. A strange situations resulted – lorries loaded with fish were detained on the Italian border, while any housewife in Rome could buy fresh local fish teeming with *A. simplex* larvae. H. Möller in Kiel, Germany convened a special workshop on the nematode problems in North Atlantic fish in 1989 (Möller 1989).

Ukraine and Byelorussia import frozen round herring from Norway. Their food inspection service demands that the herring shall be free of any parasites – which is impossible to guarantee. As the public in other countries are becoming aware of the possible presence of nematode larvae in fish, exporters, importers, and their customers are asking for basic information, such as is presented in this chapter. Also, in Spain the public is becoming aware of anisakid worms in fish, and in 1997 this subject has almost daily been given attention in the media. It is worth noting that most of the allergy work referred to above is done in Spain.

When fish die, and are left ungutted, they will become 'soft in the belly' and due to autolysis larvae may start to wriggle out of their decomposing capsules. There are several papers on the number of *A. simplex* larvae in fish viscera and in fish muscles and it is almost an axiom that they migrate from viscera into the muscles as soon as the host dies (Smith and

Woothen 1974). Thus, it may be imperative that fish should be gutted soon after capture to avoid the larvae moving into the flesh. However, in modern fisheries, large number of fish are either caught in trawls or in purse seines; and are by the ton lifted/hoisted out of the water. This means that almost all fish are exposed to heavy mechanical pressure, which very likely disrupt or damage the connective capsules surrounding the larvae. Roepstorff *et al.* (1993) handled small samples of live herring very gently and kept them ungutted for several days on ice; there was no significant post-mortem migration of *Anisakis* larvae.

The anisakid nematodes dealt with in this chapter have complicated life cycles. Knowledge of their taxonomy and biology, including their host ranges, is essential. When man interferes in the normal life cycle by consuming raw intermediate hosts, fish and squids, with live larvae, their ensuing attempt to develop in an abnormal hosts represent a zoonosis. Armed with basic biological knowledge, we should be able to minimize human cases of anisakidosis.

References

Angot, V. and Brasseur, P. (1993). European farmed Atlantic salmon (*Salmo salar* L.) are safe from anisakid larvae. *Aquaculture*, **118**: 339–344.

Bowen, W. D. (ed.) (1990). Population biology of sealworm (*Pseudoterranova decipiens*) in relation to its intermediate and seal hosts. *Can. Bull. Fish. Aquatic. Sci.*, **222**: 306.

Bouree, P., Paugam, A. and Petithory, J.-C. (1995). Anisakidosis: report of 25 cases and review of the literature. *Comp. Immun. Microbiol. Infect. Dis.*, **18**: 75–84.

Bullini, L., Arduino, P., Cianchi, R., Nascetti, G., D'Amelio, S., Mattiucci, S. *et al.* (1997). Genetic and ecological research on anisakid endoparasites of fish and marine mammals in the Antarctic and Arctic-boreal regions. – In: *Antarctic communities: species, structure and survival*, Battaglia, B., Valencia, J. and Walton, C. W. H. (eds), pp. 39–44. Cambridge University Press, Cambridge, 480pp.

Corres, L. F. de, Audicana, M., Pozo, M. D. del, Muñoz, D., Fernandes, E., Navarro, J. A. *et al.* (1996). *Anisakis simplex* induces not only anisakiasis: report on 28 cases of allergy caused by this nematode. *J. Invest. Allergol. Clin. Immunol.*, **6**: 315–319.

Ishikura, H. and Namiki, M. (eds) (1989). *Gastric anisakiasis in Japan. Epidemiology, diagnosis, treatment.* Springer-Verlag, Tokyo, vi + 144pp (ISBN: 4-431-70036-6).

Ishikura, H. and Kikuchi, K. (eds) (1990). *Intestinal anisakiasis in Japan. Infected fish, sero-immunological diagnosis, and prevention.* Springer-Verlag, Tokyo, vi + 265pp (ISBN: 4-431-70063-3).

Ishikura, H., Kikuchi, K., Nagasawa, K., Ooiwa, T., Takamiya, H., Sato, N. *et al.* (1992). Anisakidae and anisakidosis. *Prog. Clin. Parasitol.*, **3**: 43–102.

Ishikura, H., Takahashi, S., Yagi, K., Y., Nakamura, K., Kon, S., Matsuura, A., Sato, N. *et al.* (1998). Epidemiology: global aspects of anisakidosis. *Parasitology International* (Suppl.) 1998: 23. (Abstracts IX International Congress of Parasitology).

Karl, H., Roepstorff, A., Huss, H. H. and Bloemsma, B. (1995). Survival of *Anisakis* larvae in marinated herring fillets. *Int. J. Food Sci. Technol.* **29**: 661–670.

Køie, M., Berland, B. and Burt, M. D. B. (1995). Development to third-stage larvae occurs in the eggs of *Anisakis simplex* and *Pseudoterranova decipiens* (Nematoda, Ascaridoidea, Anisakidae). *Can. J. Fish. Aquat. Sci.*, **52**(Suppl. 1): 134–139.

Measures, L. N. and Hong, H., (1995). The number of molts in the eggs of sealworm, *Pseudoterranova decipiens* (Nematoda, Ascaridoidea) – an ultrastructural study. *Can. J. Fish. Aquat. Sci.*, **52**(Suppl. 1): 156–160.

Möller, H. (ed.) (1989). Nematode problems in North Atlantic fish. Report from a workshop in Kiel, 3–4 April 1989. *International Council for the Exploration of the Sea.* C. M. 1989/F:6, Mariculture Committee, 55pp.

Oshima, T. (1972). *Anisakis* and anisakiasis in Japan and adjacent area. *Prog. Med. Parasitol. Japan,* **4**: 301–393.

Paggi, L., Mattiucci, S., Gibson, D. I., Berland, B., Nascetti, G., Cianchi, R. and Bullini, L. (2000). *Pseudoterranova decipiens* species A and B (Nematoda, Ascaridoidea): nomenclatural designation, morphological diagnostic characters and genetic markers. *System. Parasitol.*, **45**: 185–197.

Pozo, M. D. del, Moneo, I., Corres, L. F. de, Audicana, M. T., Muños, D., Fernandez, E. *et al.* (1996). Laboratory determinations in *Anisakis simplex* allergy. *J. Allergy Clin. Immunol.*, **97**: 977–984.

Roepstorff, A., Karl, H., Bloemsma, B. and Huss, H. H., (1993). Catch handling and the possible migration of *Anisakis* larvae in herring, *Clupea harengus*. *J. Food Protect.*, **56**: 783–787.

Smith, J. and Wootten, R. (1974). Experimental studies on the migration of *Anisakis* sp. larvae (Nematoda: Ascaridida) into the flesh of herring, *Clupea harengus* L. *Int. J. Parasitol.*, **5**: 133–136.

Smith, J. and Wootten, R. (1978). *Anisakis* and anisakiasis. *Adv. Parasitol.*, **16**: 93–163.

Thiel, P. H. van, Kuiper, F. C. and Roskam, R. T. (1960). A nematode parasitic to herring causing acute abdominal syndromes in man. *Trop. Geograph. Ned.*, **12**: 97–113.

17 *Diphyllobothrium latum*

B. Göran Bylund

The fish tapeworm, *Diphyllobothrium latum*, occurs mainly in the temperate and subarctic regions of the Northern Hemisphere. A focus for human diphyllobothriasis has for a long time been the eastern and southern Baltic region and adjacent areas of Russia. In addition to the final host, principally man, the life cycle of this parasite includes two obligatory intermediate hosts, that is, a planktonic copepode and a fish. The infective larva developes in the muscles and visceral organs of the fish and is transmitted to man when fish products are eaten raw or without proper treatment. Thus, dietary habits are decisive for the spread of the disease, which gives diphyllobothriasis its nature of an endemic disease. Most carriers of fish tapeworm infection are symptomless or suffer very little discomfort. The most serious effect induced by the parasite is tapeworm anaemia, which is of the pernicious type. In addition to hematological changes the tapeworm anaemia is frequently accompanied by neurological symptoms. Expulsion of the parasite is easily achieved with drugs today and usually leads to complete remission. It is emphasized that campaigns aiming at reducing or eliminating diphyllobothriasis from endemic areas should primarily focus on educational work in order to prevent transmission of the infective larvae from fish to man.

Introduction

Several species of the cestode genus *Diphyllobothrium* are parasites of man. By far, the most important and the species usually associated with diphyllobothriasis or fish tapeworm disease is *D. latum* (Linnaeus 1758; Lÿhe 1910). The main distribution area of this parasite is the north temperate and subarctic regions of Eurasia and North America. It is important to recognize, however, that several other *Diphyllobothrium* species are frequently responsible for human infections. *Diphyllobothrium dendriticum* appears to be the species responsible for most human infections throughout the circumpolar region at high latitudes, beyond the range of *D. latum* (Curtis and Bylund 1991). Several species are implicated in human infections in communities bordering the Northern Pacific Ocean (*D. ursi, D. dalliae, D. nihonkaiense, D. klebanovski, D. yonagoence*, etc.). However, these species are recorded only from sporadic infections in man and apparently they are of minor importance from an epidemiologic point of view.

In 1973, Carneri and Vita (ref. von Bonsdorff 1977) estimated the number of human *Diphyllobothrium* carriers at 9 million. As routine examination for tapeworm ova is no more performed in hospitals in the Nordic countries, there is no reliable estimate of the parasite prevalence in this region today, but apparently the prevalence figures have gone down during the last decades. A well-known focus for human diphyllobothriasis has been the eastern and southern Baltic region and adjacent areas of Russia. In Finland, for example, 20–25%

of the population harboured the fish tapeworm in 1952; 20 years later, the infection prevalence was 1.8% (Wikström 1972). There are also very sparse data available on the prevalence of *Diphyllobothrium* infection in northern native communities of the Eurasian continent and North America, that is, in the areas where human infections apparently has to be referred to other species than *D. latum*.

The life cycle and biology of *Diphyllobothrium latum*

The life cycle of *D. latum* involves the final host, principally man, and two intermediate hosts. (Figure 17.1). Although man is, without doubt, the most suitable final host for *D. latum*, it has been shown that several species of domestic and wild animals (dog, cat, pig, wolf, fox, otter seals, etc.) can harbour the parasite and release viable tapeworm eggs (von Bonsdorff 1977).

The adult parasite resides in the small intestine of the final host. The fish tapeworm is the largest parasite of man (Figure 17.2). Usually the length of the worm ranges from 5 to 10 m (maximum width 2 cm) but specimens reaching a length of 25 m have been recorded. The small (45 × 65 μm), ovoid, operculate eggs are released in enormous numbers with the faeces from the definitive host. Each worm carrier expels 20–40 million eggs a day. If the egg is discharged into fresh water the first larval stage, the *coracidium*, containing the six-hooked *oncosphere*, develops and hatches from the egg within 8–12 days at 16–20°C. The free-swimming coracidium larva hatches from the egg only in the presence of light.

The coracidium has to be ingested by an appropriate planktonic copepode in order to continue the life cycle. The host specificity is not very pronounced at this stage and about 40 different species of freshwater copepodes have been recorded as first intermediate hosts of *D. latum* (von Bonsdorff 1977). Calanoids of the genera *Diaptomus* and *Eudiaptomus* and cyclopoids of the genus *Cyclops* are highly liable to infection and many species in these groups

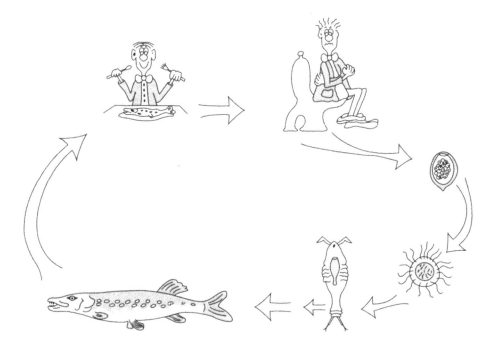

Figure 17.1 The life cycle of the human fish tapeworm, *D. latum*.

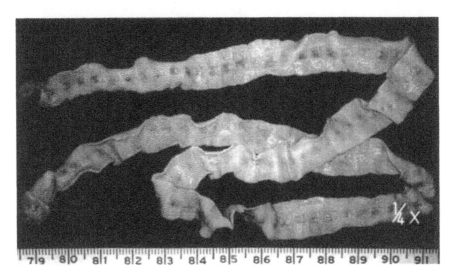

Figure 17.2 Segment of adult *D. latum* expelled from man.

have a circumpolar distribution. The larva penetrates into the hemocoel of this first intermediate host, where the next larval stage, the *procercoid* measuring 300–500 μm, develops. This development takes 2–6 weeks, depending on the water temperature.

If the copepode is ingested by a plankton-eating fish of a particular species the larva penetrates the intestinal wall and migrates to the muscles or various visceral organs of the fish and develops into the *plerocercoid* larvae, the stage infective for the definitive host (Figure 17.3). The whitish plerocercoids are on average 5–15 mm long but may reach even 40–50 mm in length (Figure 17.4). They may occur in almost any organ of the fish, frequently also free in the abdominal cavity. Usually, the plerocercoids lie unencysted in the tissues of the fish but they may sometimes be enclosed in thin connective tissue cysts. From an epidemiological point of view, the occurrence of larvae in the muscle tissue and gonads of the fish are of main importance. If the plankton-eating fish is eaten by an appropriate predatory fish, the larvae may migrate to different organs of this fish and retain their infectivity. The larvae may survive and retain their viability for years in the fish.

Numerous species of freshwater fish may serve as second intermediate hosts for *D. latum* (von Bonsdorff 1977). In the endemic regions in the northwestern Europe the plerocercoids are mainly recorded from northern pike (*Esox lucius*), perch (*Perca fluviatilis*), burbot (*Lota lota*) and ruff (*Acerina cernua*) and these fish species must be considered the most important transmitters of fish tapeworm infection to man. Plerocercoids are rarely recorded from salmonid fish and when occurring in these fish they are encysted on the intestinal tract of the fish and thus are without significance for transmission of the infection to man as the entrails are discarded when the fish is cleaned. Contrary to previous statements it seems to be firmly established that coregonid fish do not harbour plerocercoids of *D. latum* although they frequently harbour larvae of other *Diphyllobothrium* species. *Diphyllobothrium latum* larvae are sporadically reported from some other freshwater fish species too but the validity of these findings is questionable; erroneous identification of larvae of this parasite group previously caused a considerable confusion concerning the true fish hosts for *D. latum* larvae (Bylund 1975).

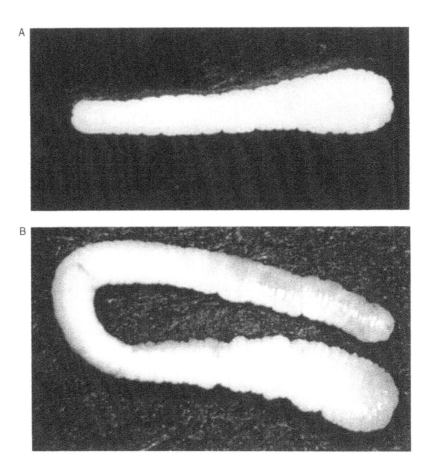

Figure 17.3 (A) Plerocercoid larva of *D. latum* (~8 mm). (B) Pleroceroid larva of *D. latum* (~15 mm).

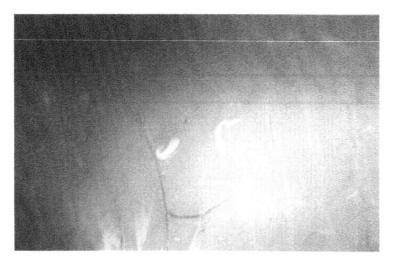

Figure 17.4 Diphyllobothrium latum larvae under peritoneum of host fish (size of larvae ~8 mm).

In North America, the infection is referred to pike (*Esox lucius*), wall-eyed pike (*Stizostedeon vitreum*), sand pike (*S. canadense*), burbot (*Lota maculosa*), yellow perch (*Perca flavescens*), and possibly also *Onchorhynchus* species.

When fish tissues containing viable larvae are eaten by man (or another suitable host), the parasite establishes itself in the small intestine. Within 3–4 weeks it grows and develops into an adult, egg-producing worm. The growth rate during this phase is very fast, up to 22 cm per day, that is, almost 1 cm per hour.

Epidemiology

The complicated life cycle of *D. latum* is completed and maintained if:

- the feeding habits of the human population include ingestion of raw or inadequately prepared fish products of the species harbouring plercoid larvae;
- sanitary habits and sewage disposal facilitate the spread of worm eggs to appropriate natural waters;
- the water biotope available is suitable for development of the larval stages and has the plankton and fish species appropriate for the larvae.

The infection is established in an area only if all these prerequisites are present and, consequently, the life cycle and transmission of the parasite are broken if only one of these prerequisites is eliminated (von Bonsdorff and Bylund 1982).

Raw fish and fish products are highly appreciated dietary components in many population groups. Slightly salted or marinated filets of several freshwater fish species are traditional components in the meals in many regions in the Baltic area. The fillets are salted for one or a few days only and are eaten without further preparation. In addition, slightly salted hard roe is a delicacy served at home as well as in the most pretentious restaurants in the same area, a hazardous delicacy, however, as the roe of pike and burbot, for example, may contain larvae of *D. latum*. Dietary habits are usually local; thus, the infection is attributable to different fish dishes in different regions. Raw or insufficiently heated fish products, however, are always the components responsible for transmission of the infection to man.

The present hygenic practice of flushing faecal products through water closets and sewage systems into lakes and rivers, greatly facilitates the spread of tapeworm infection, especially as the sewage purification plants in many smaller communities and villages are imperfect or absent.

Water biotopes suitable for development of the larval stages exist in large parts of the temperate and subarctic areas. The embryonic development in the egg and hatching of the first larval stage normally takes place in fresh waters at temperatures between 4 and 25°C. However, the parasite life cycle is not strictly bound to a freshwater biotope; the larval development can take place in brackish water with a salinity up to about 0.5–0.6%. Thus, the larvae of *D. latum* was previously frequently encountered in coastal areas of the Baltic Sea where the salinity is low. Shallow littoral zones with water temperatures of 15–20°C provide the most favourable conditions for the parasite development. As the host specificity is not very pronounced, suitable species of intermediate hosts are present in most parts of the temperate and subarctic regions. Fish species like pikes, perches and burbots, most susceptible to the infection, have a more or less circumpolar range in the northern Hemisphere.

As the ecological conditions of the water biotopes in very large areas fulfil the requirements of the parasite, the dietary habits mainly are decisive for the spread of the disease; this also gives diphyllobothriasis its nature of an endemic disease. The reproductive potential of the parasite is very high. Due to this even sporadic worm carriers can give rise to high prevalence of larvae in a fish population. There are numerous examples where population transfers or immigrants retaining their dietary habits have induced new endemic foci of *D. latum* infection (Almer 1974; von Bonsdorff 1977).

The role of domestic and wild animals in the epidemiological patterns of diphyllobothriasis is somewhat controversial. Although it has been firmly shown that a number of animals can harbour the parasite and release viable eggs, all available data indicate that the parasite is absent in regions where all ecological prerequisites for the parasite are fulfilled, including the presence of susceptible fish eating mammals, but human population is absent. Thus it seems reasonable to assume that animals are of minor or no importance for the dissemination of the infection.

Clinical manifestations

The clinical manifestations and pathogenesis involved in tapeworm anaemia were carefully studied in Finland by von Bonsdorff and colleagues already some decades ago when the infection prevalence in the country were still very high (von Bonsdorff 1977).

Most carriers of fish tapeworm infection are symptomless or suffer very little discomfort from the infection. Vague abdominal and gastrointestinal discomfort, fatigue and weakness, dizziness, diarrhoea, and numbness of the extremities are symptoms experienced by some worm carriers. A sensation of hunger as well as a 'craving for salt' can be associated with the infection (von Bonsdorff 1977). Symptoms from the central nervous system (paresthesia, disturbed co-ordination, impairment of the deep sensibility, optic neuropathy, etc.) frequently accompany the disease in worm carriers suffering serious illness from the infection, that is, patients with manifest tapeworm anaemia (Björkenheim 1951; von Bonsdorff 1977).

The most serious effect induced by the parasite is tapeworm anaemia, a disease almost exclusively recorded in connection with *D. latum* infection. The tapeworm anaemia is megaloblastic, of the pernicious type characterized by reduced red cell counts, increased diameter of the red cells, poikilocytosis, anisocytosis, occurrence of nucleated erythrocytes, megaloblasts and normoblasts in the peripheral blood, etc. (von Bonsdorff 1977). The morphology of the cellular components of the, mostly hyperplastic, bone marrow is significantly affected with increased frequency of megaloblasts and promegaloblasts as the predominating change.

Expulsion of the worm after drug treatment usually leads to a rather rapid, more or less complete remission of the neurological symptoms. It takes the patients about two months to recover from the anaemia; the return to a normal serum B_{12} level is much slower and can take more than 1 year (Nyberg and Saarni 1964).

The parasite induces anaemia by competing with the host intestine for dietary B_{12} vitamin. The tapeworm can establish itself in different levels of the intestine and its site has a decisive influence on the development of anaemia. If the parasite resides high up in the intestine, in the jejunum, it absorbs and utilizes large quantities of the dietary vitamin before it reaches the ileum, the absorbtive zone of the intestine.

The B_{12} vitamin is of essential importance for the human body and is involved, for example, in the haematopoiesis. The serum levels of the vitamin is normally 150–900 pg/ml. When below 100 pg/ml, significant tendency towards cytomegaly in blood cells, bone marrow and epithelial cells are recorded. In the surveys carried out in Finland in the 1950s about 50% of the worm carriers reached this critical level while 2% suffered from manifest anaemia. Surprisingly, in a more recent survey (1978–1979) comprising about 350 worm carriers in Finland, significant reduction of the B_{12} vitamin levels were not observed and none of the worm carriers had developed anaemia (Bylund, unpublished data). An explanation for this might have been that the population utilized improved diets with higher vitamin contents compared to the situation some decades earlier.

Diagnosis and therapy

The diagnosis of *Diphyllobothrium* infection is based upon detection of the characteristic egg in the stool of the worm carrier. Due to the enormous production of eggs the diagnosis is usually rapid and egg concentrations techniques are usually not needed. There is evidence for short-term periodicity, however, in the egg production by the parasite and the egg release may temporarily cease (Kamo *et al.* 1986). In suspected cases, therefore, faecal samples should be taken and examined with intervals.

As the morphology of the *Diphyllobothrium* egg is very characteristic, there is no problem with differential diagnosis from other human parasites. It is impossible, however, to distinguish between different *Diphyllobothrium* species on the basis of egg size and morphology.

The tapeworm anaemia is diagnosed from changes in the blood picture and bone marrow. For differential diagnosis genuine pernicious anaemia has to be considered.

Expulsion of the parasite is easily achieved in diphyllobothriasis. Efficient drugs against the parasite, with minor side effects, have been available for decades (desaspidine-derivatives, niclosamide, Praziquantel®, etc.).

Substitution therapy with B_{12} vitamin is needed in cases with pronounced anaemia.

Prevention and control

The aim of preventive and control measures must be to break the life cycle of the parasite. Theoretically, any point of the life cycle can be attacked. In practice the measures must be focused on the following links: (a) elimination of the infection from persons harbouring the parasites; (b) preventing contamination of lakes and rivers with viable tapeworm eggs through appropriate sewage treatment and disposal; and finally (c) preventing transmission of infective larvae from fish to man. As stated earlier, elimination of the parasite from worm carriers is easily achieved today once these are traced.

In endemic areas, elimination of the tapeworm eggs from waste waters and sewage systems is an urgent need. If properly dimensioned, modern purification plants eliminate 95–99% of the tapeworm eggs (Bylund *et al.* 1975) from wastewater. Overloading the plant, however, very rapidly reduces its retention effect and tapeworm eggs passing the plant remain viable. Tapeworm eggs are sensitive to dehydration, low temperatures ($< -5°C$) and also to chemicals (chlorine, formaldehyde). They are rapidly destroyed in garbage piles and latrines but application of this knowledge is difficult or impossible in communities flushing the faecal products into sewage systems.

It seems obvious that preventive measures should most efficiently focus on preventing the transmission of infective larvae from fish to man. Infected fish must not necessarily be avoided but should be treated in manners rendering the larvae innocuous before the fish is eaten. The plerocercoid do not survive temperatures over $+56°C$ or below $-10°C$ (Pesonen and Wikgren 1960; Salminen 1970). Thus, the infection risk is efficiently eliminated if the fish is fried, boiled, or adequately smoked. The risk is also eliminated and the fish can be eaten even without proper heat treatment (raw, salted, marinated, and hard roe) if the fish is kept in household freeze ($-18°C$) for a day or two.

Our basic knowledge on all aspects of the life cycle and epidemiology of *D. latum* is rather complete today. From this knowledge, it is easily emphasized that campaigns aiming at reducing or eliminating diphyllobothriasis from endemic areas should primarily focus on educational work, through public health services, in order to prevent transmission of the infective larvae from fish to man.

References

Almer, B. (1974). Controlling the broad tapeworm, *Diphyllobothrium latum* (L.) (in Swedish). *Limnological Surveys from the National Swedish Environmental Protection Board SNV PM*, **460**, 1–30.

Björkenheim, G. (1951). Neurological changes in pernicious tapeworm anaemia. *Acta Medica Scand. Suppl.*, **260**, 1–125.

von Bonsdorff, B. (1977). *Diphyllobothriasis in man*. London: Academic Press, pp. 1–189.

von Bonsdorff, B. and Bylund, G. (1982) The ecology of *Diphyllobothrium latum*. *Ecology of Disease*, **1**, 21–26.

Bylund. G. (1975). *Delimitation and characterization of European* Diphyllobothrium *species*. Thesis submitted to the Åbo Akademi University, Åbo, Finland, pp. 1–23.

Bylund, G., Wikström, M. and Penttinen, K. (1975). Sewage purification plants – barriers against spread of the fish tapeworm (in Swedish). Information from the Institute of Parasitology, Åbo Akademi University, Åbo, Finland, **14**, 1–9.

Curtis, M. A. and Bylund, G. (1991). Diphyllobothriasis: fish tapeworm disease in the circumpolar North. *Arctic Med. Res.*, **50**, 18–25.

Kamo, H., Yazaki, S., Fukumoto, S., Maejima, J. and Kawasaki, H. (1986). Evidence of egg discharging periodicity in experimental human infection with *Diphyllobothrium latum* (in Japanese). *Jpn J. Parasitol.*, **35**, 53–57.

Nyberg, W. and Saarni, M. (1964). Calculations on the dynamics of vitamin B_{12} in fish tapeworm carriers spontaneously recovering from vitamin B_{12} deficiency. *Acta Medica Scand. Suppl.*, **412**, 65–71.

Pesonen, T. and Wikgren, B.-J. (1960). Bandmasklarvernas salt- och temperaturtolerans. *Memoranda Societas Fauna & Flora Fennica*, **35**, 112 – 118.

Salminen, K. (1970). The effect of high and low temperature treatments on the infestiveness of *Diphyllobothrium latum* with regard to public health. *Acta Vet. Scand.*, Supplementum, **32**, 1–29.

Wikström, M. (1972) The incidence of fish tapeworm, *Diphyllobothrium latum*, in the human population of Finland. *Commentationes Biologicae Societas Scientiarum Fennica*, **58**, 3–11.

18 *Echinococcus* spp.

Inga-Lill Gustavsson Moringlane

Introduction

The cestodes (tape worms) of the genus *Echinococcus* (family Taenidae) may give rise to infection – echinococcosis or hydatidosis in the human host. Echinococcal disease was apparently known to Hippocrates (460–370 BC), who described water-filled cysts in the liver of infected patients. The aetiology of the disease was, however, not identified until the seventeenth century. It was first in 1808 that Rudolphi used the term echinococcus to name the parasite. The word 'echinococcus' is derived from the Greek and means hedgehog berry. In the beginning of this century Dévé described the asexual cycle of the worm and furthermore demonstrated that the scolex plays a major role both in cyst formation and parasitic life cycle.

The life cycle involves two mammalian hosts. Definitive hosts of the Northern Biotype are always carnivores, most often members of the dog (Canidae) or cat (Felidae) families. The animals harbour the adult tapeworm in the small intestine and become infected by ingesting the larval (metacestode) form of the parasite with intermediate host tissue.

The reindeer and tile elk (*Alces alces*) are the principal intermediate hosts in northern regions. In Eurasia, the range of the elk extends southward to about latitude 50° N. Intermediate hosts are mainly herbivorous, but also omnivorous mammals that become infected by ingesting eggs passed in the faeces of infected definite hosts, which may contaminate grass or other vegetation or drinking water. Humans are incidental intermediate hosts of the larval form (Figure 18.1).

Four species of *Echinococcus* are recognized, three of which cause distinctive forms of disease: *Echinococcus granulosus* (CHD), *E. multilocularis* (AHD), and *E. vogeli* (polycystic hydatid disease, PHD). *Echinococcus oligarthrus*, has so far been recognised only in South America where only a couple of cases of human infection have been reported to date (Lopera *et al.* 1989). The adult cestodes are small (2–10 mm) and they consist of a scolex with four suckers and a double row of rostellar hooks and two to five proglottids. Taxonomically significant morphologic differences between adult *Echinococcus* species include the form of the strobila, the position of the genital pore in mature and gravid proglottids, the size of rostellar hooks, the number and distribution of testes and the form of the gravid uterus. The eggs are secreted into the faecal mass and pass into the environment via host faeces. Despite the fact that the definitive host may harbour many thousands of these small worms, infections are rarely symptomatic in this host (Schantz 1989).

Pathogenesis

Echinococcus granulosus

Ingested eggs hatch and release embryos (onchospheres) into the small intestine of the host. These embryos penetrate the intestinal mucosa, enter the blood circulation of the host and

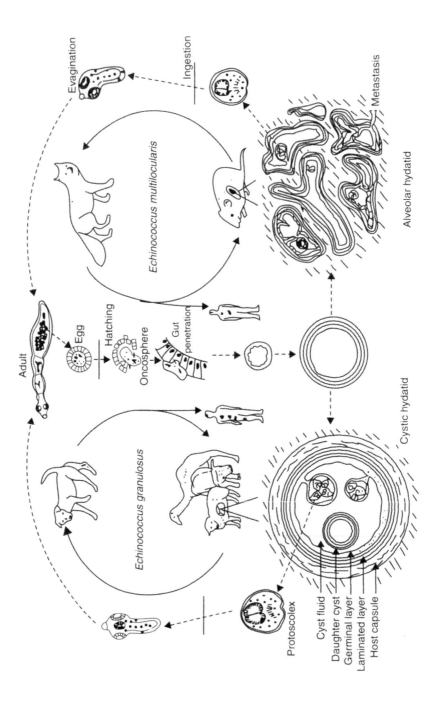

Figure 18.1 Life cycle of *Echinococcus*, *E. granulosus*, and *E. multilocularis*. Reproduced with permission. Wen *et al.* (1993) Diagnosis and treatment of human hyatidosis. *Br. J. Clin. Pharmacol.*, **35**, 567–574.

are carried to the liver and other sites where cystic development (metacestode) starts (Figure 18.1). Spherical well-delineated, primary cysts, containing living scolices, are formed in the human host. These cysts are most often found in the liver (65% of cases) followed by the lungs (25%) and other organs such as kidney, spleen, reproductive organs, heart, and seldom in bone. The space-occupying process causes repression or displacement of vital host tissue, vessels or organs. Thus, the clinical manifestations of the disease are determined by the site, the size and the number of cysts (Gottstein and Hemphill 1997).

Echinococcus multilocularis

The most common life cycle of *E. multilocularis* involves transmission between foxes (host to adult worm) and rodents as intermediate hosts (larval worm). Domestic dogs and cats can also become infected when eating infected rodents. Human infection occurs as a result of ingestion of tapeworm eggs by contamination of fox, dog, or cat faeces. The natural hosts are generally separated ecologically from humans. Infection of humans by *E. multilocularis* is less common than infections by *E. granulosus*.

The primary localization of *E. multilocularis* larvae in humans, as well as in the natural intermediate hosts (rodents) is the liver. Local expansion of the lesion and metastases to the lungs and brain may follow. The typical lesion is a dispersed mass of fibrous tissue with a lot of scattered cavities ranging from a few millimetres to centimetres in size. In chronic cases, a central necrotic cavity can be formed, containing a viscous, yellowish to brown fluid, which may be superinfected. Metacestode proliferation is usually accompanied by a granulomatous host reaction and the induction of intense cellular inflammatory infiltration of the periparasitic hepatic area. The larval mass proliferates by exogenous budding of the germinative membrane, thus producing an alveolar like pattern of microvesicles. The invasion of surrounding tissues resembles cancer in behaviour and appearance. An entire hepatic lobe may be replaced by the larval lesion before the disease becomes clinically manifest. Scolices are rarely observed in infections of *Echinococcus multilocularis* in human (Gottstein and Hemphill 1997).

Epidemiology

Geographic distribution of Echinococcus granulosus

The distribution of *E. granulosus* is cosmopolitan including North and Central Europe, the Mediterranean, Central Asia, north and north-western China, Australia, East Africa, and parts of South America.

In a recent publication, the infection by *E. granulosus* is regarded as the most frequent parasitosis in Europe (Seiferth *et al.* 1993). Information on cases of echinococcosis in Western Europe is relatively complete in comparison with other geographic regions where reporting systems are generally lacking and the quality of information varies.

The northern biotype of Echinococcus granulosus

Studies of populations of *E. granulosus* from different regions have demonstrated intra-specific variations in final and intermediate host assemblages together with other characteristics. According to Rausch (1986), two main biological forms of *E. granulosus* have been recognized, namely the 'European Form' and the 'Northern Form', which is considered to be ancestral to the 'European Form'. The 'Northern Form' occurs in Holarctic zones and under favourable conditions also at lower latitudes in North America (north of 45° N latitude) and in Eurasia.

In the larval stage, the northern biotype of *E. granulosus* occurs in ungulate animals of the family Cervidae, 'cervid strain'. There is no evidence that its development takes place in domestic ungulates, excluding the domesticated reindeer. The dog, *Canis lupus*, may replace the wolf as a final host. The extent of involvement of other canids in the cycle in Eurasia is not clear. The reindeer and elk are the principal intermediate hosts in northern regions. In Eurasia, the range of the elk extends southward to latitude 50° N.

The Nordic countries

There is evidence that this form (the cervid strain) occurs in Norway, Sweden, and Finland in a dog–domestic reindeer cycle. In northern Norway, the prevalence of *E. granulosus* cysts in the lungs of reindeer is reported to have decreased after introduction of annual praziquantel treatment of dogs and their exclusion from slaughter places (Kummeneje 1982).

Echinococcus granulosus cysts have also been found in the lungs of slaughtered reindeer in northern Sweden (Roneus 1966) and also lately, in 1996 (A. Uggla, National Veterinary Institute, Uppsala, Sweden, personal communication). In the period in between there were no veterinary observations of *E. granulosus* cysts in reindeer. The latest human case of autochthonous *E. granulosus* in Sweden was detected in 1984 in a 20-year-old woman of Lappis origin, with a large liver cyst. Imported cases are annually diagnosed in Sweden especially from the Mediterranean region, Central Europe and the Near East (Czechowski *et al.* 1992).

There are no recent epidemiological reports from Finland, but in 1969 1.2% of reindeer were reported to be infected with cysts in areas bordering northern Sweden. Iceland is free of *E. granulosus*, and the last human case of cystic echinococcosis was diagnosed in 1960. *Echinococcus* sp. is not present in Greenland possibly due to the unsuitability of the local rodents as intermediate hosts.

The Baltic states

In Estonia, Latvia, and Lithuania, the notification of echinococcal disease is not mandatory and the prevalence is generally not well known.

In Estonia, echinococcosis has not been a reported disease for a long time. There are no official data on prevalence, however, single cases of *E. granulosus* in animals in 1993–1994 were seen. In Latvia, 10 human cases with *E. granulosus* in the liver were notified and one case of cerebral echinococcus since 1984. The latest cases were from 1995. There are also veterinary data from 1994 to 1995, which show a very low incidence in cattle (0.001–0.002%) (Ludmila Jurevica epidemiologist, personal communication).

In Lithuania, the epidemiological situation is also uncertain. During the same period, 1992–1996, nine cases of human echinococcosis were recorded, two of which were fatal. From the same period, positive serology for echinococcosis has been reported in 108/533 (20%) patients with suspected disease. Furthermore, in 1997, 152 persons were examined during 10 months, 40 of whom turned out with a positive serology of echinococcus (personal communication, A. Laiskonis, Laboratory of Parasitology of State Public Health Centre, Kaunas, 1998).

Russian Federation and adjacent countries of Central Asia

In north-eastern Siberia, rates of infection have been high (25–70%) in domesticated reindeer. The sheep strain, and perhaps other pastoral strains of *E. granulosus*, occur throughout most of Russia and the newly independent nations of the former Soviet Union. (Martynenko *et al.* 1988). High rates of infection in several domestic intermediate hosts, including sheep, cattle, camels, and pigs have been recorded.

Geographic distribution of *Echinococcus multilocularis*

Alveolar hydatid disease is a highly lethal zoonotic infection caused by the larval stage of *E. multilocularis*. The life cycle of *E. multilocularis* involves, most often, transmission between foxes (host to adult worm) and rodents as intermediate host (larval form). *Echinococcus multilocularis* is found in the Northern Hemisphere with a high prevalence in northwestern Canada, Alaska, Eurasia (former USSR), Japan, and China. Hunters, trappers, and people who work with fox fur are at greater risk. In the northern tundra zone, *E. multilocularis* occurs in foxes, mainly the Arctic fox, *Apollo lagopus* and the rodents that they prey on. *Echinococcus multilocularis* is also present on some sub-arctic islands including St Lawrence Island. The cestode is found on some islands of the Canadian Arctic Archipelago but is not present on the northernmost islands of Canada or Greenland, possibly due to the unsuitability of the local rodents as intermediate hosts (Rausch 1986).

However, *E. multilocularis* has, prior to the 1960s, spread from the northern tundra zone and become established in central North America. Only two human cases have been reported to date within the central North American focus. No other human cases are known from this region.

Hyperendemic foci has been described in some Eskimo villages in the North American tundra zone where local dogs feed upon infected commensal rodents and AHD have been diagnosed in Eskimos of a limited number of communities in Alaska (Anonymous 1991; Wilson *et al.* 1995). There are two other species within the genus *Echinococcus*, which are considered valid, *E. vogeli* and *E. oligarthrus*. Infection in humans with *E. vogeli* or *E. oligarthrus* results in a polycystic form of disease. Only a couple of cases with *E. oligarthrus* have been reported.

Role of Echinococcus *strains*

The understanding of the nature and diversity of variation within the genus *Echinococcus* has increased rapidly in the last years through successful characterisation of the nuclear and mitochondrial genomes of representative isolates of the strains (Thompson and Lymbery 1990). The new data, based on genome patterns, generally support previous characterisations based on morphological and biological criteria. The current data suggest that about seven genetically distinct populations exist concerning *E. granulosus*. Concerning the different populations of *E. multilocularis* from Europe, Alaska, and North America few data are up to now available. However, large morphological studies and experimental infection trials concluded that the cestode in central North America appeared to be indistinguishable from that of the northern tundra zone. The only exception was indications of differences in the ability of the larvae to develop in rodents of various species. It is now important to recognize that biological differences may exist between populations identified as *E. granulosus* and that these differences may account for local differences in patterns of transmission and public health significance of the disease.

Diagnosis and treatment in human echinococcosis

Clinical diagnosis of cystic or alveolar echinococcosis in human is based on clinical findings, epidemiological data, morphological changes revealed by ultrasonography, computerized tomography (CT), magnetic resonance imaging (MRI), and immunological and other laboratory tests.

Surgical intervention is the basic form of treatment of cystic echinococcosis but chemotherapy with benzimidazole carbamates can be regarded as an alternative treatment in patients not suitable for operation or most often as a supplementary treatment pre- and postoperative to prevent relapses of the disease (Czeckowski *et al.* 1992). Concerning treatment of alveolar echinococcosis, surgical resection of the involved liver segment and of metacestode lesions from other affected organs is indicated. Post-operative chemotherapy with benzimidazole carbamates is often given for several years. However, long-term prognosis is generally bad. Liver transplantation has been performed in a limited number of patients with alveolar echinococcosis.

Prophylactic measures

Vaccination of intermediate hosts may in the future reduce the prevalence of human echinococcosis. In immunisation trials of sheep with onchosphere secretions or with an onchosphere homogenate 97–98% resistance to a challenge infection of the sheep with *E. granulosus* eggs have been achieved. The vaccine has the potential to be used as a tool for control of transmission of *E. granulosus* through its natural intermediate hosts (particularly domestic ungulates).

Traditional methods of hydatid control (antihelminthic treatment of dogs) include regular treatment within the prepatent period for prolonged periods (20–30 years) to achieve satisfactory control (Bowles and McManus 1993; Heath and Holeman 1997).

References

Anonymous, (1991). A collaborative study of echinococcosis by physicians of Alaska and the Soviet Far East. A progress report. *Alaska Medicine*, **33**, 34–35.

Bowles, J. and McManus, D. P. (1993). Molecular variation in Echinococcus. *Acta Tropica*, **53**, 291–305.

Craig, P. S., Eckert, J., Jenkins, D. J., Macpherson, C. N. L., and Thakur, A. (1995). In: Epidemiology and control of hydatid disease. In: *Echinococcus and hydatid disease* (R. C. Thompson and A. J. Lymbery, eds) Cab International, Oxon, pp. 233–244, 286–288.

Czechowski, J., Gustavsson Moringlane, I.-L., and Aust-Kettis, A. (1992). Ultrasound investigation during treatment of liver hydatid disease with benzimidazole-carbamates. *Ultrason. Pol.*, **23**, 29–39.

Gottstein B. and Hemphill, A. (1997). Immunopathology of echinococcosis. *Chem. Immunol.*, **66**, 177–208.

Heath, D. D. and Holeman, B. (1997). Vaccination against echinococcus in perspective. *Acta Tropica*, **67**, 37–41.

Kummeneje, K. (1982). Control of echinococcosis (hydatidosis) in reindeer and dogs. Abattoir statistics from Kautokeino, West Finmark, northern Norway. *Norsk Veterinaertidskrift*, **94**, 419.

Lopera, R. D., Melendez, R. D., Fernandez, I., Sirit, J., and Perera, M. P. (1989). Orbital hydatid cyst of *E. oligarthrus* in a human in Venezuela. *J. Parasitol.*, **75**, 467–470.

Martynenko, V. B. *et al.* (1988). Echinococcus distribution in the USSR. *Medisinskaia Parazitologiia Prazitarnye*, **3**, 84–88.

Rausch, R. L. (1986). Life cycle patterns and geographic distribution of Echinococcus species. In: *The biology of Echinococcus and hydatid disease* (R. C. A. Thompson, ed.). George Allen and Unwin, London, pp. 44–80.

Roneus, O. (1966). Studies on the aetiology and pathogenesis of white spots in the liver of pigs. *Acta Vet. Scand.*, **7** (Suppl 16), 1–112.

Schantz Peter, M. (1986). Echinococcosis. In: *Tropical medicine and parasitology* (Goldsmith and Heyneman, eds). Prentice-Hall, p. 503.

Seiferth, T., Endsberger, G., and Stolte, M. (1993). Echinococcosis – current status of diagnosis and therapy. *Leber Magen Darm*, **4**, 161–164.

Thompson, R. C. and Lymbery, A. J. (1990). Echinococcus: biology and strain variation. *Int. J. Parasitol.*, **4**, 457–470.

Wilson, J. F. *et al.* (1995). Alveolar hydatid disease. Review of the surgical experience in 42 cases of active disease among Alaskan Eskimos. *Ann. Surgery*, **221**, 315–323.

19 *Enterobius vermicularis*

Kjell Alestig and Johan Carlson

Enterobiasis, pinworm or threadworm disease, is caused by a small nematode, *Enterobius vermicularis*, also named *Oxyuris vermicularis*. This parasite infects man only and animals have no place in the life cycle of the worm. There are related species, however, that can infect animals such as chimpanzees and gibbons. Enterobiasis is a very old disease. *Enterobius vermicularis* eggs have been found in a coprolite carbon in Utah, United States, and been dated to 7837 BC. The worm was described and characterized in 1758 by Linnaeus (Cook 1994).

Epidemiology

Enterobiasis is probably the most common parasite infection in the world. While other worms are often nearly eradicated mainly due to an increase in community hygiene standards in many developed countries, enterobiasis is the only worm infection which still is rather prevalent in the population. Enterobiasis is common in all countries and infects people of all ages but infections are especially frequent in children of 5–10 years of age (Juckett 1995) and the number of infected subjects globally has been calculated to be in the range of one billion. When children live in crowded conditions, the infection is easily spread and infection rates up to 80–90% have been reported from institutions in developing countries (Makhlouf *et al.* 1994). In infected families, adults and children usually share the infection and re-infections after drug therapy are common. This infection may also spread sexually, especially between homosexuals. In institutions, the disease is spread in a similar way as many bacteria (e.g. streptococci and pneumococci). Commonly, viable eggs will transmit the disease via direct and indirect contacts from bed-linen, clothes, toys, etc. Therefore, enterobiasis is a disease that is rather difficult to eradicate.

Life cycle

Adult *E. vermicularis* inhabit the lumen of the caecum and appendix. The female is about 10 mm long and 0.5 mm broad. The male is smaller, only 2–5 mm long. Males have a curved tail and a single large copulation spicule. Besides *E. vermicularis*, a variant, *E. gregorii*, has been described from France and from the United Kingdom (Hugot and Tourte-Schaefer 1985; Chittenden and Ashford 1987; Ahn *et al.* 1992; Hasegawa *et al.* 1998). The life-cycle of this variant is identical, but the *E. gregorii* male has a shorter copulation spicule.

After ingestion of eggs, these are hatched in the stomach and the small intestine. The larvae then migrate downwards the intestine and after moulting twice the larvae become adults. The number of worms may vary considerably from a few up to several hundreds of adults, but their life span is short and adult worms do not live for more than 1–3 months.

Oviposition starts after approximately 5 weeks of age and a gravid female may contain 10 000 eggs. The gravid worm migrates via the colon to the anus where egg deposition occurs. The worm will then disintegrate and die, often disrupting during itching. The larvae in the eggs laid around the anus become infective already within 6 h and auto-infection via fingernails/finger sucking is common (Herrström et al. 1997). Hatching larvae near the anus may also migrate back to the intestine. Eggs are also found in high numbers in beds and on the floor.

Clinical signs

Most pinworm infections are asymptomatic. When symptoms occur, it is usually in the form of itching from the anus area. Itching is usually worse at night when migration of females and egg deposition occurs. The degree of symptoms are related to the parasite load. Pruritus at night may disturb the sleep for affected children and certainly are an annoyance in the daily life of the affected family. The medical importance of the disease is otherwise small. Some worms may find their way into adjacent orifices, most commonly the female genito-urinary tract, producing irritative symptoms such as vulvovaginitis and cystourethritis. Case reports of extra-intestinal infestation by enterobius in the peritoneal cavity, endometrium, lungs, liver, and other organs occur, but mainly from developing countries where high para-site loads could be expected in many patients (Arora et al. 1997; al-Rufaie et al. 1998). Secondary bacterial infections after itching have been reported in children with atopic der-matitis. The worms probably do not cause appendicitis, even if some studies have suggested such a connection (Dahlstrom and Macarthur 1994; Makni et al. 1998). Thus, itching, sleep-ing problems, and possible psychological effects because of fear of worms remain the main clinical features of this infection.

Diagnosis

For the proper diagnosis of an enterobius infection, eggs from the peri-anal area should be visualized. In the morning, a strip of tape should be applied with the sticky side to the skin. The strip can then be transferred onto a glass slide for direct microscopy. Eggs have a typical appearance. Repeated investigations are of value; three consecutive specimens will detect 90% of the infections. Other tests, for example, a standard stool examination, are less valuable and eosinophilia is generally not present.

Treatment

Treatment is most important for children with heavy infections and with clinical symptoms. The aim of treatment is to reduce the number of worms so symptoms will disappear. Piperazine, pyrantel embonate, and mebendazole are some of the drugs recommended for treatment (Cook 1994; Gan et al. 1994; Conners 1995; Venkatesan 1998). Mebendazole is usually preferred since the drug has a low intestinal absorption allowing for equal doses to be used in children and adults with a satisfactory effect on the worm load. In addition, there are seldom any significant side effects. Albendazole is perhaps even more potent and is some-times recommended as a second line drug. Neither albendazole nor mebendazole should be given during pregnancy while pyrantel embonate could be used safely in pregnant women.

Most often, treatment is given with single doses of mebendazole, 100 mg, to all members of a family and an additional dose 1 week later is often recommended. To remove eggs from the surroundings, a change of bed- and night-clothes may be performed. However, hygienic measures should not be used in excess. Whatever measures are taken, re-infections are common, especially in children attending nurseries or other institutions. A recent Swedish study reported a prevalence of 27% pin worm infections in children in three different nurseries (Kjellberg and Heyman 1993). With rigorous hygienic measures in the form of intense cleaning of floors, of walls and of toys, the prevalence could be reduced only down to 7%. As no children had any symptoms the usefulness of these strong efforts, was questioned by the authors.

Thus, physicians should remember that enterobiasis is usually an innocent infection, but often difficult to eradicate. Too much attention with rigorous hygiene and repeated treatments may not lead to eradication, but to increasing psychological problems in the affected family. Drug treatment shall mainly be given when symptoms occur and after treatment the worm should be given no further attention.

References

Ahn, Y. K., Chung, P. R., and Soh, C. T. (1992). *Enterobius gregorii* Hugot, 1983 recovered from school children in Kangwon-Do, Korea. *Kisaengchunghak Chapchi*, **30**, 163–167.

al-Rufaie, H. K., Rix, G. H., Perez Clemente, M. P., and al-Shawaf, T. (1998). Pinworms and postmenopausal bleeding. *Journal of Clinical Pathology*, **51**, 401–402.

Arora, V. K., Singh, N., Chaturvedi, S., and Bhatia, S. (1997). Fine needle aspiration diagnosis of a subcutaneous abscess from *Enterobius vermicularis* infestation. A case report. *Acta Cytologica*, **41**, 1845–1847.

Chittenden, A. M. and Ashford, R. W. (1987). *Enterobius gregorii* Hugot, 1983 first report in the U.K. *Annals of Tropical Medicine and Parasitology*, **81**, 195–198.

Conners, G. P. (1995). Piperazine neurotoxicity: worm wobble revisited. *Journal of Emergency Medicine*, **13**, 341–343.

Cook, G. C. (1994). *Enterobius vermicularis* infection. *Gut*, **35**, 1159–1162.

Dahlstrom, J. E. and Macarthur, E. B. (1994). *Enterobius vermicularis*: a possible cause of symptoms resembling appendicitis. *Australian and New Zealand Journal of Surgery*, **64**, 692–694.

Gan, Y., Wu, Q., Ou, F., Li, G., Huang, J., Wei, Q., Nong, C., and Nong, L. (1994). Studies on the efficacy of albendazole candy for treatment of intestinal nematode infections. *Chung-Kuo- Chi-Sheng- Chung-Hsueh-Yu- ChiSheng-Chung-Ping-Tsa-Chih*, **12**, 147–149.

Hasegawa, H., Takao, Y., Nakao, M., Fakuma, T., Tsuruta, O., and Ide, K. (1998). Is *Enterobius gregorii* Hugot, 1983 (Nematoda: Oxyuridae) a distinct species? *Journal of Parasitology*, **84**, 131–134.

Herrström, P., Friström, A., Karlsson, A., and Högstedt, B. (1997). *Enterobius vermicularis* and finger sucking in young Swedish children. *Scandinavian Journal of Primary Health Care*, **15**, 146–148.

Hugot, J. P. and Tourte-Schaefer, C. (1985). Etude morphologique des deux Oxyures parasites de l'homme: *Enterobius vermicularis* et *E. gregorii*. *Annales de Parasitologie Humaine et Comparee*, **60**, 57–64.

Juckett, G. (1995). Common intestinal helminths. *American Family Physician*, **52**, 2039–2048, 2051–2052.

Kjellberg, G. and Heyman, B. (1993). Springmaskinfektion hos förskolebarn (Enterobiasis in preschool children). *Läkartidningen*, **90**, 2993–2995.

Makhlouf, S. A., Sarwat, M. A., Mahmoud, D. M., and Mohamad, A. A. (1994). Parasitic infection among children living in two orphanages in Cairo. *Journal of the Egyptian Society of Parasitology*, **24**, 137–145.

Makni, S., Makni, F., Ayadi, A., and Jlidi, R. (1998). L'oxyurose appendiculaire. A propos de 205 cas. *Annales de Chirurgie*, **52**, 668.

Venkatesan, P. (1998). Albendazole. *Journal of Antimicrobial Chemotherapy*, **41**, 145–147.

20 *Toxocara canis*

Inger Ljungström

Toxocara canis is a parasitic nematode and has a cosmopolitan distribution. The definitive hosts are dogs, foxes, and less frequently other canids, and the life cycle is very complex in the hosts. The paratenic hosts, mammals and birds, acquire the infection by ingesting eggs excreted by infected canids. The larvae hatch in the small intestine, penetrate the mucosa, and after migration the larvae can be found in virtually all organs of the paratenic hosts. In humans the parasite causes visceral, ocular, and covert toxocarosis. The clinical signs and symptoms are a consequence of the extent and frequency of infection, larval migration, and the host response.

Toxocara excretory–secretory (ES) products are highly antigenic and useful in serodiagnosis, the antigen is produced by prolonged *in vitro* cultivation of the larvae. The antigen is composed of five major groups of molecules, which are heavily glycosylated proteins. The major host responses to these antigens include a marked eosinophilia and hypergammaglobulinemia.

History

Toxocara spp. are parasitic nematodes belonging to the order *Ascaridida*. In 1782, *Toxocara canis* was described as *Ascaris canis* and the clinical illness named visceral larval migrans (VLM) was ascribed in 1947 (Perlingiero and Gyorgy 1947) to *Ascaris lumbricoides*. However, Beaver *et al.* (1952) described a series of children who had eosinophilia and suffered severe multisystem diseases, and by serial sections of a larva from a patient, they identified the larva as *T. canis*. At about the same time, Wilder (1950) discovered lesions of the eyes containing nematode larvae, now known as ocular larva migrans (OLM). It was several years later before the larvae were identified as *Toxocara* (Nichols 1956). During the same period, studies were initiated and directed at serological diagnosis of VLM and demonstrated the use of indirect haemagglutination (Jung and Pacheco 1958). Most of the early attempts at serodiagnosis were disappointing, partly due to the use of antigens from adult stage worms because we now know that most of the larval antigens are not shared with the adult worms (Glickman *et al.* 1986). The breakthrough for serodiagnosis was the report of de Savigny (1975) describing the production of ES antigens by prolonged *in vitro* culture of *T. canis* larvae in serum-free medium, followed by the development of a diagnostic test (de Savigny *et al.* 1979).

Three different kinds of syndromes are associated with *T. canis* infection; visceral and ocular toxocarosis and the third, covert toxocarosis, which is a milder form of the syndrome VLM (Glickman *et al.* 1987; Taylor *et al.* 1987; Nathwani *et al.* 1992). However, most people do not develop any clinical symptoms after infection (Bass *et al.* 1983).

VLM and OLM are in general associated with *T. canis*, but other ascarids can also cause some form of larva migrans, such as *T. cati* (definite host: cats), *Baylisascaris procyonis* (definite host: racoons), and *Toxascaris* (definite host: dogs and cats).

Biology of *Toxocara canis*

The definitive hosts for *T. canis* (Werner 1782) are dogs, foxes, and less frequently other canids, and the paratenic hosts are mammals and birds. The life cycle of *T. canis* is very complex in the definitive hosts (Lloyd 1998). The paratenic hosts acquire the infection by ingesting eggs excreted by infected canids. The eggs are usually found in soil, but as they are very sticky, they may adhere to fingers, foodstuffs, insects, etc. Thus, risk factors include geophagia (pica); contact with dogs, particularly pups; poor hygiene and contaminated food, for example, raw snails (Romeu *et al.* 1991). In food animals, which also act as paratenic hosts, larvae have been found in all tissues, with the highest number in liver especially after superinfection and/or reinfection. Consumption of raw liver (Nagakura *et al.* 1989; Salem and Schantz 1992), rabbit giblets and 'lightly grilled' meat (Strüchler *et al.* 1990) and raw cattle meat (Espana *et al.* 1993) have been related to the infection.

Following ingestion of infected eggs, the larvae hatch in the small intestine, penetrate the mucosa, and within a week after infection migrate through the liver and lungs to other tissues; the larvae (350–450 × 14–21 μm) can be found in virtually all organs. Some of the larvae may remain in the liver. The larvae do not mature during the migration, even if they make their way back into the intestine. Most larvae survive for many months or even years, although their distribution in the tissue may continue to shift.

Eggs (90 × 75 μm) of *T. canis* are not embryonated when passed in the stool of the canids and are thus not directly infective for humans. To develop, the eggs need a temperature above 10°C and at a temperature of 15–25°C, the infective larvae will develop within 2–7 weeks. In most cold conditions, the eggs can survive for years, even over extremely cold winters (Ghadirian *et al.* 1976). It has also been shown that *T. canis* eggs resist both the anaerobic and aerobic conditions of sewage processing, and that the surviving eggs are viable (Black *et al.* 1982). Eggs stored for years in sludge also remained both viable and infective (O'Donnell *et al.* 1994). Ozone treatment has no effect on the viability of the embryonated second stage larvae (Ooi *et al.* 1998) but heat, more than 35°C, desiccation, and microwaves (Bouchet *et al.* 1986) will kill the larva within the eggs.

Antigens

Toxocara excretory–secretory (TES) products are highly antigenic. The total production per larva has been estimated at 8 ng/day and is constant over time (Badley *et al.* 1987). TES is composed of five major groups of molecules as determined by sodium dodecyl sulphate-polyacrylamide gel electrophoresis (SDS-PAGE), having relative molecular weights of 32, 55, 70, 120, and 400 kDa (Maizels *et al.* 1987, 1993; Maizels and Robertson 1991). By using immunoelectron-microscopy with monoclonal antibodies the internal sites containing antigens has been localized to the oesophageal gland and the midbody secretory gland (Page *et al.* 1992). The antigens are heavily glycosylated proteins, possessing 40% carbohydrate by weight, with *N*-acetyl-galactosamine and galactose as the dominant sugar (Meghji and Maizels 1986). In contrast to somatic extracts, TES do not contain phosphoryl-choline (Sugane and Oshima 1983a), a cross-reactive determinant produced by many helminths (Sugane and Oshima 1983b). However, cross-reaction has been observed with other helminths (Nicholas *et al.* 1986; Lynch *et al.* 1988; Kennedy *et al.* 1989; Smith 1991). Lately, Loukas *et al.* (1998) have identified a cystein protease migrating at 30 kDa, in somatic extracts of *T. canis* larvae. The recombinant protease was expressed in bacteria, used to immunize mice and the subsequent antiserum reacted specifically with the 30 kDa native protease in the larval extract. Further studies are in progress to investigate the efficacy of the recombinant protease as a diagnostic antigen.

Disease and pathogenesis

The clinical signs and symptoms are a consequence of the extent and frequency of infection, larval migration, and the host response. VLM or visceral toxocarosis seems to be associated with the host immune response to a large number of larvae migrating in the tissue. It is characterized by high eosinophilia, usually more than 30%, fever, hepatosplenomegaly, lymphadenopathy, hypergamma-globulinemia, respiratory signs, abdominal pain, and elevated titres of blood group isohaemagglutinins. Neurological manifestations have also been observed, such as convulsions, and larvae have been found in brains of patients at postmortem (Hill *et al.* 1985; Nelson *et al.* 1990). However, the results in a case-control study by Magnaval *et al.* (1997) suggest that migration of the larvae in human brain does not frequently induce a recognizable neurological syndrome.

OLM or ocular toxocarosis is believed to be caused by a lower infective dose, that is, fewer larvae migrate in the tissue. Persistent eosinophilia, leukocytosis, and elevated isohaemagglutinin levels commonly associated with VLM usually do not occur with ocular disease. Many patients may not have any clinical events except when a larva enters the eye. The patients are usually also older compared to patients presenting VLM (Ljungström and van Knapen 1989; Gillespie *et al.* 1993). Visual loss is usually unilateral and the degree is variable from dimness of vision through to blindness. Initially, granulomas, due to the inflammatory response, were the most frequently reported complication, but with improved diagnosis lesions other than the classical posterior pole granuloma have been reported (Shields 1984). OLM is an important differential diagnosis of retinoblastoma in children (Shields *et al.* 1991).

Patients with covert toxocarosis (CT) may or may not present eosinophilia and when present it is generally not very high (Taylor *et al.* 1987, 1988). This syndrome is associated with positive serology and the clinical signs are common and non-specific but together they form a recognisable symptom complex. Recurrent abdominal and limb pain, lack of energy or tiredness are often recognized in covert toxocarosis but the same symptoms are also observed in patients suffering VLM.

In asymptomatic toxocarosis, positive serology with or without eosinophilia can occur in the absence of other symptoms or signs. That asymptomatic toxocarosis is common is indicated by the relatively high seroprevalence of the general population in all countries of the world, although to some extent cross-reactions cannot be ruled out.

Immunopathology

Studies have shown that TES products exhibit an overlapping but not identical epitopes repertoire to those antigens expressed at the larval surface. The glycocalyx outside the epicuticle is recognized as a dynamic structure, which turns over quite rapidly and serves as a renewable source of large quantities of antigens. The major host responses to these antigens include a marked eosinophilia and hypergamma-globulinemia. Both eosinophils and IgE antibodies are associated with the expansion of the Th2 subset of T helper cells that secrete the cytokines, interleukin (IL)-4 and IL-5. There are reasons to believe that the antigens released from *T. canis* larvae favour the induction of Th2 cell population. On the other hand, the proportion of Th1 subset cells is less then normal during the infection, resulting in limited or no secretion of IL-2 and IFN-gamma (Del Prete *et al.* 1991, 1995).

Besides the immune responses mentioned earlier, the infective larvae elicit a production of specific IgG, IgM, IgA, and IgE isotype antibodies to TES. Interestingly, the specific IgM responses do not diminish over time, which may indicate a failure of isotype switching function or something antigenically unique about TES, or both (Bowman *et al.* 1987; Smith 1993).

Despite the paratenic host's response with both non-specific inflammatory and specific immune reactions, there is a prolonged survival of the larvae in tissues and granulomas. This phenomenon can, to some degree, be explained by immune evasion. In short-term acute infections, using immunohistochemistry Parson *et al.* (1986) demonstrated TES deposition in sinuous patterns suggestive of larval migration. In chronic infections, TES was localized within granulomas, both within the core of those containing larvae, or extracellularly within the inner rim of the collagen capsule of those in which neither larvae nor visible larval remnants were evident. These observations seem consistent with active shedding of TES as it is formed in the surface coat of the parasite. In this way the larvae are able to shed the host immune response, escape and re-migrate.

Several reports have demonstrated that in animals, which have been previously sensitized, a large proportion of migrating larvae will remain within the liver rather than continuing to migrate. Larval numbers imply that larvae are trapped in the liver rather than delayed in their migration. However, liver trapping does not protect the eyes or brain of sensitized mice from larval migration, nor does it result in larval killing (Parsons and Grieve 1990). Eosinophils seem not to be necessary for liver trapping as treatment of mice with antibody to IL-5 prevents both blood and tissue eosinophilia but does not affect liver trapping (Parsons *et al.* 1993). The biological significance of larval trapping is still unknown.

Epidemiology

Toxocara canis has a cosmopolitan distribution. Although it seems more common in wet tropical areas, the parasite is found both in countries with severe winters and those with dry summers.

In Sweden, the first serologically confirmed case of toxocarosis was published in 1979 (Carlson *et al.* 1979) and in 1996 the first report of VLM in an adult Norwegian was published (Lund-Tønnesen 1996). A seroepidemiological survey of young healthy Swedish adults ($n = 323$) gave the prevalence of 7% indicating that subclinical toxocarosis occurs in healthy Swedes. In the sera of patients ($n = 175$), suspected of having contracted toxocarosis, 25% were seropositive suggesting that clinically covert toxocarosis exists in Sweden (Ljungström and van Knapen 1989). In Lithuania, a seropositive reaction was observed in 11.5% of sera ($n = 739$) from patients with clinical symtoms suggesting VLM (Bajoriniene and Balkjawiczius 1988). At Toronto's Hopital for Sick Children a retrospective search for cases during the period 1952 through 1978 was performed. Only 18 VLM and three possible OLM cases were identified, indicating that VLM poses little risk to the health of children in the Toronto area (Fanning *et al.* 1981).

Toxocara canis infection has also been reported in 28% of wild red foxes ($n = 192$) collected all over Sweden, with 14% of foxes above latitude 67° N infected (Christensson 1983, 1988).

However, no *Toxocara* infection was observed in Arctic foxes taken on Banks Island, The Northwest Territories (Eaton and Secord 1979). In the metropolitan area of Copenhagen 68 red foxes were collected and *Toxocara* eggs could be detected in 23.5% and *T. canis* worms were recovered from 17/21 (81%) of these foxes (Willingham *et al.* 1996).

During the years 1971–1981 the National Veterinary Institute, Sweden, analysed the occurrence of *T. canis* in Swedish dogs by autopsy ($n = 90$) and routine stool examination ($n = 19\,044$). The prevalence was found to be 5 and 6.5%, respectively (Christensson 1983). In routine stool samples ($n = 426$) from dogs sent to the laboratory of The Norwegian College of Veterinary Medicine, Norway, during 1981–1982, *Toxocara* was diagnosed in 4%. In hospitalized dogs ($n = 2\,290$), during the 5-year period, 1978–1982, only 0.7% had *Toxocara* eggs in their stool (Tharaldsen 1983). In Halifax, Nova Scotia, the prevalence of *T. canis* eggs in stray dogs ($n = 474$), was found to be 26.6% and the prevalence was noticably greater in pupies, 56.1%, than in mature dogs, 11.9% (Malloy and Embil 1978).

In Stockholm, 86 sandboxes were investigated for *Toxocara* eggs and 32% were found to be infected (Christensson 1983, 1988). A similar figure was also observed in Oslo. Of 13 sandpits, 38.5% were found infected with *Toxocara* spp.

References

Badley, J. E., Grieve, R. B., Bowman, D. D., Glickman, L. T. and Rockey, J. H. (1987). Analysis of *Toxocara canis* larval excretory–secretory antigens: physico-chemical characterization and antibody recognition. *Journal of Parasitology*, **73**, 593–600.

Bajoriniene, D. and Balkjawiczius, B. (1988). The problem of toxocarosis in Lithuania. *Wiadomosci Parazytologiczne*, **34**, 233–238.

Bass, J. L., Mehta, K. A. Glickman, L. T. and Eppes, B. M. (1983). Clinically inapparent *Toxocara* infection in children. *New England Journal of Medicine*, **308**, 723–724.

Beaver, P. C., Snyder, C. H., Caerrera, G. M., Dent, J. H. and Lafferty, J. N. (1952). Chronic eosinophilia due to visceral larva migrans. *Pediatrics*, **9**, 7–19.

Black, M. I., Scarpino, P. V., O'Donnell, C. J., Meyer, K. B., Jones, J. V. and Kaneshiro, E. S. (1982). Survival rates of parasite eggs in sludge during aerobic and anaerobic digestion. *Applied and Environmental Microbiology*, **44**, 1138–1143.

Bouchet, F., Boulard, Y., Baccam, D. and Leger, N. (1986). Ultrastructural studies of alterations induced by microwaves in *Toxocara canis* eggs: prophylactic interest. *Zeitschrift für Parasitenkunde*, **72**, 755–764.

Bowman, D. D., Mika-Grieve, M. and Grieve, R. B. (1987). Circulating excretory–secretory antigen levels and specific antibody responses in mice infected with *Toxocara canis*. *American Journal of Tropical Medicine and Hygiene*, **36**, 75–82.

Carlson, M. G., Grabell, I., Lindahl, L. and Tonell, U. (1979). Toxocariosis – a neglected diagnosis. *Läkartidningen*, **76**, 2691–2692.

Christensson, D. (1983). *Toxocara* and related ascarid roundworms in some Swedish animals – occurrence and prevention. In: *Proceedings of the XI Symposium of the Scandinavian Society for Parasitology. Information*, **17**, pp. 14–15, Åbo Akademi, Finland.

Christensson, D. (1988). High-latitude *Toxocara*. *Parasitology Today*, **4**, 322.

de Savigny, D. H. (1975). *In vitro* maintenance of *Toxocara canis* larvae and a simple method for the production of *Toxocara* ES antigens for use in serodiagnostic tests for visceral larava migrans. *Journal of Parasitology*, **61**, 781–782.

de Savigny, D. H., Voller, A. and Woodruff, A. W. (1979). Toxocariasis: serological diagnosis by enzyme immunoassay. *Journal of Clinical Pathology*, **32**, 283–288.

Del Prete, G. F., De Carli, M., Almerigogna, F., Daniel, C. K., D'Elios, M. M., Zancuoghi, G. *et al.* (1995). Preferential expression of CD30 by human CD4+ T cells producing Th2-type cytokines. *FASEB Journal*, **9**, 81–86.

Del Prete, G. F., De Carli, M., Mastromauro, C., Biagiotti, R., Macchia, D., Falagiani, P. *et al.* (1991). Purified protein derivative of *Mycobacterium tuberculosis* and excretory–secretory antigen(s) of *Toxocara canis* expand *in vitro* human T cells with stable and opposite (type 1 T helper or type 2 T helper) profiles of cytokine production. *Journal of Clinical Investigation*, **88**, 346–350.

Eaton, R. D. and Secord, D. C. (1979). Some intestinal parasites of Arctic fox, Banks Island, N.W.T. *Canadian Journal of Comparative Medicine*, **43**, 223–230.

Espana, A., Serna, M. J., Rubio, M., Redondo, P. and Quintanilla, E. (1993). Secondary urticaria due to toxocariasis: possibly caused by ingesting raw cattle meat? *Journal of Investigational Allergology and Clinical Immunology*, **3**, 51–52.

Fanning, M., Hill, A., Langer, H. M. and Keystone, J. S. (1981). Visceral larva migrans (toxocariasis) in Toronto. *Canadian Medical Association Journal*, **124**, 21–26.

Ghadirian, E., Viens, P., Strykowsik, H. and Dubreuil, F. (1976). Epidemiology of toxocariasis in the Montreal area. Prevalence of *Toxocara* and other helminth ova in dogs and soil. *Canadian Journal of Public Health*, **67**, 495–498.

Gillespie, S. H., Dinning W. J., Voller, A. and Crowcroft, N. S. (1993). The spectrum of ocular toxocariasis. *Eye*, **7**, 415–418.

Glickman, L. T., Schantz, P. M. and Grieve, R. B. (1986). Toxocariasis. In: *Immunodiagnosis of parasitic diseases*, K. W. Walls and P. M. Schantz (eds) . New York, Academic Press, pp. 201–231.

Glickman, L. T. *et al.* (1987). Visceral larva migrans in French adults: a new disease syndrome? *American Journal of Epidemiology*, **125**, 1019–1034.

Hill, I. R., Denham, D. A. and Scholtz, C. L. (1985). *Toxocara canis* larvae in the brain of a British child. *Transactions of the Royal Society of Tropical Medicine and Hygiene*, **79**, 351–354.

Jung, R. C. and Pacheco, G. (1958). The use of intradermal and indirect haemagglutination tests for the diagnosis of visceral larva migrans. *Proceedings of the 6th International Congress of Tropical Medicine and Malaria*, **2**, 5–13.

Kennedy, M. W., Qureshi, F., Fraser, E. M., Haswell-Elkins, M. R., Qureshi, F. and Smith, H. V. (1989). Antigenic relationships bettween the surface-exposed, secreted and somatic materials of the nematode parasites *Ascaris lumbricoides*, *Ascaris suum*, and *Toxocara canis*. *Clinical and Experimental Immunlogy*, **75**, 493–500.

Ljungström, I. and van Knapen, F. (1989). An epidemiological and serological study of *Toxocara* infection in Sweden. *Scandinavian Journal of Infectious Diseases*, **21**, 87–93.

Lloyd, S. (1998). Toxocarosis. In: *Zoonoses*, S. R. Palmer, E. J. L. Soulsby and D. I. H. Simpson (eds). Oxford University Press, Oxford, pp. 841–854.

Loukas, A., Selzer, P. M. and Maizel, R. M. (1998). Characterisation of Tc-CPL-1, a cathepsin L-like cysteine protease from *Toxocara canis* infective larvae. *Molecular and Biochemical Parasitology*, **92**, 275–289.

Lund-Tønnesen, S. (1996). Visceral larva migrans. An unusual cause of eosinophilia in adults. *Tidsskrift for Den Norske Lægeforening*, **116**, 2660–2661.

Lynch, N. R., Wilkes, L. K., Hodgen, A. N. and Turner, K. J. (1988). Specificity of *Toxocara* ELISA in tropical populations. *Parasite Immunology*, **10**, 323–337.

Magnaval, J. F., Galindo, V., Glickman, L. T. and Clanet, M. (1997). Human *Toxocara* infection of the central nervous system and neurological disorders – a case-control study. *Parasitology*, **115**, 537–543.

Maizels, R. M., Gems, D. H. and Page, A. P. (1993). Synthesis and secretion of TES antigens from *Toxocara canis* infective larvae. In: *Toxocara and Toxocariasis: clinical, epidemiological and molecular perspectives*, J. W. Lewis, and R. M. Maizels (eds). The Institute of Biology and British Society of Parasitology, London, pp. 141–150.

Maizels, R. M., Kennedy, M. W., Meghji, M., Robertson, B. D. and Smith, H. V. (1987). Shared carbohydrate epitopes on distinct surface and secreted antigens of the parasitic nematode *Toxocara canis*. *Journal of Immunology*, **139**, 207–214.

Maizels, R. M. and Robertson, B. D. (1991). *Toxocara canis*: secreted glycoconjugate antigens in immunobiology and immunodiagnosis. In: *Parasitic nematodes – antigens, membranes and genes* M. W. Kennedy (eds). Taylor and Francis Ltd, London, pp. 95–115.

Malloy, W. F. and Embil, J. A. (1978). Prevalence of *Toxocara* spp. and other parasites in dogs and cats in Halifax, Nova Scotia. *Canadian Journal of Comparative Medicine*, **42**, 29–31.

Meghji, M. and Maizels, R. M. (1986). Biochemical properties of larval excretory-secretory glycoproteins of the parasitic nematode *Toxocara canis*. *Molecular and Biochemical Parasitology*, **18**, 155–170.

Nagakura, K., Tachibana, H., Kaneda, Y. and Kato, Y. (1989). *Toxocariasis* possibly caused by ingesting raw chicken. *Journal of Infectious diseases*, **160**, 735–736.

Nathwani, D., Laing, R. B. and Currie, P. F. (1992). Covert *toxocariasis* – a cause of recurrent abdominal pain in childhood. *British Journal of Clinical Practice*, **46**, 271.

Nelson, J., Frost, J. L. and Schochet, S. S. Jr. (1990). Unsuspected cerebral *Toxocara* infection in a fire victim. *Clinical Neuropathology*, **9**, 106–108.

Nicholas, W. L., Stewart, A. C. and Walker J. C. (1986). Toxocariasis; a serological survey of blood donors in the Australian Capital Territory together with observations on the risks of infection. *Transaction of the Royal Society of Tropical Medicine and Hygiene*, **80**, 217–221.

Nichols, R. L. (1956). The etiology of visceral larava migrans. I. Diagnostic morphology of infective second-stage *Toxocara* larvae. *Journal of Parasitology*, **42**, 349–362.

O'Donnell, C. J., Meyer, K. B., Jones, J. V. Benton, T. Kaneshiro, E. S. Nichols, J. S. *et al.* (1984). Survival of parasite eggs upon storage in sludge. *Applied and Environmental Microbiology*, **48**, 618–625.

Ooi, H. K., Lin, C. L. and Wang, J. S. (1998). Effect of ozone treatment on *Toxocara canis* eggs. *Journal of Veterinary Medical Science*, **60**, 169–173.

Page, A. P., Hamilton, A. J. and Maizels, R. M. (1992). *Toxocara canis*: monoclonal antibodies to carbohydrate epitopes of secreted (TES) antigens localize to different secretion-related structures in infective larvae. *Experimental Parasitology*, **75**, 56–71.

Parson, J. C., Bowman, D. D. and Grieve, R. B. (1986). Tissue localization of excretory–secretory antigens of larval *Toxocara canis* in acute and chronic murine toxocariasis. *American Journal of Tropical Medicine and Hygiene*, **35**, 974–981.

Parsons, J. C. and Grieve, R. B. (1990). Kinetics of liver trapping of infective larvae in murine toxocariasis. *Journal of Parasitology*, **76**, 529–536.

Parsons, J. C., Coffman, R. L. and Grieve, R. B. (1993). Antibody to interleukin 5 prevents blood and tissue eosinophilia but not liver trapping in murine larval toxocariasis. *Parasite Immunology*, **15**, 501–508.

Perlingiero, J. and Gyorgy, P. (1947). Chronic eosinophilia: report of a case with necrosis of the liver, pulmonary infiltrations, anemia and ascaris infestation. *American Journal of Diseases in Children*, **73**, 34–43.

Romeu, J., Roig, J., Bada, J. L., Riera, C. and Muñoz, C. (1991). Adult human toxocariasis acquired by eating raw snails. *Journal of Infectious Diseases*, **164**, 438.

Salem, G. and Schantz, P. (1992). Toxocaral visceral larva migrans after infection of raw lamb liver. *Clinical Infectious Diseases*, **15**, 743–744.

Shields, J. A. (1984). Ocular toxocariasis. A review. *Survey of Ophthalmology*, **28**, 361–381.

Shields, J. A., Parson, H. M., Shields, C. L. and Shah, P. (1991). Lesions simulating retinoblastoma. *Journal of Pediatric Ophthalmology and Strabismus*, **28**, 338–340.

Smith, H. V. (1991). Immune evasion and immunopathology in *Toxocara canis* infection. In: *Parasitic Nematodes – Antigens, Membranes and Genes*, M. W. Kennedy (eds). Taylor and Francis Ltd, London, pp. 116–139.

Smith, H. V. (1993). Antibody reactivity in human toxocariasis. In: *Toxocara and Toxocariasis: Clinical, epidemiological and molecular perspectives*, J. W. Lewis and R. M. Maizels (eds), The Institute of Biology and British Society of Parasitology, London, pp. 91–109.

Strüchler, D., Weiss, N. and Grassner, M. (1990). Transmission of toxocariasis. *Journal of Infectious Diseases*, **162**, 571–572.

Sugane, K. and Oshima, T (1983a). Purification and characterization of excretory and secretory antigen of *Toxocara canis* larvae. *Immunology*, **50**, 113–120.

Sugane, K. and Oshima, T (1983b). Activation of complement in C-reactive protein positive sera by phosphorylcholine bearing component isolated from parasite extract. *Parasite Immunology*, **5**, 385–395.

Taylor, M. R., Keane, C. T., O'Connor, P., Girdwood, R. W. and Smith, H. (1987). Clinical features of covert toxocariasis. *Scandinavian Journal of Infectious Diseases*, **19**, 693–696.

Taylor, M. R., Keane, C. T., O'Connor, P., Mulvihill, E. and Holland, C. (1988). The expanded spectrum of toxocaral disease. *Lancet*, **1**, 692–695.

Tharaldsen, J. (1983). On the prevalence of *Toxocara* in Norway. In: *Proceedings of the XI Symposium of the Scandinavian Society for Parasitology. Information*, **17**, pp. 12–13, Åbo Akademi, Finland.

Wilder, H. C. (1950). Nematode endophthalmitis. *Transaction of American Academy of Ophthalmololoy and Otolaryngology*, **55**, 99–104.

Willingham, A. L., Ockens, N. W:, Kapel, C. M. and Monrad, J. (1996). A helmintho-logical survey of wild red foxes (*Vulpes vulpes*) from the metropolitan area of Copenhagen. *Journal of Helminthology*, **70**, 259–263.

21 *Trichinella* spp.

Johan Lindh and Inger Ljungström

Trichinella spiralis is a parasitic nematode and the causative agent of trichinellosis. It has been found in almost all warm-blooded vertebrates and completes its entire lifecycle within the same host. The lifecycle of *T. spiralis* can be divided into two phases. The first phase which lasts for around one month starts when the parasite enters its host and undergoes three larval stages to form the adult stage in the intestine. The adults then mate, reproduce, and release the newborn larvae into the bloodstream. The second phase, which can be decades long, starts with an active penetration of the skeletal muscle cells and the development of the first larval stage. The nematode induces the muscle cell to undergo dedifferentiation and a parasite/muscle cell complex is formed, a nurse cell.

This chapter will describe the changes that occur during infection, both in the host and in the parasite. Furthermore, the human immune pathology together with the most abundant antigens will also be described.

History

Trichinella spiralis is a parasitic nematode belonging to the order Enoplida (Figure 21.1). The suborder Dorylaimina is small with *Trichuris* species as the closest relative (Storer *et al.* 1973).

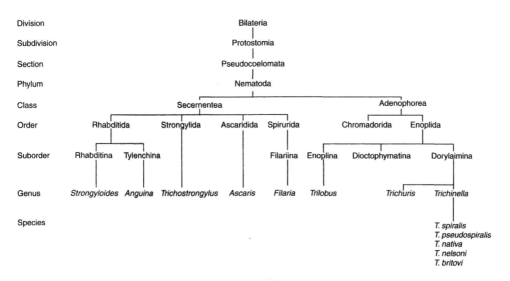

Figure 21.1 Classification of *Trichinella*. These data were obtained from Storer *et al.* (1973) and Lichtenfels *et al.* (1994).

Table 21.1 The distribution of *Trichinella* species/phenotypes and some characteristics

Species/phenotype	Distribution	Characteristics
Trichinella spiralis s.str. (Owen 1835)	Cosmopolitan	High infectivity to pigs and rats Highest female fertility of the genus No resistance to freezing of muscle larvae
Trichinella nativa (Britov and Boev 1972)	Arctic and subarctic areas	High resistance to freezing of muscle larvae Widespread in wildlife
Trichinella britovi (Pozio *et al.* 1992)	Temperate areas of Palaearctic regions	Low infectivity to pigs and rats Low resistance to freezing of muscle larvae Muscle larvae can survive freezing for some months
Trichinella pseudospiralis (Garkavi 1972)	Australian and Palaearctic regions	Non-encapsulation in host muscle Low infectivity for swine
Trichinella nelsoni (Britov and Boev 1972)	In Africa, south of the Sahara	Slow nurse cell development Low infectivity to pigs and rats No resistance to freezing of muscle larvae Widespread in wildlife
Trichinella T5	Temperature areas of North America (USA)	Considered a phenotype of *T. britovi*
Trichinella T6	Montana and Pensylvania in USA	Considered a phenotype of *T. nativa* Less resistant to freezing compared to *T. nativa* muscle larvae
Trichinella T8	In Africa, south of the Sahara	Considered a phenotype of *T. britovi*

Trichinella spiralis was first discovered in 1835 when a medical student, James Paget, at St Bartholomew's Hospital, London, saw white particles in a cadaver (reviewed by Campbell and Denham 1983). The particles were analyzed by microscopy and the results published as a description of a new parasite species named *Trichina spiralis* (Owen 1835). It then took another 25 years before Friedrich Albert von Zenker, 1860, established the lifecycle and documented that transmission could occur to humans from infected pig meat (Zenker 1860). Zenker's association of a defined pathogen with a defined disease was a milestone in medical microbiology, though it rarely receives the recognition it deserves, being overshadowed by later discoveries in bacteriology.

The name of the parasite was later changed to *Trichinella spiralis* since the name *Trichina* had already been used for a genus of a fly (Railliet 1896). Over the years it was believed that only one species of *Trichinella* occurred. Today five sibling species and three phenotypes of uncertain taxonomic level have been identified (Table 21.1; Pozio *et al.* 1992). Of the various species *T. spiralis* has been most studied, both in animals and in humans, as being the most significant for causing disease and economic losses.

Biology of *Trichinella spiralis*

Trichinella spiralis is known to have a broad vertebrate host range and can infect all vertebrates tested, except Chinese hamsters. *Trichinella spiralis* has also been documented in all continents of the world (reviewed by Despommier 1983). The lifecycle of *T. spiralis* is characteristic of nematodes and includes an adult stage followed by four larval stages (Figure 21.2). *Trichinella spiralis* spends the majority of its life at larval stage 1 (L1) which is also the infective stage. When any warm-blooded vertebrate consumes an infected muscle tissue the consumer can get trichinellosis.

Following ingestion of infected muscle tissue, the acidic environment together with pepsin in the stomach, releases the L1 larvae within minutes. No visible changes occur in the larvae during this phase. The free larvae migrate rapidly to the small intestine and enter the columnar epithelial cells; this process may take as little as 10 min (Despommier *et al.* 1978; Wright 1979). At this stage the larvae are approximately 1 mm long and have a diameter of approximately 0.040 mm. Since each columnar cell measures 0.032 × 0.0085 mm, the nematode occupies about 117 columnar cells. After penetration, a fusion of the columnar cell membranes occurs to form a syncytium (Wright 1979). The mechanism(s) for the penetration and generation of the syncytium is (are) not known.

The L1 larvae then rapidly undergo three molts to become L4 larvae after 30 h of infection (Wright 1979). Moulting in *T. spiralis* seems to be similar to that seen in other worms, at least from the L1 to L2 stages where the skin is physically removed and the worms literally crawl out of it. Similarly, *T. spiralis* probably needs a physical barrier to rub against in order to remove parts of its outer cuticle.

Sexually mature adults are observed at 1 h, to up to several months, after the L4 stage is reached (with a peak at five days). In the adult stage, a difference in size between the females and males is seen. The females, which now occupy 415–425 columnar cells, are approximately 3 mm long, while the males are still approximately 1 mm long (Wright 1979; Stewart and Giannini 1982). Other important changes which occur are the development of the male genitalia and the female cloacal opening which are not present in the L1 stage. Changes can be seen in the L2 stage and the organs are developed in the adult stage (Burnham and Despommier 1984). The two outer layers of the cuticle also disappear in the L2 stage

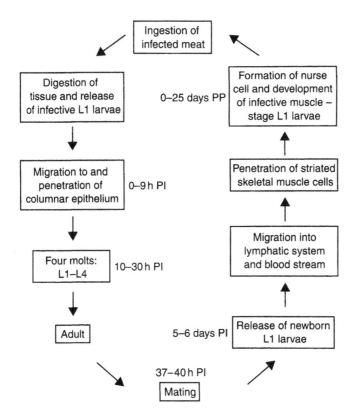

Figure 21.2 The lifecycle of *Trichinella spiralis*. Digestion of infected muscle tissue (top) results in
release of the first larval stage. The larva then migrates and penetrates the columnar
epithelial cells (left). They mate as adults after four molts, 5–6 days post-infection (PI)
(bottom). The newborn larvae migrate to the lymphatic system and are passively
transported to the skeletal muscle cells which they mechanically penetrate (right).
Approximately 20 days after the penetration (PP) of the skeletal muscle cell, a nurse
cell is formed, which can support the parasite for several years or until the tissue gets
digested again (top, reviewed by Despommier 1983).

(Stewart *et al.* 1987), which mediates a change in the surface coat. Different patterns of surface molecules are seen in all stages of the nematode (Philipp *et al.* 1980; Parkhouse and
Ortega-Pierres 1984). Other changes occurring include development of hypodermal gland
cells. These cells are organized in four rows, two dorsal and two ventral, which run almost
for the whole length of the nematode. The function of these subcuticular cells is unknown
(Wright 1979).

Mating occurs 37–40 h after infection but has a peak at day 5. It is believed that
a pheromone system is in operation when the nematodes are finding each other (Belosevic
and Dick 1980). Embryogenesis takes approximately 90 h and the release of the newborn
larvae can occur as early as five days, but with a peak between days 8 and 9 after infection.
Fecundity varies depending on the infectivity of the *Trichinella* isolate, the number of infecting nematodes, and the immunological status of the infected animal. Values in mice vary
between 14 and 600 newborn larvae per L1 larva (reviewed by Despommier 1983).

The newborn larvae migrate from the lumen of the columnar epithelial cells to the lamina propria of the villus tissue. They then can penetrate the lymphatic system and travel via the thoracic duct to the blood system (Harley and Gallicchico 1971; Despommier *et al.* 1978) where they are passively transported to their target cells, the striated skeletal muscle cells (Dennis *et al.* 1970). It is not known how the nematode recognizes these cells, or how penetration occurs but it is thought that the nematode mechanically penetrates the muscle cells, then migrates several worm lengths away from its point of entry (Despommier 1993). The worm now starts a visible period of growth and development without moulting to become a mature infective L1 larva. The volume increases by more than 270 times from day 5 to day 20, when its differentiation to L1 is complete. The main organ that is developed during this stage is the stichosome, an organ that has been proposed to be used for secretion. Each stichosome consists of cells, stichocytes, which contain secretory granules. The number of stichocytes in the stichosome increases from about 20 on day 1 after entering the cell to a final 50–55 at day 20. The stichocytes differentiate into either cells containing 10–15 alfa-granules, found in the posterior region, or cells containing 35–40 beta-granules, found within the anterior region (Despommier and Muller 1976; Takahashi *et al.* 1989). Mature stichocytes are about 0.025 mm in diameter and posses, a single nucleus, mitochondrion, Golgi complex, rough endoplasmic reticulum, and either approximately 2 000 alfa-granules or approximately 3 000 beta-granules (Despommier and Muller 1976). The granules are surrounded by a single membrane and contain molecules which are recognized by antibodies raised against surface antigens from the nematode, suggesting a secretory role for the stichocytes (Despommier and Muller 1976; Grencis *et al.* 1986; Takahashi *et al.* 1989).

The muscles most frequently involved in trichinellosis are the fibres nearest to the site of attachment of the tongue, larynx, diaphragm, and neck. Once the parasite has invaded the muscle cell, the vascular network surrounding the muscle increases, a phenomenon which is thought to be induced by the parasite (Baruch and Despommier 1991). After 20 days of nematode development in the muscle cell, the parasite has developed a stable environment, a nurse cell (Despommier *et al.* 1975). The nurse cell continues to grow for another 30 days and the thickness of the capsule increases until day 50 after penetration of the nematode into the muscle cell (Teppema *et al.* 1973). Between days 10 and 14 after penetration of the muscle cell, the final two layers of the nematode cuticle, 3 and 4, can be observed for the first time.

It is in the first 20 days during exponential growth of the larvae that most changes occur, both in macromolecules and in the microanatomical structure observed in the infected muscle cell (Jasmer 1990). There is a disappearance of myofilaments in the infected muscle cells. This is correlated with a decrease in the content of muscle-specific proteins, such as myosin heavy chain, α-actin, and tropomysin (Teppema *et al.* 1973; Jasmer 1990). There is also an increase in the number of organelles, including smooth and rough endoplasmic reticulum, free ribosomes, mitochondria which appear vacuolated, Golgi apparatus, and lysosomes (Maier and Zaiman 1966; Despommier 1975). The DNA content in the infected muscle cell nuclei increases approximately two fold compared with the non-infected muscle cell (Jasmer 1993). These data indicate that the infected muscle cell has left the normal differentiated state and re-entered the cell cycle in order to undergo S-phase in which the DNA content multiplies for the later division into two cells. Since no division of the infected muscle cell has been observed, it is thought that the cell stops at the G2/M-phase (Jasmer 1993). The levels of specific RNA transcripts encoding several muscle-specific proteins are decreased to <0.1% when compared with non-infected muscle cells (Jasmer *et al.* 1991). Transcript levels of the myogenic regulatory proteins MyoD and myogenin are reduced to <1% and <10%, respectively, while the level of the negative regulator Id is increased to approximately 250% (Jasmer 1993).

The decreased expression of the muscle-specific proteins has been suggested to be due to the down regulation of the myogenic regulatory proteins, MyoD and myogenin, in the infected muscle cell (Jasmer 1993; Lindh 1996). The changes in the expression patterns of the myogenic regulatory proteins could be due to a secreted molecule from the nematode, which might interact with the host molecules and be responsible for their different expression pattern (Ko *et al.* 1992, 1994a,b; Jasmer 1993; Lindh 1996). The increase in the number of different organelles together with the increase in DNA synthesis to approximately 4N has led to the following suggestion: the differentiated muscle cell has left the differentiated state because of secreted products from the nematode and entered the cell cycle again (Stewart 1983; Jasmer 1993). Evidence that the dedifferentiation process seen in the infected muscle cells is due to ES products comes from experiments in which ES was collected from L1-stage larvae and then reintroduced into healthy muscle cells (Ko *et al.* 1992). The reorganization seen in these muscle cells was similar to those seen in muscle cells infected with infective larvae (Ko *et al.* 1992, 1994a,b; Jasmer *et al.* 1994). A possible negative regulator to muscle differentiation has also been identified and found to be secreted from the L1 larva into the muscle cell (Lindh 1996).

Antigens

Approximately 20 different proteins have been identified as antigenic in the *T. spiralis* L1 infective stage, ranging in size from 11 to 105 kDa (Despommier and Lacetti 1981). Together, the major immunodominant epitopes are found to be glycoproteins in the range of 40–50 kDa, and are collectively known as TSL-1 (Appleton *et al.* 1991; Ortega-Pierres *et al.* 1996). These antigens all share the same immunodominant carbohydrate epitope, tyvelose, which is unique for *Trichinella* (Ellis *et al.* 1994). Antibodies raised towards TSL-1 recognize one major 43-kDa protein (Gp45) which is stage specific and localizes to the stichosome and cuticle of the mature larvae (Despommier *et al.* 1990; Vassilatis *et al.* 1992; Jasmer *et al.* 1994). However, the function of TSL-1 is not fully understood yet.

Pathogenesis

The pathogenesis and symptoms associated with trichinellosis vary with the number of viable larvae ingested. The human disease can be divided into two main phases. The first phase occurs approximately between the first week after infection until the eighth week after infection. The first symptoms to appear are fever and in most cases diarrhoea together with abdominal pain. These early symptoms are probably due to an inflammatory reaction in the small intestine that is parasitized by the adult nematodes. The symptoms then gradually move into the second phase which can continue for several years depending on the infection rate. The fever from the first phase decreases and the pain from the abdominal area is moved to the muscles most frequently used, such as the tongue, facial, diaphragm, neck muscles, and flexor muscles of the extremities. Movement of the muscles increases pain and the patients feel weak. These symptoms are due to the migration of the newborn larvae to the muscle cells and the penetration of these cells (reviewed by Pawlowski 1983). Once the larvae begin to encapsulate, symptoms in the patients subside, and eventually the cyst wall and larvae calcify. Modern medical treatment, such as different derivatives of benzimidazoles, helps in most cases, especially if they are given in an early stage of the infection, before the nematodes enter the muscle cells. Once the nematode has become encapsulated in the muscle cell, the treatment is less likely to be successful (reviewed by Campbell and Denham 1983).

Immunopathology

As described above the parasite causes several pathological changes. The majority of the clinical features of trichinellosis are observed during the origin of infection. These features can be related to the capacity of *Trichinella* to induce allergic responses, a property shared by many species of worm parasites. The molecular basis of this allergenicity is still undefined, although clearly it must reflect particular characteristics of worm antigens and the manner in which they are presented to the host. Studies in mouse models show that *Trichinella* infections preferentially stimulate cells of the T helper 2 (Th2) subset of CD4+ T lymphocytes (Grencis *et al.* 1991). These cells release the cytokines necessary for the development of many of the allergic components of the disease, and it can reasonably be assumed that a similar situation exists during human infections. During the intestinal phase there is a marked infiltration of inflammatory cells, including neutrophils, eosinophils, and mast cells into the gut mucosa. Significant changes take place in mucosal architecture (e.g. villous atrophy); fluid flux across the mucosa is disturbed, mucus production is increased, and intestinal transit time is decreased. All of these changes are the result of T cell activity and all can be related to the symptoms appearing during the intestinal phase of infection, of which diarrhoea is the most characteristic. Eosinophilia is a consequence of T cell response to both the intestinal adult worms and to the muscle larvae and is dependent upon release of the cytokine IL-5 (Herndon and Kayes 1992). It has been shown in rodents that infection stimulates both parasite-specific IgE and total IgE antibodies as well as IgG isotypes (IgG1) that are involved in hypersensitivity reactions (e.g. Gabriel and Justus 1979), and this is consistent with the dominance of the T cell response by the Th2 subset. Although *Trichinella*-specific IgE responses have been detected in humans by some workers (e.g. Bruschi *et al.* 1990) others have not been able to find such responses (Ljunström *et al.* 1988). This may be due to different assays used and/or time elapsed after infection when measured. It does seem probable that many of the systemic allergic symptoms of the disease do reflect type-1 hypersensitivity responses.

Invasion of the muscles and formation of the characteristic cysts are accompanied by the development of intense inflammatory responses. These are again T cell-dependent, involve accumulation of eosinophils and other leucocytes, and are maintained by continuous release of antigenic material from the larvae. This inflammation is the direct cause of the myositis that occurs at this stage of infection and which contributes, together with the physical disruption of muscle fibres, to the mechanical and electrophysiological disturbances associated with severe trichinellosis.

Epidemiology

The epidemiology of trichinellosis is characterised by two main cycles, a synanthropic-domestic cycle (*T. spiralis*) and a sylvatic cycle (all species and phenotypes, including *T. spiralis*). As shown in Table 21.1, *T. spiralis*, *T. britovi*, and *T. nativa* are distributed in the Northern Hemisphere. Table 21.2 summarizes the distribution of the various species in nine countries (Kapel 1996).

Table 21.2 The distribution of *Trichinella* species in
northern countries

Species/country	T. spiralis	T. nativa	T. britovi
Estonia	x	x	x
Denmark	?	?	?
Finland	x	x	
Greenland		x	
Latvia	x		x
Lithuania	x		x
Norway		x	x*
Russia	x	x	x
Sweden	x	x	x

Note
* T. britovi, personal communication, J. Tharaldsen.

References

Appleton, J. A., Bell, R. G., Homan, W., and Van Knapen, F. (1991). Consensus on *Trichinella spiralis* antigens and antibodies. *Parasitology Today*, **7**, 190–192.

Baruch, A. M. and Despommier, D. D. (1991). Blood vessels in *Trichinella spiralis* infections: a study using vascular casts. *Journal of Parasitology*, **77**, 99–103.

Belosevic, M. and Dick, T. A. (1980). *Trichinella spiralis*: comparison with an Arctic isolate. *Experimental Parasitology*, **49**, 266–276.

Bruschi, F., Tassi, C., and Pozio, E. (1990). Parasite-specific antibody response in *Trichinella* sp. 3 human infection: a one year follow-up. *American Journal of Tropical Medicine & Hygiene*, **43**, 186–193.

Burnham, J. C. and Despommier, D. D. (1984). Development of the male genitalia of *Trichinella spiralis* during the enteral phase of infection in the mouse: an SEM study. *Journal of Parasitology*, **70**, 310–311.

Campbell, W. C. and Denham, D. A. (1983). In: *Trichinella and trichinosis* (W. C. Campbell, ed.). Plenum Press, New York, pp. 335–366.

Dennis, D. T., Despommier, D. D., and Davis, N. (1970). Infectivity of the newborn larva of *Trichinella spiralis* in the rat. *Journal of Parasitology*, **56**, 974–977.

Despommier, D. D. (1975). Adaptive changes in muscle fibers infected with *Trichinella spiralis*. *American Journal of Pathology*, **78**, 477–496.

Despommier, D. D. (1983). In: *Trichinella and trichinosis* (W. C. Campbell, ed.). Plenum Press, New York, pp. 75–151.

Despommier, D. D. (1993). *Trichinella spiralis* and the concept of niche. *Journal of Parasitology*, **79**, 472–482.

Despommier, D. D., Aron, L., and Turgeon, L. (1975). *Trichinella spiralis*: growth of the intracellular (muscle) larva. *Experimental Parasitology*, **37**, 108–116.

Despommier, D. D., Gold, A. M, Buck, S. W., Capo, V., and Silberstein, D. (1990). *Trichinella spiralis*: secreted antigen of the infective L1 larva localizes to the cytoplasm and nucleoplasm of infected host cells. *Experimental Parasitology*, **71**, 27–38.

Despommier, D. D. and Laccetti, A. (1981). *Trichinella spiralis*: proteins and antigens isolated from a large-particle fraction derived from the muscle larva. *Experimental Parasitology*, **51**, 279–295.

Despommier, D. D. and Muller, M. (1976). The stichosome and its secretion granules in the mature muscle larva of *Trichinella spiralis*. *Journal of Parasitology*, **62**, 775–785.

Despommier, D. D., Sukhdeo, M., and Meerovitch, E. (1978). *Trichinella spiralis*: site selection by the larva during the enteral phase of infection in mice. *Experimental Parasitology*, **44**, 209–215.

Ellis, L. A., Reason, A. J., Morris, H. R., Dell, A., Iglesias, R., Uberia, F. M., and Appleton, J. A. (1994). Glycans as targets for monoclonal antibodies that protect rats against *Trichinella spiralis*. *Glycobiology*, **4**, 585–592.

Gabriel, B. W. and Justus, D. E. (1979). Quantitation of immediate and delayed hypersensitivity responses in *Trichinella*-infected mice. Correlation with worm expulsion. *International Archives of Allergy and Applied Immunology*, **60**, 275–285.

Grencis, R. K., Crawford, C., Pritchard, D. I., Behnke, J. M., and Wakelin, D. (1986). Immunization of mice with surface antigens from the muscle larvae of *Trichinella spiralis*. *Parasite Immunology*, **8**, 587–596.

Grencis, R. K., Hultner, L., and Else, K. J. (1991). Host protective immunity to *Trichinella spiralis* in mice: activation of Th subsets and lymphokine secretion in mice expressing different response phenotypes. *Immunology*, **74**, 329–332.

Harley, J. P. and Gallicchico, V. (1971). *Trichinella spiralis*: migration of larvae in the rat. *Experimental Parasitology*, **30**, 11–12.

Herndon, F. J. and Kayes, S. G. (1992). Depletion of eosinophils by anti-IL-5 mAb treatment of mice infected with *Trichinella spiralis* does not alter parasite burden of immunological resistance to reinfection. *Journal of Immunology*, **149**, 3642–3647.

Jasmer, D. P. (1990). *Trichinella spiralis*: altered expression of muscle proteins in trichinosis. *Experimental Parasitology*, **70**, 452–465.

Jasmer, D. P. (1993). *Trichinella spiralis* infected skeletal muscle cells: arrest in G2/M is associated with the loss of muscle gene expression. *Journal of Cell Biology*, **121**, 785–793.

Jasmer, D. P., Bohnet, S., and Prieur, D. J. (1991). *Trichinella* spp: differential expression of acid phosphatase and myofibrillar proteins in infected muscle cells. *Experimental Parasitology*, **72**, 321–331.

Jasmer, D. P., Yao, S., Vassilatis, D., Despommier, D., and Neary, S. M. (1994). Failure to detect *Trichinella spiralis* p43 in isolated host nuclei and in irridated larvae of infected muscle cells which express the infected cell phenotype. *Molecular and Biochemical Parasitology*, **67**, 225–234.

Kapel, C. M. O. (1996). Trichinellosis in the Nordic region and the Baltic countries. *Bulletin of the Scandinavian Society for Parasitology*, **6**, 7–10.

Ko, R. C., Fan, L., and Lee, D. L. (1992). Experimental reorganization of host muscle cells by excretory/secretory products of infective *Trichinella spiralis* larvae. *Transactions of the Royal Society of Tropical Medicine and Hygiene*, **86**, 77–78.

Ko, R. C., Fan, L., and Lee, D. L. (1994a). Experimental reorganisation of host muscle cells by excretory/secretory products of infective *Trichinella spiralis* larvae. *Transactions of the Royal Society of Tropical Medicine and Hygiene*, **86**, 77–78.

Ko, R. C., Fan, L., Lee, D. L., and Compton, H. (1994b). Changes in host muscles induced by excretory/secretory products of larval *Trichinella spiralis* and *Trichinella pseudospiralis*. *Parasitology*, **108**, 195–205.

Lichtenfels, J. R., Pozio, E., Dick, T. A., and Zarlenga, D. S. (1994). Workshop on systematics of *Trichinella*. In: *Trichinellosis* (W. C. Campbell, E. Pozio, and F. Bruschi, eds). Istituto Superiore di Sanita Press, Rome, Italy, pp. 619–623.

Lindh, J. (1996). *Molecular cloning and functional analysis of muscle stage proteins in Trichinella spiralis*. PhD Thesis, University of London.

Ljungström, I., Hammarström, L., Kociecka, W., and Smith, C. I. E. (1988). The sequential appearance of IgG subclasses and IgE during the course of *Trichinella spiralis* infection. *Clinical Experimental Immunology*, **74**, 230–235.

Maier, D. M. and Zaiman, H. (1966). The development of lysosomes in rat skeletal muscle in trichinous myositis. *Journal of Histochemistry and Cytochemistry*, **14**, 396–400.

Ortega-Pierres, M. G., Yepez-Mulia, L., Homan, W., Gamble, H. R., Lim, P. L., Yakahashi, Y., *et al.* (1996). Workshop on a detailed characterization of *Trichinella spiralis* antigen: a platform for future studies on antigens and antibodies to this parasite. *Parasite Immunology*, **18**, 273–284.

Owen, R. (1835). Description of a microscopic enteozoon infecting the muscles of the human body. *Transactions of the Zoological Society of London*, **1**, 315–324.

Parkhouse, R. M. and Ortega-Pierres, G. (1984). Stage-specific antigens of *Trichinella spiralis*. *Parasitology*, **88**, 623–630.

Pawlowski, Z. S. (1983). In: *Trichinella and trichinosis* (W. C. Campbell, ed.). Plenum Press, New York, pp. 367–401.

Philipp, M., Parkhouse, R. M., and Ogilvie, B. M. (1980). Changing proteins on the surface of a parasitic nematode. *Nature*, **287**, 538–540.

Pozio, E., La Rosa, G., Rossi, P., and Murrell, K. D. (1992). Biological characterizations of *Trichinella* isolates from various host species and geographic regions. *Journal of Parasitolology*, **78**, 647–653.

Railliet, A. (1896). Quelques rectifications a la nomenclature des parasites. *Recueil de Medicine Veterinaire*, **3**, 157–161.

Stewart, G. L. (1983). In: *Trichinella and Trichinosis* (W. C. Campbell, ed.). Plenum Press, New York, pp. 75–151.

Stewart, G. L., Despommier, D. D., Burnham, J., and Raines, K. M. (1987). *Trichinella spiralis*: behavior, structure, and biochemistry of larvae following exposure to components of the host enteric environment. *Experimental Parasitology*, **63**, 195–204.

Stewart, G. L. and Giannini, S. H. (1982). *Sarcocystis, Trypanosoma, Toxoplasma, Brugia, Ancylostoma*, and *Trichinella* spp.: a review of the intracellular parasites of striated muscle. *Experimental Parasitology*, **53**, 406–447.

Storer, T. I., Usinger, R. L., Stebbins, R. C., and Nybakken, J. W. (1973). *General zoology*, 6th edn. McGraw-Hill Book Company, New York.

Takahashi, Y., Uno, T., Mizuno, N., Yamada, S., and Araki, T. (1989). Ultrastructural localization of antigenic substances in *Trichinella spiralis*. *Parasitology Research*, **75**, 316–324.

Teppema, J. S., Robinson, J. E., and Ruitenberg, E. J. (1973). Ultrastructural aspects of capsule formation in *Trichinella spiralis* infection in the rat. *Parasitology*, **66**, 291–296.

Vassilatis, D. K., Despommier, D., Misek, D. E., Polvere, R. I., Gold, A. M., and Van der Ploeg, L. H. (1992). Analysis of a 43-kDa glycoprotein from the intracellular parasitic nematode *Trichinella spiralis*. *Journal of Biological Chemistry*, **267**, 18459–18465.

Wright, K. A. (1979). *Trichinella spiralis*: an intracellular parasite in the intestinal phase. *Journal of Parasitology*, **65**, 441–445.

Zenker, F. A. (1860). Über die Trichinen-Krankheit des Menschen. *Virchows Archiv für Pathologische Anatomie und Physiologie und für Klinische Medizin*, **188**, 561–572.

22 Swimmers' itch

Cecilia Thors, Ewert Linder, and Göran Bylund

Introduction and history of the disease

Schistosomal dermatitis, cercarial dermatitis or swimmers' itch is caused by penetration of larvae, cercariae, of blood flukes (schistosomes) into human skin which are released from infected snails (Figure 22.1). Being a global problem, the disease is widespread, for example, in northern and central Europe, in Russia, and in North America.

Depending on the geographic location, it is variously known as 'swimmers' itch', 'bather's dermatitis', 'cercarien dermatitis', 'sawah', 'rice paddy itch', 'lake side disease', 'clam-diggers itch', etc. (Hunter 1975).

The cause of swimmers' itch in temperate and northern regions is usually avian schistosomes. Swimmers' itch is also an initial symptom of schistosomiasis caused by several species of tropical, human schistosomes like *Schistosoma mansoni, S. haematobium,* and *S. japonicum.* In addition, swimmers' itch can be caused by schistosomes infecting mammals other than man, such as cattle.

The symptoms of cercarial dermatitis were described already more than 150 years ago when Fujii (1847) described the initial symptoms of the 'Katayama disease'. Cort (1928) made the first description of cercarial dermatitis caused by an avian schistosome. While collecting molluscs (*Lymnaea emarginata-angulata*) at Douglas lake in Michigan, US, he accidentally noticed that cercariae of *Cercaria elvae* [first described by La Valette St George (1855), mentioned as *Trichobilharzia ocellata* in Europe (McMullen and Beaver 1945)] produced a severe prickly sensation on the wrists and that this was followed by the appearance of papules which evolved into a pustular eruption with intense itching within 48 h. Since then, the dermatitis produced by cercariae of bird and mammalian schistosomes has been described by several workers over the years (e.g. Christenson and Green 1928; Matheson 1930; Vogel 1930; Wesenberg-Lund 1934; Olivier 1949; Cort 1950; Pirilä and Wikgren 1957; Berg and Reiter 1960; Hoeffler 1974, Hoeffler 1977; Kirschenbaum 1979; Shimizu *et al.* 1981; Kimmig and Meier 1985; Sevcova *et al.* 1987; Beer and German 1993; Kullavanijaya and Wongwaisayawan 1993; Thune 1994; Kolárová *et al.* 1999).

The life cycle and biology

At least eight genera of cercariae of non-human schistosomes causing swimmers' itch have been recorded (see Table 22.1) (Farley 1971; Blair and Islam 1983; Kolárová 2000), most of them from freshwater snails but several also from marine snails (Penner 1950; Chu and Cutress 1954; Bearup and Langsford 1966). Much of the scientific work carried through in order to clarify the biology and epidemiology of these parasites has focused on *Trichobilharzia* spp, the species, together with *Bilharziella polonica*, considered to be most frequently involved

Figure 22.1 (1) Infectious cercaria giving rise to swimmers' itch. (2) Typical severe cercarial dermatitis with maculo-papular rash caused by *Trichobilharzia* spp. swimmers' itch of 'accidental host' and the definite host for the parasite, a seabird. (3) Adult *Trichobilharzia* spp. in seabird. (4) Egg of *Trichobilharzia* spp. which gives rise to the miracidial stage which infects the intermediate host, a *Lymnaea* snail. (5) The infectious cercaria (1) arise from sporocysts developing in the snail. (See Colour Plate IV.)

in outbreaks of swimmers' itch in Europe and North America. Apparently the basic biology and life cycles of other species of avian/mammalian schistosomes are rather similar although the species of intermediate and final hosts are different (McMullen and Beaver 1945; Hoeffler 1974; Goff and Ronald 1981; Wojcinski *et al.* 1987; McKown *et al.* 1991; Loken *et al.* 1995).

Various water fowls (usually mallards, *Anas platyrhinchos*) are the final hosts of the genus *Trichobilharzia*, while snails of the genera *Lymnaea*, *Stagnicola*, *Radix*, *Planorbis*, and *Planorbarius*, are the most common intermediate hosts (Pirilä and Wikgren 1957; Beer and German 1994; Kolárová *et al.* 1997). The adult worms are located in the vessels of the gut and mesenteries of the final host and produce eggs, which are transported to the intestinal lumen and are

Table 22.1 The family *Schistosomatidae* [from Farley (1971) and Kolárová (2000)]

Subfamily	Genus	Final host	Agent of SI	References
Schistosomatinae	Austrobilharzia[a] (Microbilharzia)	Birds	Yes	(Stunkard and Hinchcliffe 1952; Appleton and Lethbridge 1979)
	Bivitellobilharzia	Elephants		(Vogel and Minning 1940)
	Heterobilharzia	Mammals	Yes	(Malek and Armstrong 1967; Goff and Ronald 1981; McKown et al. 1991)
	Macrobilharzia	Birds		(Kohn 1964)
	Orientobilharzia	Mammals	Yes	(Sahba and Malek 1979)
	Ornithobilharzia[a]	Birds		(Witenberg and Lengy 1967; Morales et al. 1971)
	Schistosoma	Mammals	Yes	(Kullavanijaya and Wongwaisayawan 1993)
	Schistosomatium	Mammals	Yes	(Swartz 1966; Malek 1977; Loker 1983)
Bilharziellinae	Bilharziella	Birds	Yes	(Beer et al. 1995)
	Trichobilharzia	Birds	Yes	(Matheson 1930; Müller and Kimmig 1994; Pilz et al. 1995)
Gigantobilharziinae	Dendritobilharzia	Birds		(Vande Vusse, 1980; Canaris et al. 1981; Wojcinski et al. 1987)
	Gigantobilharzia[b]	Birds	Yes	(von Dönges 1965; Matsumura et al. 1984; Appleton and Randall 1986)
Griphobilharziinae	Griphobilharzia	Crocodiles		

Notes
a Cercariae developing in marine water snails.
b Cercariae developing in marine and freshwater snails.

shed with the faeces into the water. Some species seem to be located in the nasal blood vessels of the host and accordingly eggs may be shed in the nasal secretions (Fain 1955; Palmer and Ossent 1984; Horák *et al.* 1998). The ciliated miracidium, after hatching from the egg, penetrates into the snail (Figure 22.2). A complicated process of asexual replication in the snail tissues, through sporocyst and daughter sporocyst generations, generates the characteristic free-swimming, brevifurcate cercaria larvae, infective for the final host (Islam 1986). The cercariae are phototactic, swimming close to the water surface and when finding a suitable host bird penetrate the skin of the feet and migrate via the heart and lungs to their final localization, the tissues of the gut or the mesenteric veins (Neuhaus 1952; Bourns *et al.* 1973; Ellis *et al.* 1975).

Shallow, warm coastal waters are the biotopes preferred by the intermediate hosts as well as by the final hosts. Such biotopes are usually also much frequented by humans for recreational purposes. Although humans are not appropriate hosts for further development of the parasite, cercariae, coming in contact with human skin, attach to and penetrate the skin (Haas and van de Roemer 1998). The mechanisms for attachment to and penetration of the human skin are essentially analogous to those of the tropical schistosomes capable of utilizing man as final hosts. The bird schistosomes, however, are usually unable to penetrate deeper into the human tissues and are destroyed in the skin (von Dönges 1964). However, some authors have shown that cercariae of the genera *Trichobilharzia* and *Bilharziella* are able to migrate to the lungs in mammals (Oliver 1953; Appleton and Brock 1986; Haas and Pietsch 1991; Horák and Kolárová 2000). Recently Horák and Kolárová (2001) reported

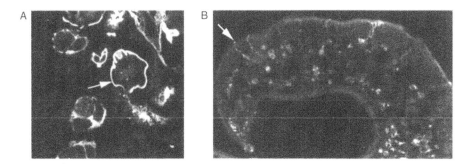

Figure 22.2 Serum antibodies from patient with cercarial dermatitis strains *Trichobilharzia* cercariae in infected *Lymnaea stagnalis* snail (A) and delicate structures connecting subtegumental cells with surface of male schistosomes in (B).

central nervous system (CNS) injury in ducklings and mice caused by the nasal schistosome *Trichobilharzia regenti*. The swimmers' itch is the immune response involving an allergic reaction induced by antigenic components released from the penetrating larvae or, more importantly, from dead, decomposing cercariae in the human skin (Macfarlane, 1949).

Taxonomy

As mentioned earlier, there are a large number of species of bird and non-human, mammalian schistosomes inducing allergic reactions and creating public health problems when invading human skin. In different regions, species of the following genera have been considered responsible for swimmers' itch: *Austrobilharzia (Microbilharzia)*, *Bilharziella*, *Gigantobilharzia*, *Heterobilharzia*, *Orientobilharzia*, *Schistosoma*, *Schistosomatium*, and *Trichobilharzia* (Stunkard and Hinchcliffe 1952; Leigh 1955; Sahba and Malek 1979; Appleton and Randall 1986; McKown *et al.* 1991). The taxonomy of these parasites is not completely understood today (Farley 1971). Morphological and functional criteria applied for identification of the cercariae are swimming patterns and resting posture in the water, spination, size of the cercariae and proportional size of different organs of the larvae, number and organization of the penetration glands, structures of the excretory system (number of flame cells), localization of the sensory papillae, etc. (Nasir and Erasmus 1964; Farley 1971; Richard 1971; Blair and Islam 1983).

Signs and symptoms of cercarial dermatitis

Cort (1950), in his excellent review of swimmers' itch, presented a description of the signs and symptoms of swimmers' itch (in already sensitized persons) which essentially covers what has been repeated in the literature over the years:

> The penetration of the cercariae into the skin produces a prickling, or itching sensation resembling that caused by the bites of small insects. This may continue for an hour or more but usually subsides earlier. During this time maculae, about 1 to 2 millimeters in diameter, appears at the site of the penetration of each cercaria. These maculae usually soon disappear, but in some cases they may persist for several hours, or until they are replaced by papules. Diffuse erythema may occur instead of the macular reaction, and

in few cases, a local urticaria, may develop near the penetration site. Usually about 10 to 15 hours after penetration, discrete papules, 3 to 5 millimeters in diameter, replace the maculae. The development of the papules is accompanied by an intense itching. They are distinct, indurate and surrounded by a zone of erythema, the size of which depends usually on the amount of rubbing or scratching induced by the pruritus. The area surrounding them may be edematous especially after rubbing or scratching. Not infrequently in heavy exposures, the papules become confluent, and the whole area is swollen and edematous. On the second or third day vesicles form on them and these are often ruptured by rubbing or scratching. The papules, usually disappear in about a week after infection, leaving small pigmented spots on the skin. The itching, except in very severe cases, is sporadic and intermittent and usually disappears after several days. The lesions may become pustular from secondary bacterial infection, and intense scratching sometimes produce severe complications from secondary infection and injury to the skin.

It is evident that the response in persons without prior exposure to the cercariae are relatively inconspicuous and may pass unnoticed, while only persons sensitized by previous exposure develop the typical symptoms.

Many of the early scientists who produced cercarial dermatitis experimentally commented on the fact that certain individuals showed little or no reaction to the penetration of the cercariae. Also it was frequently noted that some of the people exposed on bathing beaches failed to develop any signs of dermatitis. Early studies in Canada and the United States (Macleod and Little 1942; Macfarlane 1949; Olivier 1949) showed striking differences in the reaction to the penetration of the cercariae after the first as compared with subsequent infections and suggested that the degree of skin reactivity reflects the degree of immune response induced as the result of sensitization by previous exposure.

It is clear and well documented that sensitization, due to repeated infections, is the factor releasing the typical symptoms of swimmers' itch, that is, papulae and weal formation, pruritus. There is evidence that the sensitization may persist for several years, since persons who had been exposed to cercariae up to 12 years previously, but not since then, developed severe dermatitis even when exposed to very small numbers of cercariae.

Pathology and histopathology

Several descriptions are published on the pathogenesis and histopathology of swimmers' itch, both from spontaneous and experimental infections. Although old, some of these descriptions are fully valid today.

Vogel (1930) described sections of 24-h lesions in an individual who apparently had been previously exposed, but was only partially sensitized. He found the cercariae and some cellular exudate in the burrows. There was oedema and a mild leucocytic reaction, and the cells of the cercariae were beginning to disintegrate. Brackett (1940) made biopsies of lesions in an individual who had a history of repeated schistosome dermatitis over a period of 6 years. At 29 h after infection, he found no cercariae but marked serous and cellular exudates both in epidermis and in dermis. These studies show that cercariae had been destroyed within 29 h by the intensive cellular reaction. At 50 h the dermis was more intensely infiltrated with eosinophils and lymphocytes.

The itching is evidently of two types: the prickling sensation experienced immediately after exposure to cercariae appears to be induced by cercariae penetrating through and burrowing in the skin. The latter, intensive histamine-related itch is of a diffuse type and is increased by scratching and rubbing. This occurs in the early and late papular stages of the pathogenesis and results from the antigen–antibody reaction.

In previously non-exposed individuals, the erythematous reaction to the cercariae may appear only eight days after the penetration of the cercariae. Apparently, the reaction varies depending on the species of parasite and the sensitivity of the individual affected.

Topical application of antihistamines and anti-inflammatory substances on affected skin areas will relieve the itching but has little effect on the lesions.

Differential diagnostics

A correct diagnosis of swimmers' itch is not always very easy. Similar external symptoms may occur after insect bites and in some communicable diseases, such as chickenpox, measles, and some venereal diseases. Similar symptoms may also sometimes be induced by plants in sensitive persons. The localization of the affected skin areas is important to consider for the differential diagnosis. A history of bathing or immersion of affected skin areas in water is of course most important for a presumptive diagnosis. However, allergic reactions after exposure to blue-green algae (cyanobacteria) have also been recorded (Solomon and Stoughton 1978).

It should be kept in mind that the cercariae causing swimmers itch in northern regions are released from the snails and swarm in the waters only during short periods in summer and during warm, sunny days.

Cnidarians (coelentarates) is a group of marine animals that contain stinging structures known as nematocysts, and include jellyfish, corals, sea anemones, hydroids, etc. Contacts with these animals can induce dermatitis (Black and Szmant 1992). In northern areas, however, there are few representatives of these animals able to produce symptoms in humans.

There is evidence for an antibody response in swimmers' itch patients and serology can be used for diagnostics (Kimmig and Meier 1985; Kolárová *et al.* 1994; Pilz *et al.* 1995).

Control and prevention

As with tropical human schistosomes, much of the attempts aiming at control of swimmers' itch have been focused on the control of the snail populations in the vicinity of beaches and coastal areas used for recreation. As most of the chemicals tested for this purpose are non-selective and poisonous for many groups of organisms (including man) they are not approved today in most countries. Harvesting of snails from beach areas is a theoretical possibility, whereas prevention of growth of vegetation may have an effect. Recently, attempts have been made to reduce the infection pressure in an area by adding the schisto-somicide, praziquantel, to bird feeds and thereby eliminating the infection from the final host (Müller *et al.* 1993; Reimink *et al.* 1995).

References

Appleton, C. C. and Brock, K. (1986). The penetration of mammalian skin by cercariae of *Trichobilharzia* sp. (Trematoda: *Schistosomatidae*) from South Africa. *Onderstepoort Journal of Veterinary Reseach*, **53**, 209–211.

Appleton, C. C. and Lethbridge, R. C. (1979). Schistosome dermatitis in the Swan Estuary, Western Australia. *Medical Journal of Australia*, **1**, 141–145.

Appleton, C. C. and Randall, R. M. (1986). Schistosome infection in the kelp gull, *Larus dominicanus*, from Port Elizabeth, Republic of South Africa. *Journal of Helminthology*, **60**, 143–146.

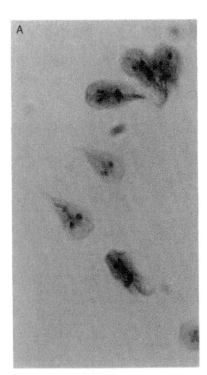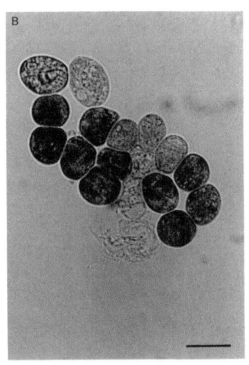

Colour Plate I Giardia intestinalis: (A) trophozoites stained with Giemsa stain and (B) iodine-stained cysts. Bar = 10 μm. (See Figure 10.1, page 103.)

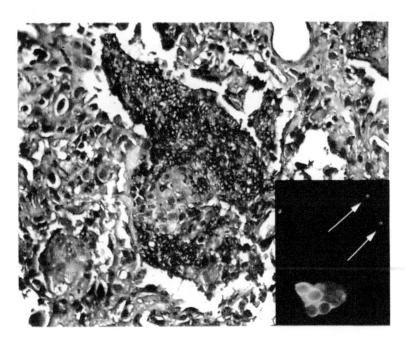

Colour Plate II Pneumocystis carinii organisms identified in histological section of infected lung tissue and in broncho-alveolar lavage fluid obtained from a patient with fulminant PCP. Immunoperoxidase staining of intra-alveolar masses of *Pneumocystis carinii* is seen. Using fluorescence microscopy of lavage fluid both asci (cysts) and trophozoites are seen using monoclonal anti-pneumocystis antibody 3F6 as the marker (100). (See Figure 12.3, page 121.)

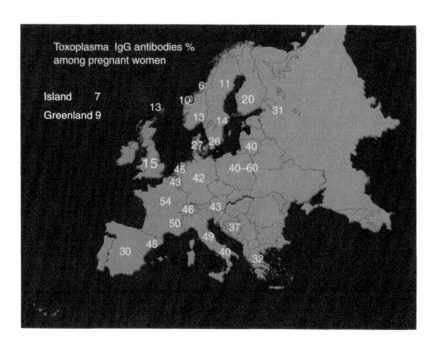

Colour Plate III The prevalence of *Toxoplasma* IgG antibodies in pregnant populations in different countries in Europe. (See Figure 14.1, page 152.)

Colour Plate IV (1) Infectious cercaria giving rise to swimmers' itch. (2) Typical severe cercarial dermatitis with maculo-papular rash caused by *Trichobilharzia* spp. swimmers' itch of 'accidental host' and the definite host for the parasite, a seabird. (3) Adult *Trichobilharzia* spp. in seabird. (4) Egg of *Trichobilharzia* spp. which gives rise to the miracidial stage which infects the intermediate host, a *Lymnaea* snail. (5) The infectious cercaria (1) arise from sporocysts developing in the snail. (See Figure 22.1, page 206.)

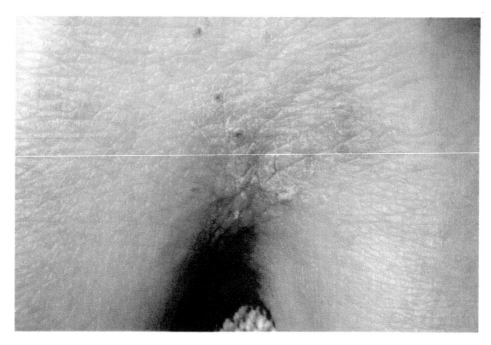

Colour Plate V Scabies: eczematous lesions of the finger webs are common, but not always observed. (See Figure 24.1, page 246.)

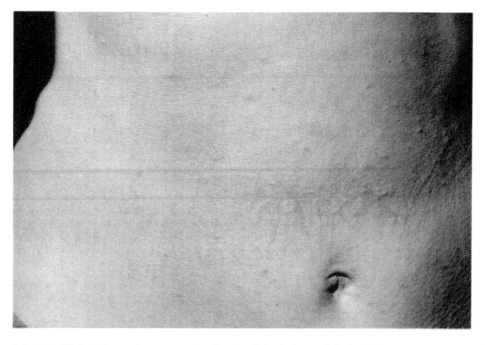

Colour Plate VI Scabies: an important clue for the clinical diagnosis is that 'the symptoms are much worse than the clinical signs' – intense itch, but minimal skin lesions – urticarial or excoriated papules. (See Figure 24.2, page 246.)

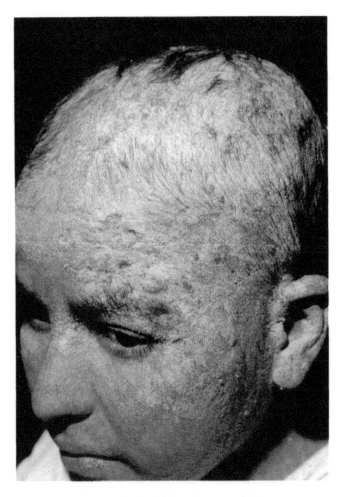

Colour Plate VII Crusted scabies: while in otherwise healthy persons the infestation usually means only 10–20 mites, an immuno-compromised patient (HIV, tumor diseases, etc.), might develop quite another clinical disease with widespread hyperkeratotic lesions: crusted scabies or 'Norwegian scabies'. This patient suffered from systemic lupus erythematosus and was on high doses of cortisone since a long time. (See Figure 24.5, page 248.)

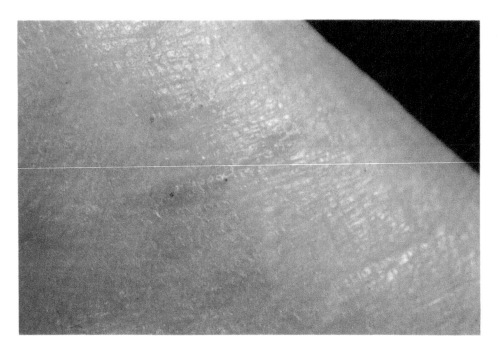

Colour Plate VIII Scabies: burrows are typical and can be described as tortuous lines with a slight scaling, most often found on the flexor surface of the wrist, between the fingers, and sometimes on the genitals. At the end of the burrow often a dark spot can be seen. (See Figure 24.3, page 247.)

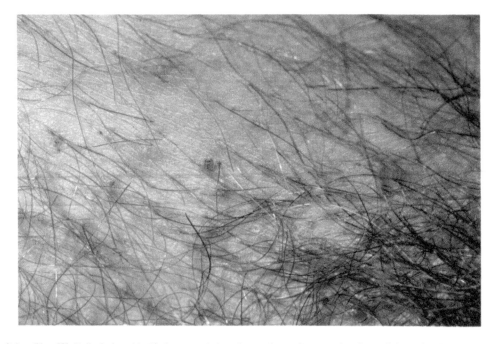

Colour Plate IX Pediculosis pubis: if the examining doctor has a low grade of suspicion, the diagnosis might easily be missed. The lice are often difficult to see and are often mistaken for small crusts. (See Figure 24.6, page 249.)

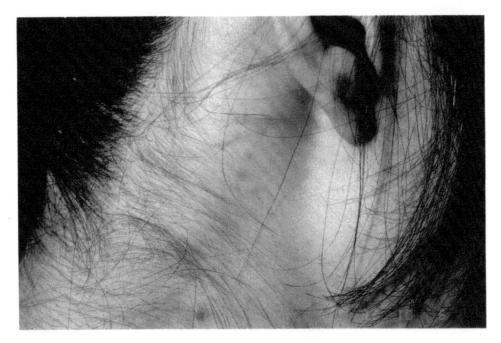

Colour Plate X Pediculosis capitis: usually in young people the lice cause itching bite marks, especially in the nape region. The lice themselves are often difficult to discover, but the nits (Swedish: gnetter) should always be possible to identify. (See Figure 24.7, page 250.)

Colour Plate XI Pediculosis corporis or *vestimenti*: the lice do not live on the body but in the clothes, but they bite and feed on human skin. The *clothing* should be treated by thorough cleaning, not the patient! (Photo: Leif Strandberg). (See Figure 24.8, page 250.)

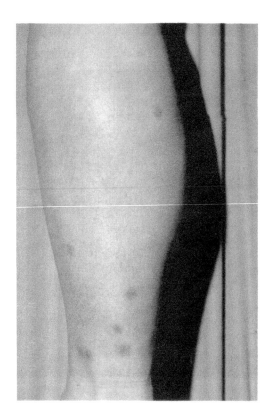

Colour Plate XII Flea bites: various species of fleas, each species specific for a special kind of animal, or human, are rare nowadays. Fleas live in the moist areas of floors, but they search for 'blood meals' from animals or humans. Flea bites are most common on the lower legs. (See Figure 24.9, page 251.)

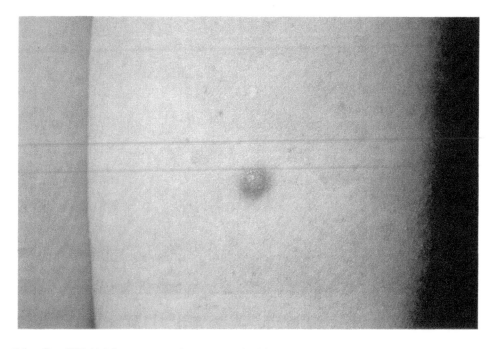

Colour Plate XIII Nodulus cutaneous after a mosquito bite: a small (approximately 5 mm), very hard papule, with a characteristic ring of pigmentation. (See Figure 24.10, page 252.)

Bearup, A. J. and Langsford, W. A. (1966). Schistosome dermatitis in association with rice growing in the northern territory of Australia. *Medical Journal of Australia*, **1**, 521–525.

Beer, S. A. and German, S. M. (1993). The ecological prerequisites for a worsening of the cercariasis situation in the cities of Russia (exemplified by the Moscow region). *Parazitologiia*, **27**, 441–449.

Beer, S. A. and German, S. M. (1994). The ecological prerequisites for the spread of schistosomal dermatitis (cercariasis) in Moscow and the Moscow area. *Meditsinskaia Parazitologiia i Parazitarnye Bolezni*, 16–19.

Beer, S. A., Solonets, T. M., Dorozhenkova, T. E. and Zhukova, T. V. (1995). Human cercariasis caused by schistosomatid larvae from aquatic birds in the Narochanka River recreational area of Byelarus. *Meditsinskaia Parazitologiia i Parazitarnye Bolezni*, 8–11.

Berg, K. and Reiter, F. H. (1960). Observations on schistosome dermatitis in Denmark. *Acta Dermato-Venereologica*, **40**, 369–380.

Black, N. and Szmant, A. (1992). Larval thimble jellyfish (*Linuche unqui culata*) as a possible cause of seabather's eruption. *1992 Symposium on Flonda Keys Regional Ecosystem*, 18 November, Miami FL.

Blair, D. and Islam, K. S. (1983). The life-cycle and morphology of *Trichobilharzia australis* n. sp. (Diginea: *Schistosomatidae*) from the nasal blood vessels of the black duck (*Anas superciliosa*) in Australia, with a review of the genus *Trichobilharzia*. *Systematic Parasitology*, **5**, 89–117.

Bourns, T. K., Ellis, F. C. and Rau, M. E. (1973). Migration and development of *Trichobilharzia ocellata* (Trematoda: *Schistosomatidae*) in its duck hosts. *Canadian Journal of Zoology*, **51**, 1021–1030.

Brackett, S. (1940). Pathology of schistosome dermatitis. *Archives of Dermatology and Syphilis*, **42**, 410–418.

Canaris, A. G., Mena, A. C. and Bristol, J. R. (1981). Parasites of waterfowl, from southwest Texas: III. The green-winged teal, *Anas crecca. Journal of Wildlife Diseases*, **17**, 57–64.

Christenson, R. O. and Green, W. P. (1928). Studies on biological and medical aspects of 'swimmers itch'. Schistosome dermatitis in Minnesota. *Minnesota Medicine*, **11**, 573–575.

Chu, G. W. T. C. and Cutress, C. E. (1954). *Austrobilharzia variglandis* (Miller and Northup, 1929) Penner, 1953 (Trematoda, *Schistosomatidae*) in Hawaii with notes on its biology. *Ibidem*, **40**, 515–524.

Cort, W. W. (1928). Schistosome dermatitis in the United States (Michigan). *Journal of the American Medical Association*, **90**, 1027–1029.

Cort, W. W. (1950). Studies on schistosome dermatitis XI. Status of knowledge after more than twenty years. *American Journal of Hygiene*, **52**, 251–307.

von Dönges, J. (1964). Hautreaktionen bei Schistosomeninvasion. *Deutsche Medizinische Wochenschrift*, **32**, 1512–1516.

von Dönges, J. (1965). *Gigantobilharzia suebica* n. sp. (Trematoda). Ein Dermatitiserreger beim Menschen. *Zeitschrift für Parasitenkunde*, **24**, 65–75.

Ellis, J. C., Bourns, T. K. and Rau, M. E. (1975). Migration, development, and condition of *Trichobilharzia ocellata* (Trematoda: *Schistosomatidae*) in homologous challenge infections. *Canadian Journal of Zoology*, **53**, 1803–1811.

Fain, A. (1955). Une nouvelle bilharzoise des oiseaux: La trichobilharziose nasale. Remarque sur l' importance des schistosomes d'oiseaux en pathologie humaine. Note préliminaire. *Annales de la Société belge de Mèdecine tropicale*, **35**, 323–327.

Farley, J. (1971). A review of the family *Schistosomatidae*: excluding the genus *Schistosoma* from mammals. *Journal of Helminthology*, **45**, 289–320.

Fujii, D. (pen name Yoshinao). (1909). An account of a journey to Katayama. (Originally published in 1847). In Japanese. Translated in: Kean B. H., Mott J. E., Rusell A. J. (eds). *Tropical medicine and parasitology. Classic investigations*, Cornell University Press, Ithaca, NY two volumes, pp. 677, 1978. *Chugai Iji Shinpo*, **691**, 55–56.

Goff, W. L. and Ronald, N. C. (1981). Certain aspects of the biology and life cycle of *Heterobilharzia americana* in east central Texas. *American Journal of Veterinary Research*, **42**, 1775–1777.

Haas, W. and Pietsch, U. (1991). Migration of *Trichobilharzia ocellata* schistosomula in the duck and in the abnormal murine host. *Parasitology Research*, **77**, 642–644.

Haas, W. and van de Roemer, A. (1998). Invasion of the vertebrate skin by cercariae of *Trichobilharzia ocellata*: penetration processes and stimulating host signals. *Parasitology Research*, **84**, 787–795.

Hoeffler, D. F. (1974). Cercarial dermatitis. Its ethiology, epidemiology and clinical aspects. *Archive for Environmental Health*, **29**, 225–229.

Hoeffler, D. F. (1977). Swimmer's itch (cercariae dermatitis). *Cutis*, **19**, 461–465.

Horák, P. and Kolárová, L. (2000). Survival of bird schistosomes in mammalian lungs. *International Journal of Parasitology*, **30**, 65–68.

Horák, P. and Kolárová, L. (2001). Bird schistosomes: do they die in mammalian skin? *Trends in Parasitology*, **17**, 66–69.

Horák, P., Kolárová, L. and Dvorak, J. (1998). *Trichobilharzia regenti* n. sp. (*Schistosomatidae, Bilharziellinae*), a new nasal schistosome from Europe. *Parasite*, **5**, 349–357.

Hunter, G. W. (1975). *Schistosome cercarial dermatitis and other rare schistosomes that may infect man.* Krieger, New York.

Islam, K. S. (1986). Development of *Trichobilharzia australis* Blair & Islam, 1983 in the snail, *Lymnaea lessoni* Deshayes and in an experimental definitive host, the Muscovy duck. *Journal of Helminthology*, **60**, 301–306.

Kimmig, P. and Meier, M. (1985). Parasitologische Untersuchungen, Diagnose und Klinik der Zerkariendermatitis – Hygienische Bedeutung für Badegewässer gemässigter Zonen. *Zentralbl. Bakteriol. Mikrobiol. Hyg.*, **181**, 390–408.

Kirschenbaum, M. B. (1979). Swimmer's itch. A review and case report. *Cutis*, **23**, 212–216.

Kohn, A. (1964). On the genus *Macrobilharzia* Travassos 1922 (Trematoda, Schistosomatoidea). *Memorias do Instituto Oswaldo Cruz*, **62**, 1–6.

Kolárová, L. (2000). *Schistosomatidae*: overview of medically important pathogenic agents. http://www.natur.cuni.cz/~horak/.

Kolárová, L., Horák, P. and Sitko, J. (1997). Cercarial dermatitis in focus: schistosomes in the Czech Republic. *Helminthologia*, **34**, 127–139.

Kolárová, L., Skirnisson, K. and Horak, P. (1999). Schistosome cercariae as the causative agent of swimmer's itch in Iceland. *Journal of Helminthology*, **73**, 215–220.

Kolárová, L., Sykora, J. and Bah, B. A. (1994). Serodiagnosis of cercarial dermatitis with antigens of *Trichobilharzia szidati* and *Schistosoma mansoni*. *Central European Journal of Public Health 2*, **1**, 19–22.

Kullavanijaya, P. and Wongwaisayawan, H. (1993). Outbreak of cercarial dermatitis i Thailand. *International Journal of Dermatology*, **32**, 113–115.

La Valette St George, A. J. H. (1855). *Symbolae ad Trematodum Evolutionis Historian.* Berolini, pp. 1–40.

Leigh, W. H. (1955). The morphology of *Gigantobilharzia huttoni* (Leigh 1953), an avian schistosome with marine dermatitis-producing larva. *Journal of Parasitology*, **41**, 262–269.

Loken, B. R., Spencer, C. N. and Granath, W. O., Jr. (1995). Prevalence and transmission of cercariae causing schistosome dermatitis in Flathead Lake, Montana. *Journal of Parasitology*, **81**, 646–649.

Loker, E. S. (1983). A comparative study of the life-histories of mammalian schistosomes. *Parasitology*, **87**, 343–369.

Macfarlane, W. V. (1949). Schistosome dermatitis in New Zealand. Part II. Pathology and immunology of cercarial lesions. *American Journal of Hygiene*, **50**, 152–167.

Macleod, J. A. and Little, G. E. (1942). Continued studies on cercarial dermatitis and the trematode family *Schistosomatidae* in Manitoba. Part I. *Canadian Journal of Research (Sect. D)*, **20**, 170–181.

Malek, E. A. (1977). Geographical distribution, hosts, and biology of *Schistosomatium douthitti* (Cort, 1914) Price, 1931. *Canadian Journal of Zoology*, **55**, 661–671.

Malek, E. A. and Armstrong, J. C. (1967). Infection with *Heterobilharzia americana* in primates. *American Journal of Tropical Medicine and Hygiene*, **16**, 708–714.

Matheson, C. (1930). Notes on *Cercaria elvae*, Miller, as probable cause of an outbreak of dermatitis at Cardiff. *Transactions of the Royal Society of Tropical Medicine and Hygiene*, **23**, 421–424.

Matsumura, T., Sawayama, T., Honda, M. and Asada, S. (1984). Avian schistosomiasis (paddy field dermatitis) in a rural city of Hyogo Prefecture, Japan – seasonal emergence of *Gigantobilharzia sturniae* cercariae from an intermediate host snail, *Polypylis hemisphaerula*. *Kobe Journal of Medical Sciences*, **30**, 17–23.

McKown, R. D., Veatch, J. K. and Fox, L. B. (1991). New locality record for *Heterobilharzia americana*. *Journal of Wildlife Diseases*, **27**, 156–160.

McMullen, D. B. P. and Beaver, C. (1945). Studies on schistosome dermatitis. IX. The life cycles of three dermatitis-producing schistosomes from birds and a discussion of the subfamily *Bilharziellinae* (Trematoda: *Schistosomatidae*). *American Journal of Hygien*, **42**, 128–154.

Morales, G. A., Helmboldt, C. F. and Penner, L. R. (1971). Pathology of experimentally induced schistosome dermatitis in chickens: the role of *Ornithobilharzia canaliculata* (Rudolphi, 1819) Odhner 1912 (Trematoda: *Schistosomatidae*). *Avian Diseases*, **15**, 262–276.

Müller, V. and Kimmig, P. (1994). *Trichobilharzia franki* n. sp. – a causative agent of swimmer's itch in south-western Germany. *Applied Parasitology*, **35**, 12–31.

Müller, V., Kimmig, P. and Frank, W. (1993). The effect of praziquantel on *Trichobilharzia* (Diginea, *Schistosomatidae*), a cause of swimmer's dermatitis in humans. *Applied Parasitology*, **34**, 187–201.

Nasir, P. and Erasmus, D. A. (1964). A key to the cercariae from British freshwater molluscs. *Journal of Helminthology*, **38**, 245–268.

Neuhaus, W. (1952). Biologie und Entwicklung von *Trichobilharzia szidati* n. sp. (Trematoda, *Schistosomatidae*), einem Erreger von Dermatitis beim Menschen. *Zeitschrift für Parasitenkunde*, **15**, 203–266.

Oliver, L. (1953). Observations on the migration of avian schistosomes in mammals previously unexposed to cercariae. *Journal of Parasitology*, **39**, 237–246.

Olivier, L. (1949). Schistosome dermatitis, a sensitization phenomenon. *American Journal of Hygiene*, **49**, 290–302.

Palmer, D. and Ossent, P. (1984). Nasal schistosomiasis in mute swans in Switzerland. *Revue Suisse de Zoologie*, **91**, 709–715.

Penner, L. R. (1950). *Cercaria littorinalinae* sp. nov., a dermatitis-producing schistosome larva from the marine snail, *Littorina planaxis* Philippi. *Journal of Parasitology*, **36**, 466–472.

Pilz, J., Eisele, S. and Disko, R. (1995). Cercaria dermatitis (swimmer's itch). Case report of cercaria dermatitis caused by *Trichobilharzia* (Diginea, *Schistosomatidae*). *Hautarzt*, **46**, 335–338.

Pirilä, P. and Wikgren, B.-J. (1957). Cases of swimmers itch in Finland. *Acta Dermato-Venereologica*, **37**, 140–148.

Reimink, R. L., DeGoede, J. A. and Blankespoor, H. D. (1995). Efficacy of praziquantel in natural populations of mallards infected with avian schistosomes. *Journal of Parasitology*, **81**, 1027–1029.

Richard, J. (1971). La chetotaxie des cercaires. Valeur systematique et phyletique. *Memoires du Museum National d'Histoire Naturelle, Nouvelle Serie, Serie A. Zoologie*, **67**, 1–179.

Sahba, G. H. and Malek, E. A. (1979). Dermatitis caused by cercariae of *Orientobilharzia turkestanicum* in the Caspian Sea area of Iran. *American Journal of Tropical Medicine and Hygiene*, **28**, 912–913.

Sevcova, M., Kolarova, L. and Gottwaldov, A. (1987). Cercarial dermatitis. *Ceskosloven ská Dermatologie*, **62**, 369–374.

Shimizu, M., Matsuoka, S. and Katsuhiko, A. (1981). Cercaria dermatitis. *The Journal of Dermatitis*, **8**, 117–124.

Solomon, A. E. and Stoughton, R. B. (1978). Dermatitis from purified sea algae toxin (debromoaplysiatoxin). *Archives of Dermatology*, **114**, 1333–1335.

Stunkard, H. W. and Hinchcliffe, M. C. (1952). The morphology and life history of *Microbilharzia variglandis* (Miller and Northrup, 1926), Stunkard and Hinchcliffe, 1951, avian blood-flukes whose larvae cause 'swimmers-itch' of ocean beaches. *Journal of Parasitology*, **38**, 248–265.

Swartz, L. G. (1966). An occurrence of *Schistosomatium douthitti* (Cort, 1914) Price, 1931, in Alaska in a new natural definitive host, *Clethrionomys rutilus* (Pallas). *Canadian Journal of Zoology*, **44**, 729–730.

Thune, P. (1994). Cercariadermatitt eller svømmekløe – et lite kjent, men hyppig forekommende sykdomsbilde i Norge. *Tidsskrift for den Norske Lægeforening nr 15*, **114**, 1694–1695.

Vande Vusse, F. J. (1980). A review of the genus *Dendritobilharzia* Skrjabin and Zakharow 1920 (Trematoda: *Schistosomatidae*). *Journal of Parasitology*, **66**, 814–822.

Vogel, H. (1930). Hautveränderungen durch *Cercaria ocellata*. *Dermatologisch Wochenschrift*, **90**, 577–581.

Vogel, H. and Minning, W. (1940). Bilharziose bei Elephanten. *Arch Schiffs, Tropen-Hyg*, **44**, 562–574.

Wesenberg-Lund, C. (1934). Contributions to the development of the Trematoda Diginea. Part II. The biology of the freshwater cercariae in Danish freshwaters. *Mem de l'Acad Roy des Sc et des Lettres de Danemark, Copenhague. Section des Sciences*, 9me serie t V no. 3 pp. 1–233 *(Kgl danske Vid Selsk Skr nat afd 9 5:1)*, **3**, 1–233.

Witenberg, G. and Lengy, J. (1967). Redescription of *Ornithobilharzia canaliculata* (Rud.) Odhner with notes on classification of genus *Ornithobilharzia* and the subfamily schistosomatinae (Trematoda). (The second Israel south Red sea expedition 1965 report nr 4.). *Israel Journal of Zoology*, **16**, 193–204.

Wojcinski, Z. W., Barker, I. K., Hunter, D. B. and Lumsden, H. (1987). An outbreak of schistosomiasis in Atlantic brant geese, *Branta bernicla hrota. Journal of Wildlife Diseases*, **23**, 248–255.

23 Medically important parasitic arthropods (insects, ticks, and mites) of the northern Holarctic region

Thomas G. T. Jaenson

Medical entomology is the science about the arthropods, that is, insects, arachnids, and related evertebrates, which cause human disease. It is a very vast and important biomedical discipline encompassing subjects such as entomology, ecology, epidemiology, infectious disease control, and environmental science including the science about the conservation of nature and natural resources. In the following brief account, I will focus on the main features concerning the biology and medical importance of the taxa (taxonomic groups) of parasitic arthropods which are of outstanding significance to human health in the northern parts of the Holarctic (Palaearctic and Nearctic) zoogeographical region.

A parasite can be defined as an organism living in or on another organism, its host, from which the parasite obtains food. Parasites living inside the host are denoted as endoparasites (e.g. myiasis-causing fly larvae, and *Sarcoptes* and *Demodex* mites) while parasites living on the host are termed ectoparasites. Ectoparasites are arthropods which spend the whole or a great part of the life-cycle on the host. They may be subdivided into host ectoparasites, nest ectoparasites, and field ectoparasites.

Host ectoparasites, for example, blood-sucking lice (Anoplura), spend the whole life-cycle on the host. Nest ectoparasites (e.g. soft ticks (Argasidae), fleas (Siphonaptera), and bed bugs (Heteroptera: Cimicidae)) spend all stages in the vicinity of the host on which one or several of the active stages take blood. Most species of hard ticks (Ixodidae) can be denoted as field ectoparasites; they spend most of their life away from the host but when a host has been encountered the tick usually spends several days attached to its skin until fully engorged (Nelson *et al.* 1975).

Due to limitation of space, medically important blood-sucking Diptera such as mosquitoes (Culicidae), black-flies (Simuliidae), biting midges (Ceratopogonidae), horse-flies, deer flies and clegs (Tabanidae), stable-flies and horn-flies (blood-sucking Muscidae), and louse-flies (Hippoboscidae) are not treated in this chapter. In a broad parasitological sense, the blood-sucking adult sex(es) of these Diptera may be regarded as temporary, blood-feeding ectoparasites. However, in a stricter sense, they may not be regarded as ectoparasites. Further information on ectoparasitic and endoparasitic arthropods of medical and veterinary importance can be obtained from Smart (1948), Gordon and Lavoipierre (1972), Smith (1973), Harwood and James (1979), Alexander (1984), Jaenson (1985), Lane and Crosskey (1993), Kettle (1995), and Wall and Shearer (1997) and from references provided in these books. Very useful are also the publications and unpublished documents on vector biology and control from the World Health Organization (WHO) and the Pan American Health Organization (PAHO). In order to follow the most recent research on ectoparasitic and endoparasitic arthropods the following monthly journals are highly recommended: *The Journal of Medical Entomology*, *Medical and Veterinary Entomology*, and *The Review of Medical and Veterinary Entomology*.

Insects

Flies (Diptera) causing myiasis

The larvae of several different families of the order Diptera (flies) cause a condition termed myiasis, which can be defined as the infestation by fly larvae of living tissues or organs of vertebrates. From a parasitological point of view there are two types of myiasis-causing flies. One type of fly species are obligatory parasites which, for their complete development, need tissues of live hosts. The other category, that is, the facultatively parasitic fly species usually develop in dead, decaying animals or other decomposing organic materials (faeces, rotting vegetation, etc.) but sometimes deposit eggs or larvae on infected wounds or on mucous membranes of live hosts. Different terms are used to indicate which part of the body that has become infested: rectal, gastric, enteric or intestinal; urinary or urogenital; auricular; ophthalmic; dermal, subdermal, or cutaneous; and nasopharyngeal (Zumpt 1965; Harwood and James, 1979). Hall and Smith (1993) classify myiasis into three main groups: cutaneous, body cavity and accidental myiasis. Cutaneous myiasis may be subdivided into: (i) blood-sucking or sanguinivorous myiasis, that is, when fly larvae attach to the skin and suck blood (e.g. the Congo floor maggot *Auchmeromyia*); (ii) furuncular myiasis, that is, when larvae stay in the skin and make boil-like swellings, for example, the mango fly *Auchmeromyia* and the human bot-fly *Dermatobia hominis*; and (iii) creeping myiasis (larva migrans) when larvae burrow in human skin but cannot complete development in man (Hall and Smith, 1993). Body cavity myiasis may be subdivided into nasopharyngeal, auricular, pulmonary, and ophthalmomyiases, that is, when eggs or larvae are deposited in or remain in nose, sinuses and pharyngeal cavities, and in ear, lung, or eye, respectively. Examples of accidental myiases are intestinal (enteric, rectal) when larvae are ingested by accident or reach the intestine via the anus. Another type of accidental myiasis is urogenital myiasis when flies are attracted to infected tissues or dirty clothes. Depending on fly species, number of eggs or larvae deposited, site of oviposition or larviposition, and the health status of the host myiasis may range from an asymptomatic or benign conditions to severe and even fatal infestations.

Hypoderma *spp. (cattle and deer warble flies)*

The warble fly (family Oestridae, subfamily Hypodermatinae, genus *Hypoderma*) larvae normally infest cattle and deer, but occasionally horses and humans. The gravid females of the cattle warble flies (*H. bovis*, *H. lineatum*) usually glue their eggs to the hairs on the legs of cattle and wild bovines. The newly hatched larva burrows either directly into the skin or into the hair follicles. The larva then gradually works its way through the tissues and eventually reach the skin on the host's back where it makes a small breathing opening in the skin. The inflammation of the surrounding tissues causes a swelling denoted as a warble. When mature the larvae leave the host by dropping to the ground in which they pupate. The life-cycle and biology of the reindeer or caribou warble fly, *Hypoderma* (= *Oedemagena) tarandi*, is similar to that of the cattle-infesting warble flies. *Hypoderma tarandi* is a common parasite of reindeer and caribou in the northern parts of the Holarctic region. It is a well-known cause of human ophthalmomyiasis in, for example, Norway and Sweden.

Infestation of man by *Hypoderma* larvae may result from direct oviposition by gravid female flies as well as by handling cattle having the fur infested by recently hatched larvae. Creeping myiasis, skin abscesses, serious ophthalmic myiasis, and even intracerebral myiasis due to *Hypoderma* have been recorded in humans (Zumpt 1965; Hall and Smith 1993). In some cases, the larva can be squeezed out of an open wound, but in most instances surgery is necessary. The infestation of *Hypoderma* in abnormal hosts often causes serious tissue destruction by the larval migrations in the body (eyes, spinal cord, and brain) causing blindness, paralyses or even fatal damage.

Oestrus ovis *(the sheep nasal bot fly)*

Oestrus ovis (family Oestridae, subfamily Oestrinae) is considered to be originally a Palaearctic species which has subsequently spread to most sheep-farming areas of the world (Kettle 1995). The main hosts of *O. ovis* are sheep and goats. When intending to oviposit the female sheep nostril fly normally hovers in front of the nostrils of sheep or goats. The eggs hatch immediately during the oviposition. The newly hatched larvae are 'sprayed' into the nose of the host by the hovering adult female fly. The larvae then develop in the nasal cavity and later move into the frontal and maxillary sinuses where larval development is completed. When fully mature, the larvae move forward and are sneezed out by the host and drop to the ground where they burrow into the soil and pupate. Clinical symptoms in sheep and goats may range from mild discomfort, nasal discharge or head shaking to allergic and inflammatory responses followed by secondary bacterial infection and sometimes death. Larvae may occasionally enter the brain, causing ataxia, circling, and head pressing (Wall and Shearer 1997). Infestation causing ocular myiasis, and sometimes invasion of the brain followed by death, also occurs in dogs and humans. *Oestrus ovis* is, however, not capable to complete its development in these 'abnormal' hosts.

Cephenemyia *spp. (deer nostril flies)*

The genus *Cephenemyia* (family Oestridae, subfamily Oestrinae) is restricted to the Holarctic, and the larvae develop only in Cervidae (deer). The developmental biology of *Cephenemyia trompe* (a parasite of reindeer in the Palaearctic), of *C. auribarbis* (a parasite of caribou in the Nearctic) and related *Cephenemyia* species is very similar to that of *O. ovis*. The female fly deposits first-stage larvae in the nostrils of the host and the larvae subsequently move to the pharyngeal and nasal cavities. The mature larvae move to the pharyngeal and nasal cavities, leave the host and pupate in the ground. *Cephenemyia* spp. are, like *O. ovis*, often the cause of ocular myiasis in humans.

Gasterophilus *spp. (stomach bots of equines)*

Nine species of *Gasterophilus* (family Oestridae, subfamily Gasterophilinae) have been recorded from equines; six species parasitize domestic horses and donkeys. *Gasterophilus intestinalis* is the most important and most widely distributed of the horse bot flies. It was originally a Palaearctic species, but is now common also in North America and in Australia. The *Gasterophilus* larvae are normally parasites of equines. In the northern temperate region there is one generation per year. Typically, the eggs are laid on the host, licked into the mouth by the host so that the first instar larvae can develop in the tissues of the oral cavity. The second and third instar larvae attach themselves to the intestinal mucosa where they feed for several months. The prepupae are expelled with the faeces and then pupariate.

Occasionally the 1–2 mm long, first instar *Gasterophilus* larvae will penetrate the human skin and cause a creeping myiasis (cutaneous larva migrans) in which the larvae move in the epidermis up to 20 mm per day, causing considerable irritation (Kettle 1995). Similar symptoms are caused by certain tropical human-parasitizing nematodes. There are records that *Gasterophilus* larvae can cause ocular myiasis.

'Accidental' myiasis

Ingestion of food or liquid contaminated with living eggs or larvae of fly species belonging to a number of different families, but most often Calliphoridae (blow flies, in particular *Calliphora* and *Lucilia*), Muscidae (including the house-fly *Musca domestica*), and Sarcophagidae (flesh-flies), cause enteric (intestinal) myiasis. Usually, the condition passes unnoticed without any signs or symptoms. However, enteric myiasis may cause severe disturbances including malaise, abdominal pain, and diarrhoea with bloody discharge. Living or dead larvae may be present in stool or vomit.

Rectal myiasis may occur when gravid female flies are attracted to soiled clothes or skin on which they deposit eggs or larvae. The larvae then gain entrance to the intestine via the anus. A number of species in several different families of Diptera can be involved in rectal myiasis. Species most often encountered in rectal myiasis are excrement feeders such as certain species of Muscidae and Sarcophagidae, and the rat-tailed larvae of the drone fly *Eristalis tenax* (Syrphidae).

Many cases of wound or traumatic myiasis, in which fly larvae develop in skin wounds or lesions may be included as an accidental type of myiasis. There are, for instance, numerous reports from hospitals and similar institutions where fly larvae have been encountered in leg sores of geriatric patients. Species of flies infesting such sores usually belong to Calliphoridae, Sarcophagidae, Fanniidae, Muscidae, and Phoridae.

Myiasis-causing flies commonly imported from the tropics

Dermatologists, infectious disease specialists, general practitioners, and other physicians should be acquainted with the fact that people who have recently arrived from the subtropics or tropics can sometimes have their signs and symptoms in skin or other tissues explained by infestation of myiasis-causing fly larvae.

Dermatobia hominis *(the human bot-fly, tórsalo)*

This relatively large, purplish fly parasitises in the larval stage man, cattle, dogs, and other species of domestic and wild mammals, and birds. It is widely distributed in Latin America where it is an important cattle parasite. The larvae can complete their development to the pupal stage in humans. The larvae may sometimes cause a serious, painful, and lengthy (six weeks) form of myiasis. The condition can be fatal, particularly in children. In view of the lengthy developmental period this myiasis-causing species is one of the more common ones to be diagnosed in clinics in North America and northern Eurasia among people who have recently visited Latin America.

Cochliomyia hominivorax *(the New World screw-worm fly)*

This calliphorid, called the New World screw-worm fly, has its main area of distribution in Latin America. The adult *C. hominivorax* is a relatively large, bluish green fly with an orange-coloured head. The gravid female lays her eggs in wounds and onto mucous membranes of the natural body orifices. Main hosts are cattle but there are numerous records of infestation also on humans. The larvae can cause serious tissue destruction; even fatal cases are known, for example, when the umbilicus of a newly born baby has become infested. *Cochliomyia hominivorax* has been eradicated by the mass-release of laboratory-reared, gamma-irradiated, sterile male flies from the southern United States and northern Mexico. Following an accidental introduction of *C. hominivorax* by infested livestock from South America to Libya the fly was eradicated recently by the use of the mass-release of sterile male flies.

Chrysomyia bezziana

Chrysomyia bezziana is an obligate agent of myiasis and a close relative of *C. hominivorax*. The biology of the two fly species is similar but *C. bezziana*, the Old World screw-worm fly, has its main area of distribution in the tropics of Africa and Asia.

Cordylobia anthropophaga *(the tumbu fly)*

The tumbu fly or mango fly (*Cordylobia anthropophaga*) is distributed in Africa, south of the Sahara. The adult blow-fly (family Calliphoridae) is yellowish and about 1 cm long. The eggs are deposited in dry soil or sand or on clothes laid on the ground to dry, especially if there is a smell of urine or faeces. If a larva comes into contact with the skin of a potential host (humans, pigs, dogs, rodents, etc.) the small larva penetrates the skin, usually unnoticed, to reside in the subcutaneous tissues. Within a week, a boil-like swelling develops. It is often painful and is often secondarily infected with bacteria. The fully developed larva leaves the host and pupates in the ground. Infestation of a single person by dozens of larvae at the same time is sometimes seen. In hospitals in northern countries, the infestation is relatively commonly recorded in patients who have just returned from tropical Africa. The larva can be partly forced out of the skin by covering the spiracles (breathing openings) by vaseline. Then the larva can be gently pulled out by a forceps.

Fleas (Siphonaptera)

The insect order Siphonaptera, that is, fleas, comprises about 2 000 species and subspecies (Lewis 1995). The adult flea is a blood-sucking wingless ectoparasite with a laterally flattened body. Many fleas have well-developed legs adapted for jumping. Like the Diptera (flies, mosquitoes, and midges) the Siphonaptera has a so-called complete development (egg, larva, pupa, and adult). The larva is a worm-like creature without legs. Its food is organic material which it encounters in the lair or nest of its host. The pupa is enclosed in a cocoon. The fully developed flea can lie for several months inside the cocoon waiting for a potential host. This explains why people can be attacked by fleas in houses which have been standing unoccupied for a long time.

Most flea species are ectoparasites on mammals. A lesser number feed on birds. However, in contrast to the Anoplura, the fleas are not strictly host species specific. This implies that fleas sometimes can transmit microparasitic infections from one host species to a different host species. Fleas will abandon the bodies of dying or dead hosts. This is of importance in the epidemiology of plague: fleas leaving dying, plague-infected rodents will seek new hosts and will thereby increase the potential for transmission of the disease. An adult flea may live for a year, but plague-infected fleas have a reduced life-span.

In a country like Sweden, about 50 flea species have been recorded. Some of these, in particular the cat flea (*Ctenocephalides felis*), the dog flea (*C. canis*), and bird fleas (*Ceratophyllus* spp.) will occasionally attack man (Figure 23.1).

The human flea (*Pulex irritans*) is nowadays a rarity in Swedish houses and apartments. The modern homes are generally too dry and clean for fleas to thrive there. When *P. irritans* is found in Sweden, it is usually as a parasite of pigs or of wild burrow-dwelling mammals

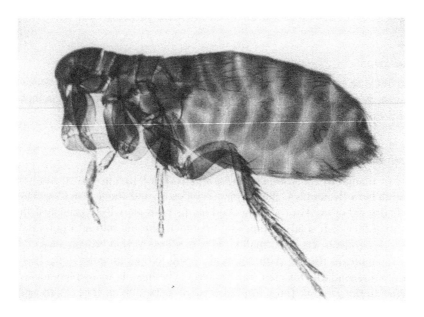

Figure 23.1 The bird flea *Ceratophyllus*. Natural size *c*.2–3 mm. (Photo: T. G. T. Jaenson©.)

like the red fox (*Vulpes vulpes*). The human flea, however, is still a common parasite of humans in many other, less developed parts of the world. It thrives in dirty localities.

Plague

Plague or pest is a bacterial infection caused by *Yersinia pestis*. The infection is primarily a zoonosis among wild rodents. It is transmitted within these rodent populations by fleas of different genera including *Xenopsylla*. Occasionally, the infection is transmitted by fleas from the enzootic hosts to plague-susceptible hosts, for example, squirrels and prairie dogs, which may die in large numbers (epizootic plague). Urban, rat-borne plague may occur when synanthropic (peridomestic) rat species come into contact with plague-infected enzootic rodents or plague-infected epizootic rodents in or near urban areas. When the synanthropic rodents become infected, the risk for a human plague epidemic is great.

Transmission of plague bacteria to man is mainly by infected rat fleas (*Xenopsylla* spp.). The infection can follow after the bite of an infected flea; after having crushed an infected flea and then contaminated a wound, for example, the bite wound, with infectious material; after having crushed an infected flea between ones teeth; or by inhalation of dry flea faeces or other material containing plague bacteria. The global incidence of human plague is about 1 500–5 000 cases per year. Enzootic, asymptomatic foci are present in North and South America, North, West, Southern and East Africa and Madagascar, and in several countries in Asia. Human plague is usually a very serious disease. About 30–90% of untreated cases of bubonic plague are fatal. Pulmonic plague, which is nearly always fatal, occurs when the primary infection gains entry by inhalation of infectious material via the lungs.

Wherever plague in an urban or peri-urban area is diagnosed, control of fleas and rats should be carried out without any delay. Flea control is mainly done by treating flea- and rat-infested localities with suitable insecticides. Control of rats is mainly by removing potential food sources for rats, by baited traps, and by poisonous baits.

DDT is one of the compounds which is usually used for controlling epizootics and epidemics of plague. This serious, but presently relatively rare, disease is caused by a bacterium which naturally occurs in certain populations of rodents inhabiting areas where, in general, human cases of plague rarely occur. However, from these natural plague foci the infection may spread. Surveillance of the infection in the natural plague foci should therefore be carried out on a permanent, routine basis. Epidemics of plague can occur in any area of the world where the sanitary and environmental conditions favour the breeding of rats and their fleas in close association with man. To avoid human cases of plague in urban and suburban areas, surveillance and control of fleas and rodents are the main measures to rely on. The monitoring of resistance against chemical insecticides and rodenticides among flea and rodent populations, respectively, should, therefore, be carried out routinely, particularly in countries or regions where plague is enzootic. Environmental methods including the reduction of potential food sources for rodents, rodent trapping by baited traps, and poisonous baits to kill rodents are among the main methods recommended for control of domestic and peridomestic rodent populations. A moderately effective vaccine against plague is available for use by persons potentially becoming exposed to the infection, for example, people living within or near plague enzootic foci. Valuable information relevant for the control of plague, fleas, and rodents is provided in PAHO (1982) and WHO (1973, 1974, 1988a,b, 1991). In view of the detrimental effects caused by DDT in non-target organisms, the occurrence of high levels of resistance to DDT in a number of populations of plague vectors (WHO 1992), and the availability of relatively cheap and apparently less harmful, alternative chemicals, for example, deltamethrin, it is considered inappropriate to use any of the persistent organochlorine compounds in attempts to control flea vectors of plague (Jaenson 1996).

Flea-borne typhus

Flea-borne typhus, endemic typhus, murine typhus, murine spotted fever, or rodent typhus is an infection due to *Rickettsia mooseri* (*R. typhi*) which is an asymptomatic infection among rodents in many parts of the world, particularly in subtropical and tropical areas. Human infection is usually acquired indoors when infectious flea faeces or an infected flea is crushed and rubbed into a (bite-) wound. Inhalation of infectious flea faeces can also lead to infection. *Rickettsia mooseri* is mainly transmitted to humans via rodent fleas, in particular *Xenopsylla cheopis*. The rickettsia can also be transmitted by the body louse (*Pediculus humanus*). Infection with *R. mooseri* is, in a similar manner as Q-fever (*Coxiella burnetii*) and the tick-borne typhuses (*R. africae*, *R. conorii*, etc.) occasionally recorded in Northern hospitals among patients who have recently returned from, for example, the Mediterranean area or Africa.

Blood-sucking bugs (Heteroptera)

Cimex lectularius *(bed-bugs)*

The bed-bugs belong to the family Cimicidae within the order true bugs (Heteroptera). All cimicids feed on blood only. Two species feed primarily on human hosts, that is, *Cimex lectularius* (mainly in temperate areas) and *C. hemipterus* (mainly in the tropics). A few decades ago, the bedbugs were almost eradicated from northern and western Europe, but at present they seem to be increasing in abundance. The bed-bugs undergo so-called hemimetabolous

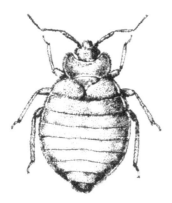

Figure 23.2 The bed-bug *Cimex lectularius*. Natural size *c*.6 mm. (Drawing by Inga Thomasson after Smart (1948).)

development, that is, the development occurs gradually and there is no intervening pupal stage. In common with the lice (Anoplura) the body of the bed-bugs is dorso-ventrally flattened and seen from above it is more or less oval (Figure 23.2). The young bedbugs resemble the adults but are yellowish white and smaller. The adults are usually reddish brown in colour and about 6 mm long. The wings are reduced to small scales. Thus, the bedbugs cannot fly but they are capable of running quite rapidly.

The human-parasitizing bed-bugs are active mainly at night, particularly between 2 and 4 in the morning when most people are in deep sleep. At 15°C they suck blood about once every week but at 25°C they feed about once every night. However, they can survive starvation for long periods. After the blood meal, the bed-bugs hide in a secluded place, for example, under mattresses, behind pictures on walls or behind loose wallpaper, in cracks and crevices in walls or on the floor, in furniture, etc. A bed-net impregnated with an insecticide, for example, permethrin, provides good protection against bed-bugs if one is forced to sleep in a bed-bug infested room. Transfer of bed-bugs from one place to another is believed to take place by the transportation of infested furniture, etc. Eradication of bed-bugs from infested houses is preferably done by thorough cleaning followed by insecticidal spraying of the infested rooms, particularly the bed-rooms. In particular, the hiding-places of the bed-bugs, where also the eggs are laid, should be treated with a suitable insecticide (Schofield and Dolling 1993).

During their blood-ingestion, the bed-bugs inject saliva. The proteins in the saliva can cause allergic reactions with oedema and itching at the site of the bite. Continuous, massive attacks by bed-bugs will eventually cause symptoms due to the disturbed sleep and occasionally also anaemia (iron deficiency).

Blood-sucking lice (Anoplura)

All known species of blood-sucking lice (Anoplura) are obligate parasites spending their whole life-cycle on mammals. Transmission of lice from one host to another is usually by close inter-host contact. Lice move slowly. They have no wings and cannot fly or jump. The development from egg to adult is of the so-called hemimetabolous type. This means that the morphology of the young, newly hatched louse very much resembles that of the older louse, only that the latter one is bigger. Thus, in contrast to so-called holometabolous insects, which have a 'complete' development with a pupal stage, the lice do not pass through a pupal stage.

The lice are dorso-ventrally flattened insects with powerful extremities each one ending with a strong, simple claw. Depending on the stage and species they range in length from 0.5 to 8 mm (Ibarra 1993). The lice move about quite slowly and are usually seen clinging to the hairs of the host's fleece. They are highly host species specific and adapted to the normal skin temperature of the host. Healthy lice do not normally leave their host. If a louse is removed from its host the louse will usually die within one or two days. There are three species of Anoplura on man, that is, the crab louse or pubic louse (*Pthirus pubis*), the head louse (*Pediculus capitis*), and the clothing louse or body louse (*Pediculus humanus*). The three species of human lice, all have a world-wide distribution. The two species of *Pediculus* are also recorded from New World monkeys, gibbons, and the great apes while *Pthirus pubis* is only recorded from gorillas and man (Ibarra 1993; Kettle 1995). Thus, lice from, for instance, domestic animals cannot survive on humans and vice versa.

Pediculus capitis *(the head louse)*

Although the morphology of the head louse, *P. capitis*, is very similar to that of the clothing louse, their ecology and medical importance are quite different. The adult female *P. capitis* is about 3 mm long. The male is slightly smaller and has dark transverse bands on the dorsal part of the abdomen. The general colour of the body is dirty yellowish; recently blood-fed lice can be reddish, but later the blood-fed louse becomes much darker (black-brown). Also, louse populations living in dark hair are usually darker than those living in light hair. The form of the body is elongated, more or less oval, and the three pairs of legs are, in contrast to those of the crab louse, almost identically shaped (Figure 23.3). The yellowish eggs are about 0.8 mm long and are glued firmly, usually close onto the base of the hairs of the scalp. The glue is so strong that it cannot be dissolved without destroying the hair. The eggs will hatch after about a week. When the first stage nymph begins to hatch the egg fills with air

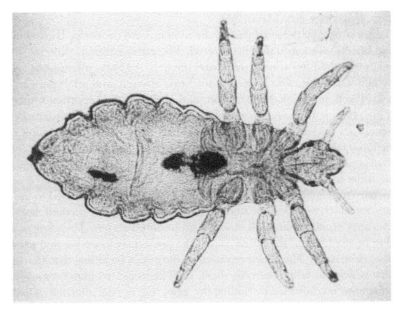

Figure 23.3 The head louse *Pediculus capitis*. Natural size *c.*1–2 mm. (Photo: T. G. T. Jaenson©.)

and becomes glistening white. The egg-shells, sometimes also the eggs, are popularly called nits. The white, empty egg-shells are usually located further away from the base of the hairs. Since the hairs grow about 1 cm per month, it is possible to roughly estimate for how long the infestation has been going on. The young louse resembles the adult louse but is smaller. It sucks blood from the skin of the scalp about five times per day and changes skin three times before being fully developed. The life-cycle from egg to egg takes usually three weeks. The maximum life-span of a female louse is about three weeks (Ibarra 1993).

Head lice infestations are common among all age groups, although most common among children, in all socio-economic classes throughout the world. During the 1970s in northern and western Europe, head louse infestations increased markedly. In North America, next to infections with common cold viruses head louse infestations may be the most common 'disease' affecting school children (Ibarra 1993).

Transmission of head lice is almost invariably by direct bodily contact, usually by head to head contact. The close bodily contact among playing children in nursery and primary schools may, at least partly, explain why head louse epidemics are most commonly recorded among children about 5–10 years old. Until a head louse epidemic is detected it has usually been going on 'silent' during several weeks or months. Light infestations can occur even under good sanitary conditions. Infestations of head lice can occur even in clean hair (MPA 1999).

The only food of the head louse is blood. It will rapidly desiccate to death unless it has a regular supply of blood. Thus, a head louse away from the host's head will not survive for longer than about a day. This is one main reason why we consider head-to-head contact to be the main mode of transmission.

However, since a louse may survive for about 24 h away from the host, it cannot be excluded that other means of transmission occasionally may occur, for example, via combs, hair brushes, caps, and helmets. If such things have been used by a louse-infested person during the last few hours delousing by washing, heating, or freezing is recommended. Children should be urged not to exchange caps, helmets, combs, brushes, etc. between each other. In the control of head louse epidemics, it is not necessary to clean other clothes, bedclothes, carpets, playthings, etc. (MPA 1999).

A louse infestation is usually detected due to the itching from the scalp. During blood-feeding, the louse injects saliva into the bite-wound. The saliva contains proteins which, when they have entered the bite-wound, will cause the itch. All louse infestations are not accompanied by itch. Children, who have recently been infested, and adult persons who have become desensitized, may lack the itch. They can, therefore, be important sources for transmission of lice. Itching bite-wounds may be severely scratched and produce further ulceration and secondary bacterial infections. A black snuff-like powder on the pillow or on the collar may be louse faeces and thus evidence of a head louse infestation.

The diagnosis of a potential head louse infestation is best done by combing the hair very thoroughly close to the scalp by using a fine-toothed 'louse comb'. A magnifying lens and good light is necessary to reliably detect any live eggs, which are firmly attached to the base of the hair, or live lice which may become dislodged by the comb. By the method described, many lice may become dislodged and fall onto the ground undetected. Therefore, a more reliable method to be used within the family is that the potentially louse-infested person is sitting naked, on a white sheet. The debris removed by the comb, including that attached to the comb and that which has fallen onto the sheet, shall be inspected under a magnifying lens. A positive diagnosis is confirmed by finding live eggs, that is, eggs attached to the base of hairs, and/or by the presence of active lice. The presence of only eggs or egg shells on hairs is not a reliable sign of an ongoing louse infestation; the eggs may be dead. Since the eggs are usually attached close to the base of the hairs which grow about 1 cm per month the location of the eggs or shells may reveal when the infestation began (MPA 1999).

The inspection including the tracing of contacts can be very time-consuming, but is necessary in order to get good results from the treatment. Parents should also be advised to inspect their children's scalp before the start of each school term and then preferably each week during the first two months of the term and thereafter, once a month (MPA 1999).

The eggs are so firmly 'cemented' to the hair that they cannot usually be removed or destroyed by ordinary louse-combs made of plastic. Therefore, the stronger, more fine-toothed steel combs are preferred. It is considered likely that complete elimination of relatively light head louse infestations can be achieved by thorough combing using a fine-toothed steel comb.

In order to achieve complete eradication, especially of dense infestations, it is likely that this can be achieved more rapidly and easily by combining the use of a fine-toothed steel comb with insecticidal treatment of the hair. The choice of insecticide depends partly on the resistance status of the louse population in the geographical area concerned. Therefore, it is necessary to get information about the sensitivity of the local louse population to the potentially preferable insecticides.

There is no indication that head lice, under natural conditions, transmit any human-pathogenic viruses, bacteria, or other microparasites.

Pediculus humanus *(the human clothing louse, body louse)*

The body louse is morphologically very similar to the head louse and the two species were until recently treated as two subspecies, *P. humanus humanus* (synonym: *P. humanus corporis*) and *P. humanus capitis*. However, the biology and medical importance of the two species are distinctly different.

The primary microhabitat for all stages of the body louse is the clothes of humans. The blood-feeding stages (i.e. all stages except the eggs) only leave the clothes of its host in order to blood-feed on the host's skin. Each individual louse feeds about four times in 24 h. The gravid female body louse usually glues her eggs to fibres in the seams. On rare occasions, she may attach eggs onto body hairs.

The body louse is present in all parts of the world on persons who live under poor hygienic conditions and who cannot wash or change their clothes regularly. Thus, infestations with body lice are associated with poverty, very poor hygienic conditions, and (often) a cold climate where people need to wear clothes. In the tropics, the presence of the body louse is concentrated to relatively cold areas such as the highlands of Ethiopia and the South American Andes where people wear clothes on the greater part of the body. In such areas, the heaviest infestations usually occur during the coldest part of the year. In Europe and the former Soviet Union, during the Second World War, the body louse populations had excellent opportunities to increase due to the extremely unhygienic conditions in concentration camps, prisons, refugee camps, etc. Then, the body louse became a relatively rare species. However, with the vast economic and social changes that have now taken place in, for example, eastern Europe and Russia, an increasingly greater part of the human population in these areas have become extremely poor and even homeless. Therefore, the possibilities for the body louse to find optimal living conditions have greatly increased. In a recent investigation involving 300 homeless men in Moscow, 19% were positive for body lice (or eggs); 3–25 lice per person were recorded from their clothes; 12% of 268 louse samples were positive by polymerase chain reaction (PCR) for *Bartonella quintana* (Rydkina *et al.* 1999).

In contrast to the head louse, the body louse is an extremely important vector of human-pathogenic microorganisms. In Russia, louse-transmitted diseases have caused more deaths than any other infectious disease in recent centuries (Rydkina *et al.* 1999). Louse infestations are promoted by wars and natural disasters when people are forced into crowded, unhygienic conditions. Thus, outbreaks of louse-borne diseases is a constant threat to people living under primitive conditions such as those prevailing during conditions of war, in refugee camps and in prisons in poor countries. The body louse is the vector of three medically important infections, namely, epidemic typhus caused by *Rickettsia prowazekii*, trench fever caused by *Bartonella quintana*, and epidemic relapsing fever caused by *Borrelia recurrentis*.

Louse-borne epidemic typhus

Epidemic typhus may by some people have been considered a disease of the past. Yet, it is a constant threat during each major war. During the Second World War, an outbreak of epidemic typhus caused disease in more than 20 million people in Russia. Thereafter, the largest outbreak occurred some years ago in the over-crowded prisons in Burundi (Raoult *et al.* 1997). *Rickettsia prowazekii* can persist in a latent form in people who have long ago had overt epidemic typhus. In people under immunological stress the latent form can become apparent as Brill–Zinsser disease. Recently, there have been outbreaks every year of Brill–Zinsser disease in all regions of the former Soviet Union (Rydkina *et al.* 1999).

Trench fever

Trench fever is a louse-borne infection caused by the bacterium *B. quintana*. The latest great epidemics of trench fever took place during First World War in Europe (Maurin and Raoult, 1996). A large outbreak of trench fever accompanied by epidemic typhus took place recently in Burundi (Raoult *et al.* 1998). Cases of *B. quintana* infection have been reported during the last decade from Europe and North America, mainly in HIV-infected persons, homeless persons, and persons with chronic alcoholism (Rydkina *et al.* 1999). *Bartonella quintana* can cause bacteremia in immunocompromised patients. In healthy patients, *B. quintana* can cause bacillary angiomatosis, lymphadenopathy, endocarditis, and central nervous system (CNS) infections (Rydkina *et al.* 1999).

Louse-borne relapsing fever

Louse-borne relapsing fever is also denoted as epidemic relapsing fever. The causative agent is the spirochaete *B. recurrentis*. The disease is endemic in the highlands of Africa, particularly Ethiopia, and in scattered foci in other parts of the world. Serious epidemics have prevailed during or shortly after outbreaks of epidemic typhus, often in connection with wars. The disease was prevalent in Russia during First and Second World Wars.

Pthirus pubis *(the crab louse, pubic louse)*

The crab louse or pubic louse, *Pthirus pubis*, is rounder, that is, shorter (1–2 mm) and broader than the two, more oval *Pediculus* species (Figure 23.4). In the crab louse the first pair of legs is much more slender than the second and third pairs. In the microscope the crab louse somewhat resembles a 'miniature crab'. Since they are so small, they are difficult to find even with a magnifying glass. In contrast to the clothing louse, the pubic louse including its eggs only occur on the hairy parts of the body. *Pthirus pubis* mainly occurs on hairs in the pubic

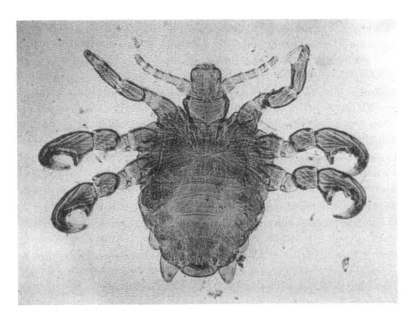

Figure 23.4 The public louse *Pthirus pubis*. Natural size 1–2 mm. (Photo: T. G. T. Jaenson©.)

and perianal regions, and less frequently in the axillae, eyebrows, and beard. Since *P. pubis* prefers thick hairs, infestations of children are relatively rare, but if *P. pubis* are found on children it is usually on their eyebrows. However, lice on the eyebrows are often proved to be head lice. The pubic louse is a parasite of humans in all parts of the world. The prevalence of *P. pubis* appears to have been increasing in the human population in recent years (Kettle 1995). Pthiriasis or pediculosis pubis, as an infestation with *P. pubis* is denoted, is claimed to be the most contagious of the sexually transmitted diseases (Felman and Nikitas, cited in Kettle 1995). It is likely that most pubic louse infestations are not diagnosed by physicians. In other words, pthiriasis is most likely quite underdiagnosed and may be nearly as prevalent as head louse infestations.

A 'normal-sized' pubic louse population on a single infested person is about 10 lice. The pubic louse feeds more or less continuously during the day and night. Like the *Pediculus* species, the pubic louse is strongly dependent on its host and will die within 24 h if it is dislodged from the host.

There is no indication that *P. pubis*, under natural conditions, transmits any human-pathogenic microorganism including viruses. The main physical symptom of pthiriasis is itching which, however, is relatively benign. The psychological suffering of pthiriasis may be more severe. Since the lice, during blood-feeding, injects an enzyme (which transforms haemoglobin to biliverdin) with the saliva, bluish spots can sometimes be seen in the skin. The faeces of pubic lice can be seen as a brownish powder on the skin or in the underclothes.

Like the other two species of human-parasitizing lice, the pubic louse is transmitted from one person to another by bodily contact, often by sexual activity or simply by sharing a bed with an infested person. It should be emphasized that transfer of pubic lice can occur by ordinary (non-sexual) bodily contact. However, transmission of *P. pubis* via toilet seats, towels, etc. is presumably very uncommon.

Acari (mites and ticks)

Non-insect arthropods of great medical importance in the northern Holarctic region include several groups of ectoparasitic mites, for example, the minute chigger mite larvae (*Leptotrombidium*, *Neotrombicula*) as well as endoparasitic mites (e.g. *Sarcoptes scabiei*) and the larger ticks (Ixodidae, Argasidae). Mites and ticks (Acari) belong to the class of arthropods denoted as Arachnida. The scorpions, harvest-men, and true spiders also belong to the arachnids. About 30 000 species of mites and 800 species of ticks have been described, but about half a million species of mites are believed to exist (Varma 1993). Most Acari, except the nymphs and adults of most ticks, are very minute animals and should preferably be studied with a compound or scanning electron microscope. Nearly all mite species lay eggs. The egg develops to a six-legged larva, which then becomes an eight-legged nymph. Depending on the species there may be one or several nymphal stages. The life-cycle in some species can be as short as one week but in other species (argasid ticks) can be a decade or more.

Infestation by mites is termed acariasis. Mites can be medically important because they ingest blood, lymph, or dermal tissues, because they may cause dermatitis or other tissue damages, or cause serious allergic reactions, and/or because they can transmit pathogenic microparasites (from wild or domesticated vertebrates) to man. In the following sections, the Acari has been denoted as: (i) ectoparasitic mites; (ii) endoparasitic mites; (iii) soft ticks; and (iv) hard ticks. It should be emphasized that the first two groups are not taxonomic entities, but only an artificial way of treating a number of taxonomically diverse mite taxa into two groups based on their way of living on or inside the host, respectively.

Ectoparasitic mites

Dermanyssus gallinae *(the chicken mite, red poultry mite)*

The chicken mite, *Dermanyssus gallinae*, has a world-wide distribution and is a nocturnally blood-feeding ectoparasite on domestic and wild birds. During the summer when synanthropic birds, for example, house sparrows, starlings or doves, have their nests inside or on the outside of human habitations, the mites will feed on the nestlings and their avian parents. When the birds leave the house the mites will try to find substitute hosts, which often happen to be people inhabiting the house. The bites can be painful and irritating. The length of the mite is about 0.7 mm (unfed) to 1 mm (fed). The recently fed mite is bright red in colour. Poultry houses can be infested by thousands of *D. gallinae* which may lead to serious blood-loss, reduced egg laying, and even death of the birds (Varma 1993). Infestations in schools and other buildings (often at the end of the summer) usually originates from birds' nests under eaves or in attics. The problem can usually be solved by removal of the nests, and by vacuum cleaning and insecticidal treatment of the attic and infested rooms.

Liponyssoides sanguineus *(the house-mouse mite)*

The house-mouse mite, *Liponyssoides sanguineus*, is a nest-dweller and comes to its hosts (mice, rats, and other rodents) only to feed. *Liponyssoides sanguineus* is a vector of rickettsial pox (caused by *Rickettsia akari*) in man. The infection has occurred as outbreaks in the US, Ukraine, and in Africa.

Ornithonyssus sylviarum *(the northern fowl mite)*

The general appearance and biology of the northern fowl mite, *Liponyssoides sanguineus*, is similar to those of *D. gallinae*. The northern fowl mite parasitizes domestic and wild birds but will also attack mammals, including humans. It is common in the northern temperate regions of North America and Europe but occurs also in southern Africa and southern Australia.

Cheyletiella *spp.*

There are several species of *Cheyletiella* mites which feed on the keratin layer of the epidermis of cats, dogs, and rabbits. People handling infested animals can become infested and develop an itching dermatitis.

Neotrombicula *and* Eutrombicula *chigger mites*

More than 1200 species of trombiculid mites (family Trombiculidae) have been described. About 20 of these are known to attack man, thereby causing chigger dermatitis (trombidiosis) or transmitting disease (scrub typhus or tsutsugamushi fever).

In several parts of Europe, including the British Isles and Scandinavia, microscopic chigger mites (*Neotrombicula autumnalis*) also denoted as harvest mites, are present in relatively distinct, focal localities. In the New World, species of the genus *Eutrombicula* have a biology and medical importance similar to those of *Neotrombicula*. The larval mites (about 0.25 mm long) of *N. autumnalis* feed mainly on small mammals and birds. Only the larval stage is parasitic on vertebrates. The larvae are not blood-feeders but rather feed on the enzymatically liquefied skin tissues of the host. People visiting *Neotrombicula*-infested localities in the late summer may be attacked by hordes of the creamy white to bright red larvae. The individual larvae are difficult to see, but larvae attacking in clusters can easily be detected. People who have previously been bitten by chigger mites will develop an allergic condition, chigger mite dermatitis, scrub itch or trombidiosis, presumably caused by allergens in the saliva of the mites. In the typical case of trombidiosis there is no reaction until after the second time of exposure. About 2–3 weeks are needed between the first and second exposure in order for a typical allergic reaction to develop. Usually, the dermatitis is not serious but is often intensely itchy and starts a few hours after exposure. Small, red pustules (3–4 mm) can be seen, mainly on areas where the skin is moist, where the clothes are tight to the skin, or under the watch-bracelet (Varma 1993; Kettle 1995).

Personal protection from larval chigger mites is of course most optimally done by avoiding to visit infested habitats. If this is not possible, boots and trousers (well tucked into the boots) can be used, particularly if treated with an effective mite-repellent or acaricide, for example, diethyl-toluamide (DEET), benzyl-benzoate, or permethrin.

Scrub typhus, tsutsugamushi fever

Larval chigger mites of the genus *Leptotrombidium* (relatives to *Neotrombicula*; family Trombiculidae) are also ectoparasitic on various small mammals including rodents and birds. In southern and eastern Asia a serious human disease called scrub typhus or tsutsugamushi fever is transmitted by *Leptotrombidium* larvae. The larval mites acquire the infection (*Orientia tsutsugamushi* = *Rickettsia orientalis*) by feeding on infected rodents, particularly *Rattus* spp. The infection can presumably be spread to new areas by infected mites being transported on birds. Since the infection in the mite population is transmitted transovarially, via the eggs, to the new generation the infection can be transmitted further to susceptible veretebrates. Man is an unnatural host for the mite larvae (and for *O. tsutsugamushi*). Human infections can, however, occur when people visit the mite-infested 'typhus islands' and are attacked by hordes of infected chigger larvae. The infection is usually present in small and scattered foci in many different biotopes from flooded alluvial plains in Japan, scrub and disturbed forests in southeastern Asia, to semi-deserts in Pakistan and alpine areas in the Himalayas (Varma in WHO 1989b). *Leptotrombidium deliense* is the principal vector over most of the distributional area of tsutsugamushi fever. In Japan, *L. akamushi* is the main vector. In the Far East of Russia *L. pavlovskyi* is an important vector (Varma 1993).

A few days after the infection, in about 40% of cases, a small nodule can be seen at the site of the mite bite. The nodule will become necrotic and suppurate. About a week after the appearance of disease symptoms (general malaise, headache, fever, regional lymphadenitis, etc.) a spotted fever exanthema usually appears on the trunk, on the extremities and in the face. There is often bronchitis and encephalitis. Pneumonia, encephalitis, and heart failure are serious complications and may indicate fatal cases. During serious epidemics the mortality rate of untreated cases can exceed 60%. Scrub typhus has been, and may continue to be, of great importance in military medicine.

Endoparasitic mites

Sarcoptes scabiei *(scabies mites)*

The scabies mite, *Sarcoptes scabiei*, is the cause of scabies in humans and other mammals. If man is infested by any of the forms of the scabies mite which normally parasitize animals the infestation usually disappears after some time. This is because each one of the different scabies mite populations are relatively host specific.

The scabies mite is whitish to nearly transparent, almost round, and due to its size (1/3 mm) hardly visible to the naked eye (Figure 23.5). The female scabies mite makes ducts in the epidermis in which the eggs are laid. The eggs hatch after about one week. After the larval and nymphal instars the adult stage is reached. The adult mites will mate on the skin surface (Gordon and Lavoipierre 1972). After copulation, the female burrows in the skin,

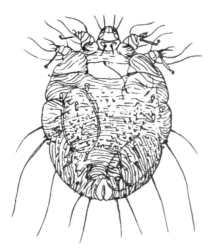

Figure 23.5 Female scabies mite *Sarcoptes scabiei* with two eggs. Natural size *c*.0. 3 mm. (Drawing by Inga Thomasson after Smith (1973).)

particularly where it is thin. Particularly favoured sites are the skin between the fingers, the underside of the wrists, the elbows and knees, navel, breasts, shoulders, buttocks, scrotum, and penis. The face and scalp are not usually infested, except in children. The length of the ducts varies between a few millimetres to several centimetres. The number of ducts and adult females can sometimes exceed 100. The average number of ducts per infected person may be about 10–15. The itching, which is an immunological response to mites and their faeces, is particularly serious during night-time. In newly infested persons the itching begins not until about a month after the initial infestation. The itching will often indirectly cause secondary bacterial infections of the skin with purulent ulcers. Infested persons will develop an extensive rash with erythema and follicular papules that can cover areas where no mites can be found (Alexander 1984; Varma 1993). Scabies-infected individuals with immunological defects, for example, HIV-infected people, can develop a very serious and very contagious type of scabies, denoted as crusted or Norwegian scabies (after a Norwegian scientist who first described the condition). In persons with crusted scabies the mite population increases to extremely high abundances (>1000 mites per person).

The diagnosis is preferably made by using a thin needle to catch a (female) mite in a burrow in the skin. A captured mite can thereafter be identified in the microscope. By pouring ink onto supposedly scabies-infested skin and then washing away surplus ink from the skin a positive scabies diagnosis is made by finding ink remaining in the mite ducts.

Human infestation with *S. scabiei* has a world-wide distribution but appears to be most prevalent in relatively poor subtropical and tropical areas. Persons of all ages are affected, although prevalence rates appear to be the highest in children. Thus, investigations of pre-school children in the tropics have revealed point prevalences between 30 and 80%.

Transmission of scabies mites is presumably mainly by prolonged close personal contact, for example, by holding hands or by resting or sleeping in the same bed as a scabies-infested person. However, it is likely that the mites can also be transmitted via clothes, towels, bed linen, etc. Since the mites are most active during the night, transmission is also most likely to occur during this time.

Control of scabies infestation is primarily by acaricidal (chemical) treatment of the infested skin. In similarity to the situation in louse infestations, the choice of scabicide to be recommended depends partly on the resistance status of the mite population in the geographical area concerned. The resistance status can change within a few months in a particular area depending on the types of acaricides that have been used there. Thus, it is not possible to give any general recommendation here about the choice of acaricide to be used. Rather, it is necessary to get up-to-date information about which acaricide is recommended in a particular area. The most commonly used anti-scabies products contain as the active ingredient(s) benzyl-benzoate (with disulfiram), crotamiton (*N*-ethyl-*O*-crotontoluide), lindane or malathion. Benzyl-benzoate is not recommended for usage on children. Lindane is one of the environmentally harmful persistent organic pollutants (POPs) the use of which should be abandoned (Jaenson 1996).

When scabies has been detected in a family, all its members and other close contacts should be treated. The treatment with a suitable acaricide, usually obtainable in the local pharmacy, should be preceded by thoroughly washing the body. The duration of the treatment, that is, the time that the active substance shall remain on the skin depends on the type of acaricide and how it is formulated. After the acaricidal treatment, the body should be thoroughly washed with soap and water. Clothes, towels, bed linen, caps, etc. should now be washed or, if this is not possible, an alternative is to place, for example, gloves, shoes, etc., to air for 5–7 days.

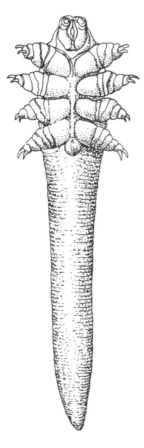

Figure 23.6 The hair follicle mite *Demodex folliculorum*. Natural size *c*.8 mm. (Drawing by Inga
Thomasson after Smart (1948).)

Demodex *spp. (hair follicle mites)*

There are two species of mites in the genus *Demodex* parasitic on humans, namely,
D. folliculorum (in the hair follicles) (Figure 23.6 and *D. brevis* (in the sebaceous glands). Most
people harbour these mites in the face, particularly around the nose. Older persons are more
often infested than younger ones. The medical importance of the presence of *Demodex* is
difficult to assess but most infestations appear to be benign and of no or little pathological
significance. In rare cases, these mites are thought to be involved in the loss of eye-lashes and
in granulomatous acne (Alexander 1984; Varma 1993).

Soft ticks (Argasidae)

Ticks are large blood-sucking Acari, sometimes >1 cm in length. There are three families of
ticks, Argasidae, Ixodidae, and Nutalliellidae, consisting of a total of about 800 species. The
latter family contains only one species of no known medical significance. The development
of ticks is from an egg via one (Ixodidae) or more (Argasidae) nymphal stages to the adult
stage. The larva has six legs while the nymph and adult have eight legs. The soft ticks

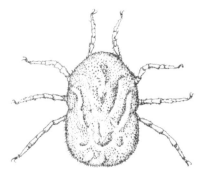

Figure 23.7 The soft tick *Ornithodoros*. Natural size *c.*8 mm. (Drawing by Inga Thomasson after Smith (1973).)

(Argasidae) differ from the hard ticks (Ixodidae) by lacking a dorsal shield (scutum). Moreover, in the soft ticks the mouth-parts are not visible when the tick is observed from above. Between the larval stage and the adult stage, soft ticks can have up to five nymphal stages. The eggs are laid when the female has ingested a blood meal (Sonenshine 1991, 1993; Varma 1993).

Ornithodoros *spp.*

Ticks of the genus *Ornithodoros* can be distinguished from other soft ticks (*Argas, Otobius*) by the presence of mamillae (minute, regular, usually hemispherical elevations) of the integument and the absence of a distinct lateral margin to the body (Varma 1993) (Figure 23.7). Most species of *Ornithodoros* inhabit animal burrows, nests, dens, and caves where they feed on wild mammals or seabirds. These ticks are the main vectors and reservoirs of human relapsing fever (*Borrelia* spp.) also denoted as tick-borne (endemic) relapsing fever. The foci of this infection are very localized and restricted to huts, caves, etc. The spirochaetes are transmitted both transstadially and transovarially in the tick population. To the ticks' blood hosts the spirochaetes are transmitted by tick bite but sometimes also by contaminated coxal fluid from the ticks. Human infections play no role in the dynamics of this zoonosis among the natural hosts of the ticks, that is, small mammals. In the US, three *Ornithodoros* species (*O. hermsi, O. parkeri,* and *O. turicata*) are vectors of relapsing fever *Borrelia* spp. (*B. hermsi, B. parkeri,* and *B. turicatae*). From Israel to western China *O. tholozani* is the vector of relapsing fever. In Africa and Latin America, other *Ornithodoros* species transmit other species of relapsing fever borreliae. A typical infection in a human patient includes three to seven attacks at 4–7-day intervals. The clinical picture of the disease is similar to the one exhibited by louse-borne relapsing fever. *Ornithodoros*, parasitizing seabirds, are also vectors of several arboviruses which can cause human disease.

Hard ticks (Ixodidae)

The mouth-parts of the hard ticks are visible if the tick is inspected from above (in soft ticks the mouth-parts are hidden under the body) (Figure 23.8). The adult male has a hard, dorsal shield (scutum) which covers the whole dorsal part of the body. In the adult female, nymph and larva only the anterior part of the dorsum is covered by the scutum. The ixodids

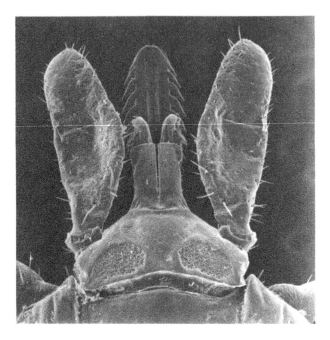

Figure 23.8 Anterior portion with the mouth-parts of a female hard tick *Ixodes ricinus*. (Photo: G. Wife and T. G. T. Jaenson©.)

have only one nymphal stage between the larval and adult stages. While the argasids ingest their blood meals relatively rapidly and may take several blood meal during each active stage (larva, nymphs, adult), the ixodids only ingest one (large) meal during each active stage. In view of their roles as vectors of viroses, rickettsioses, ehrlichioses, tularemia, and babesioses the ixodids are of great medical importance in the tropical, subtropical, and temperate regions. From a veterinary point of view, the Ixodidae is the most important of all arthropod families. Some of the medically most important ixodid species are found in the genera *Amblyomma*, *Dermacentor*, *Haemaphysalis*, *Hyalomma*, *Ixodes*, and *Rhipicephalus* (Sonenshine 1993; Varma 1993).

Tick-borne encephalitides

With few exceptions, the medically important arboviruses are transmitted by either mosquitoes or by ixodid ticks. Humans are in most cases a dead end host for the virus, and are therefore not necessary for the continuing survival of the virus in nature. In the temperate Palaearctic region, there are two types of encephalitides transmitted among wild animals by *Ixodes* ticks. The central European or western tick-borne encephalitis (TBE) is transmitted by *I. ricinus* and occurs in scattered foci in Europe and western Russia. The eastern form is called Russian spring–summer encephalitis (RSSE), is transmitted by *I. persulcatus* and occurs from eastern Europe to the Far East and into northeastern China. Most infected people get a subclinical infection. In typical clinical cases, the disease has a bi-phasic course. After an incubation period of about 10 days (after the tick bite) influenza-like symptoms appear, which last for a few days. But after about a week with few or no symptoms the patient's condition worsens. Signs and symptoms of meningoencephalomyelitis may develop, including fever, neck rigidity, and headache, sometimes with paralyses of one or more extremities. The second phase of the disease usually lasts for 2–3 weeks. Fatigue and remaining signs of paralysis may persist for a year or more. The RSSE type is, in general, more severe than the TBE type. Also, the symptoms are more severe in the elderly than in children. Effective vaccines are available.

Crimean-Congo haemorrhagic fever

The tick *Hyalomma marginatum* is common in southern, central, and eastern Europe and in the former USSR. Migratory birds are important hosts for the subadult stages of *H. marginatum*. Such birds carry the ticks between Africa and Europe. *H. marginatum* is a major vector of a serious arboviral disease, Crimean-Congo haemorrhagic fever. The virus is known to be enzootic in dry areas of central and eastern Europe, most of the former USSR and in many parts of Africa.

Colorado tick fever

Dermacentor andersoni is the main vector of the arbovirus causing Colorado tick fever in the northwestern US and in western Canada. The vertebrate reservoir consists of various small mammals. In human cases encephalitis may develop, particularly in children.

Tick-transmitted rickettsioses

Rickettsia species within the so-called spotted fever group have been demonstrated among ticks in Europe, for example, *R. conorii* in Spain, France, and Italy and *R. slovaca* in eastern Europe. In 1979, Burgdorfer *et al.* demonstrated the presence of a rickettsia, later named *R. helvetica*, in *Ixodes ricinus* collected in Switzerland. Recent studies have demonstrated that *R. helvetica* is present in about 20% of nymphal (Figure 23.9) and adult *I. ricinus* in southern Sweden (Nilsson *et al.* 1998). Nilsson *et al.* (1999) have also demonstrated the presence of *R. helvetica* in heart muscle tissue and adjacent tissues in two men who had suddenly died during sporting activities. Even though it is not proven that *R. helvetica* was the cause of the death of these young men, this hypothesis needs to be tested.

Rickettsia rickettsii, the etiological agent of Rocky Mountain spotted fever, is widely distributed in both North and Latin America. The vectors in eastern and western US are *Dermacentor variabilis* (the American dog tick) and *D. andersoni* (the Rocky Mountain wood tick), respectively (Burgdorfer 1975). The infection in humans appears now to be most prevalent in the eastern US. Untreated *R. rickettsi* infection is a serious disease. However, with the advent of antibiotics the mortality rate has now decreased considerably.

Rickettsia conorii is the etiological agent of Mediterranean tick typhus or boutonneuse fever in the Mediterranean area, in Israel and in southern and south-eastern Asia. It is so-called because it is associated with a button-like lesion at the site of the tick bite. The main vector is the brown dog tick, *Rhipicephalus sanguineus*.

Rickettsia sibirica is distributed in northern Asia from the Far East westwards to Armenia. *Dermacentor marginatus*, *D. silvarum*, *Hyalomma concinna*, and *R. sanguineus* are the main vectors. The symptoms of *R. sibirica* infection resemble those of mild Rocky Mountain spotted fever with a rash, fever, and headache.

The rickettsioses mentioned are all transmitted via tick bite but some of the infections can presumably also be contracted, for instance, by picking infected ticks from dogs,

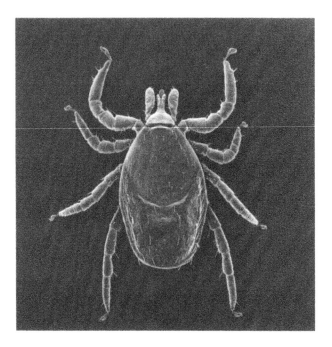

Figure 23.9 Nymph of the hard tick *Ixodes ricinus*. Natural size *c*.1.5 mm. (Photo: G. Wife and T. G. T. Jaenson©.)

crushing the ticks between the fingers, and then inadvertently contaminating wounds or mucous membranes with infective material. The *Rickettsia* organisms are usually transmitted, both trans-stadially and transovarially, in the vector populations.

Q-fever

This infection is caused by the bacterium *Coxiella burnetii* which occurs in both wild and domesticated animals. *Coxiella burnetii* is enzootic in sheep, goats, and cattle in nearly all parts of the world, particularly in hot and dry areas. The organism has been associated with anorexia and abortion in domestic animals. Ixodid ticks of the genera *Ixodes* and *Amblyomma* are important vectors of *C. burnetii*, particularly in nature. Human infection, however, is primarily through inhalation of infective material and by consumption of unpasteurized milk and milk products from cattle, sheep, and goats. Pneumonia is the main symptom in the human disease, but occasionally the disease is very severe with hepatitis or endocarditis.

Lyme borreliosis

Lyme disease or Lyme borreliosis is the most prevalent arthropod-borne human infection in the temperate region of the Northern Hemisphere (Jaenson 1991a). This area includes the major part of the US, Europe, Russia, and Japan. In Sweden, which is a country with one of the highest incidences, the annual number of human Lyme disease cases is estimated to be about 10 000. The disease is caused by infection with one of several *Ixodes*-transmitted spirochaetal species within the *Borrelia burgdorferi* complex. In Europe, *Ixodes ricinus* is the main vector of the *Borrelia* species from the mammal reservoir (rodents, insectivores, hares, and birds) to humans (Jaenson 1988, 1998). In European Russia the main vectors are

I. ricinus (in the west) and *I. persulcatus* (towards the east). In Asia (Russia, Japan, and China) *I. persulcatus* is the primary vector. In the US, there are two main vectors, that is, *I. scapularis* in the east and *I. pacificus* in the west. In the US and in Europe the nymphal stage of the main vector species is the most important stage as a link vector of the spirochaetes from the enzootic focus, usually maintained by small mammals, to man. Roe deer and other cervids are important blood hosts for the females, and therefore indirectly for the maintenance of the whole tick population. However, there is no or very little evidence that cervids play a significant role as reservoirs of the spirochaetes. Thus, ticks feeding on cervids are usually not infected with spirochaetes from these cervids (Jaenson 1991a, 1998).

In most cases of human Lyme borreliosis, the cardinal sign denoted as erythema migrans occurs on the skin at the site of the tick-bite a few days to several weeks after the bite. Sometimes, several skin rashes occur on the same patient, presumably due to haematogenous spread of the spirochaetes. Apart from the skin erythema the early symptoms are usually influenza-like. The infection may disappear spontaneosly (or by antibiotic treatment) or proceed to become more serious with neurologic symptoms, arthritis, and sometimes myocarditis. The signs and symptoms of Lyme disease are numerous. The appearance of the disease is very variable and may, in one patient be completely different from that in another.

Other tick-transmitted bacterial infections of both medical and veterinary importance are the ehrlichioses, caused by rickettsia-like organisms of the genus *Ehrlichia* spp. and tularemia, caused by the bacterium *Francisella tularensis*. There are several *Ehrlichia* species of medical importance transmitted both in the Nearctic and in the Palaearctic areas. Tularaemia is confined to the Holarctic region. Mixed infections, with *Borrelia burgdorferi* s.l., *Ehrlichia* spp., *Francisella tularensis* and TBE-virus may occur in the same individual tick.

Babesioses

Babesioses are tick-transmitted infections of great veterinary importance. They are caused by protozoa of about 100 different species in the genus *Babesia*. Two species are known to cause human disease, that is, *B. microti* transmitted by *Ixodes scapularis* (*I. dammini*) in northeastern US and *B. divergens* transmitted by *I. ricinus* in Europe.

The natural host of *B. microti* in the US is the white-footed mouse, *Peromyscus leucopus*. Infection with *B. microti* in humans is usually self-limiting. Many people have an asymptomatic infection, and the severity of the disease symptoms usually increases with age. Severe disease due to *B. microti* is usually seen only in the elderly or other people with a weakened immune defence. In young people the infection usually only causes sero-conversion without any disease symptoms. *B. microti* also occurs in Europe where it is transmitted mainly by *I. trianguliceps* among rodents. Since *I. trianguliceps* is relatively host specific and (almost) never feeds on humans the risk for *B. microti* infection is almost non-existent in Europe (Jaenson 1988).

Most patients who have contracted human babesiosis in Europe had been splenectomized some time before the appearance of symptoms of this disease. The course of the disease resembles that of malignant malaria. The mortality rate has been about 50%. The causative organisms in all cases of European human babesiosis seems to have been *B. divergens* which is naturally transmitted by *I. ricinus* among cattle.

Tick paralysis

Tick paralysis is caused by injection of a neurotoxin by blood-feeding female ixodid ticks and by some *Argas* species. The neurotoxin, which may be different in different species, affect synapses in the spinal cord and blocks the neuro-muscular junctions. The first symptoms appear 5–7 days after attachment of the female tick. One single female tick is sufficient to kill an adult human. In the beginning, the paralysis affects the legs and then ascends to the trunk, arms, and head within a few hours. The most important tick species involved in human tick paralysis are *Ixodes holocyclus* (in Australia), *Dermacentor andersoni* (in western North America) and *D. variabilis* (in eastern North America). Removal of the tick, including its mouth-parts, usually leads to quick and complete recovery, except in cases involving *I. holocyclus* which may become worse by removal of the tick. Thus, in Australia an anti-toxin is injected intravenously and allowed to circulate in the body before removing the tick (Kettle 1995).

Tick bites

The bites of most species of soft ticks are painful. The bite of the pigeon tick, *Argas reflexus*, can cause severe reactions including unconsciousness. In contrast, the hard ticks usually attach to the human body unnoticed. This is presumably an adaptation to their long duration of feeding (several days). A tick attached to the skin should be removed as soon as possible. The capitulum (front-part) of the tick should be grasped with a very fine-pointed forceps and then pulled backwards, at a right angle to the skin surface. By this method the whole tick is usually removed intact. If a portion of the mouth-parts remains in the skin it is usually possible to remove it by the forceps. The remaining bite-wound should be washed in clean water and thereafter, with an antiseptic solution.

Control of ticks and tick-borne infections

Pasture spelling and pasture rotation may be used to control tick species attacking and transmitting diseases to farm animals. Pasture spelling is widely used in Australia (Wilkinson 1957). Certain breeds, races, or species of livestock are more tolerant than others to ticks and tick-borne infections. Such livestock should be used for traction, and meat and milk production rather than less disease-resistant breeds. There are vaccines available against some important tick-borne diseases. Anti-tick vaccines may become available for veterinary use in the near future. For personal protection in tick-infested areas the use of appropriate clothing, possibly with the additional use of a chemical tick-repellent, and prompt removal of attached ticks are measured that can be recommended. Further information on tick control methods is available in Jaenson *et al.* (1991b), Mwase *et al.* (1991), and references therein. A number of plants are traditionally used for their tick-repelling properties (Secoy and Smith 1983; Curtis 1991, 1992). There is an urgent need to investigate the pharmaceutical and arthropod-repellent potentials of many of these plant species.

General considerations about methods for control of disease-causing arthropods

Environmental measures for control of ectoparasites, mosquitoes, and other arthropods of medical and veterinary importance almost disappeared with the advent of chemical, synthetic insecticides (WHO 1982). However, arthropod vectors of disease soon started to develop resistance to the pesticides. Moreover, concern over the environmental side-effects from pesticides has contributed to a decrease in the development of new public health pesticides. The need to use an integrated approach including both old and new methods of mosquito control is now generally accepted (Curtis 1984, 1991; Jaenson 1991b). Biological control agents and environmental management measures are effective components of such strategies (WHO 1982). Many of the arthropod-related problems in poor urban areas are similar to those appearing under situations of war and natural disasters. The PAHO's publication on 'Emergency vector control after natural disaster' (PAHO 1982), the WHO's (1982) 'Manual on environmental management for mosquito control,' and the WHO's (1991) 'CAP guide for insect and rodent control through environmental management' provide much valuable information.

Some advantages of environmental management measures are that they are effective and have long-term effects; their long-range costs are relatively low; the additional benefits may be considerable; better housing and recreational and sanitary facilities in urban areas can contribute to social development and higher standards of living; the protection of the people from the hazards associated with the use of some chemical pesticides is not required; they can effectively contribute to the prevention and control of several vector-borne and water-associated diseases; the disadvantages of environmental management operations are mainly their capital cost, the length of time required for completion, and complexity of important works (WHO 1982). Specific measures that can be undertaken by individuals or the community against problems with bed-bugs, flies, lice, mosquitoes, rodents, solid wastes, wastewater, sanitation, and house design are given in the CAP guide (WHO 1991).

Arthropod-related problems in poor urban and rural settings

The rapid growth of densely populated, low-income settlements – mainly in the Third World but even in the North – has come to constitute one of the most serious threats to health (WHO 1988a). The urban poor can be regarded as the interface between underdevelopment and industrialization, and their disease pattern often reflects the problems of both. From the first they receive a number of infectious diseases and malnutrition, and from the second a wide range of chronic and social diseases (WHO 1988a). The importance of water-associated diseases can be seen in many urban areas where, what were previously primarily rural diseases, for example, malaria and dengue, are now becoming endemic in the cities (WHO 1988a). A great number of infectious diseases are transmitted, or directly caused, by insects, mites, and other arthropods to humans in urban areas.

Most vector-borne human diseases are, or can be transmitted in the home environment which offers two key features: (i) food from man, his animals and products; (ii) shelter from extremes of climate. Apart from most flies and some mosquitoes, nearly all medically important arthropods in the domestic environment are active mainly at night. During daytime these insects, ticks, and mites are resting in dark crevices and similar protected shelters. Thus, reduction of dark undisturbed hiding-places and restricted access to sources of blood or other food at night can decrease the prevalence of these arthropods. Schofield and White (1984) and Schofield *et al.* (1991) have written informative overviews on house design and domestic ectoparasites and vectors of disease, from which the following information has been extracted.

The floor in many types of primitive houses is of beaten earth. Cracked or even mud floors may provide refuges for flea larvae, bed-bugs, soft ticks (vectors of relapsing fever), and house dust mites. Replacement of mud floor with cement reduces the infestation with these arthropods. Houses raised on stilts to avoid flooding have a space underneath, often used for storage of firewood etc., which may become infested with rodents, snakes, spiders, cockroaches, bugs, etc.

The wall is usually made of locally available material, for example, corrugated sheets, split logs, timber planks, mud and sticks, or woven thatch. Plywood, hardboard, or even cardboard from old boxes is often used in slum areas where better material cannot be afforded. Most types of walls having cracks or crevices or spaces between planks, stones, or logs provide excellent undisturbed refuges for a number of ectoparasites, vectors, and pests. The few, if any, windows ensure a dark, poorly ventilated, usually humid environment where the arthropods will thrive.

The roof: Thin roofs exposed to direct sunlight become very hot during sunny summer days which discourage insects. The tendency of many insects to fly upwards when they encounter an obstruction such as a house wall, means that the eaves under a roof are the main points of entry of house-entering insects such as endophagic flies and mosquitoes. If the eaves cannot be blocked or screened an alternative solution is to fit a ceiling. However, the roof-space above a ceiling may become a favoured hiding place for blood-sucking bugs, fleas and mites, and for the ectoparasites' hosts, such as rodents, bats, and birds.

Restriction of food sources for rodents, birds, bats, and arthropods is important since these animals can serve as reservoirs and/or vectors for infections. Protection of the food may be achieved by using screened food stores. Restriction of other food sources for rats, flies, cockroaches, ants, etc. include the provision of closed refuse containers and well-maintained toilets. Domestic populations of blood-sucking bugs, fleas, and mites can be controlled or eradicated by the removal of domestic animals (dogs, cats, pigs, goats, sheep, cattle, etc.) to suitable outhouses or enclosures. In Sweden and other parts of northern and western Europe, improved housing and the construction of animal shelters separate from the domestic habitations was probably the single most important factor in diverting the malaria mosquitoes (*Anopheles* spp.) from feeding on humans to feeding on animals. In this way, a gradual decline and eventual disappearance of malaria took place (Jaenson 1983).

The use of mosquito nets over beds at night gives relatively effective protection, not only against mosquitoes and other night-flying insects, but also gives some protection against fleas and bed-bugs. The protective effect is considerably enhanced if the netting is treated with an appropriate insecticide, for example, permethrin or deltamethrin at the recommended dosages. However, an unimpregnated poorly maintained or incorrectly used net does not provide satisfactory protection. Bed-nets which have not been treated with an insecticide will hardly provide any protection against blood-sucking bugs, fleas, ticks, and mites which actually inhabit the bed (WHO 1989a).

Breeding sites: Exposed standing water is likely to serve as a breeding site for dipterous insects (flies, mosquitoes, biting midges, etc). Thus, it is unwise to locate houses within the flight range (usually about 2 km) of these insects, unless the houses are adequately screened or the breeding sites controlled. On-site sanitation systems, such as pit-latrines or septic tanks and cess pits, are much cheaper than a system of sewers but have the drawback that under warm weather conditions they may provide suitable breeding of the nuisance mosquito *Culex pipiens* and a number of potentially myiasis-causing flies, for example, many blow-fly species (Calliphoridae). If cess pits are carefully constructed, kept free of cracks and apertures and screened, mosquito and fly problems can be avoided, but enforcement of such standards in low-income areas may be difficult (Curtis 1984). Almost all gravid blow-flies attempting to enter a pit latrine do so via the vent pipe, so a screen could effectively minimize the input of blow-fly eggs. This is less reliable because some enter via the latrine hole (Curtis 1984).

Flies of several families, including many species of facultatively myiasis-causing flies, have their primary breeding sites in animal and human wastes. Human faeces is among the most dangerous substances with which people can come into contact; it is the principal source of the pathogenic organisms of many communicable diseases, particularly infections of the intestinal tract (WHO 1988a,b). The house fly *Musca domestica* and several other flies in the families Muscidae, Calliphoridae, and Sarcophagidae are very efficient 'mechanical' vectors of pathogens (viruses, bacteria, protozoa, and helminth eggs) from human faeces to human food. These flies are often most serious pests in warm weather when they can reach high population densities. They breed in decaying food and garbage and are attracted to human food, human faeces, dead animals, garbage, rotting vegetation, and other organic material. Removal, destruction or protection of potential fly breeding sites is likely to effectively reduce the incidence of a number of diseases such as cholera, hepatitis A, and amoebiasis and other intestinal parasites.

Arthropod resistance to synthetic and natural insecticides and acaricides

Resistance is defined by WHO (1992) as 'an inherited characteristic that imparts an increased tolerance to a pesticide, or group of pesticides, such that resistant individuals survive a concentration of the compound(s) that would normally be lethal to the species'. In practice, the operational criterion is usually taken as 20% or more survivors of individual arthropods tested to the normally used diagnostic concentration of the pesticide, using WHO test kits in the field.

The appearance of insecticide-resistant vector or pest insect population is practically inevitable provided that the particular insecticide is used above a certain level for a certain duration of time.

The genes providing for susceptibility of the arthropod vector populations to particular groups of pesticides may be considered a non-renewable resource that should optimally be maintained by all people engaged in vector or pest control operations. Since there are only a very limited number of pesticides available for use in vector control programmes, these chemicals should be regarded as a valuable resource and protected accordingly (WHO 1992).

According to Miller (1988), the types and mechanisms for insecticide resistance can be grouped within four categories:

1   Behavioural resistance, where insect behaviour becomes modified so that the insect no longer comes into contact with the insecticide.
2   Penetration resistance, where the composition of the insect exoskeleton becomes modified in ways that inhibit insecticide penetration.
3   Site-insensitivity, where the chemical site of action for the insecticide becomes modified to have reduced sensitivity to the active form of the insecticide.
4   Metabolic resistance, where the metabolic pathways of the insect become modified in ways that detoxify the insecticide, or disallow metabolism of the applied compound into its toxic form. The most important mechanisms of metabolic resistance involve multi-function oxidases, glutathione-S-transferase, and esterases in the case of pyrethroids (which are almost all esters). In general, site-insensitivity or metabolic detoxification are the main resistance mechanisms, and physiological resistance can be seen as resulting from an interplay of these factors (Miller 1988).

Management of pesticide resistance: Based on the previous statements all efforts should be done to prevent or at least delay the appearance of insecticide resistance among important disease vector arthropods and other pest populations. To extend the time of effective use of pesticides involves a reliable system of surveillance, early detection, and accurate monitoring should be used (WHO 1992). This is even more obvious when considering that cross-resistance and multiple resistance, to chemicals to which an arthropod has never been exposed, will often evolve. Resistance management may be achieved by (WHO 1992; Roberts and Andre 1994):

- The use of non-chemical, rather than chemical control, if feasible.
- Agricultural pesticides can select for resistance in mosquitoes. In general, the resistance selected for is so broad that only by banning whole classes of chemicals could the insecticides against mosquitoes be fully safeguarded. For instance, to prevent resistance to malathion in malaria vectors breeding in rice field one would have to ban the agricultural use of all organophosphates and carbamates (Lines 1988).
- Agricultural use of pesticides should be drastically limited and taken into consideration regarding the promotion of development of resistant arthropods of public health importance. For instance, pyrethroid-impregnated bed-nets may become less effective because of the widespread use of pyrethroids in agriculture, especially in paddy fields where the malaria vector may breed.
- The use of mixtures of chemically unrelated pesticides. The basic assumption is that vectors resistant to pesticide A will be killed by pesticide B and therefore less likely to produce any resistant offspring. The absence so far among medical vectors of resistance to *Bacillus thuringiensis israelensis* is believed to be because it contains several toxic proteins.
- Application against a single life stage (adult females rather than both sexes or all life stages) of the target species.
- The use of pesticides of relatively low persistence (since pesticides which degrade rapidly are less likely, compared to more persistent chemicals, to produce pesticide resistance.
- Use of pesticides that produce resistance only within a single class of pesticides.
- Varying the dose or frequency of pesticide application.
- Application of pesticides in limited areas, with high levels of disease transmission, rather than area-wide; the former method will reduce the selection pressure.
- Limitation of pesticide application to periods of the year when the vector problem is important.
- Alternating or rotating among different (unrelated) pesticides. If two pesticides, A and B, are used sequentially the resistance level to A will decrease when B is used.
- Use of the most appropriate pesticide formulations.
- Use of synergists if feasible. These chemicals inhibit specific detoxification enzymes and thereby reduce the selective advantage of arthropods that have such enzymes.
- Integrated control, including biological control and environmental management should be used wherever it is appropriate. With regard to pesticide usage, environmental management, and biological control will reduce the selection pressure and prolong the effective duration of life of the pesticide used.
- The development of biological control agents which are resistant to certain pesticides used in agriculture and/or public health.
- Intensified research on identification of new pesticides with different metabolic pathways than the ones presently available.

References

Alexander, J. O'Donel. (1984). *Arthropods and human skin*. Springer-Verlag, Berlin, Germany, 422 pp.

Burgdorfer, W. (1975). A review of Rocky Mountain spotted fever (tick-borne typhus), its agent and its tick vectors in the United States. *J. Med. Entomol.* **12**: 269–278.

Curtis, C. F. (1984). Low cost sanitation systems and the control of flies and mosquitoes. *Trans. R. Soc. Trop. Med. Hyg.* **78**: 298.

Curtis, C. F. (ed.) (1991). *Control of disease vectors in the community*. Wolfe Publ. Ltd., London, 233 pp.

Curtis, C. F. (1992). Personal protection methods against vectors of disease. *Rev. Med. Vet. Entomol.* **80**: 543–553.

Gordon, R. M. and Lavoipierre, M. M. J. (1972). *Entomology for students of medicine*, 3rd edn. Blackwell, Oxford, UK.

Hall, M. J. R. and Smith, K. G. V. (1993). Diptera causing myiasis in man. In: Lane, R. P. and Crosskey, R. W. (eds), *Medical insects and arachnids*. Chapman & Hall, London, UK, pp. 429–470.

Harwood, R. F. and James, M. T. (1979). *Entomology in human and animal health*, 7th edn. Macmillan Publ. Co., New York, 548 pp.

Ibarra, J. (1993). Lice (Anoplura). In: Lane, R. P. and Crosskey, R. W. (eds), *Medical insects and arachnids*. Chapman & Hall, London, UK, pp. 517–528.

Jaenson, T. G. T. (1983). Malaria in Sweden – entomological aspects (in Swedish). *Läkartidningen* **80**: 2418–2421.

Jaenson, T. G. T. (1985). *Medical entomology. Insects and mites which cause disease in man* (in Swedish). Liberförlag, Malmö/Almqvist & Wiksell International, Stockholm, Sweden, 95 pp.

Jaenson, T. G. T. (1988). The ecology of tick-borne infections in Fennoscandia – an overview (in Swedish). *Läkartidningen* **25**: 2329–2331.

Jaenson, T. G. T. (1991a). Epidemiology of Lyme borreliosis. *Parasitol. Today* **7**: 39–45.

Jaenson, T. G. T. (1991b). Urban insect vectors with particular reference to households in poor countries. Unpubl. document for Stockholm Environment Institute, Stockholm, June 1991, 30 pp.

Jaenson, T. G. T. (1996). Alternatives to POP pesticides for control of arthropods of medical and veterinary importance, chapter 1. In: Johnson, A. (ed.), *Alternatives to persistent organic pollutants*, 327 pp. KEMI Report Series 4/96, Swedish National Chemicals Inspectorate & Swedish Environmental Protection Agency, Stockholm, Sweden, pp. 21–77.

Jaenson, T. G. T. (1998). The tick *Ixodes ricinus* as a disease vector in Northern Europe (in Swedish). Information from The Medical Products Agency, Uppsala, Sweden, April 1998.

Jaenson, T. G. T., Fish, D., Ginsberg, H. S., Gray, J. S., Mather, T. N. and Piesman, J. (1991). Methods for control of tick vectors of Lyme borreliosis. *Scand. J. Infect. Dis.* **7**: 151–157.

Kettle, D. S. (1995). *Medical and veterinary entomology*, 2nd edn. CAB International, Wallingford, Oxon, UK, 725 pp.

Lane, R. P. and Crosskey, R. W. (1993). *Medical insects and arachnids*. Chapman & Hall, London, UK, 723 pp.

Lewis, R. E. (1995). Fleas (Siphonaptera). In: Lane, R. P. and Crosskey, R. W. (eds), *Medical insects and arachnids*. Chapman & Hall, London, UK, pp. 529–575.

Lines, J. D. (1988). Do agricultural insecticides select for insecticide resistance in mosquitoes? A look at the evidence. *Parasitol. Today* **4**: S17–S20.

Maurin, M. and Raoult, D. (1996). *Bartonella (Rochalimaea) quintana* infections. *Clin. Microbiol. Rev.* **9**: 273–292.

Miller, T. A. (1988). Mechanisms of resistance to pyrethroid insecticides. *Parasitol. Today* **4**: 58–59.

MPA (1999). The treatment of head lice infestation – recommendations. Information from the Medical Products Agency (MPA), Uppsala, Sweden. **10**(5): 75–78.

Mwase, E. N., Pegram, R. G. and Mather, T. N. (1991). New strategies for controlling ticks. In: Curtis C.F. (ed.), *Control of disease vectors in the community*. Wolfe Publ. Ltd., London, pp. 93–102.

Nelson, W. A., Keirans, J. E., Bell, J. F. and Clifford, C. M. (1975). Host–ectoparasite relationships. *J. Med. Entomol.* **12**: 143–166.

Nilsson, K., Lindquist, O., Liu, A. J., Jaenson, T. G. T., Uhnoo, I., Friman, G. and Påhlson, C. (1998). *Rickettsia helvetica* in *Ixodes ricinus* ticks in Sweden. *J. Clin. Microbiol.* **37**: 400–403.

Nilsson, K., Lindquist, O. and Påhlson, C. (1999). Association of *Rickettsia helvetica* with chronic perimyocarditis in sudden cardiac death. *Lancet* **354**: 1169–1173.

PAHO (1982). Emergency vector control after natural disaster. Pan American Health Organization Scient. Publ. No. 419. PAHO, Washington, DC, USA.

Raoult, D., Ndihokubwayo, J. B., Tissot-Dupont, H., Roux, V., Faugere, B., Abegbinni, R. *et al.* (1998). Outbreak of epidemic typhus associated with trench fever in Burundi. *Lancet* **352**: 353–358.

Raoult, D., Roux, V., Ndihokubwayo, J. B., Bise, G., Baudon, D., Martet, G. *et al.* (1997). Jail fever (epidemic typhus) outbreak in Burundi. *Emerg. Infect. Dis.* **3**: 357–360.

Roberts, D. L. and Andre, R. G. (1994). Insecticide resistance issues in vector-borne disease control. *Am. J. Trop. Med. Hyg.* **50**: 21–34.

Rydkina, E. B., Roux, V., Gagua, E. M., Predtechenski, A. B., Tarasevich, I. V. and Raoult, D. (1999). *Bartonella quintana* in body lice collected from homeless persons in Russia. *Emerg. Infect. Dis.* **5**: 176–178.

Schofield, C. J. and Dolling, W. R. (1993). Bedbugs and kissing-bugs (bloodsucking Hemiptera). In: Lane, R. P. and Crosskey, R. W. (eds), *Medical insects and arachnids*. Chapman & Hall, London, UK, pp. 483–516.

Schofield, C. J. and White, G. B. (1984). House design and domestic vectors of disease. *Trans. R. Soc. Trop. Med. Hyg.* **78**: 285–292.

Schofield, C. J. *et al.* (1991). The role of house design in limiting vector-borne disease. In: Curtis C. F. (ed.), *Control of disease vectors in the community*. Wolfe Publ. Ltd, London, pp. 187–212

Secoy, D. M. and Smith, A. E. (1983). Use of plants in control of agricultural and domestic pests. *Econ. Bot.* **37**: 28–57.

Smart, J. (1948). *A handbook for the identification of insects of medical importance*, 2nd edn. British Museum (Natural History), London, UK.

Smith, K. G. V. (ed.) (1973). *Insects and other arthropods of medical importance*. British Museum (Natural History), London, UK, 561 + XI pp.

Sonenshine, D. E. (1991). *Biology of ticks*, vol. 1. Oxford University Press, New York.

Sonenshine, D. E. (1993). *Biology of ticks*, vol. 2. Oxford University Press, New York.

Varma, M. G. R. (1993). Ticks and mites (Acari). In: Lane, R. P. and Crosskey, R. W. (eds), *Medical insects and arachnids*. Chapman & Hall, London, UK, pp. 597–658.

Wall, R. and Shearer, D. (1997). *Veterinary entomology*. Chapman & Hall, London, UK, 439 pp.

WHO (1973). Technical guide for a system of plague surveillance. *Weekly Epidemiol. Rec.* **48**: 149–160. WHO, Geneva, Switzerland.

WHO (1974). Ecology and control of rodents of public health importance. WHO Techn. Rep. Ser. 553, WHO, Geneva, Switzerland, 42 pp.

WHO (1982). Manual on environmental management for mosquito control. With special emphasis on malaria vectors. WHO Offset Publ. no. 66. WHO, Geneva, Switzerland, 283 pp.

WHO (1988a). Urbanisation and its implications for child health. Potential for action, WHO, Geneva, Switzerland, 80 pp.

WHO (1988b). Urban vector and pest control. WHO Techn. Rep. Ser. 767. WHO, Geneva, Switzerland, 77 pp.

WHO (1989a). The use of impregnated bednets and other materials for vector-borne disease control. WHO/VBC/89.981, WHO, Geneva, Switzerland, 45 pp.

WHO (1989b). Geographical distribution of arthropod-borne diseases and their principal vectors. WHO/VBC/89.967. WHO, Geneva, Switzerland, 131 pp.

WHO (1991). Insect and rodent control through environmental management. A community action programme. WHO, Geneva, Switzerland, 107 pp. +62 cards.

WHO (1992). Vector resistance to pesticides. WHO Techn. Rep. Ser. 818. WHO, Geneva, Switzerland, 59 pp.

Wilkinson, P. R. (1957). The spelling of pasture in tick control. *Aust. J. Agric.* **8**: 414–423.

Zumpt, F. (1965). *Myiasis in animals and man in the Old World*. Butterworths, London, 267 pp.

24 Dermatological manifestations of parasitic arthropods

Johan Landegren

In Scandinavia today we are confronted with four kinds of ectoparasites, that is, parasites that will stay on the body or in the skin for a relatively long period of time. These are as follows: Scabies, *Pediculosis pubis*, *Pediculosis capitis*, and *Pediculosis corporis*.

Short commentaries will be made about 'temporary' parasites, which sometimes cause considerable diagnostic problems: animal scabies – especially fox scabies, *Cheyletiella* infestation, fleas, mosquito bites and nodulus cutaneous, bee stings and wasp stings, caterpillar dermatitis. Some important comments will be made about a very special dermatological problem which is psychiatric in nature, *delusion of parasitosis*.

Scabies

Symptoms

The symptoms are much more relevant for the diagnosis than the signs and a preliminary diagnosis of scabies can be made. Typical symptoms are: an incubation time of about three weeks after a possible source of transmission, often not admitted. Localized itching is common.

Diagnosis

Lesions and/or symptoms in the following areas could help in diagnosis:

a the thighs – not laterally, not medially, not frontal but the frontal–medial aspect;
b the axillae – not in the vaults but in the anterior axillary fold;
c the buttocks;
d the mid-abdominal area – the patient will usually describe the area as a belt (Figure 24.2);
e the volar surfaces of the lower parts of the arms, and sometimes the hands, especially the finger webs (Figure 24.1); and
f the breasts in women and the penis in men.

The back, the distal parts of the legs, and the face (except in children), are usually not affected.

The tentative diagnosis of scabies (Figure 24.2) should always be confirmed by the identification of burrows (Figure 24.3) and preferably by the identification of the parasite. This can usually be achieved through the use of a blunt sewing needle top (Figure 24.4). The parasite, that is, the female mite, measures about 0.4 mm, and can easily be identified in the microscope.

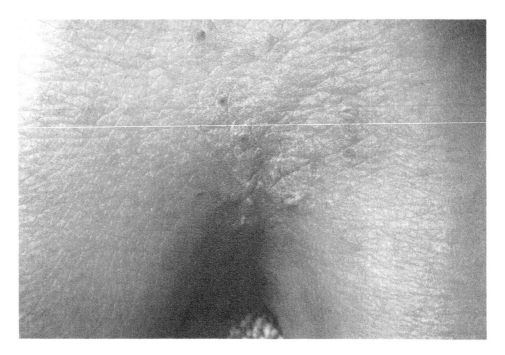

Figure 24.1 Scabies: eczematous lesions of the finger webs are common, but not always observed. (See Colour Plate V.)

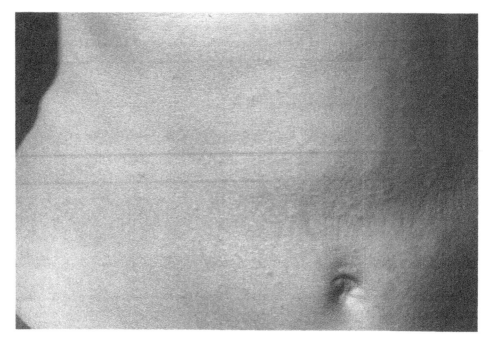

Figure 24.2 Scabies: an important clue for the clinical diagnosis is that 'the symptoms are much worse than the clinical signs' – intense itch, but minimal skin lesions – urticarial or excoriated papules. (See Colour Plate VI.)

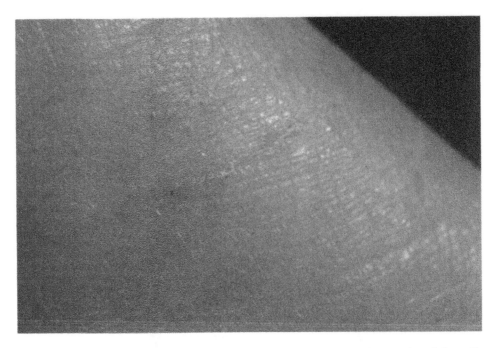

Figure 24.3 Scabies: burrows are typical and can be described as tortuous lines with a slight scaling, most often found on the flexor surface of the wrist, between the fingers, and sometimes on the genitals. At the end of the burrow often a dark spot can be seen. (See Colour Plate VIII.)

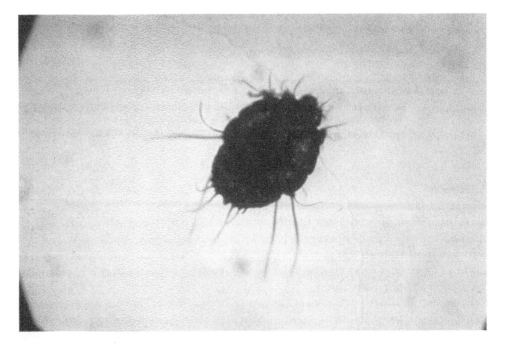

Figure 24.4 The scabies mite: the 'dark spot' represents the mite, which can usually be 'catched' by a blunt needle and identified microscopically with low-grade magnification (×100).

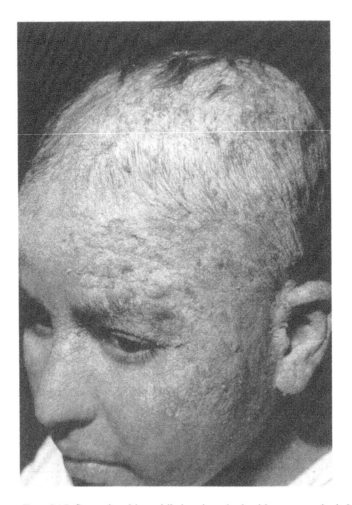

Figure 24.5 Crusted scabies: while in otherwise healthy persons the infestation usually means only 10–20 mites, an immuno-compromised patient (HIV, tumor diseases, etc.), might develop quite another clinical disease with widespread hyperkeratotic lesions: crusted scabies or 'Norwegian scabies'. This patient suffered from systemic lupus erythematosus and was on high doses of cortisone since a long time. (See Colour Plate VII.)

Treatment

A single application of 24 h of Tenutex™ (benzylbensoate 22.5%, disulfiram 2% in a cream base) has proven 100% effective (Landegren *et al.* 1979). This is properly done by generous application of the cream all over the body below the neck. A single dose of ivermectin is reported to be effective (Tzenow *et al.* 1997). As the infection is transferred by close body contact, spouse and children should be treated, even when no symptoms have occurred (incubation period = three weeks).

Fox scabies is caused by a variant of the same agent that causes human scabies, *Sarcoptes scabiei*. Handling infested foxes may cause a transient, local infestation of man, but the parasite will not be able to multiply in human skin; thus no active antiparasitic treatment is necessary.

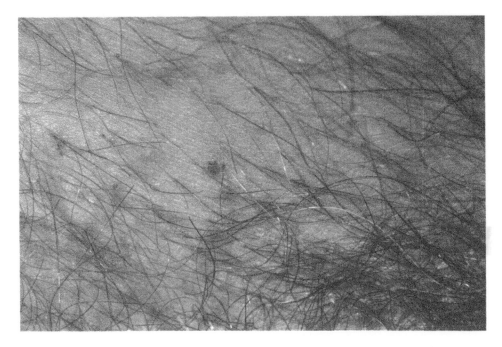

Figure 24.6 Pediculosis pubis: if the examining doctor has a low grade of suspicion, the diagnosis might easily be missed. The lice are often difficult to see and are often mistaken for small crusts. (See Colour Plate IX.)

Pediculosis pubis

A well-known disorder almost always transferred through sexual contact. The diagnosis may, however, be difficult if not suspected (Figure 24.6) as the examining doctor may interpret the small lice as superficial crusts.

Treatment

There are presently several highly effective antiparasitic remedies for external use (malation, pyrethrum derivates, etc.).

Pediculosis capitis

Head lice are found world wide, often causing small epidemics in schools in northern Europe. The louse itself is often very difficult to find, but the typical nits can usually be found and identified by the microscope (Figure 24.7).

Treatment

As for *Pediculosis pubis*, but normally only the scalp area has to be treated. Spread to grown-up relatives is rather uncommon, but brothers, sisters, and schoolmates are very often affected.

Pediculosis corporis (or vestimenti)

Although during the 1940s this parasite was the cause of great problem among refugees and other people living in crowded communities, this variant is now rare. Sporadic cases

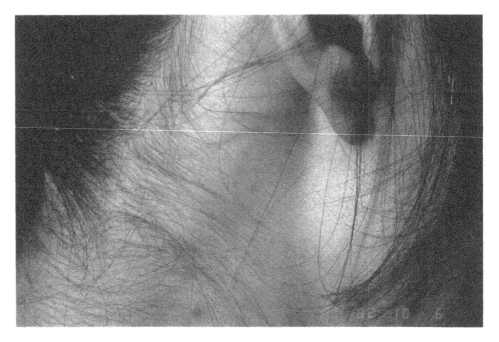

Figure 24.7 Pediculosis capitis: usually in young people the lice cause itching bite marks, especially in the nape region. The lice themselves are often difficult to discover, but the nits (Swedish: gnetter) should always be possible to identify. (See Colour Plate X.)

Figure 24.8 Pediculosis corporis or *vestimenti*: the lice do not live on the body but in the clothes, but they bite and feed on human skin. The *clothing* should be treated by thorough cleaning, not the patient! (Photo: Leif Strandberg). (See Colour Plate XI.)

are, however, seen among 'tramps' and other persons living in very poor situations (Figure 24.8).

Treatment

The clothes should be treated (i.e. cleaned) – not the patient!

Cheyletiella

This is a mite that may occur in great numbers in the fur of dogs, especially boxers. Close contact with the dog will cause the transfer of mites to human skin, where they will bite and cause red itchy papules. These papules tend to be great in numbers.

Fleas

Fleas are not a great problem in Scandinavia today, with one important exception of diagnostic importance: every spring (April–May) several patients seek medical advice because of intensely pruritic papules on their skin, almost always on their legs (Figure 24.9). The history is often the same: handling of a bird's nesting-house, especially if taking it into the warmth in-door for cleaning, and a few days later intensely itching big papules appear.

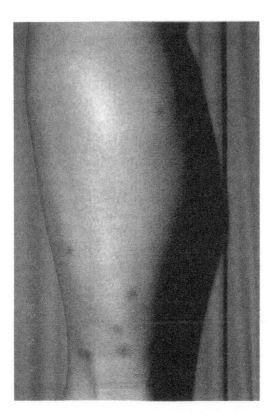

Figure 24.9 Flea bites: various species of fleas, each species specific for a special kind of animal, or human, are rare nowadays. Fleas live in the moist areas of floors, but they search for 'blood meals' from animals or humans. Flea bites are most common on the lower legs. (See Colour Plate XII.)

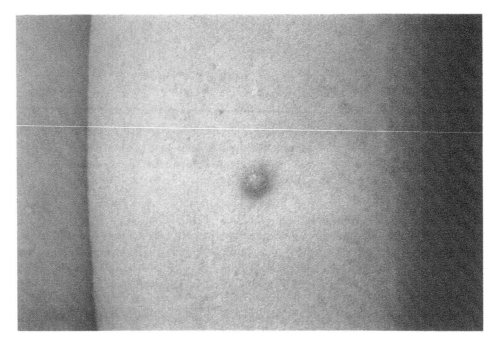

Figure 24.10 Nodulus cutaneous after a mosquito bite: a small (approximately 5 mm), very hard papule, with a characteristic ring of pigmentation (see text). (See Colour Plate XIII.)

Treatment

The fleas will survive for approximately three weeks – they will not be able to multiply. A potent topical corticosteroid, such as clobetasol proprionate (Dermovat™), will give relief.

Mosquito bites

Mosquito bites rarely – if ever – cause diagnostic problems. However, a late effect of mosquito bites, which very often causes dermatological consultations, is nodulus cutaneous (skin nodule), which is a very common type of skin lesion, usually on the leg. It is usually a consequence of a previous mosquito bite and is made up of a dermal proliferation of collagen. It is hard and adheres to the skin surface. It is completely benign but constitutes an important differential diagnostic item towards pigment cell tumors such as malignant melanoma! (Figure 24.10).

Bee stings and wasp stings

These are rather common during summer time. They result in a rather intense and painful local reaction. A general reaction of an allergic type, usually urticaria, may occur, but is rare. The acute reaction can be treated with epinephrine and internal corticosteroids. Desensitization is possible.

Caterpillar dermatitis

Please see legend of Figure 24.11.

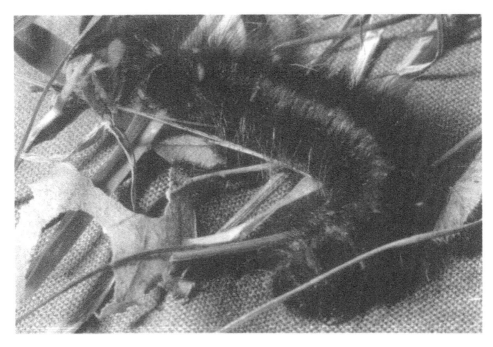

Figure 24.11 Caterpillar (dermatitis): in Scandinavia, every autumn (September–October) some children will ask for medical advice because of intensively pruritic papules or vesicles of their palms. After a directed questioning, the doctor might find out that the child has been playing with – or cared for – a caterpillar, a larva, usually of the species *Macrothilatio rubi* which has noxious hairs as a protection against bird attack. This benign disorder will resolve by itself within a few days.

Delusion of parasitosis (parasitophobia)

This disorder is not very common, but is very often misunderstood and misinterpreted.

The patient may complain of parasites in the skin – sometimes with itch, but more often without – and often has consulted several other doctors and received repeated treatment procedures. The disorder is not a 'phobia' in strict psychiatric sense: thus 'delusion of parasitosis, is preferred.

For appropriate handling of the patient (often a female above 60 years of age) the doctor should understand that the patient does not *believe*, but she *knows* that she has parasites in her skin! Sometimes the husband/spouse also has the same pathological ideas.

Therapy

A kind understanding of the patient's problems is most important. Trying to convince her that her ideas are 'wrong' is not fruitful. Neuroleptics like pimozid (Orap®) or haloperidol (Haldol™) will often be of great help, but most patients will not accept a 'psychic' drug.

References

Landegren, J., Borglund, E., and Storgårds, K. (1979). Treatment of scabies with disulfiram and benzyl bensoate emulsion: a controlled study. *Acta Dermatol. Venereol.*, **59**, 274–276.

Tzenow, I., Melnik, B., and Wehmeier, M. (1997). Orale Behandlung der Scabies mit Ivermectin. *Der Hautarzt*, **48**, 2–4.

Part 4

Clinical presentation and laboratory investigations

25 Clinical symptoms of toxoplasmosis – an overview

Birgitta Evengård

Clinical symptoms of this parasitic infection, depend on the immune status of the individual. Clinical pictures will thus differ between the immune-competent and the immune-incompetent individual. This chapter will therefore be divided into these two main groups.

It is established that cell-mediated immunity is, to a high degree, responsible for the immune reactions against *Toxoplasma gondii* infection. B-cells are stimulated to antibody production, which may be used for the purpose of diagnosis.

Thus, individuals with impaired cell-mediated immune response of $CD4^+$ cells and of $CD8^+$ cells might have a more severe, and sometimes lethal, outcome while the immuno-competent individual might hardly notice the infection. A special situation occurs when pregnant women suffer a primary infection. The foetus lacks immuno-competence, which develops gradually during the course of pregnancy. Thus maternal toxoplasmosis may lead to foetal infection, the severity of which is dependent on the degree of intra-uterine development and the time of infection.

Clinical symptoms in the immuno-competent individual

The incubation time varies, but is usually between 10 and 14 days depending on the dose, the route of entry of the parasite, and individual factors (Krick and Remington 1978). The primary *Toxoplasma* infection is not commonly recognized, since the symptoms described are non-specific. Low-grade fever, sore throat, night sweats, myalgia, fatigue, malaise, and a maculopapular rash may occur, as well as, enlargement of the liver and spleen. Initially a lymphadenopathy may occur, often in the cervical region, and the nodes are usually discrete and non-tender and do not suppurate. This condition can clinically be mistaken for mononucleosis or a cytomegalovirus infection. A low number of atypical lymphocytes may be present in the blood.

These nodes usually disappear within a month but can persist for a longer period in some individuals. Enlarged nodes in the abdomen can cause abdominal pain. Retinochoroiditis, usually unilateral, may occur in cases of an acute, acquired infection. Pneumonitis, myocarditis, pericarditis, hepatitis, and central nervous system involvement may occur, but is rare (Sherman and Nozik 1992). Pulmonary toxoplasmosis has been described in immuno-competent patients (Pomeroy and Filice 1992). Shortness of breath and cough are the most common symptoms, together with fever and rales. Lymphadenopathy and hepatosplenomegaly are also common, and chest roentgenographs may show bilateral interstitial infiltrates. Mortality rates tend to be low.

Toxoplasmic myocarditis and polymyositis in an adult with primary toxoplasmosis has been described (Montoya *et al.* 1997) and it is recommended that a toxoplasmic serological profile is performed on patients with myocarditis and/or polymyositis of unknown origin. There are case reports describing recurrent toxoplasmosis in immuno-competent hosts (Norrby and Eilard 1976; Candolfi *et al.* 1993). Conventional immunological tests proved no abnormalities. In one case *Toxoplasma* parasitemia was detectable in blood using polymerase chain reaction although immunological parameters, including antibody responses, were normal (O'Connell *et al.* 1993). This is, however, a rare condition and should be followed, as it might be an early sign of a defect in the immune-defense. Treatment can be given.

Eye manifestations

Ocular toxoplasmosis is perhaps the most common manifestation of *Toxoplasma* infection in the immuno-competent individual. It has been considered as being of congenital origin but there is new evidence indicating that it can also be the manifestation of an acquired infection (Glasner *et al.* 1992; Ronday *et al.* 1994; Montoya and Remington 1996).

Retinochoroiditis is commonly found and lasts on an average for four months. The natural course usually takes weeks to months. The patient suffers in general from two to three attacks in a lifetime (Friedman and Knox 1969). It typically involves one eye at a time. Both eyes may, however, have chorioretinal scars. Initial symptoms include floaters and decreased visual acuity. Pain and redness are unusual findings but may occur if a secondary iridocyclitis develops. The bulbar conjunctiva, cornea, and anterior chamber are usually normal. From the active focus of the retina, a dense vitreous reaction with cellular material exudes, creating a 'headlight in the fog' appearance typical for this condition. Reactivated retinal lesions are usually found at the border of old atrophic scars. Retinal vessels may be sheathed with inflammatory cells, and a localized serous detachment of the retina may develop. The cysts may be identified not only in previously active spots but also in normal retinal tissue. Reactivation may be triggered by factors such as stress, pregnancy, trauma, and conditions causing immuno-incompetence (Sherman and Nozik 1992).

Symptoms of intra-uterine infection

The clinical symptoms in pregnant women correspond to those described above. If a pregnant woman is infected for the first time during pregnancy *Toxoplasma* tachyzoites may cross the placenta and infect the foetus. During the acute stage with parasitemia, parasites may colonize the placenta which may serve as a reservoir, supplying viable organisms to foetus, if the mother is not treated (Stray-Pedersen 1993). The infection is mostly asymptomatic (in approximately 90%) in the mother (Wong and Remington 1994). If the maternal infection is acquired 1–2 months before conception transmission might occur. The risk for transmission increases with length of gestation, with almost 100% transmission in the last month. The incidence and severity of congenital infection depend on which trimester the infection was acquired in. When contracted during the first trimester, 10–15% of the foetuses are infected and the result may be spontaneous abortion, stillbirth, or severe disease in the newborn. During the second trimester the incidence of foetal infection is 30%, and during the third trimester, 60%, (Wong and Remington 1994). About 85% of infected infants show no clinical sign of infection at birth.

The clinical manifestations in infected newborns are varied. Sequale may not occur or may develop at various times. Most signs and clinical findings are non-specific, and testing for *Toxoplasma* infection should be performed on wide indications. Although infections during the first trimester result in a low number of infected infants, the damages are more severe. The classical triad of retinochoroiditis (usually bilateral), microcephaly, and intracranial calcifications is fortunately a rare finding. Other signs are strabismus, blindness, hearing loss, epilepsy, psychomotoric or mental retardation, anaemia, jaundice, rash, petechiae, encephalitis, pneumonitis, and hydrocephalus. If clinical signs are found at birth the sequalae are usually severe. Approximately 85% of sub-clinically infected newborns will subsequently develop chorioretinitis; 10–30%, hearing loss; and 20–75%, developmental delay (Wilson *et al.* 1980).

When the transmission occurs later in pregnancy, the frequency of chorioretinitis increases which may cause a symptomatic retinochoroiditis up til the age of 20 (Koppe *et al.* 1986). Ultra-sound findings combined with maternal serology have been found to be significantly related to clinical outcome (Virkola *et al.* 1997).

Infection in immuno-compromised hosts

AIDS patients

Infection with *T. gondii* in these patients can involve the brain, the lung, and the eye. A multi-organ involvement with acute respiratory failure and haemodynamic abnormalities, as in septic shock, have been described (Lucet *et al.* 1993; Al-Kassab *et al.* 1995). The association of high fever, acute dyspnea, recent onset of thrombocytopenia, and a very high level of lactate dehydrogenase activity are suggestive of disseminated toxoplasmosis.

Toxoplasma encephalitis (TE) is the most common manifestation of infection in AIDS patients and is the most frequent cause of focal lesions in the central nervous system (Luft and Remington 1992). More than 95% of the cases are due to reactivation of a chronic infection and the incidence is thus proportional to the prevalence in the particular region. TE usually manifests itself when CD4 count falls below $100/mm^3$.

In Europe and Africa 10–50% of HIV-infected patients sero-positive to *Toxoplasma gondii* will develop TE. Initial symptoms are often headache, confusion, and fever while focal symptoms will develop depending on the localization of the infected focus. Hemiparesis and/or abnormalities of speech are also major initial manifestations. Brainstem involvement causes neurological symptoms such as cranial nerve lesions, ataxia, palsies, and dysmetria. Non-focal symptoms can predominate and include weakness, disorientation, psychosis, confusion, or coma. Frequently multiple lesions occur. The onset can be insidious over weeks but can also be acute. Meningismus is rare. The cerebrospinal fluid may be normal or may have an increased number of cells and an increased protein level.

During the past decade the incidence of *Toxoplasma* pneumonia has increased in immuno-compromised patients, probably due to improved diagnosis (Mariuz *et al.* 1997).

Toxoplasmic pneumonia has similar non-specific clinical manifestations as seen in pneumonia caused by *Pneumocystis carinii*. The onset tends, however, to be more rapid. Initial symptoms are usually dyspnea, fever, and a non-productive cough, which may rapidly progress to acute respiratory failure (Derouin *et al.* 1989). Chest roentgenographs commonly show bilateral intestinal infiltrates, with or without nodules, and hilar adenopathy. The mortality can be as high as 35% and extra-pulmonary disease is present in about 50% of patients with toxoplasmic pneumonitis (Beaman *et al.* 1995). Diagnosis is made by isolation or identification of the tachyzoite forms in broncho-alveolar lavage (BAL) fluid (Catterall *et al.* 1986). In a study conducted in France, 5% of BAL specimens were positive (Derouin *et al.* 1990).

Severe infections of the entire gastrointestinal tract and pancreas have also been described (Luft 1989) as have cardiac tamponade or biventricular failure. Retinochoroiditis can be found. This manifestation caused by *T. gondii* is less haemorrhagic than retinochoroiditis caused by cytomegalovirus and is associated with a heavy vitriol haze and severe involvement of the vitreous and anterior uvea.

Non-AIDS immuno-compromised patients

There is a need for clinical alertness of *Toxoplasma* infection in these patients as mortality is high if treatment is not given (Israelski *et al.* 1993). Infection may be due to primary infection or re-activation of a latent infection. There is therefore a need for testing of recipient and donor when organs are transplanted so that mismatches for *Toxoplasma* immunity can be correctly managed.

Transfer of Toxoplasma with transplants

Heart

Severity ranging from asymptomatic sero-conversion to myocardial infiltration or disseminated neurological disease or death has been described following heart transplantation (Michaels *et al.* 1992; Gallino *et al.* 1996). Onset of the disease was within the first 6.5 months after transplantation. This occurred if the recipient was sero-negative and the donor sero-positive. Prophylactic strategies are recommended.

Kidney

Re-activation of a latent infection has been described but the most common is a primary infection if a sero-negative donor receives an organ from a sero-positive donor (Renoult *et al.* 1997). Within three months following transplantation fever, neurological disturbances, and pneumonia due to toxoplasmosis occurred.

Liver

Sero-conversion after transplantation has been reported (McCabe and Chirurgi 1993), so has a case of fulminant disseminated infection (Mayes *et al.* 1995). *Toxoplasma* infection should be considered in the differential diagnosis in multi-organ failure in the early period following transplantation. Retinochoroiditis following liver transplantation has also been reported (Blanc-Jouvan *et al.* 1996).

Bone marrow

Re-activation of a latent infection is a serious problem in transplantation of bone marrow (Derouin *et al.* 1992; Chandrasekar and Momin 1997). The re-activation occurs within the first six months after marrow transplant, with the highest incidence in the second and third months. Cerebral or a disseminated infection has been described in addition to pulmonary toxoplasmosis (Saad *et al.* 1996) and retinochoroiditis (Peacock *et al.* 1995). *Toxoplasma* infection was considered to contribute to death in at least 40% of the cases (Slavin *et al.* 1994). Primary infection in the sero-negative recipient has also been reported. Diagnosis is mainly made by the demonstration of parasites in body fluids or tissues.

Malignancies

Patients with Hodgkin's disease and other lymphomas have been reported especially to be vulnerable to *Toxoplasma* infection. Manifestations of central nervous system, the myocardium, and the lungs are described (Israelski and Remington 1993).

References

Al-Kassab, A. K., Habte-Gabr, E., Mueller, W. F., and Azher, Q. (1995). Case report. Fulminant disseminated toxoplasmosis in an HIV patient. *Scand. J. Infect. Dis.*, **27**, 183–185.

Beaman, M., McCabe, R. E., Wong, S., and Remington, J. S. (1995). *Toxoplasma gondii.* In: *Principles and practice of infectious diseases* (G. L. Mandell and J. E. Bennett, eds).

Blanc-Jouvan, M., Boibieux, A., Fleury, J., Fourcade, N., Gandilhon, F., Dupouy-Camet, J., Peyron, F., Ducerf, C. (1996). Chorioretinitis following liver transplantation:detection of *Toxoplasma gondii* in aqueous humor. *Clin. Infect. Dis.*, **22**(1),184–185.

Chandrasekar, P. H. and Momin, F. (1997). Disseminated toxoplasmosis in marrow recipients: a report of three cases and a review of the literature. Bone marrow transplant team. *Bone Marrow Transplant*, **19**(7), 685–689.

Candolfi, E., de Blay, F., Rey, D., Christmann, D., Treisser, A., Pauli, G., and Kien, T. (1993). A parasitologically proven case of *Toxoplasma* pneumonia in an immunocompetent pregnant woman. *J. Infect.*, **26**, 79–81.

Catterall, J. R., Hofflin, J. M., and Remington, J. S. (1986). Pulmonary toxoplasmosis. *Am. Rev. Respir. Dis.*, **133**, 704–711.

Derouin, F., Sarfati, C., Beauvais, B., Iliou, M. C., Dehen, L., and Larrivière (1989). Laboratory diagnosis of pulmonary toxoplasmosis in patients with AIDS. *J. Clin. Microbiol.*, **27**, 1661–1663.

Derouin, F., Sarfati, C., Beauvais, B. *et al.* (1990). Prevalence of pulmonary toxoplasmosis in HIV-infected patients. *AIDS*, **4**, 1036.

Derouin, F., Devergie, A., Auber, P., Gluckman, E., Beauvais, B., Garin, Y. J. F., and Lariviere, M. (1992). Toxoplasmosis in bone marrow-transplant recipients: report of seven cases and review. *Clin. Infect. Dis.*, **15**, 267–270.

Friedman, C. T. and Knox, D. L. (1969). Variations in recurrent active toxoplasmic retinochoroiditis. *Arch. Ophthalmol.*, **81**, 481–493.

Gallino, A., Maggiorini, M., Kiowski, W., Martin, X., Wunderli, W., Schneider, J., Turina, M., and Follath, F. (1996). Toxoplasmosis in heart transplant recipients. *Eur. J. Clin. Microbiol. Infect. Dis.*, **15**(5), 389–393.

Glasner, P. D., Silveira, C., Kruszon-Moran, D., Martins, M., Burnier, M., Silveira, S. *et al.* (1992). An unusualy high prevalence of ocular toxoplasmosis in Southern Brazil. *Am. J. Ophtalmol.*, **114**, 136–144.

Israelski, D. and Remington, J. (1993). Toxoplasmosis. In: *Current clinical topics in infectious diseases* (J. Remington and M. Schwartz, eds), vol. 13. Blackwell Scientific Publications, Boston, MA, pp. 322–356.

Krick, J. A. and Remington, J. S. (1978). Toxoplasmosis in the adult – an overview. *N. Engl. J. Med.*, **298**, 550–553.

Koppe, J. G., Loewer-sieger, D. H., and de Roever-Bonnet, H. (1986). Results of 20-year follow-up of congenital toxoplasmosis. *Lancet*, **101**, 254–255.

Lucet, J.-C., Bailly, M.-P., Bedos, J.-P., Wolff, M., Gachot, B., and Vachon, F. (1993). Septic shock due to toxoplasmosis in patients infected with the human immunodeficiency virus. *Chest*, **104**, 1054–1058.

Luft, B. J. (1989). *Toxoplasma gondii.* In: *Parasitic infection in the compromised host* (P. D. Walzer and R. M. Gertz, eds). Marcel Dekker, New York, pp. 179–279.

Luft, B. J. and Remington, J. S. (1992). AIDS commentary. Toxoplasmic encephalitis in AIDS. *J. Infect. Dis.*, **15**, 211–222.

Mariuz, P., Bosler, E. M., and Luft, B. J. (1997). *Toxoplasma* pneumonia. *Semin. Respir. Infect.*, **12**(1), 40–43.

Mayes, J. T., O'Connor, B. J., Avery, R., Castellani, W., and Carey, W. Transmission of *Toxoplasma gondii* infection by liver transplantation. *Clin. Infect. Dis.*, **21**(3), 511–515.

McCabe, R. and Chirurgi, V. (1993). Issues in toxoplasmosis. *Parasit. Dis.*, **7**, 587–604.

Michaels, M. G., Wald, E. R., Fricker, F. J., del Nido, P. J., and Armitage J. (1992). Toxoplasmosis in pediatric recipients of heart transplants, **14**, 847–851.

Montoya, J. G., Jordan, R., Lingamneni, S., Berry, G. J., and Remington, J. S. (1997). *Clin. Infect. Dis.*, **24**(4), 676–683.

Montoya, J. G. and Remington, J. S.(1996). Toxoplasmic chorioretinitis in the setting of acute acquired toxoplasmosis. *Clin. Infect. Dis.*, **23**(2), 277–282.

Norrby, R. and Eilard, T. (1976). Case report. Recurrent toxoplasmosis. *Scand. J. Infect. Dis.*, **8**, 275–276.

O'Connell, S., Guy, E. C., Dawson, S. J., Francis, J. M., and Joynson, D. H. M. (1993). Case report. Chronic active toxoplasmosis in an immunocompetent patient. *J. Infect.*, **27**, 305–310.

Peacock, J. E., Greven, C. M., Cruz, J. M., and Hurd, D. D. (1995). Reactivation toxoplasmic retinochoroiditis in patients undergoing bone marrow transplantation: is there a role for chemoprophylaxis? *Bone Marrow Transplant*, **15**(6), 983–987.

Pomeroy, C. and Filice, G. A. (1992). Pulmonary toxoplasmosis: a review. *Clin. Infect. Dis.*, **14**(4), 863–870.

Renoult, E., Georges, E., Biava, M. F., Hulin, C., Frimat, L., Hestin, D., and Kessler, M. (1997). Toxoplasmosis in kidney transplant recipients: report of six cases and review. *Clin. Infect. Dis.*, **24**(4), 625–634.

Ronday, M. J. H., Stilma, J. S., Barbe, R. F., Kijlstra, A., and Rothova, A. (1994). Blindness from uveitis in a hospital population in Sierra Leone. *Br. J. Ophtalmol.*, **78**, 690–693.

Saad, R., Vincent, J. F., Cimon, B., de Gentile, L., Francois, S., Bouachour, G., and Ifrah, N. (1996). Pulmonary toxoplasmosis after allogeneic bone marrow transplantation: case report and review. *Bone Marrow Transplant*, **18**(1), 211–212.

Sherman, M. D. and Nozik, R. A. (1992). Other infections of the choroid and retina. *Ocular. Infect.*, **6**(4), 893–908.

Slavin, M. A., Meyers, J. D., Remington, J. S., and Hackman, R. C. (1994). *Toxoplasma gondii* infection in marrow transplant recipients: a 20 year experience. *Bone Marrow Transplant*, **13**(5), 549–557.

Stray-Pedersen, B. (1993). Toxoplasmosis in pregnancy. In: *Clinical obstetrics and gynaecology*, Baillière Tindall, pp.107–137.

Virkola, K., Lappalainen, M., Valanne, L., and Koskiniemi, M. (1997). Radiological signs in newborns exposed to primary *Toxoplasma* infection in utero. *Pediatr. Radiol.*, **27**(2), 133–138.

Wilson, C. B., Remington, J. S., Stagno, S. *et al.* (1980). Development of adverse sequalae in children born with subclinical congenital *Toxoplasma* infection. *Pediatrics*, **66**, 767–774.

Wong, S. and Remington, J. S. (1994). Toxoplasmosis in pregnancy. *Clin. Infect. Dis.*, **18**, 853–862.

26 Treatment of toxoplasmosis

Babill Stray-Pedersen

Toxoplasma infections in immune-competent individuals are usually asymptomatic and unrecognized, and treatment is not discussed. In cases with symptoms infection is most often self-limiting and treatment is unnecessary. Only in the few patients where the infection affects vital organs, for example, brain, lung, liver, and heart, should therapy be given. Also, in cases with prolonged lymphadenopathy, fever, malaise, and fatigue, treatment may lead to improvement of the condition, although most patients will recover without therapy.

In pregnant women, however, *Toxoplasma* infection acquired at any time during pregnancy should be treated in order to reduce infection of the placenta and to prevent or to modify the infection in the foetus (Stray-Pedersen 1993). Two other groups should also be considered for treatment: patients with ocular toxoplasmosis and immuno-compromised individuals with infections. In these cases re-activation of a pre-existent latent infection may occur and therapy is needed in those with cerebral or disseminated toxoplasmosis (McCabe and Oster 1989; Joss 1992).

Drugs

Until today the drugs used regularly for treatment of toxoplasmosis act by slowing down the multiplication of tachyzoites, either by interfering with protein synthesis or by inhibition of vital enzymes (Derouin 2000). Only four drugs have been used to a great extent: pyrimethamine, sulfonamide, spiramycin, and clindamycin (WHO 1990). These drugs do not eliminate the *Toxoplasma* tissue cysts, nor do they prevent new tissue cyst formation. Today in the AIDS era there is an urgent need for developing and trying new drugs which may affect the cystic stage. Azithromycin and atovaquone seem promising in animal models, and have shown remarkable acitivity against the tissue cysts *in vitro* (Huskinson *et al.* 1991). These drugs are now on trial in individuals with AIDS and ocular toxoplasmosis.

Pyrimethamine

Pyrimethamine is an antimalarial drug that is used in combination with sulfonamides (sulfadiazine or triple sulfonamide). The drug has also been used in combination with the long-acting sulfaderivates sulfadoxine (Fansidar) or sulfametopyrazine (Dapsone) which can be given as a weekly once dosage. Pyrimethamine has been accepted as the drug of choice for treatment of toxoplasmosis since the beginning of the 1950s (WHO 1990). Hitherto, no other drug, alone or in combination, has been shown to be superior to the pyrimethamine–sulfa combination. Both drugs are folic acid antagonists, which act synergistically to suppress the proliferation of tachyzoites by impairment of DNA synthesis. Their combined activity is 10 times greater than expected if their effects were merely additive (Sheffield and Melton 1975; Derouin and Chastang 1989).

Pyrimethamine is capable of penetrating the blood brain and the blood retina barriers and concentrates in both the brain and the retinal tissue (Remington *et al.* 1995).

The half-life of pyrimethamine is 4–5 days. The oral dose is usually 0.5–1 mg/kg per day. The tablet may be crushed and administered in a suspension with juice or food. The most common side effects are those related to suppression of the bone marrow. Peripheral white blood cells and platelet count should be regularly monitored during treatment. Adding folinic acid (not folic acid) 3–10 mg every third day may prevent the haematological complication. Other less-serious side effects include nausea, headache, and an unpleasant taste in the mouth.

In humans pyrimethamine has no proven teratogenic effect. In 1978 WHO approved the free use of Fansidar (pyrimethamine 25 mg + sulfadoxine 500 mg) as malaria prophylaxis during pregnancy, while an editorial in Lancet (1983) stated that pyrimethamine was safe even during early pregnancy – if folinic acid is added as a supplement. However, as a precaution in the first trimester pyrimethamine is usually replaced by spiramycin (WHO 1988).

Sulfonamides

Sulfadiazine or trisulfapyrimidines have similar activity in combination with pyrimethamine. Most other sulfonamides are inferior. The usual oral dose is 50–100 mg/kg daily with a double loading dose. The general recommendation is at least 4 g/day. The patient must maintain a good urine output to avoid crystalluria and oliguria. Alkalization of the urine with sodium bicarbonate may also reduce the chance of crystalluria in AIDS patients. Today no evidence exists that sulfonamides may cause foetal abnormalities (Briggs *et al.* 1994).

Spiramycin

Spiramycin is a macrolide antibiotic which is capable of killing free and intracellular parasites by interfering with protein synthesis. It concentrates markedly in tissues, but the penetration to the cerebrospinal fluid (CSF) is poor, so it has no logical role in the treatment of cerebral toxoplasmosis. It does accumulate in the placenta, and the concentration is up to five times greater than that of the corresponding maternal serum. Spiramycin crosses the placental barrier, but the foetal blood level is only half of the corresponding maternal serum levels (Forestier *et al.* 1987). A dose of 3 g/day (9 MIU) is recommended during pregnancy. The main advantage of spiramycin lies in the complete safety of the foetus. It has no teratogenic effect whatsoever, and is well tolerated.

Clindamycin

Clindamycin acts on protein synthesis. It is highly lipid soluble, and penetrates both the eye and dense tissue. It has been used in the treatment of ocular toxoplasmosis and as an alternative treatment of cerebral toxoplasmosis (Luft and Hafner 1990). However, its penetration of CSF is poor. The recommended dose is 600 mg, four times a day. The drug is used together with pyrimethamine because by itself it merely suppresses and does not fully inhibit the spread of the infection in tissues (Piketty *et al.* 1990). Adverse reaction to clindamycin includes vomiting and diarrhoea. The drug is not teratogenic and can be used during pregnancy, but the effect on prenatal toxoplasmosis is lacking.

Macrolides

Three new macrolides: azithromycin, roxithromycin, and clarithromycin, have all shown activity in animal models of toxoplasmosis (Chang and Pechere 1988; Derouin 2000). The AIDS Clinical Trials Group has recently completed a prospective multicentre study of Azithromycin 1 200 mg/day in combination with 50–75 mg pyrimethamine. The results were favourable (Mariuz *et al.* 1994). In another case report Farthing *et al.* (1992) describes a dramatic response to azithromycin in a patient with Toxoplasma encephalitis who was allergic to sulfonamide and in whom clindamycin and doxycycline were ineffective. The advantage of azithromycin is the high tissue penetration and the concentration within fibroblast and phagocytic cells. This could be useful in eradicating intracellular parasites and parasites in the cystic stage. Azithromycin has been used during pregnancy, the concentration in the placenta is up to 90-fold higher than the plasma level of the mother (Stray-Pedersen 1996).

Immunotherapy

Administration of immune mediators such as gamma-interferon have been shown to enhance survival of *Toxoplasma*-infected mice (Suki *et al.* 1990). Inflammatory foci in the brains of mice with encephalitis decreased, but the effect was transient. It has been proposed that immunotherapy should be used as adjunctive treatment of life-threatening toxoplasmosis in AIDS patients with severely impaired immune function (Mariuz *et al.* 1994).

Cotrimoxazole

Trimethoprim and sulfamethxazole is a combination of drugs with comparable half-life. Experimental data and animal studies indicate that administration of cotrimoxazole to *Toxoplasma gondii*-infected pregnant mice gives a better *in utero* control of congenital toxoplasmosis than does spiramycin (Nguyen and Stadtsbader 1985). In humans the efficacy has been established for the prophylaxis of *Pneumocystis carinii* and *T.gondii* infections in patients with AIDS and has resulted in a marked reduction of the incidence of cerebral toxoplasmosis (Berlanger *et al.* 1999). In addition, for the treatment of Toxoplasma encephalitis the combination (trimethoprim 10 mg/kg/day and sulfamethxazole 50 mg/kg/day) was as effective and better tolerated than the standard pyrimethamine/sulfadiazine regimens (Torre *et al.* 1998).

Tetracyclines

Doxycycline has shown inhibitory activity in mice models (Chang *et al.* 1990). The best results have been obtained when a combination with pyrimethamine has been used. Doxycycline is lipid soluble and capable of penetrating the brain, especially if the meninges are inflamed. However, trials in AIDS patients with encephalitis has not been too successful (Joss 1992).

Atovaquone

Atovaquone is a hydroxynaphthoquinone with potent activity against malaria, *P. carinii*, and *T. gondii* in *in vitro* studies and in laboratory animals (Fausto *et al.* 1991). The drug was shown to be effective on the cystic stages of *Toxoplasma* in mice models (Derouin 2000). The effect of treatment has been greatly enhanced when administered in combination with pyrimethamine or sulfadiazine (Aranjo *et al.* 1993). The drug has been well tolerated in HIV-infected adults and infants and seems to be promising (Hugues *et al.* 1998; Torres *et al.* 1997).

Eight AIDS patients with cerebral toxoplasmosis who were intolerant or non-responders to standard therapies were treated with oral atovaquone 750 mg four times daily; seven patients showed clinical and radiological improvement (Kovacs 1992).

Treatment regimens

Acute acquired infection in the immune-competent patient

These infections are not treated unless significant organ dysfunction occurs (myocarditis, encephalitis) or unless systemic symptoms are severe or prolonged. When treatment is indicated, the combination of pyrimethamine and sulfonamides with folinic acid is recommended for a period of 3–4 weeks. Thereafter, a reassessment of the patient's condition should be performed. In severe cases, the usual daily dose of 25 mg pyrimethamine may be increased to 50 mg or more. In less-serious cases drugs of lower toxicity, such as spiramycin (3 g/day) or clindamycin (2 g/day), are recommended. Infection acquired by laboratory accidents or transfusion of blood products are potentially severe, and these patients should probably be treated regardless of signs or symptoms of infection (Haverkos 1987).

Ocular toxoplasmosis

The combination of pyrimethamine and sulfonamide has been recommended for treatment of active chorioretinitis. Therapy is usually administered for at least 1 month, and improvement is generally noticed after 10 days. Longer treatment may be given if active infection persists. Due to side effects ophthalmologists have tried to replace pyrimethamine with clindamycin (300 mg orally, four times a day) (Lackanpal *et al.* 1983). Systemic cortiocsteroids are added when lesions involve the macula, optic nerve head, or papillomacular bundle. Overall, the benefits of treating Toxoplasma chorioretinitis are not too favourable. A European controlled trial by Rothova *et al.* (1989) found that none of the treatment regimes reduced the duration of inflammatory activity in comparison to untreated patients. Pyrimethamine and sulfadiazine significantly reduced the *size* of the retinal lesions by 52% compared to 25% in controls, while clindamycin, together with sulfadiazine, showed marginal improvement (32%) and cotrimoxazol were ineffective. In all regimes corticosteroides were added. Therapy is today advocated only if the central visual function is threatened either by local or general inflammation, while peripheral retinal pathology is usually allowed to resolve spontaneously under observation (Dutton 1989a,b). At present, the choice of treatment is between sulfonamides and pyrimethamine or clindamycin. Previously pyrimethamine was the drug of choice, but most recent studies are in favour of the clindamycin combination (Dutton 1989a,b). Sub-conjunctival treatment has shown to be effective (Jeddi *et al.* 1997).

Toxoplasmosis in the immuno-compromised patient

Acute infection

Immuno-compromised patients should always be treated when the infection is acute. The therapy should be given for at least three weeks, but up to six weeks or more may be required for severely ill patients who have not achieved a complete response. Pyrimethamine combined with sulphadiazine and folinic acid is a therapy of choice for AIDS patients with toxoplasmosis. This regimen has been associated with clinical response in 68–95% of cases with encephalitis, but unfortunately up to 40% of the patients develop side effects, requiring discontinuation of therapy (Leport and Raguin 1990; Luft *et al.* 1993). Substitution of clindamyacin for sulphadiazine while continuing with pyrimethamine is an acceptable solution. However, there is little evidence to support regular first use of clindamycin in preference to sulphadiazine–pyrimethamine. Cotrimoxazole has also, on occasions been successfully used as an alternative to sulphadiazine (Solbreux *et al.* 1990). However, its regular use is questionable as encephalitis has developed in patients receiving cotrimoxazole for *P. carinii* infections (Haverkos 1987). Of the newer agents azithromycin and atoquavone are now on trial (Torres *et al.* 1995).

Maintenance and prophylaxis

When suppressive therapy is discontinued, surviving tissue cysts may induce Toxoplasma encephalitis and other clinical manifestations in up to 80% of AIDS patients. Since the standard regimen does not eliminate the tissue cysts, it is generally agreed that AIDS patients will require life-long maintenance therapy. Many clinicians favour continuing therapy with the same drug combinations in the same daily doses as used for acute infection. Others believe that both doses and frequency can be reduced. One regimen of pyrimethamine (25 mg) and sulphadiazine (75 mg/kg) twice a week plus folinic acid (10 mg daily) was fully effective (Pedrol 1990). In contrast, pyrimethamine plus clindamycine twice weekly was less effective. A recent trial recommends cotrimoxazole as prophylactic therapy, while in high-risk patients with <100 CD4 cells/mm^3 pyrimethamine (50 mg) should be given three times weekly (Leport *et al.* 1996). Other agents under study include Fanzidar and Dapsone, which have a long half-life. In the future, there is a possibility that drugs which eliminate the tissue cysts may become available. These drugs would be of enormous benefit for AIDS patients and congenitally infected infants.

Prevention of toxoplasmosis in transplant recipients

Seronegative recipients of organ transplants (heart, kidney, liver, and bone marrow) from a seropositive donors often develop toxoplasmosis. Some centres have a programme to prevent this form of toxoplasmosis by treating the recipient with anti-toxoplasma drugs for 4–6 weeks (Joss 1992).

Primary toxoplasma infection in pregnant women

Antiparasitic treatment should be offered to every primary infected woman. The aim is to prevent foetal infection if it has not already occurred, and to treat and reduce tissue damage in an already infected foetus. In France, spiramycin has been used since the beginning of the 1970s. In a study of 542 pregnant women, Desmonts and Couvreur (1984) reported that parasites were detected less frequently in the placentas of spiramycin-treated mothers (19%) than in untreated mothers (50%). Maternal therapy reduced the placental infection from 25 to 8% in the first trimester, from 54 to 19% in the second trimester, and from 65 to 44% in the third trimester. Moreover, a significant reduction in the incidence of congenitally infected infants of the treated (23%) versus untreated (61%) mothers were noted, but the proportion of babies with clinical disease was the same in both groups. Thus, spiramycin appeared to reduce the incidence of foetal infection, but had little effect on an already infected foetus. The critics focus on the fact that the untreated mothers do not represent a real control group, but more or less historical controls.

Later studies, primarily from France, have shown that pyrimethamine–sulfadiazine treatment in pregnancy is more effective that spiramycin in the eradication of parasites from placenta. In a retrospective study of congenitally infected infants, Couvreur *et al.* (1988) detected parasites in 89% of placentas of untreated women, in 75% of spiramycin-treated women, and in only 50% of women treated with pyrimethamine–sulfadiazine. Furthermore, pyrimethamine–sulfadiazine is shown to lead to a significant reduction in the number of severely affected babies and a shift to less severe and sub-clinical forms (Hohlfeld *et al.* 1989). In a recent multicentre European study (Foulon *et al.* 1999) a comparison was made between 144 seroconverting mothers, of whom 119 received prenatal treatment (mainly spiramycin) and 25 mothers who were not treated at all. For the first time the effect of gestational age at maternal infection on transmission risk was taken into account. It was shown that prenatal anti-parasitic treatment had no impact on the maternal–foetal transmission rate, but a significant beneficial effect on sequelae, especially on severe sequelae in the infected infants at 1 year of age. Early start of treatment in pregnancy resulted in the significant reduction in the number of infants with severe complications (Foulon *et al.* 1999). A similar non-beneficial effect on the transmission rate, but probably a clinically important effect on complications, has been reported in a French study (Gilbert *et al.* 2001). A combination of pyrimethamine and the long acting sulfadoxine (Fansidar) has been used in pregnancies with proven infected offspring (Maisonneuve *et al.* 1984; Piene and Garin 1989). All these studies conclude that treatment influences the long-term sequelae of infected infants and that Fansidar is well tolerated. These drugs can be given once a week, and thus the compliance is probably better than for daily pyrimethamine.

The present guidelines recommend that anti-parasitic treatment should be offered to every primary infected woman (Stray-Pedersen and Foulon 2000). Initially, spiramycin can be given until the status of the foetus is settled by prenatal diagnosis (Table 26.1). If the foetus is infected, pyrimethamine–sulfadiazine–folinic acid should be offered in three-week courses alternating with spiramycin. Alternative regimen is pyrimethamine–sulphadoxine (Fansidar) two tablets every week, sometimes alternating with spiramycin daily until term.

Treatment of neonates

It is today agreed that every case of congenital toxoplasmosis should be treated whether or not the newborn infant displays clinical manifestation. Treatment should be instituted as soon as possible after birth, and given for a minimum of 6 months, usually 1 year. Therapy beyond 12 months is only recommended in cases where the infection is still active. In infants with symptomatic congenital toxoplasmosis prospective studies from the United States using pyrimethamine–sulfadiazine continuously for a whole year showed striking effects on the ophthalmologic follow-up (Mets *et al.* 1996), neuro developmental outcome (McAuley *et al.* 1994), and hearing loss (McGee *et al.* 1992).

The key question is whether or not treatment is necessary in sub-clinical cases with no symptoms at all. Controlled trials with long prospective follow-ups do not exist. There is, however, historical evidence that early treatment protects against development of late ocular lesions and cerebral symptoms (Wilson *et al.* 1980; Piene and Garin 1989). Pyrimethamine and sulfadiazine in combination have proven more effective than spiramycin alone.

The present guidelines recommend that *symptomatic* infants should be treated for at least 1 year with pyrimethamine–sulfadiazine. In cases with active chorioretinitis or cerebral infection, corticosteroids should be added.

Table 26.1 Treatment of Toxoplasma infection in pregnant women and newborn infants

Drugs	Adults	Newborn
Spiramycin	$3\,g = 9\,MIU/day$	$150\,000–300\,000\,IU/kg/day$
or		
Pyrimethamine (P)	25 mg/day	0.5–1 mg/kg/day
Sulfadiazine (S)	50–100 mg/kg/day (initially: double dose)	50–100 mg/kg/day
or		
Fansidar		
(1 tablet = 25 mg pyrimethamine + 500 mg sulfadoxine)	2 tablets/week	1 tablet/20 kg/week
During pyrimethamine therapy:		
Folinic acid (F)	15 mg twice weekly	3–5 mg, twice weekly
Blood cells counts (platelet, white cells) every 1–2 weeks		
If ocular or CNS toxoplasmosis:		
Corticosteroids		1–2 mg/kg/day
Indications		
Pregnant women with acquired infection		
Before conception:	No need for treatment	
Suspected cases:	Spiramycin	
Proven cases:		
1st trimester:	Spiramycin continuously	
2nd–3rd trimester:	P + S + F (3 weeks) + Spiramycin (4–6 weeks) Alternative: Fansidar (2 tablets/week)	
Evidence of foetal infection:	P + S + F (3 weeks), then Spiramycin (3–6 weeks). Repeat until delivery	
Newborns with congenital infection		
Suspected infection:	Spiramycin until diagnosis	
Subclinical congenital infection:	P + S + F (4 weeks) + Spiramycin (4–6 weeks) Repeat until 12 months of age Alternative: (a) P + S + F (3 months) (b) Fansidar 1 tablet/20 kg/week	
Overt congenital infection:	P + S + F (from birth until 6–12 months), Alternative after 6 months of age: Spiramycin (4 weeks) + P + S + F (4 weeks)	

In *asymptomatic* newborns the regimen varies. Traditionally, spiramycin is recommended during the first months of life, thereafter four-week courses of pyrimethamine–sulfadiazine alternating with six-week courses of spiramycin – altogether four courses during the first year of life (McLeod *et al.* 2000). In Denmark a single three-months course with pyrimethamine–sulfadiazine–folinic acid is recommended (Petersen and Eaton 2000). In France and Switzerland, Fansidar and folinic acid are given weekly during the first year.

Still, after 30 years of experience, the optimum schedules and duration of therapy during pregnancy and in newborns remain unknown. Future international multicentre studies with long-term follow-up observations are needed to provide sufficient data.

References

Aranjo, F. G., Lin, T, and Remington, J. S. (1993). The activity of atovaquone in murine toxoplasmosis is markedly augmented when used in combination with pyrimethanine or sulphadiazine. *J. Infect. Dis.*, **167**(2), 494–497.

Berlanger, F., Derouin, F., Grangeot-Keros, L., *et al.* (1999). Incidence and risk factors of toxoplasmosis in a cohort of HIV infected patients 1988–1996. *Clin. Infect. Dis.*, **28**, 575–581.

Briggs, G. G, Freeman, R. K., and Yaffe, S. J. (eds) (1994). *Drugs in pregnancy and lactation.* Williams & Wilkins, Baltimore, MD.

Couvreur, J., Desmonts, G., and Thulliez, Ph. (1988). Prophylaxis of congenital toxoplasmosis. Effects of spiramycin on placental infection. *J. Antimicrob. Chemother.*, **22**(Suppl. B), 193–200.

Derouin, F. and Chastang, C. (1989). In vitro effects of folate inhibitors on *Toxoplasma gondii*. *Antimicrob. Agents Chemother.*, **33**, 1753–1759.

Derouin, F. (2000). Drugs effective against *Toxoplasma gondii*. Present status and future perspective. In: *Congenital toxoplasmosis* (P. Ambroise-Thomas and E. Petersen, eds). Springer-Verlag, France, pp. 94–110.

Desmonts, G. and Couvreur, J. (1984). Toxoplasmose congenitale. Etude prospective del'issue de la grossesse chez 542 femmes atteintes de toxoplasmose acquise en cours de gestation. *Ann. Pediat.*, **31**, 805–809.

Dutton, G. N. (1989a). Toxoplasmic chorioretinitis. *Ann. Acad. Med.*, **18**, 214–221.

Dutton, G. N. (1989b). Recent developments in the prevention and treatment of congenital toxoplasmosis. *Int. Ophtalmol.*, **13**, 407–413.

Gilbert, R. E., Dunn, D. T., Wallon, M., *et al.* (2001). Ecological comparison of the risks of mother to child transmission and clinical manifestations of congenital toxoplasmosis according to prenatal treatment protocol. *Epidemiol. Infect.*, **127**, 113–20.

Farthing, C., Rendel, M., Curriet, B., *et al.* (1992). Azithromycin for cerebral xoplasmosis. *Lancet*, **339**, 437–438.

Forestier, F., Daffos, F., Rainaut. *et al.* (1987). Suivi therapeutique foetomaternel de la spiramycine en cours de grosese. *Arch. Fr. Ped.*, **44**, 539–544.

Foulon, W., Villena, I., Stray-Pedersen, B. *et al.* (1999). Treatment of toxoplasmosis during pregnancy: a multicenter study of impact on fetal transmission and children's sequelae at one year of age. *Am. J. Obstet. Gynecol.*, **180**, 410–415.

Haverkos, H. W. (1987). Assessment of therapy for toxoplasma encephalitis. The TE study group. *Am. J. Med.*, **292**, 1108–1112.

Hohlfeld, J., Daffos, F., Thulliez, P., *et al.* (1989). Fetal toxoplasmosis: outcome of pregnancy and infant follow-up after *in utero* treatment. *J. Pediatr.*, **115**, 765–769.

Hugues, W., Dorenbaum, A., Yogev, R. *et al.* (1998). Phase I safety and pharmacokinetics study of micronized atovaquone in human immunodeficient virus infected infants and children. *Antimicrob. Agents Chemother.*, **42**, 1315–1318.

Huskinson, M. J., Aranjo, F. G., and Remington, J. S. (1991). Evaluations of the effect of *Toxoplasma gondii*. *J. Infect. Dis.*, **164**, 170–177.

Jeddi, A., Azaiez, A., Bouguila, H. *et al.* (1997). Value of clindamycin in treatment of ocular toxoplasmosis. *J. Fr. Ophtalmol.*, **20**, 418–422.

Joss, A. W. L. (1992). Treatment. In: *Human toxoplasmosis* (D. O. Ho-Yen, and A. W. L. Joss, eds). Oxford University Press, New York, pp. 119–143.

Kovacs, J. A. (1992). Efficacy of atovaquone in treatment of toxoplasmosis in patients with AIDS. *Lancet*, **340**, 637–638.

Lackanpal, V., Schocket, S. S., and Nirankari, V. S. (1983). Clindamycin in the treatment of toxoplasmic retinochoroiditis. *Am. Ophtalmol.*, **95**, 605–613.

Lancet Editorial (1983). Pyrimethamine combination in pregnancy. *Lancet*, **2**, 1005–1007.

Leport, C. and Raguin, G. (1990). Toxoplasmic encephalitis. *Curr. Opin. Infect. Dis.*, **3**, 614–619.

Leport, C., Chene, G., Morlat, P., *et al.* (1996). Pyrimethamine for primary prophylaxis of toxoplasmic encephalitis in patients with human immmunodeficiency virus infection: a double-blind, randomized trial. *J. Infect. Dis.*, **173**(1), 91–97.

Luft, B. J. and Hafner. R. (1990). Toxoplasmic encephalitis. *AIDS*, **4**, 593–595.

Maisonneuve, H., Faber, C., Piens, M.A. *et al.* (1984) Toxoplasmose congenitale. Tolerance de l'association sulfadoxine–pyrimethamine. Vingt quatre observations. *Press. Med.*, **13**(14), 859–862.

Mariuz, P., Bosler, E. M., and Luft, B. J. (1994). Toxoplasmosis in individuals with AIDS. *Infect. Dis. Clin. N. Am.*, **8**, 365–381.

McAuley, J., Boyer, K. M., Patel, D. *et al.* (1994). Early and longitudinal evaluations of treated infants and children and untreated historical patients with congenital toxoplasmosis: the Chicago collaborative treatment trial. *Clin. infect. Dis.*, **18**, 38–72.

McCabe R. E. and Oster, S. (1989). Current recommendations and future prospects in the treatment of toxoplasmosis. *Drugs*, **38**, 973–987.

McGee, T., Wolters, C., Stein, L. *et al.* (1992). Absence of sensorineural hearing loss in treated infants and children with congenital toxoplasmosis. *Otolaryngology*, **106**, 75–80.

McLeod, R., Boyer, K. and the Toxoplasmosis Study groups and Collaborators (2000). Management of and outcome for the newborn infant with congenital toxoplasmosis. In: *Congenital toxoplasmosis* (P. Ambroise-Thomas and E. Petersen, eds). Springer-Verlag, France, pp. 189–213.

Mets, M. B., Holfels, E., Boyer, K. M. *et al.* (1996). Eye manifestations in congenital toxoplasmosis. *Am. J. Ophthalmol.*, **122**(3), 309–324.

Nguyen B. T. and Stadtsbader, S. (1985). Comparative effect of cotrimoxazole and spiramycin in pregnant mice infected with *Toxoplasma gondii*. *Br. J. Pharmacol.*, **85**, 713–716.

Petersen, E. and Eaton, R. B. (2000). Neonatal screening for congenital infection with *Toxoplasma gondii*. In: *Congenital toxoplasmosis* (P. Ambroise-Thomas and E. Petersen, eds). Springer-Verlag, France, pp. 306–311.

Piens, M. A. and Garin, J. P. (1989). New perspective in the chemoprophylaxis of toxoplasmosis. *J. Chemother*, **3**, 614–619.

Piketty, C., Derouin, F., Rouveix, B. *et al.* (1990). *In vivo* assessment of antimicrobial agents against *Toxoplasma gondii* by quantification of parasites in the blood, lungs, and brain of infected mice. *Antimicrob. Agents Chemother.*, **34**, 1467–1472.

Rothova, A., Bitenhuis, H. J., Meenken, C. *et al.* (1989). Therapy of ocular toxoplasmosis. *Int. Ophtalmol.*, **13**, 415–419.

Torre, D., Casari, S., Speranza, F. *et al.* (1998). Randomised trial of trimetoprim-sulfamethxazole versus pyrimethamine–sulfadiazine for therapy of toxoplasmic encephalitis in patients with AIDS. *Antimicrob. Agents Chemother.*, **42**, 1346–1349.

Torres, R. A., Weinberg, W., Stansell, A. *et al.* (1997). Atovaquone for salvage treatment and suppression of toxoplasmic encephalitis in patients with AIDS. *Clin. Infect. Dis.*, **24**, 422–429.

Sheffield, H. G. and Melton, M. L. (1975). Effect of pyrimethamine and sulphadiazine on the fine structure and multiplication of *Toxoplasma gondii* in cell cultures. *J. Parasitol.*, **61**, 704–712.

Stray-Pedersen, B. and Foulon, W. (2000). Effect of treatment of the infected pregnant women and her fetus. In: *Congenital toxoplasmosis* (P. Ambroise-Thomas and E. Petersen, eds). Springer-Verlag, France, pp. 141–152.

Stray-Pedersen, B. (1993). Toxoplasmosis in pregnancy. In: *Infectious diseases*. Challenges for the 1990s. (G. L. Gilbert, ed.). Balliere's Clinical Obstetrics and Gynecology. Vol. 17, 1. Bailliere Tindall, London, pp. 107–137.

Stray-Pedersen, B. and The European Research Network of Congenital Toxoplasmosis Treatment Group (1996). Azitromycin levels in placental tissue, amniotic tissue and blood. In *Abstract book, of the 36th ICAAC American Society for Microbiology, New Orleans, 15–18 September 1996*.

Wilson, C. B., Remington, J. S., Stagno, S. *et al.* (1980). Development of adverse sequelae in children born with subclinical Toxoplasma infection. *Pediatrics*, **66**, 767–774.

WHO (1988). Report of the WHO consultation on public health aspects of toxoplasmosis. WHO/CDS/VPH 74, pp. 2–14.

WHO (1990). WHO model prescribing information – drugs used in parasitic diseases, pp. 53–60.

27 Clinical presentation and laboratory investigations of other parasites in HIV-infected individuals

Ib C. Bygbjerg

The acquired immunodeficiency syndrome (AIDS) is the end-stage of a progressive infection of the human host's cellular immune system, in particular the T-helper (CD4+) lymphocytes, by the human immunodeficiency virus type 1 (HIV-1). Though the closely related HIV-2 may also cause immunodeficiency and thereby give rise to opportunistic parasitic infections, HIV-related parasites in the following section will focus on those affecting HIV-1 infected individuals.

HIV-infected individuals with progressive immune deficiency have an abnormally high susceptibility to infections with non-virulent and minimally pathogenic organisms. Life-threatening infections of the respiratory, central nervous, and gastrointestinal system arise by revival of dormant infections acquired many years previously, or by exposure to new pathogens, most commonly when the CD4+ counts decrease to 10–20% of normal levels. Thus, *Pneumocystis carinii* infections typically occur in patients with CD4+ counts below 200–300 mio./l; *Toxoplasma gondii* infections with CD4+ counts below 100 mio./l; and *Cryptosporidium parvum* may be fulminant in patients with CD4+ counts below 50 mio./l, while it may be reversible if CD4+ counts are above 300–400 mio./l. Patients with *Leishmania*/HIV co-infection are virtually untreatable if CD4+ counts are low.

Besides contributing to the sufferings of HIV-infected individuals, parasitic infections are also important for the classification (stages) of HIV: several parasites are enlisted as AIDS defining events.

In the Nordic countries, the most prevalent parasites in HIV-infected patients are *P. carinii*, *T. gondii*, *C. parvum*, and *Microsporidia*. *Cyclospora cayetanensis* and *Isospora belli* are less frequently diagnosed, while *Leishmania* infections and other (sub)-tropical parasitoses are very rare. Increasing population mobility may, however, broaden the spectrum of opportunistic infections encountered, whether in Scandinavians visiting endemic areas or immigrants and refugees succeeding in crossing the borders to the cool North. In the following, emphasis is on *P. carinii*, *T. gondii*, *C. parvum*, and *Microsporidia*.

Effective chemoprophylaxis of *P. carinii* and *T. gondii* and more recently highly active anti-viral chemotherapy (HAART) of HIV have greatly affected the patterns of opportunistic infections, at least for those patients, who can afford it. Anti-parasitic prophylaxis may be stopped in HIV patients with CD4+ counts returning to levels >200 mio./l (Schneider *et al.* 1999). Resistance to HAART, however, calls for cautiousness.

Pneumocystis carinii

Clinical presentation

Though *Pneumocystis carinii* has common epitopes with fungi (Lundgren *et al.* 1992b) and may be considered a fungus rather than a protozoan (Stringer *et al.* 1989), its diagnosis and treatment are more like that of protozoa, justifying inclusion among parasites in clinical practice.

Pneumocystis carinii pneumonia (PCP) was the first opportunistic infection recognized in AIDS patients, American (Gottlieb *et al.* 1981) as well as European (Gerstoft *et al.* 1982). HIV positive individuals with decreasing CD4+ (T-helper lymphocyte) counts not taking any chemoprophylactic are at high risk of PCP.

The symptoms often start insidiously with tiredness, fever, dyspnoea, and dry cough. The incubation time may be several months, more often a few weeks. Severe chest pain, resolving spontaneously, may precede an episode of PCP (Bygbjerg, unpublished). Ambulant presentation with almost no clinical signs and even with a normal chest X-ray is common; however, the dyspnoea may suddenly worsen, and the chest X-ray's appearance turn into that of 'white lungs' (Figure 27.1). Lung stethoscopy is often unhelpful, for example, normal, but increased respiration rates, 30–40 per minute in severe cases, and clinically visible hypoxaemia (cyanosis) should alert the clinician to hospitalize and further examine the patient immediately, since manifest respiratory insufficiency, requiring mechanical ventilation may supervene, if diagnosis and treatment are delayed. In case of super- or co-infection with fungi or bacteria, the patient's dry cough may become productive; otherwise the characteristic glass-glossy/foamy sputum can only be induced by hypertonic saline inhalation or

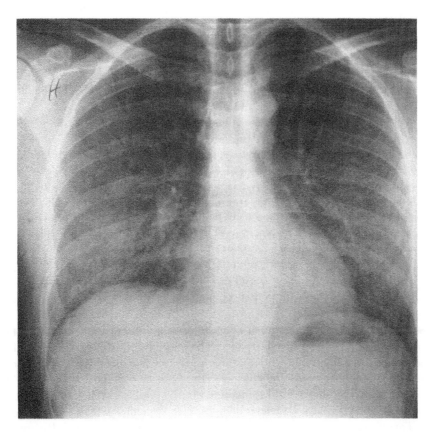

Figure 27.1 X-ray image of the lungs of an AIDS patient with *P. carinii* infection: bilateral, interstitial infiltrates ('white lungs').

brought up by bronchoalveolar lavage (BAL) (see below). The outcome of PCP in Scandinavian patients with AIDS is about 15% mortality, which may rise to 44% if mechanical ventilation is required (Lidman *et al.* 1989; Pedersen *et al.* 1989).

Extrapulmonary *P. carinii* have been described from most internal organs *post mortem* as well as *in vivo* (Cohen and Stoeckler 1991), including lymph nodes, spleen, liver, bone marrow, bones, adrenal glands, gastrointestinal tract, kidneys, thyroid gland, heart, pleura, pancreas, peritoneum, eyes, ears, and brain (Garcia and Bruckner 1993). Patients taking prophylaxis with aerosolized pentamidin, attaining only the lungs, rather than taking systemic prophylaxis with sulfamethoxazole-trimethoprim (co-trimoxazole), are at particular risk of extrapulmonary infection (Witt *et al.* 1991).

Laboratory investigations

Levels of blood haemoglobin, leukocytes, differential counts, and thrombocytes may be lowered by the HIV infection, but are normally unaffected by the *P. carinii* infection. Decreasing levels of these haematological parameters are common during high-dose therapy with co-trimoxazole, while adjunct treatment with corticosteroids – which may be life saving in patients with hypoxaemia (Bozzette *et al.* 1990) – may induce blood leukocytosis. Monitoring the levels of blood lactate dehydrogenase (LDH) is much more helpful: more than 90% of patients with PCP have elevated LDH, and if above 450 International Units, the prognosis is poor (Leoung and Hopewell 1990a).

If the patient is not manifest hypoxaemic, exercise biking may reveal latent hypoxaemia, and prompt further examination for the causative agent. As mentioned earlier, very discrete symptoms and a normal chest X-ray may change to life-threatening PCP within a day or two. Computed tomography (CT) scanning of the chest may disclose early diffuse infiltrates, but the diagnosis should be based on demonstration of *P. carinii*.

BAL +/– trans-bronchial biopsy via a flexible fibre bronchoscope was recommended as the standard procedure early on in the AIDS epidemic (Broaddus *et al.* 1985); it has a high sensitivity, about 90%, even without biopsy, which is no longer routine. Sputum induced by inhalation of hypertonic saline, in the hands of experienced clinicians and laboratory technicians, may attain almost 80% sensitivity (Leoung and Hopewell 1990b). Staining techniques, including toluidine blue, Giemsa, and methenamine silver have been described in previous chapters, as has the use of fluorescent monoclonal antibodies. Improved diagnosis can be obtained by polymerase chain reaction (PCR) on sputum (Lipschik *et al.* 1992), while PCR on serum has been disappointing (Wagner *et al.* 1997).

Demonstration of fluorescing serum antibodies to crude *P. carinii* antigens is not helpful in diagnosing PCP in AIDS (Hofmann *et al.* 1985), while IgG antibodies against gp95 antigens are increased in 66% of patients with PCP; IgM antibodies are only raised in 4% (Lundgren *et al.* 1992a).

Some patients may be too ill to co-operate to induced sputum or bronchoscopy. These may benefit from non-invasive procedures like PCR on mouth washings. However, there may not be time to await the results, and presumptive treatment with co-trimoxazole should be instituted. As mentioned above, the mortality rises from 10% to more than 40% if the patient becomes so ill that mechanical ventilation may be required. When improving, the patient may undergo BAL, since *P. carinii* is demonstrable even after several days' or weeks' treatment. Therefore, re-examination after treatment is normally not helpful. Patients not responding to standard treatment should, however, be re-examined to exclude important differential diagnostic conditions, including cytomegalovirus infection, bacterial infection (streptococci, styphylococci, *H. influenzae*), mycobacteria, fungi, and disseminated Kaposi's sarcoma, which may be indistinguishable clinically and radiologically. PCP not responding to standard treatment may be treated with alternative regimens, including trimetrexate, pentamidine, clindamycin with primaquine, atovaquone and others.

Recently, successful *in vitro* culture of *P. carinii* has been reported (Merali *et al.* 1999), but routine testing for drug resistance *in vitro* is not possible. However, mutations in dihydropteroate synthase, the enzyme essential for folate biosynthesis in *P. carinii*, which is blocked by sulphonamides, may be demonstrated by PCR and may be associated with impaired prognosis (Helweg-Larsen *et al.* 1999). Further details on the management of PCP should be sought in standard textbooks on AIDS.

Toxoplasma gondii

Clinical presentation

Toxoplasmosis is – next to PCP – the most important opportunistic tissue protozoan infection in patients with AIDS. Soon after the recognition of the AIDS syndrome, outbreaks of central nervous system (CNS) toxoplasmosis in Western Europe and North America were observed (Luft *et al.* 1983). The CNS remains the most common localization in AIDS patients, underlining the seriousness of this condition. It is found in 5–10% of North European AIDS patients (Smith *et al.* 1991), but rarely if the HIV positive patient takes regular chemoprophylaxis with co-trimoxazole (sulfamethoxazole-trimethoprim) (Gallant *et al.* 1994) or sulphone-pyrimethamine. The CD4+ count at the time of diagnosis is generally lower than for PCP, 100 or below versus 200–300 mio./l, respectively (Masur *et al.* 1989).

The clinical presentation may be acute with high fever, confusion, gross neurological deficits, seizures, and even life-threatening brain oedema, closely resembling CNS lymphoma – another common event in severely immunocompromised HIV positive patients. Or it may be insidious with tiredness, discrete mental changes, and low-grade fever, mimicking AIDS dementia.

On computed tomography (CT) scan or magnetic resonance imaging (MRI) scan of the brain, toxoplasmosis most often presents as a focal or multifocal hypodense or ring enhancing lesions (Figure 27.2). Unfortunately, CNS lymphoma may present similarly. If a brain biopsy is not wanted or possible, it is justified to treat the patient empirically with high-dose sulfadiazine plus pyrimethamine. A therapeutic response should be observed clinically within days and on CT scans after 2 weeks. Raffi *et al.* (1997) showed that 40% of AIDS patients suspected of CNS toxoplasmosis and given specific therapy, were actually not having toxoplasmosis. Thus, more sensitive and specific diagnostic tools are warranted. Even with adequate therapy and life-long maintenance therapy, the prognosis is relatively poor, and relapses not uncommon. New, potent anti-HIV drug combinations (HAART), reduces the need for life-long suppressive therapy of toxoplasmosis (Kirk *et al.* 1999), but one recent study showed that the incidence of cerebral toxoplasmosis as an AIDS-defining illness remained stable after introduction of HAART (Ives *et al.* 2001).

Other organs may be affected by toxoplasmosis, with or without simultaneous CNS involvement. Chorioretinitis may be caused by *T. gondii* in AIDS patients, but is much more often caused by cytomegalovirus (Mathiesen and Lundgren 1997). Detection of intra-ocular antibody production and PCR analysis may be helpful (Verbraak *et al.* 1996). Pulmonary toxoplasmosis has been reported with increasing frequency (Pomeroy and Filice 1992), mimicking PCP. Next to the brain and the lungs, the myocardium is the site most commonly

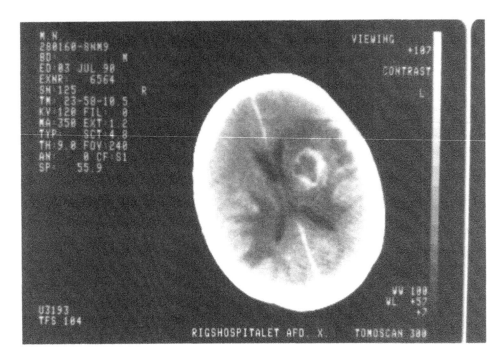

Figure 27.2 CT scan of the brain of an AIDS patient with *T. gondii* infection: a focal ring enhancing lesion is clearly visible centrally in the left hemisphere, surrounded by oedema, partially compressing and displacing the ventricle system of the brain.

involved in disseminated toxoplasmosis (Jacobs *et al.* 1991). Necrotizing pancreatitis and multiorgan failure associated with toxoplasmosis has also been described in AIDS (Ahuja *et al.* 1993).

Laboratory investigations

In practice, the diagnosis of CNS toxoplasmosis is based on typical clinical and radiological symptoms and signs, and on therapeutic response. Positive serology is present in more than 90%, but the specificity is low: in one large series of European patients, 40% of patients with non-toxoplasmic brain lesions were seropositive as well (Raffi *et al.* 1997). IgG antibodies are much more common than IgM, pointing to reactivation rather than new infections. In one Danish series of 17 AIDS patients with CNS toxoplasmosis, all were IgG positive but none IgM positive (Smith *et al.* 1991). Intrathecal antibody production may further support the diagnosis (Potasman *et al.* 1988), as may PCR on cerebrospinal fluid (Lebech *et al.* 1992). However, the diagnostic sensitivity of the latter has not been found to be satisfactory (Clinque *et al.* 1997).

The definitive diagnosis by demonstrating *T. gondii* in a brain biopsy is rarely practicable: among 17 Danish patients with CNS toxoplasmosis, only three had histologically proven diagnosis, and all three at autopsy (Smith *et al.* 1991). Pulmonary toxoplasmosis may be confirmed by microscopy of BAL or biopsies (Jacobs *et al.* 1991); the diagnosis may be supported by extreme elevations of serum LDH (Pugin *et al.* 1992). Generally, toxoplasmosis of the brain in AIDS patients responds well to treatment with high-dose sulfadiazine plus pyrimethamine. In case of adverse reactions (seen in 20–40%), clindamycin plus pyrimethamine, or atovaquone may be used (Kovacs 1994). Life-long chemotherapy/prophylaxis is hardly needed after the introduction of HAART, but resistance to HAART, however, calls for continuousness. Further details on management of toxoplasmosis should be sought in standard textbooks.

Intestinal spore-forming protozoa

Cryptosporidia, Cyclospora, Isospora, *and Microsporidia*

The AIDS epidemic has increased awareness and recognition of intestinal spore-forming protozoa hitherto considered rare pathogens. The many common characteristics of these four intestinal protozoa (reviewed by Goodgame 1996) justify a common review here, of their clinical presentation and demonstration in HIV-infected individuals.

About the same time as the AIDS epidemic was recognized in the USA, a very useful staining method for *Cryptosporidium parvum* was published in Scandinavia by veterinarians (Henriksen and Pohlenz 1981). This modified Ziehl–Neelsen acid fast stain soon became helpful to diagnose human cryptosporidiosis as well (Payne *et al.* 1983). The first cases of cryptosporidium enterocolitis in homosexual men with AIDS were reported in Denmark by Gerstoft *et al.* (1983). *Cryptosporidium parvum* may be fatal in AIDS patients. McGowan *et al.* (1993). found a median survival of 15 weeks from diagnosis. The clinical presentation in AIDS is a chronic, debilitating watery diarrhoea, with 5–10 bowel movements per day, and abdominal cramps, anorexia, and low-grade fever. Extraintestinal infection with *C. parvum* has been described in the biliary tract causing sclerosing cholangitis in patients with AIDS, and may also be associated with hepatitis, pancreatitis, and respiratory problems (Højlyng and Jensen 1988).

About 1–3% of Danish AIDS patients suffered from cryptosporidiosis (Smith and Orholm 1990) before HAART was introduced. A major nosocomial outbreak of cryptosporidiosis in a department of infectious diseases implying 18 cases – infected through an ice machine contamined by an infected patient – resulted in eight patients dying (Ravn *et al.* 1991). This underlines that infection may only require a very small inoculum, and that *C. parvum* may survive at low temperature. Since there is no effective therapy, preventative measures are vital. Unfortunately, sulphonamides (used in combination with antifolates against *P. carinii* and *T. gondii*) are not helpful for preventing *C. parvum* (Ravn *et al.* 1991), while macrolides offer some hope (Fichtenbaum *et al.* 2001). Paromomycin (White *et al.* 1994), and more recently, a 5-nitrothiazole (Nitazoxanide) have been used with some success for treatment (Doumbo *et al.* 1997), while numerous other medicaments have been disappointing, including immunotherapy with specific antibodies from cows' colostrum (Saxon and Weinstein 1987). Fortunately, HAART can restore immunity, including mucosal, and lead to eradication of opportunistic pathogens (Schmidt *et al.* 2001).

The laboratory confirmation of *C. parvum* infection can be by light microscopy of unstained faeces with or without previous concentration. Acid-fast stains are very useful, as indicated above, and *C. parvum* is clearly visible at 400 × magnification (as opposed to microsporidia). Monoclonal antibody-based fluorescent stain is also helpful. Electron microscopy of biopsies from the gut will show organisms on the brush border of the mucosal surface (Goodgame 1996). PCR and PCR-ELISA may be more sensitive than light microscopy, and may also specify genotypes (Gibbons *et al.* 2001).

Cyclospora cayetanensis, previously referred to as 'cyanobacterium-like bodies' has now been identified worldwide in the faeces of both immunocompetent and immunocompromized patients with diarrhoea (Ortega *et al.* 1993). Pape *et al.* (1994) give four reasons for the relatively few cases observed and published initially: *Cyclospora* may be confused with *Cryptosporidia*; acid-fast stain may not be ordered by the physician; co-trimoxazole chemoprophylaxis for *P. carinii* and *T. gondii* may affect *C. cayetanensis*; and cyclospora may have a low prevalence in developed countries.

In patients with low CD4+ counts, watery diarrhoea, abdominal cramps, and weight loss are the main symptoms – as in cryptosporidiosis, isosporiasis, and microsporidiosis (Goodgame 1996). When the CD4+ count is almost normal (> 400 mio./l), the patient may have no symptoms at all, and shedding of *C. cayetanensis* oocysts may resolve spontaneously after a couple of months (Schubach *et al.* 1997). A 1-week course of co-trimoxazole is effective in HIV-infected patients (Verdier *et al.* 2000).

The laboratory diagnosis can be made by trained microscopists on formolether concentrated faeces sediments, but modified acid-fast staining is more helpful. It is very important to measure the size of the oocysts. Those of cyclospora are 8–10 μm, while *Cryptosporidia* are 4–6 μm (González-Ruiz and Bendall 1995). Electron microscopy and examination of biopsies are also helpful (Goodgame 1996).

Isospora belli was recognized as an opportunistic enteric pathogen in AIDS in 1985–1986, in the USA (DeHovitz *et al.* 1986). It is most common in tropical and subtropical climates, and only single cases are observed in north European AIDS patients; in France, about 2% of chronic AIDS-related diarrhoea are associated with *Isospora* (Cotte *et al.* 1993). A negative association between isosporiasis and pneumocystosis has been related to use co-trimoxazole prophylaxis in HIV positive patients (Sorvillo *et al.* 1995). Disseminated isosporiasis has been described in AIDS (Bernard *et al.* 1997).

The identification of *I. belli* is easy in unstained wet or acid-fast stained faeces by light microscopy. The oocysts are large, 10–20 × 20–30 μm. In small-bowel biopsy specimens, *I. belli* are easily seen as oval enterocyte inclusions (Goodgame 1996). Isosporiasis in patients with AIDS can be treated effectively with a 10-day course of co-trimoxazole, and prevents by prophylactic doses of co-trimoxazole (Pape *et al.* 1989). However, refractory cases may be encountered (Bygbjerg, unpublished).

Five genera of microsporidia have been recognized in humans (Garcia and Bruckner 1993), of which two are common intestinal pathogens in patients with AIDS: *Enterocytozoon bieneusi* and *Encephalitozoon (Septata) intestinalis*; the former is by far the commonest (Goodgame 1996). *Enterocytozoon bieneusi* was the first hitherto completely unknown eukaryotic parasite discovered in man because of immunodeficiency (Canning and Hollister 1990). Its clinical significance has been reviewed by Schattenkerk *et al.* (1991). The first cases of microsporidiosis caused by *E. bieneusi* in Scandinavian patients with AIDS were reported by Højlyng *et al.* (1993). *Enterocytozoon bieneusi* is found in more than 20% of AIDS patients with permanent diarrhoea. In 1995, *E. (Septata) intestinalis* was increasingly recognized as a cause of chronic diarrhoea in patients with AIDS (Molina *et al.* 1995). Besides of chronic diarrhoea it is usually associated with fever, cholangitis, sinusitis, bronchitis, or kerato-conjunctivitis. Most patients have very low CD4+ counts.

The use of light microscopy to diagnose intestinal microsporidiosis in patients with AIDS has been outlined by Rijpstra *et al.* (1988), for example, by Giemsa staining of smears of specimens from duodenal/jejunal biopsies. Electron microscopy confirmation of the minute organisms (1–2 μm microns only) is helpful. Identification in stool is difficult (Orenstein *et al.* 1990). Invasive biopsy procedures may be avoided by PCR on stool specimens (Owen 1997; Liguory *et al.* 1997).

Albendazole – a broad-spectrum antihelmintic drug – may be a useful palliative treatment for microsporidial diarrhoea (Blanshard *et al.* 1992); however, only for *E. intestinalis*

(Molina *et al.* 1995), while no specific therapy is available for *E. bieneusi* (Goodgame 1996). Thalidomide may be useful in intractable cases of diarrhoea (Sharpstone *et al.* 1995).

Other genera of microsporidia, *Encephalitozoon cuniculi* and *Encephalitozoon hellem* have also been reported in AIDS patients with keratitis and sinusitis (Garcia and Bruckner 1993). HAART may effectively clear microsporidiosis (Martins *et al.* 2001).

Other parasites rarely encountered in HIV-infected individuals in the North

In contrast to southern Europe and other subtropical areas, AIDS patients in northern Europe rarely acquire *Leishmania*/HIV co-infection. In Denmark, a single case of untreatable visceral leishmaniasis in an AIDS patient has been described (Balslev *et al.* 1991). In southern Europe, 25–70% of adult visceral leishmaniasis cases are related to HIV infection (WHO 1997).

With increasing mobility of populations, however, co-infection with HIV and remotely acquired parasitic infections should be remembered. Thus, fulminant, systemic strongyloidiasis has been reported in AIDS (Maayan *et al.* 1987; Cahill and Shevchuk 1996).

Serious infections with *Entamoeba histolytica* would be expected to occur in AIDS patients: they do not. *Plasmodium falciparum* malaria might be expected to interact with HIV infection: it hardly does (Lucas 1990). For more details on exotic parasites occurring in – or surprisingly not occurring in – AIDS, the reader is referred to an excellent review by the British Society for Parasitology and the Royal Society of Tropical Medicine and Hygiene (1990).

References

Ahuja, S. K., Ahuja, S. S., Thelmo, W., Seymour, A., and Phelps, K. R. (1993). Necrotizing pancreatitis and multisystem organ failure associated with toxoplasmosis in a patient with AIDS. *Clinical Infectious Diseases*, **16**, 432–434.

Balslev, U., Jonsbo, F., Junge, J., and Bentsen, K. (1991). Three cases of visceral leishmaniasis (kala-azar) one of which were in an HIV-positive man. *Ugeskrift for Læger* (in Danish, with English summary), **153**, 1591–1592.

Bernard, E., Delgiudice, P., Charles, M., Boissy, C., Saint-Paul, M. C., Le Fichoux, Y. *et al.* (1997). Disseminated isosporiasis in an AIDS patient. *European Journal of Clinical Microbiology and Infectious Diseases*, **16**, 699–701.

Blanshard, C., Ellis, D. S., Tovey, D. G., Dowell, S., and Gazzard, B. G. (1992). Treatment of intestinal microsporidiosis with albendazole in patients with AIDS. *AIDS*, **6**, 311–313.

Bozzette, S. A., Sattler, F. R., Chiu, J., Wu, A. W., Gluckstein, D., Kemper, C. *et al.* (1990). A controlled trial of early adjunctive treatment with corticosteroids for *Pneumocystis carinii* pneumonia in the acquired immunodeficiency syndrome. *New England Journal of Medicine*, **323**, 1451–1457.

British Society for Parasitology and Royal Society of Tropical Medicine and Hygiene (1990). Parasitic and other infections in AIDS. *Transactions of the Royal Society of Tropical Medicine and Hygiene*, **84**, Suppl. 1, 1–39.

Broaddus, C., Dake, M. D., Stulbarg, M. S., Blumenfeld, W., Hadley, W. K., Golden, J. A. *et al.* (1985). Bronchoalveolar lavage and transbronchial biopsy for the diagnosis of pulmonary infections in the acquired immunodeficiency syndrome. *Annals of Internal Medicine*, **102**, 747–752.

Cahill, K. M. and Shevchuk, M. (1996). Fulminant, systemic strongyloidiasis in AIDS. *Annals of Tropical Medicine and Parasitology*, **90**, 313–318.

Canning, E. U. and Hollister, W. S. (1990). *Enterocytozoon bieneusi* (Microspora): prevalence and pathogenecity in AIDS patients. *Transactions of the Royal Society of Tropical Medicine and Hygiene*, **84**, 181–186.

Clinque, P., Scarpellini, P., Vago, L., Linde, A., and Lazzarin, A. (1997). Diagnosis of central nervous system complications in HIV-infected patients: cerebrospinal fluid analysis by the polymerase chain reaction. *AIDS*, **11**, 1–17.

Cotte, L., Rabodonirina, M., Piens, M. A., Perreard, M., Mojon, M., and Trepo, C. (1993). Prevalence of intestinal protozoans in French patients infected with HIV. *Journal of the Acquired Immunodeficiency Syndrome*, **6**, 1024–1029.

DeHovitz, J. A., Pape, J. W., Boney, M., and Johnson, W. A. (1986). Clinical manifestations and therapy of *Isospora belli* infection in patients with the acquired immunodeficiency syndrome. *New England Journal of Medicine*, **315**, 87–90.

Doumbo, O., Rossignol, J. F., Pichard, E., Traore, H. A., Dembele, M., Diakili, M. *et al.* (1997). Nitazoxanide in the treatment of cryptosporidial diarrhoea and other intestinal parasitic infections associated with acquired immunodeficiency syndrome in tropical Africa. *American Journal of Tropical Medicine and Hygiene*, **56**, 637–639.

Fichtenbaum, C. J., Zakin, R., Feinberg, J., Benson, C., Griffiths, J. K., and AIDS Clinical Trials Group (2001). Rifabutin but not clarithromycin prevents cryptosporidiosis in persons with advanced HIV infection. *AIDS*, **15**, 2889–2893.

Gallant, J. E., Moore, R. D., and Chaisson, R. E. (1994). Prophylaxis for opportunistic infections in patients with HIV infection. *Annals of Internal Medicine*, **120**, 932–944.

Garcia, L. S. and Bruckner, D. A. (1993). *Diagnostic Medical Parasitology*, 2nd edn, pp. 63–69, 102–103. Washington, DC: American Society for Microbiology.

Gerstoft, J., Malchow-Møller, A., Bygbjerg, I., Dickmeiss, E., Enk, C., Halberg, P. *et al.* (1982). Severe acquired immunodeficiency in European homosexual men. *British Medical Journal*, **285**, 17–19.

González-Ruiz, A. and Bendall, R. P. (1995). Size matters: the use of the ocular micrometer in diagnostic parasitology. *Parasitology Today*, **11**, 83–85.

Goodgame, R. W. (1996). Understanding intestinal spore-forming protozoa: cryptosporidia, microsporidia, isospora, and cyclospora. *Annals of Internal Medicine*, **124**, 429–441.

Gottlieb, M. S., Schroff, R., Schanker, H. M., Weisman, J. D., Fan, P. T., Wolf, R. A. *et al.* (1981). *Pneumocystis carinii* pneumonia and mucosal candidiasis in previously healthy homosexual men: evidence of a new acquired cellular immunodeficiency. *New England Journal of Medicine*, **305**, 1425–1431.

Helweg-Larsen, J., Benfield, T. L., Eugen-Olsen, J., Lundgren, J. D., and Lundgren, B. (1999). Effects of mutations in *Pneumocystis carinii* dihydropteroate synthase gene on outcome of AIDS-associated *P. carinii* pneumonia. *Lancet*, **354**, 1318–1319.

Henriksen, S. A. and Pohlenz, J. F. H. (1981). Staining of cryptosporidium by a modified Ziehl-Neelsen technique. *Acta Veterinaria Scandinavica*, **22**, 594–596.

Hofmann, B., Ødum, N., Platz, P., Ryder, L. P., Svejgaard, A., Nielsen, P. B. *et al.* (1985). Humoral responses to *Pneumocystis carinii* in patients with acquired immunodeficiency syndrome and in immunocompromized homosexual men. *Journal of Infectious Diseases*, **152**, 838–840.

Højlyng, N. and Jensen, B. N. (1988). Respiratory cryptosporidiosis in HIV-positive patients. *Lancet*, **i**, 590–591.

Højlyng, N., Nielsen, A., Wandall, J., Blom, J., Mølbak, K., Chauhan, D. *et al.* (1993). First cases of microsporidiosis in Scandinavian patients with AIDS. *Scandinavian Journal of Infectious Diseases*, **25**, 667–669.

Ives, N. J., Gazzard, B. G., and Easterbrook, P. J. (2001). The changing pattern of AIDS-defining illnesses with the introduction of highly effective antiretroviral therapy (HAART) in a London clinic. *Journal of Infections*, **42**, 134–139.

Jacobs, F., Depierreux, M., Goldman, M., Hall, M., Liesnard, C., Janssen, F., Touissant, C., and Thys, J. P. (1991). Role of bronchoalveolar lavage in diagnosis of disseminated toxoplasmosis. *Reviews of Infectious Diseases*, **13**, 637–641.

Kirk, O., Lundgren, J. D., Pedersen, C., Nielsen, H., and Gerstoft, J. (1999). Can chemoprophylaxis against opportunistic infections be discontinued after an increase in CD4-cells induced by highly active antiretroviral therapy? *AIDS*, **13**, 1647–1651.

Kovacs, J. A. (1994). Advances in the therapy of toxoplasmosis. *Annals of Internal Medicine*, **120**, 947–949.

Lebech, M., Lebech, A. M., Nelsing, S., Vuust, J., Mathiesen, L., and Petersen, E. (1992). Detection of *Toxoplasma gondii* DNA by polymerase chain reaction in cerebrospinal fluid from AIDS patients with cerebral toxoplasmosis. *Journal of Infectious Diseases*, **165**, 982–983.

Leoung, G. S. and Hopewell, P. C. (1990a) *Pneumocystis carinii* pneumonia: clinical presentation and diagnosis. In: *The AIDS Knowledge Base*, P. T. Cohen, M. A. Sande and P. A. Volberding (Eds) pp. 6.5.2.1–6.5.2.6. Waltham, MA: The Medical Publishing Group.

Leoung, G. S. and Hopewell, P. C. (1990b) *Pneumocystis carinii* pneumonia: diagnostic tissue examination and diagnostic algorithm. In: *The Aids Knowledge Base*, P. T. Cohen, M. A. Sande and P. A. Volberding (Eds). Waltham, MA: The Medical Publishing Group. pp. 6.5.3.1–6.5.3.4.

Lidman, C., Örtqvist, Å., Lundbergh, P., Julander, I., and Bergdahl, S. (1989). *Pneumocystis carinii* pneumonia in Stockholm, Sweden: treatment, outcome, one-year-follow-up and pyrimethamine prophylaxis. *Scandinavian Journal of Infectious Diseases*, **21**, 381–387.

Liguory, O., David, F., Sarfati, C., Schuitema, A. R. J., Hartskeerl, R. A., Deronin, F. *et al.* (1997). Diagnosis of infections caused by *Enterozytozoon bieneusi* and *Encephalocytozoon intestinalis* using polymerase chain reaction in stool specimens. *AIDS*, **11**, 723–726.

Lipschik, G. Y., Gill, V. J., Lundgren, J. D., Andramis, V. A., Nelsen, N. A., Nielsen, J. O. *et al.* (1992). Improved diagnosis of *Pneumocystis carinii* infection by polymerase chain reaction on induced sputum and blood. *Lancet*, **340**, 203–206.

Lucas, S. B. (1990). Missing infections in AIDS. *Transactions of the Royal Society of Tropical Medicine and Hygiene*, **84**, Suppl. 1, 49–54.

Luft, B. J., Conley, F., and Remington, J. S. (1983). Outbreak of central-nervous-system toxoplasmosis in Western Europe and North America. *Lancet*, **i**, 781–784.

Lundgren, B., Lundgren, J. D., Nielsen, T., Mathiesen, L., Nielsen, J. O., and Kovacs, J. A. (1992a) Antibody responses to a major *Pneumocystis carinii* antigen in human immunodeficiency virus-infected patients with and without *P. carinii* pneumonia. *Journal of Infectious Diseases*, **165**, 1151–1155.

Lundgren, B., Kovacs, J. A., Nelson, N. N., Stock, F., Martinez, A., and Gill, V. J. (1992b) *Pneumocystis carinii* and specific fungi have a common epitope, identified by a monoclonal antibody. *Journal of Clinical Microbiology*, **30**, 391–395.

Maayan, S., Wormser, G. P., Widerhorn, J., Sy, E. R., Kim, J. H., and Ernst, J. A. (1987). *Strongyloides stercoralis* hyperinfection in a patient with acquired immune deficiency syndrome. *American Journal of Medicine*, **83**, 945–948.

Martins, S. A., Muccioli, C., Belfort, R., Jr., and Castelo, A. (2001). Resolution of microsporidial keratoconjunctivitis in an AIDS patient treated with highly active antiretroviral therapy. *American Journal of Ophthalmology*, **131**, 378–379.

Masur, H., Ognibene, F. P., Yarchoan, R., Shelhamer, J. H., Baird, B. F., Tranvis, W. *et al.* (1989). CD4 counts as predictors of opportunistic pneumonias in human immunodeficiency virus (HIV) infection. *Annals of Internal Medicine*, **111**, 223–231.

Mathiesen, L. R. and Lundgren, J. D. (1997). Chorioretinitis. In: *HIV og AIDS* (in Danish), J. D. Lundgren, C. Pedersen, J. Gerstoft and J. O. Nielsen (Eds). Copenhagen: Munksgaard. pp. 82–83.

McGowan, J., Hawkins, A. S., and Waller, I. V. D. (1993). The natural history of cryptosporidial diarrhoea in HIV-infected patients. *AIDS*, **7**, 349–354.

Merali, S., Frevert, U., Williams, J. H., Chin, K., Bryan, R., and Clarkson, A. B., Jr. (1999). Continuous axenic cultivation of *Pneumocystis carinii*. *Proceedings of the National Academy of Sciences, USA*, **96**, 2402–2407.

Molina, J.-M., Oksenhendler, E., Beaavais, B., Safarti, C., Jaccard, A., Deronin, F. *et al.* (1995). Disseminated microsporidiosis due to *Septata intestinalis* in patients with AIDS: Clinical features and response to albendazole therapy. *Journal of Infectious Diseases*, **171**, 245–249.

Orenstein, J. M., Zierdt, V., Zierdt, C., and Kotler, D. P. (1990). Identification of spores of *Enterocytozoon bieneusi* in stool and duodenal fluid from AIDS patients. *Lancet*, **336**, 1127–1128.

Ortega, Y. R., Sterling, C. R., Gilman, R. H., Cama, V. A., and Diaz, F. (1993). Cyclospora species – a new protozoan pathogen of humans. *New England Journal of Medicine*, **328**, 1308–1312.

Owen, R. L. (1997). Polymerase chain reaction of stool: a powerful tool for specific diagnosis and epidemiological investigation of enteric microsporidia infections. *AIDS*, **11**, 817–818.

Pape, J. W., Verdier, R.-I. and Johnson, W. D. (1989). Treatment of *Isospora belli* infection in patients with the acquired immunodeficiency syndrome. *New England Journal of Medicine*, **320**, 1044–1047.

Pape, J. W., Verdier, R.-I., Boney, M., Boney, J., and Johnson, W. D. (1994). Cyclospora infection in adults infected with HIV. Clinical manifestations, treatment, and prophylaxis. *Annals of Internal Medicine*, **121**, 654–657.

Payne, P., Lancaster, L. A., Heinzman, M., and Mc Cutham, J. A. (1983). Identification of cryptosporidium in patients with the acquired immunodeficiency syndrome. *New England Journal of Medicine*, **309**, 613–614.

Pedersen, C., Lundgren, J. D., Nielsen, T., and Andersen, W. H. (1989). The outcome of *Pneumocystis carinii* pneumonia in Danish patients with AIDS. *Scandinavian Journal of Infectious Diseases*, **21**, 375–380.

Pomeroy, C. and Filice, G. A. (1992). Pulmonary toxoplasmosis: a review. *Clinical Infectious Diseases*, **14**, 863–870.

Potasman, I., Resnick, L., Luft, B. J., and Remington, J. S. (1988). Intrathecal production of antibodies against *Toxoplasma gondii* in patients with toxoplasmic encephalitis and the acquired immunodeficiency syndrome (AIDS). *Annals of Internal Medicine*, **108**, 49–51.

Pugin, J., Vanhems, P., Hirschel, B., Chave, J.-P., and Flepp, M. (1992). Extreme elevations of serum lactic dehydrogenase differentiating pulmonary toxoplasmosis from pneumocystis pneumonia. *New England Journal of Medicine*, **325**, 1226.

Ravn, P., Lundgren, J. D., Kjaedegaard, P., Hollin-Andersen, W., Højlyng, N., Nielsen, J.O. *et al.* (1991). Nosocomial outbreak of cryptosporidiosis in AIDS patients. *British Medical Journal*, **302**, 277–280.

Rijpstra, A. C., Canning, E. U., Van Ketel, R. J., Schattenkerk, J. K. M. E., and Laarinan, J. J. (1988). Use of light microscopy to diagnose small-intestinal microsporidiosis in patients with AIDS. *Journal of Infectious Diseases*, **157**, 827–831.

Saxon, A. and Weinstein, W. (1987). Oral administration of bovine colostrum anti-cryptosporidia antibody fails to alter the course of human cryptosporidiosis. *Journal of Parasitology*, **73**, 413–415.

Schattenkerk, J. K. M. E., Van Gool, T., Van Ketel, R. J., Bartelsman, J. F. W. M., Kviken, C. L., Terpstra, W. J. *et al.* (1991). Clinical significance of small-intestinal microsporidiosis in HIV-1-infected individuals. *Lancet*, **337**, 895–898.

Schneider, M. M., Borleffs, J. C., Stolk, R. P., Jaspers, C. A., and Hoepelman, A. I. (1999). Discontinuation of prophylaxis for *Pneumocystis carinii* pneumonia in HIV-1-infected patients treated with highly active antiretroviral therapy. *Lancet*, **353**, 201–203.

Schmidt, W., Wahnschaffe, V., Schafer, M., Zippel, T., Arvand, M., Meyerhans, A. *et al.* (2001). Rapid increase of mucosal CD4 T cells followed by clearance of intestinal cryptosporidiosis in an AIDS patient receiving highly active antiretroviral therapy. *Gastroenterology*, **120**, 984–987.

Schubach, T. M., Neves, E. S., Leite, A. C., Aranjo; A. Q. C., and de Moura, H. (1997). *Cyclospora cayetanensis* in an asymptomatic patient infected with HIV and HTLV-1. *Transactions of the Royal Society of Tropical Medicine and Hygiene*, **91**, 175.

Sharpstone, D., Rowbottom, A., Nelson, M., and Gazzard, B. (1995). The treatment of microsporidial diarrhoea with thalidomide. *AIDS*, **9**, 658–659.

Smith, E. and Orholm, M. (1990). Trends and patterns of opportunistic diseases in Danish AIDS patients 1980–1990. *Scandinavian Journal of Infectious Diseases*, **22**, 665–672.

Smith, E., Pers, C., Aschow, C., and Mathiesen, L. (1991). Cerebral toxoplasmosis in Danish AIDS patients. *Scandinavian Journal of Infectious Diseases*, **23**, 703–709.

Sorvillo, F. J., Lieb, L. E., Seidel, J., Kerndt, P., Turner, J., and Ash, L. K. (1995). Epidemiology of isosporiasis among persons with acquired immunodeficiency syndrome in Los Angeles County. *American Journal of Tropical Medicine and Hygiene*, **53**, 656–659.

Stringer, S. L., Stringer, J. R., Blase, M. A., Walzer, P. D., and Cushion, M. T. (1989). *Pneumocystis carinii*: sequence from ribosomal RNA implies a close relationship with fungi. *Experimental Parasitology*, **68**, 450–461.

Verbraak, F. D., Galema, M., van den Horn, G. H., Bruinenberg, M., Luyendijk, L., Danner, S. A. *et al.* (1996). Serological and polymerase chain reaction-based analysis of aqueous humour samples in patients with AIDS and necrotizing retinitis. *AIDS*, **10**, 1091–1099.

Verdier, R. I., Fitzgerald, D. W., Johnson, W. D., Jr., and Pape, J. W. (2000). Trimethoprim-sulfamethoxazole compared with ciprofloxacin for treatment and prophylaxis of Isospora belli and Cyclospora cayetanensis infection in HIV-infected patients. A randomized, controlled trial. *Annals of Internal Medicine*, **132**, 885–888.

Wagner, D., Königer, J., Kern, W. V., and Kern, P. (1997). Serum PCR of *Pneumocystis carinii* DNA in immunocompromised patients. *Scandinavian Journal of Infectious Diseases*, **29**, 159–164.

White, A. C., Chappel, C. L., Hayat, C. S., Kimball, K. T., Flanigan, T. P., and Goodgame, R. W. (1994). Paromomycin for cryptosporidiosis in AIDS: a prospective double-blind trial. *Journal of Infectious Diseases*, **170**, 419–424.

Witt, K., Nielsen, T. N., and Junge, J. (1991). Dissemination of *Pneumocystis carinii* in patients with AIDS. *Scandinavian Journal of Infectious Diseases*, **23**, 691–695.

World Health Organization (WHO) (1997). *Leishmania*/HIV co-infection. *Weekly Epidemiological Record*, **72**, 49–54.

28 Parasite infections in transplant patients

Jan Andersson and Elda Sparrelid

Introduction

Infections in transplant patients are predominantly due to pharmacologically impaired innate and adaptive immune responses. Therefore these patients suffer from infections caused by intracellular micro-organisms, while parasitic infections do not pose predominant problems. In the Nordic countries the dominating parasitic infections in transplant patients include toxoplasmosis, a limited number of imported cases of malaria, and some gastrointestinal parasite infection such as cryptosporidiosis.

Toxoplasmosis in transplant patients

Toxoplasmosis is an uncommon cause of disease in transplanted patients in the Nordic countries. Historically, it has been a problem in recipients of heart transplants and in patients with combined heart and lung transplantations. International studies show an incidence between 6 and 10% of all heart transplanted patients. The sero-prevalence in immunocompetent adults in Sweden is essentially around 15%. The vast majority of these patients who contract toxoplasmosis reject the transplanted organ despite the fact that reactivation of latent infection is a possibility. When a sero-negative recipient receives an organ from a sero-positive donor, the risk of developing primary toxolasmosis has been estimated to be approximately 50%. The clinical picture in *Toxoplasma* differs in patients transplanted with different organs. The symptoms in these patients are often nonspecific which makes *Toxoplasma* diagnosis difficult. For instance, myocarditis due to *Toxoplasma* can be interpreted as rejection. This mistake may lead to an increase in immunosuppressive therapy followed by an increased risk for dissemination of the infection from the heart to the central nervous system (CNS). More than 50% of patients with clinical manifestations of toxoplasmosis indeed have intracerebral infection. The symptoms vary and can be focal or nonfocal. Changes in mental status occurs in 75%, fever in 10–70%, epilepsy in 30%, headache in 50%, and focal neurological signs in 60% of patients with neurological infection.

Diagnosis

Diagnosis of *Toxoplasma* infection can be established by the detection of IgM- and IgG-antibodies in the serum. Detection of IgA positivity in serum is common during acute infection. After primary infection the specific IgG response persists. This may sometimes make it difficult to diagnose toxoplasmosis only by serology. Polymerase chain reaction diagnostic methods in peripheral leukocytes has recently been developed (and described elsewhere in this book). The sensitivity is, though, sometimes lower than the serological diagnostic, and in several studies it has been reported to vary between 70 and 80%. *Toxoplasma* trophozoites can be found histologically by using antigen-specific staining methods. In CNS infection, computerized tomography scan with intravenous contrast is an important instrument for diagnosis. The typical picture includes multiple lesions surrounded by contrast enhancement. Pneumonia due to *Toxoplasma* in these patients is often atypical with bilateral infiltrates of the lungs.

Most cases of pneumonia can be diagnosed by antigen detection in bronchoalveolar lavage (BAL), and myocarditis by specific antigen staining of histology sections. *Toxoplasma gondii* trophozoites, pseudocysts, and cysts can also be visualized in hematoxylin and eosin or Giemsa staining, both in BAL and in heart biopsies.

Treatment

The drugs of choice in the treatment of toxoplasmosis are pyrimethamine, initially 200 mg followed by 50–75 mg daily, combined with sulfadiazin 4–6 g daily. Leucovorin should be used to reduce the risk for bone marrow toxicity of pyrimethamine.

Prophylaxis

Cotrimoxazole prophylaxis during the first six weeks after transplantation is generally recommended in sero-negative transplanted patients receiving organs from sero-positive donors, especially those receiving heart or heart/lung transplants. This approach has led to a significant decline of clinical toxoplasmosis.

Toxoplasma infection in liver-, kidney-, or bone marrow-transplanted patients is unusual. The incidence of *Toxoplasma* infection in Sweden is below 1% in these groups of patients. Normally we do not perform *Toxoplasma* serology assessments, with the exception of heart and lung/heart donors and recipients. It is worth noting that all patients receive low-dose cotrimoxazole as prophylaxis against *Pneumocystis carinii* pneumonia during the first six months after transplantation, which may also have a prophylactic effect against *Toxoplasma*. A general recommendation to transplanted patients is that they should avoid eating raw flesh. In addition, contact with cats should be avoided due to the fact that they are the major reservoir of toxoplasmosis. The risk of being infected with *Toxoplasma* declines after the transplantation. Maximal immunosuppressive therapy is in general given during the first 1–3 months after the transplantation. Thereafter the immunosuppression is reduced to significantly lower levels.

Malaria and other imported parasitic infections in transplant patients

The classical clinical picture of malaria infection, including shivering, thrombocytopenia, and gastrointestinal symptoms, usually does not differ in organ-transplanted patients compared to immunocompetent individuals. A few cases of malarial infection following transplantation have been reported from the Nordic countries. All these cases have been patients from Nordic countries who have undergone transplantation with organs of donors from malaria-endemic regions. The diagnosis and treatment of malaria in transplanted patients are not different from that of immunocompetent individuals. This infection should always be considered when patients are transplanted in malaria-endemic regions. In addition, malaria should also be considered in individuals who have received blood transfusions in foreign countries where malaria may persist. There are no reports in the Nordic countries about increased incidence of *Giardia*, *Amoeba*, *Cryptosporidium*, or *Microsporidium* in organ-transplanted patients who have been visiting areas with increased risk for those parasitic infections.

References

Andersson, R., Sandberg, T., Berglin, E., and Jeansson, S. (1992). Cytomegalovirus infections and toxoplasmosis in heart transplant recipients in Sweden. *Scand J Infect Dis* **24**, 411–417.

Aubert, D., Foudrinier, F., Villena, I., Pinon, J. M., Biava, M. F., and Renoult, E. (1996). PCR for diagnosis and follow-up of two cases of disseminated toxoplasmosis after kidney grafting. *J Clin Microbiol* **34**, 1347.

Gallino, A., Maggiorini, M., Kiowski, W., Martin, X., Wunderli, W., Schneider, J., Turina, M., and Follath, F. (1996). Toxoplasmosis in heart transplant recipients. *Eur J Clin Microbiol Infect Dis* **15**, 389–393.

Gordon, S. M., Gal, A. A., Hertzler, G. L., Bryan, J. A., Perlino, C., and Kanter, K. R. (1993). Diagnosis of pulmonary toxoplasmosis by bronchoalveolar lavage in cardiac transplant recipients. *Diagn Cytopathol* **9**, 650–654.

Holliman, R. E., Johnson, J. D., Adams, S., and Pepper, J. R. (1991). Toxoplasmosis and heart transplantation. *J Heart Lung Transplant* **10**, 608–610.

Mayes, J. T., O'Connor, B. J., Avery, R., Casteilani, W., and Carey, W. (1995). Transmission of *Toxoplasma gondii* infection by liver transplantation. *Clin Infect Dis* **21**, 511–515.

Orr, K. E., Gould, F. K., Short, G., Dark, J. H., Hilton, C. J., Corns, P. A., and Freeman, R. (1994). Outcome of *Toxoplasma gondii* mismatches in heart transplant recipients over a period of 8 years. *J Infect* **29**, 249–253.

Renoult, E., Georges, E., Biava, M. F., Hulin, C., Frimat, L., Hestin, D., and Kessler, M. (1997). Toxoplasmosis in kidney transplant recipients: report of six cases and review. *Clin Infect Dis* **24**, 625–634.

Wreghitt, T. G., Gray, J. J., Pavel, P., Balfour, A., Fabbri, A., Sharples, L. D., and Waliwork, J. (1992). Efficacy of pyrimethamine for the prevention of donor-acquired *Toxoplasma gondii* infection in heart and heart-lung transplant patients. *Transplant Int* **5**, 197–200.

Wreghitt, T. G., Hakim, M., Gray, J. I., Baifour, A. H., Stovin, P. G., Stewart, S., Scott, J., English, T. A., and WaUwork, J. (1989). Toxoplasmosis in heart and heart and lung transplant recipients. *J Clin Pathol* **42**, 194–199.

Wreghitt, T. G., McNeil, K., Roth, C., Wallwork, J., McKee, T., and Parameshwar, J. (1995). Antibiotic prophylaxis for the prevention of donor-acquired *Toxoplasma gondii* infection in transplant patients. *J Infect* **31**, 253–254.

Part 5

Laboratory diagnostics methods

29 PCR-based diagnosis of *Toxoplasma gondii* infections

Experience from a low-prevalence area

Nancy P. Nenonen, Tomas Bergström, and Sigvard Olofsson

Polymerase chain reaction (PCR) technology has provided the clinician with a new tool, enabling rapid and improved, early and adequate diagnosis of *Toxoplasma gondii* infections. We report on the use of PCR-based diagnosis of *T. gondii* infections in the Göteborg region, a typical low-prevalence area. The PCR-based diagnosis for toxoplasma infections in Göteborg was introduced in 1993 and to date (December 2001) 842 clinical specimens have been analysed using primers based on the P30 (*SAG1*) gene. In our hands the sensitivity of the P30 system approached that obtained by using primers targeting the B1 region. The frequency of toxoplasma PCR positive samples was 1.2%, representing cases of congenital toxoplasmosis, toxoplasmosis during pregnancy, toxoplasmosis in immunocompromised patients, and retinochoroiditis. Based on reports using different PCR systems for toxoplasma diagnosis, we recommend that standardization programmes of Nordic PCR-based diagnosis of *T. gondii* infections should focus on the exchange of specimens and standards rather than striving towards a consensus PCR protocol.

Introduction

As in many other fields of laboratory diagnosis of microbiological agents, the introduction of PCR technology has provided the clinician with a new tool, enabling rapid and improved, early and adequate diagnosis of *T. gondii* infections. Despite the comparatively low prevalence of this parasite in the Nordic countries, the increased number of immunocompromised patients in medical care has extended the demand for rapid and reliable diagnosis of reactivated infection. As an example, the PCR-based diagnosis of *T. gondii* encephalitis on samples from cerebrospinal fluid (Ostergaard *et al.* 1993) was found to be the most useful test in comparison to conventional serology and antigen detection (van de Ven E *et al.* 1991). Furthermore, retinitis caused by *T. gondii* may be diagnosed by PCR assay of aqueous humour in HIV-infected patients (Danise *et al.* 1997). One caution that should be raised concerns the difficulties of interpretation of negative PCR results obtained after initiation of treatment of the infection, as was shown in a study of generalized toxoplasma infections in HIV patients (Foudrinier *et al.* 1996).

Diagnosis of congenital infections caused by *T. gondii* poses special problems, especially where transmission from the mother is the result of subclinical infection (Thulliez *et al.* 1992). Improved diagnosis is of particular importance in the newborn child, where early and safe treatment may reduce the sequelae of congenital toxoplasmosis (Thulliez *et al.* 1992). In this field, PCR has proven superior to other diagnostic methods in a large study of prenatal infections (Hohlfeld *et al.* 1994). The burden for the clinician of difficult decisions

Table 29.1 Examples of sample materials (X) selected by clinicians for PCR-based diagnosis of T. gondii infection in different clinical conditions

Diagnosis	Sample material				
	PBMC	CSF	Amniotic fluid	Biopsies	Vitreous body
Generalized infections	X			X	
Encephalitis	X	X			
Retinitis					X
Prenatal infection			X		
Congenital infection	X	X			
PCR positive/number tested[a]	5/197	3/358	0/31	1/55	2/40

Note

a Samples were referred to the laboratory for investigation of patients with clinical signs and symptoms of *T. gondii* infection.

concerning termination of pregnancy or prenatal treatment is somewhat relieved by the increased possibility of reliable diagnosis in these cases (Hohlfeld *et al.* 1994).

PCR methodology has recently been further improved by the introduction of conventional or real time quantitative PCR analysis of *T. gondii* DNA (Luo *et al.* 1997; Lin *et al.* 2000; Costa *et al.* 2001). Consistently negative findings in large sample materials from control groups suggest that positive PCR results are of clinical relevance. However, it must be borne in mind that the presence of tissue cysts in very small numbers may be due to latent rather than active infection (Ruskin and Remington 1976). Of AIDS patients who are toxoplasma-seropositive only around 30% will experience reactivation leading to severe disease (Luft and Remington 1992). It is conceivable that the passenger nature of the organism could mislead, and careful interpretation of laboratory results is required. Further diagnostic development will depend on a continuing communication between the clinician and the microbiologist concerning the relevance of PCR findings from different sample material. Examples of body fluids and specimen materials currently being used for PCR-based *T. gondii* diagnosis are given in Table 29.1.

Clinical presentation and severity of disease in the individual may vary due to factors associated with the host, and with the genotype of the infecting parasite. Despite the existence of a well-described sexual cycle in cats, *T. gondii* appears to reproduce largely clonally in nature, recombination being a rarer event. PCR-restriction polymorphism (PCR-RFLP) analyses at multiple loci indicate that three distinct lineages dominate (Howe and Sibley 1995). Although only limited data are available on strain type and infection in humans (Honore *et al.* 2000; Grigg *et al.* 2001b), recent reports based on direct PCR-RFLP of patient samples found no clear correlation between genotype, symptoms, and severity of disease (Fuentes *et al.* 2001). Direct PCR-RFLP analysis may provide new perspectives for investigation of *T. gondii* infections, and eventual refinement of diagnostic methods. However, the immediate concern remains provision of a rapid and reliable detection, and identification, of the causative organism, *T. gondii*. In what is often a potentially blinding, or fatal infection that is amenable to therapy, rapid identification of *T. gondii* can be crucial.

Technical background

One major problem associated with the laboratory diagnosis of *T. gondii* infections is that a clinical specimen containing only a few tachyzoites can reflect a life-threatening infection.

Thus, PCR-based methods for laboratory diagnosis of *T. gondii* infection should be able to detect literally single copies of the genome in such specimens, and several PCR systems with adequate sensitivity have been published (Burg *et al.* 1989; Weiss *et al.* 1991). To accomplish this, at least three genes of the *T. gondii* genome have been evaluated as target sequences for PCR, namely the B1, P30, and 18S rDNA genes (Jones *et al.* 2000).

Most systems have acceptable sensitivity and specificity, and one example of a successful PCR method is that based on the gene encoding the P30 surface membrane antigen (Savva *et al.* 1990; Weiss *et al.* 1991). The probability of detection of *T. gondii* should be increased if regions of DNA containing repeat sequences are used. As such, the B1-gene, containing 35 tandem repeats (Burg *et al.* 1989; Ho *et al.* 1992), and the ribosomal RNA-gene (Cazenave *et al.* 1992) with 110 repeats (Guay *et al.* 1993), offer a theoretical limit of 0.01 tachyzoite of *T. gondii* per sample assay (Macpherson and Gajadhar 1993). These detection systems do provide very high potential sensitivity. However as discussed below, there are some advantages in the P30 PCR method compared with the methods employing repeat DNA sequences as target, despite the inherent high theoretical sensitivity of the latter.

Developments of the PCR-based diagnosis of *T. gondii* infections now include scoring of the pathogenic properties of different *T. gondii* strains (Guo *et al.* 1997). One approach to identifying *T. gondii* strains associated with human pathogenicity was based on short term cell culture propagation and subsequent PCR, using the surface antigen *SAG2* locus (Howe *et al.* 1997). Rapid identification and discrimination of *T. gondii* types have been achieved by multilocus PCR-RFLP of *SAG3*, and *SAG4* (single copy surface antigen genes) as well as *SAG1*, *SAG2*, and B1 genes (Grigg *et al.* 2001a,b). Although chronic infections in humans and animals are said to be associated predominantly with the mouse avirulent types II or III, correlation between clinical disease, pathogenicity, and genotype in human infections is still under study (Honore *et al.* 2000; Fuentes *et al.* 2001; Grigg *et al.* 2001b). Genotyping with PCR-RFLP methods may provide new approaches for investigation and diagnosis of *T. gondii* infections, but the significance of their contribution to the laboratory diagnosis remains to be determined.

Here, we present a PCR system using the P30 gene as the target sequence, which according to our experience combines good sensitivity, 25–100 fg of *T. gondii* DNA, corresponding to one or less tachyzoite, and high specificity.

Methods

Sample preparation

a   Serum samples were pre-treated with 0.1 M NaOH for 1 h at 37°C, and thereafter neutralized with 0.1 M HCl, prior to PCR.

b   DNA was prepared from PMBC essentially as described by Landgren *et al.* (1994). After centrifugation at 4 000 × *g* for 5 min the cell pellet was lysed in buffer containing 10 mM Tris HCl (pH 8.4), 1 mM EDTA, 0.5% Tween 20, 0.5% IGEPAL (Sigma) and 400 μg/ml proteinase K for 1 h at 55°C, followed by heat inactivation at 95°C for 10 min. DNA was isolated by ethanol precipitation in the presence of salt at −20°C, washed with ethanol and acetone, and subsequently solubilized in DNAse free water.

c   Amniotic fluid was centrifuged, 5 min 5000×*g*, and the cell deposit separated from supernatant. Cell pellet was digested by treatment with proteinase K in lysis buffer prior to DNA precipitation as described for isolation of DNA from PBMC. The supernatant fluid was subjected to alkali treatment as outlined for serum samples.

d   Broncheoalveolar lavage (BAL) specimens were centrifuged as above, the cell pellets separated from supernatant, and each treated as described for amniotic fluid.

e   CSF was subjected to alkali denaturation, followed by neutralization as described for serum samples.

f   Vitreous body was treated with proteinase K in lysis buffer and the DNA precipitated as described previously, whilst vitreous humour was treated with alkali, and subsequently neutralized as described.

g   Biopsy tissues were finely chopped with a scalpel prior to digestion with proteinase K (400 µg/ml) in a lysis buffer volume of 1 ml. DNA was prepared as described previously. Where necessary gross blood contamination was removed by gentle washing of the specimen with buffered saline prior to chopping. In particular, lysates of brain tissue were extracted with phenol and chloroform prior to DNA preparation.

h   Placental tissue was prepared as described above for biopsies.

Controls

Positive control DNA was obtained from two sources: tachyzoites purified from *T. gondii* (RH strain)-infected mice, as described by Weiss *et al.* (1991), and from a cell-cultured RH strain. These DNA controls were prepared as serial dilutions in poly(dA-dT) nucleotide (20 µg/ml). Standard parasite counts subjected to freeze-thawing and subsequent dilution were also used to determine the sensitivity of the detection system.

Amplification procedures

Presence of *T. gondii* DNA was detected in a nested PCR system, where the outer primers 5'-CACACGGTTGTATGTCGGTTTCGCT-3' (forward, nt 149–172; Gene Bank Accession Number S63900) and 5'-TCAAGGAGCTCAATGTTACAGCCT-3' (reverse, nt 520–497), were designed by Weiss *et al.* (1991), from the published P30 sequence (Burg *et al.* 1988). This system gave an outer product of 372 base pairs. The inner nesting primers were 5'-TGACGAGTATGTTTCCGAAGGC-3' (corresponding to nt 203–224) and 5'-TGGGCAGATTTGCCTGTTGGGT-3' (nt 468–447), amplifying a region of 266 base pairs. The specificity of the method was ascertained: (i) by using material from autopsies and PMBC of *T. gondii*-infected and uninfected mice, as described (Weiss *et al.* 1991); (ii) by cross-checking with our standard DNA positive controls for CMV, HSV, EBV, and VZV PCR (reviewed by Bergström *et al.* 1995); and (iii) by oligonucleotide hybridization of the PCR products using the 5'-biotin labelled probe (5'-biotin-CTCACACCGACGGAGAAC-CACTTC3') (nt 370–393) described by Weiss *et al.* (1991).

Master mixes were prepared using the thermostable DNA polymerase (AmpliTaq, Applied Biosystems) at a concentration of one unit TAQ per reaction mix of 50 µl, 10 mM Tris HCl (pH 8.3 at 20°C), 1.5 mM $MgCl_2$, 50 mM KCl, 200 µM each dNTP, and with 0.2 µM of the forward and reverse primers in the outer or inner PCR reaction mixes.

Sample DNA was added to the reaction mixes as 5, 8, or 10 µl of H_2O-solubilized, concentrated DNA. In the case of placenta, biopsies, and brain tissue, DNA preparations were also tested in dilution to reduce problems of inhibition or DNA overload. To detect and avoid the problems of cross-contamination between specimens, water controls were set between each sample preparation and precautions taken as described by Kwok and Higuchi (1989). Target DNA was amplified in the 9 600 thermocycler (Applied Biosystems) with an initial cycle programme of 30 cycles followed by a nesting round of 40 cycles. In the first

PCR, a denaturation step of 94°C for 2 min 30 s was followed by a 30 cycle programme of denaturation at 94°C for 15 s, primer annealing at 65°C for 10 s, and primer elongation at 72°C for 15 s, with a time extension at 72°C of 1 s per cycle. Five microlitres of the first-round PCR were 'nested' into the inner primer system mixes and subjected to similar cycling conditions, with the exception that amplification was continued for 40 cycles. PCR products were detected by gel electrophoresis of 15 μl of reaction mix on a 1.8% agarose with ethidium bromide (0.2 μg/ml) giving fluorescent staining of nucleic acid amplicons on exposure to UV light.

To define the problem of false negative PCR reactions caused by inhibitors present in the sample DNA, for example, iron from haemoglobin, repeat PCR reactions were performed using β-globin primers and specimen DNA. 'Spiking' of repeat toxoplasma primer PCR mixes with the specimen DNA sample plus control *T. gondii* positive DNA was also used to detect false negative reactions caused by sample inhibition.

Toxoplasma positive samples were confirmed by hybridization of the amplified products with a specific biotin labelled probe, indicated above (Weiss *et al.* 1991). In some instances, the PCR products were sequenced, using the ABI FS dd NTP D-rhodamine cycle sequencing kit, and the labelled products analysed on the ABI 310 (Applied Biosystems).

Results and discussion

Our experience of the use of PCR for diagnosis of infections with *T. gondii* is that the method has improved the service of the laboratory in this field, and that the detection of *T. gondii*-specific DNA is a valuable complement to our serological assays. Although *T. gondii* infections are rare in the Göteborg area, the clinician often requests diagnostic analyses for this pathogen due to availability of effective treatment.

Performance of P30 and B1 PCR

The current PCR method using the P30 gene as target sequence, detected 25–100 fg *T. gondii* DNA, corresponding to the DNA of one tachyzoite (Cornelissen *et al.* 1984) compared to the 10–100 fg obtained with a typical B1 assay (a slight modification of the system described by Burg *et al.* (1989), performed on the same occasion. These results are in line with the comparison between P30 and B1 PCR made by Wastling *et al.* (1993). The nested P30 PCR product (266 bp) is more easily discernible on agarose gels than the low molecular sized product (90 bp) of the nested B1 PCR. Also, with the given primers and PCR conditions, use of the P30 system avoids the problems of obscuring smears and extra bands, which have been associated with the B1 method. Although disconcerting, the problems of the B1 system are said to be less disturbing in clinical specimen analysis than in studies of conventional positive standards prepared from infected mice (Mats Olsson, personal communication). Our own laboratory results do not agree here, and this difference in binding performance between the two PCR systems probably reflects the lower annealing temperature of most published B1 methods (Jones *et al.* 2000). In one report, the B1 PCR was used for screening purposes whereas P30 PCR was used for confirmation (Hohlfeld *et al.* 1994). When we analysed suspensions with pre-defined numbers of intact tachyzoites it was found that the P30 PCR, as well as a B1 PCR, was able to produce a positive PCR reaction for one or less than one tachyzoite present in the sample volume processed for *T. gondii* detection (Figure 29.1). The P30 PCR method as described here, therefore, combines sensitivity with excellent technical performance.

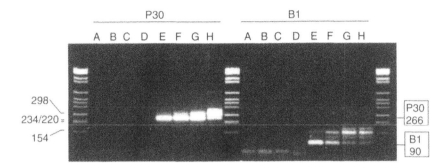

Figure 29.1 Sensitivity of the P30 and B1 toxoplasma PCR systems. Detection of DNA from a known number of parasites inoculated into master mixes as follows: (lane A) master mix only; (B) water; (C) 0.0025 parasites; (D) 0.025 parasites; (E) 0.25 parasites; (F) 2.5 parasites; (G) 25 parasites; (H) 250 parasites per master mix. DNA molecular weight marker, consisting of a mixture of pBR329, cleaved with Bgl I and Hinf I is included. Electrophoresis was run in 1.8% agarose and the gels were stained with ethidium bromide.

Experience of the P30 PCR in routine diagnosis

As in most parts of Scandinavia, the Gothenburg region is considered to be a low-endemic area with respect to *T. gondii* infections (Andersson *et al.* 1992; Bergström *et al.* 1998). The seroprevalence of IgG antibodies to *T. gondii* in blood donors in Göteborg is approximately 10% (unpublished observation). The PCR-based diagnosis of *T. gondii* infections in Göteborg was introduced in 1993 and 842 clinical specimens have been analysed to date (December 2001). The types of specimen received included EDTA blood for PBMC preparation (199 samples), serum (45 samples), vitreous body and fluid (40 samples), cerebrospinal fluid (356 samples), biopsies from various tissues (57 samples), placenta (81 samples), amniotic fluid (32 samples). Various other body fluids were examined such as urine, broncheoalveolar lavage, ascites and pericardial fluid (43 samples). The frequency of toxoplasma PCR positive samples was 1.2%, representing positive samples obtained from the 10 patients presented in Table 29.2. In contrast to PBMC or CSF samples no positive PCR signals were obtained from serum. The rapid laboratory confirmation of *T. gondii* infection was important. Besides identifying the etiological agent for a severe medical condition under investigation, it should be noted that in at least six of these cases, the early PCR diagnosis of *T. gondii* infection was of significance for the clinical management of the patient.

The specificity of the P30 PCR products was confirmed by hybridization of the positive samples obtained from each patient (10 cases). In addition, PCR amplicons from three cases (patient nos. 8, 9, 10; Table 29.2) were control sequenced, and the 375 bp outer products of the P30 target gene were shown to vary in just one nucleotide, nt 293 (Accession no. S85174). At this position, the HIV positive case (patient no.10, from Lebanon) read *GTT* coding for valine, whilst the two congenital cases (patient nos. 8 and 9) sequenced as *GCT* for alanine. GenBank Blast analyses of the P30 sequences derived from the HIV positive case of cerebral toxoplasma indicated that it was similar to the RH strain at nt position 293. The two cases of congenital infection that resulted in neonatal death, gave P30 toxoplasma sequences identical to strains P and/or C described by Bulow and Boothroyd (1991). These findings emphasise the P30 target region as well conserved between the cases studied, despite

Table 29.2 PCR-positive cases with *T. gondii* infections diagnosed in Göteborg 1993–2001[a]

Patient No.	Gender, age	Serology Toxo IgM / IgG	Toxoplasma PCR-positive specimen	Diagnosis	Underlying disease
1	Neonate M, 9 days	Neg/Pos	PBMC	Congenital toxoplasmosis	
2	F, 21 years	Pos/Pos	PBMC	Toxoplasmosis during pregnancy, mother of pat. 1.	
3	M, 14 years	Not available	Liver biopsy	Generalized toxoplasmosis with hepatitis	Liver transplanted
4	M, 39 years	Neg/Pos	Vitreous body	Retinochoroiditis	
5	F, 52 years	Neg/Neg	CSF	Cerebral toxoplasmosis	Liver transplanted
6	F, 57 years	Neg/Pos	Vitreous body	Retinochoroiditis	
7	F, 66 years	Pos/Pos	PBMC	Generalized toxoplasmosis	Malignant lymphoma
8	M, 3 days[†]	Pos/Pos	PBMC	Congenital toxoplasmosis	
9	M, 1 day[†]	No data	PBMC, CSF	Congenital toxoplasmosis	
10	M, 29 years[†]	Neg/Pos	PBMC, CSF	Cerebral toxoplasmosis	Untreated HIV-1 infection

Note
a A total of 842 specimens were analysed.

their widely different geographical sources of infection. Therefore, the P30 primer based PCR provides a good detection system for diagnosis of toxoplasma infections, confirming previous observations.

Access to rapid toxoplasma diagnosis is of special relevance for paediatricians engaged in the diagnosis of congenital infections. This is exemplified by a case of congenital toxoplas-mosis where diagnosis by PCR amplification of *T. gondii* DNA from peripheral blood led to early treatment of the infant, and seemingly normal brain development on examination at age 2 years, despite presence of intra-cranial calcifications at birth (Bergström *et al.* 1998). The laboratory diagnosis of *T. gondii* infection was instigated by findings detected on careful ophthalmic investigation of the neonate. Rapid detection of *T. gondii* DNA in the baby's PBMC prompted immediate early treatment that was continued for 12 months. At 2 years of age, the boy showed normal development for his age, the opalescence of the vitreous body detected at birth had resolved bilaterally, and the chorioretinitic areas were demarcated. The child appeared to have visual impairment in the form of loss of visual fields in at least one eye. The mother, who experienced sub-clinical infection during pregnancy, was PCR-positive for toxoplasma DNA in a sample of peripheral blood taken nine days after delivery. This case of child and mother illustrates the usefulness of the PCR method for diagnosis once the clinical suspicion has been raised, indicating prompt treatment of the infection. Our PCR observations as well as results obtained in Stockholm (Fahnehjelm *et al.* 2000) using PKU cards for neonatal antibody screening, emphasize the importance of careful ophthalmic examination and rapid laboratory diagnosis of congenital toxoplasmosis of neonates, as treatment has been shown to reduce long-term sequelae.

In all, we detected three cases of congenital toxoplasmosis (1993–2001) by examining neonatal PBMC and/or CSF by the toxoplasma specific P30 PCR (Table 29.2). This should be viewed in the light of the Swedish study based on PKU sample testing (Evengard *et al.* 2001), giving a prevalence of 0.7 cases per 10 000 births, and the 3.3 per 10 000 (11 cases) found in Norway (Jenum *et al.* 1998). The two additional cases of congenital toxoplasmosis diagnosed by PCR in the Göteborg region exhibited the classical features of very severe congenital toxoplasma infection and resulted in neonatal death. Although the exposure of pregnant women to toxoplasma is rare in this area, the condition is regarded with informed concern among women attending the maternity clinics. These cases illustrate the significant contribution PCR can make in confirming the diagnosis of congenital toxoplasmosis, or past maternal infection. Speedy confirmation of the diagnosis is very important, not only in prompting rapid management of the baby, but also in providing reassurance for the women that subsequent pregnancies are unlikely to be affected (Chatterton 1992).

In our experience, serology is of limited value in the diagnosis of toxoplasma-induced retinochoroiditis. Levels of IgM antibodies to *T. gondii* are often too low to allow detection, and IgG antibodies also tend to be of low titre. In this field, PCR diagnosis based on samples of vitreous body/vitreous humour seems especially promising. Although the material presented is very small, the finding of two positive samples out of 40 examined (5%) indicates that vitreous body is a valuable test sample in cases of uveitis as has been observed previously (de Boer *et al.* 1996). These two positive findings were probably not due to the irrelevant detection of dormant parasites, since control material, vitreous body specimens taken on surgery in the absence of clinical suspicion of *T. gondii* infection ($n = 10$), was found to be negative (unpublished observation).

The remaining four toxoplasma PCR positive cases in our study were found among immunosuppressed patients. It is interesting to note that following liver transplantation, one patient developed cerebral toxoplasmosis, probably after a primary infection as judged by the seronegative status at onset, and the other transplant case showed signs of toxoplasma-induced hepatitis. One elderly patient with acute myeloic leukaemia was a typical case of generalized toxoplasmosis where prolonged fever had resisted all therapeutic efforts until the PCR analysis of PBMC gave the correct diagnosis. This patient lived in very close contact with many stray cats. It should also be noted that only one case of AIDS-related toxoplasmosis was diagnosed, which may reflect the low prevalence, not only of *T. gondii* but also of HIV infection, in the Göteborg area. In this case, severe cerebral toxoplasmosis was detected in a newly arrived immigrant from the Lebanon. On admission to hospital, the young man was comatose, and despite rapid laboratory diagnosis of HIV positive antibodies, and strongly positive toxoplasma PCR signals in the CSF and PBMC, the patient succumbed to the infection.

These 10 cases describe our experience of the laboratory diagnosis of toxoplasma infection over 10 years. They typify the opportunistic character of *T. gondii* infection in humans, where a subtle shift in the balance between the immune competence of the host and the quiescent passenger state of the protozoa, can result in overwhelming infection. Accurate diagnosis requires alertness in the clinic and the laboratory.

PCR-based diagnosis of Toxoplasma gondii – strategies and tactics

A general problem with *T. gondii* diagnosis is that active disease may be reflected by the presence of only a few tachyzoites in some types of specimens. Although PCR methods facilitate the detection of as little as one or a few parasites present in such specimens, it could be erroneous to rely only on the high sensitivity of the test. First, the Poisson distribution of tachyzoites in a fluid specimen implies that at low tachyzoite densities there is a significant probability that negative results are obtained due to an uneven distribution. Moreover, special care is needed to ensure that the few tachyzoites contained in such specimens are not non-specifically adsorbed or degraded during transport to the laboratory. With this background, a safer strategy would be to include examination of specimens from locations expected to contain higher concentrations of *T. gondii*, such as lymph nodes, or to analyse several portions, preferably over a time interval, from easily obtained body fluids such as PBMC, before ruling out the diagnosis. As these infections may persist and cause symptoms over many weeks and months, repeated diagnostic efforts are recommended in analogy with the search for bacterial pathogens by repeated blood culture sampling from patients with prolonged fever of unknown origin.

The introduction of several PCR methods in different laboratories for *T. gondii* diagnosis raises the important question of which method represents the optimal technology for the Nordic countries. However, with the documented high sensitivity and specificity of different methods, we see several advantages in diversity within this diagnostic field (Lebech and Petersen, 1992; Stray-Pedersen and Jenum, 1992; Ostergaard *et al.* 1993; Lappalainen *et al.* 1995; Guy *et al.* 1996). Just as for PCR-based detection of other microbial agents, where we have some experience from the field of herpesvirus infections, reviewed in (Bergström *et al.* 1995), there is probably no such thing as a 'gold standard' set-up. Thus, methods appearing as excellent in one laboratory can perform poorly in another, and vice versa. In addition, protozoa, as other microbial agents, are under constant evolution, and therefore the use of different independent PCR primer systems could be an advantage and help in discovering emerging strains. Although recent population genetic studies identified a very limited number of *T. gondii* genotypes in nature, recombination may occur even though a rare event (Grigg *et al.* 2001a,b). Studies of genotype, associated human clinical disease, and immunocompetence in different patient groups, may provide new perspectives for investigation of toxoplasma disease, and eventual refinement of diagnostic methods. But the essential requirement of a reliable, well-proven detection system, such as the P30 or B1 PCR, for rapid laboratory diagnosis of *T. gondii* infections, and an awareness of the importance of maintenance of good liaison between clinician and laboratory, cannot be over emphasized.

Standardization programmes of Nordic PCR-based diagnosis of *T. gondii* infections should therefore focus on the exchange of specimens and standards rather than striving towards a consensus PCR protocol. Furthermore, this information exchange should include sharing of experiences concerning suitable sample materials as well as optimal time points and conditions for sampling. To conclude, a strategy combining the advantages of PCR methods with a sharing of experience between different laboratories and the clinicians seems especially suitable for the improvement of *T. gondii* diagnosis in a low-prevalence area with a well-organized health care system.

Acknowledgements

We thank Drs Jean Chatterton and Alec Joss, Microbiology Laboratory, Inverness, UK, and Drs Mats Olsson and Anders Magnusson, Department of Parasitology, SMI, Stockholm, for valuable advice and the kind gifts of quantified *T. gondii* suspensions which were used as standards. We also wish to thank Dr Edward Trybala, Ms Mona Jensen, and Ms Birgitta Hägg, Department of Clinical Virology, University of Göteborg, for advice and assistance.

References

Andersson, R., Sandberg, T., Berglin, E. and Jeansson, S. (1992). Cytomegalovirus infections and toxoplasmosis in heart transplant recipients in Sweden. *Scand. J. Infect. Dis.* **24**, 411–417.

Bergström, T., Olofsson, S., Studahl, M., Kyllerman, M., Darin, N., Martinell, J. and Ricksten, A. (1995). Gendetektion med PCR vid virala CNS-infektioner ger snabb och specifik diagnostik av herpesgruppens virus. *Lakartidningen* **92**, 427–432.

Bergström, T., Ricksten, A., Nenonen, N. P. and Olofsson, S. (1998). Congenital *Toxoplasma gondii* infection diagnosed by PCR amplification on peripheral mononuclear blood cells from the child and the mother. *Scand. J. Infect. Dis.* **30**, 202–204.

Bulow, R. and Boothroyd, J. C. (1991). Protection of mice from fatal *Toxoplasma gondii* infection by immunization with p30 antigen in liposomes. *J. Immunol.* **147**, 3496–3500.

Burg, J. L., Grover, C. M., Pouletty, P. and Boothroyd, J. C. (1989). Direct and sensitive detection of a pathogenic protozoan, *Toxoplasma gondii*, by polymerase chain reaction. *J. Clin. Microbiol.* **27**, 1787–1792.

Burg, J. L., Perelman, D., Kasper, L. H., Ware, P. L. and Boothroyd, J. C. (1988). Molecular analysis of the gene encoding the major surface antigen of *Toxoplasma gondii*. *J. Immunol.* **141**, 3584–3591.

Cazenave, J., Forestier, F., Bessieres, M. H., Broussin, B. and Begueret, J. (1992). Contribution of a new pcr assay to the prenatal diagnosis of congenital toxoplasmosis. *Prenat. Diagn.* **12**, 119–127.

Chatterton, J. M. (1992). Health promotion. In: D. Ho-Yen and A. W. L. Joss (eds). *Human toxoplasmosis*. Oxford: Oxford University Press, pp. 174–175.

Cornelissen, A. W., Overdulve, J. P. and van der Ploeg, M. (1984). Determination of nuclear DNA of five eucoccidian parasites, *Isospora* (*Toxoplasma*) *gondii, Sarcocystis cruzi, Eimeria tenella, E. acervulina* and *Plasmodium berghei*, with special reference to gamontogenesis and meiosis in *I. (T.) gondii. Parasitology* **88**, 531–553.

Costa, J. M., Ernault, P., Gautier, E., and Bretagne, S. (2001). Prenatal diagnosis of congenital toxoplasmosis by duplex real-time PCR using fluorescence resonance energy transfer hybridization probes. *Prenat. Diagn.* **21**, 85–88.

Cristina, N., Darde, M. L., Boudin, C., Tavernier, G., Pestre-Alexandre, M. and Ambroise-Thomas, P. (1995). A DNA fingerprinting method for individual characterization of *Toxoplasma gondii* strains: combination with isoenzymatic characters for determination of linkage groups. *Parasitol. Res.* **81**, 32–37.

Danise, A., Cinque, P., Vergani, S., Candino, M., Racca, S., De Bona, A., Novati, R., Castagna, A. and Lazzarin, A. (1997). Use of polymerase chain reaction assays of aqueous humor in the differential diagnosis of retinitis in patients infected with human immunodeficiency virus. *Clin. Infect. Dis.* **24**, 1100–1106.

de Boer, J. H., Verhagen, C., Bruinenberg, M., Rothova, A., de Jong, P. T., Baarsma, G. S., Van der Lelij, A., Ooyman, F. M., Bollemeijer, J. G., Derhaag, P. J. and Kijlstra, A. (1996). Serologic and polymerase chain reaction analysis of intraocular fluids in the diagnosis of infectious uveitis. *Am. J. Ophthalmol.* **121**, 650–658.

Evengard, B., Petersson, K., Engman, M. L., Wiklund, S., Ivarsson, S. A., Tear-Fahnehjelm, K., Forsgren, M., Gilbert, R. and Malm, G. (2001). Low incidence of toxoplasma infection during pregnancy and in newborns in Sweden. *Epidemiol. Infect.* **127**, 121–127.

Fahnehjelm, K. T., Malm, G., Ygge, J., Engman, M. L., Maly, E. and Evengard, B. (2000). Ophthalmological findings in children with congenital toxoplasmosis. Report from a Swedish prospective screening study of congenital toxoplasmosis with two years of follow-up. *Acta. Ophthalmol. Scand.* **78**, 569–575.

Foudrinier, F., Aubert, D., Puygauthier, T., Rouger, C., Beguinot, I., Halbout, P., Lemaire, P., Marx, C. and Pinon, J. M. (1996). Detection of *Toxoplasma gondii* in immunodeficient subjects by gene amplification: influence of therapeutics. *Scand. J. Infect. Dis.*, **28**, 383–386.

Fuentes, I., Rubio, J. M., Ramirez, C. and Alvar, J. (2001). Genotypic characterization of *Toxoplasma gondii* strains associated with human toxoplasmosis in Spain: direct analysis from clinical samples. *J. Clin. Microbiol.* **39**, 1566–1570.

Grigg, M. E., Bonnefoy, S., Hehl, A. B., Suzuki, Y. and Boothroyd, J. C. (2001a). Success and virulence in *Toxoplasma* as the result of sexual recombination between two distinct ancestries. *Science* **294**, 161–165.

Grigg, M. E., Ganatra, J., Boothroyd, J. C., and Margolis, T. P. (2001b). Unusual abundance of atypical strains associated with human ocular toxoplasmosis. *J. Infect. Dis.* **184**, 633–639.

Guay, J. M., Dubois, D., Morency, M. J., Gagnon, S., Mercier, J. and Levesque, R. C. (1993). Detection of the pathogenic parasite *Toxoplasma gondii* by specific amplification of ribosomal sequences using comultiplex polymerase chain reaction. *J. Clin. Microbiol.* **31**, 203–207.

Guo, Z. G., Gross, U. and Johnson, A. M. (1997). *Toxoplasma gondii* virulence markers identified by random amplified polymorphic DNA polymerase chain reaction. *Parasitol. Res.* **83**, 458–463.

Guy, E. C., Pelloux, H., Lappalainen, M., Aspock, H., Hassl, A., Melby, K. K., Holberg Pettersen, M., Petersen, E., Simon, J. and Ambroise Thomas, P. (1996). Interlaboratory comparison of polymerase chain reaction for the detection of *Toxoplasma gondii* DNA added to samples of amniotic fluid. *Eur. J. Clin. Microbiol. Infect. Dis.* **15**, 836–839.

Ho, Y., Joss, A. W., Balfour, A. H., Smyth, E. T., Baird, D. and Chatterton, J. M. (1992). Use of the polymerase chain reaction to detect *Toxoplasma gondii* in human blood samples. *J. Clin. Pathol.* **45**, 910–913.

Hohlfeld, P., Daffos, F., Costa, J. M., Thulliez, P., Forestier, F. and Vidaud, M. (1994). Prenatal diagnosis of congenital toxoplasmosis with a polymerase-chain-reaction test on amniotic fluid. *N. Engl. J. Med.* **15;331**, 695–699.

Honore, S., Couvelard, A., Garin, Y. J., Bedel, C., Henin, D., Darde, M. L. and Derouin, F. (2000). Genotyping of *Toxoplasma gondii* strains from immunocompromised patients. *Pathol. Biol. (Paris)* **48**, 541–547.

Howe, D. K. and Sibley, L. D. (1995). *Toxoplasma gondii* comprises three clonal lineages: correlation of parasite genotype with human disease. *J. Infect. Dis.* **172**, 1561–1566.

Howe, D. K., Summers, B. C. and Sibley, L. D. (1996). Acute virulence in mice is associated with markers on chromosome VIII in *Toxoplasma gondii*. *Infect. Immun.* **64**, 5193–5198.

Howe, D. K., Honore, S., Derouin, F. and Sibley, L. D. (1997). Determination of genotypes of *Toxoplasma gondii* strains isolated from patients with toxoplasmosis. *J. Clin. Microbiol.* **35**, 1411–1414.

Jenum, P. A., Stray-Pedersen, B., Melby, K. K., Kapperud, G., Whitelaw, A., Eskild, A. and Eng, J. (1998). Incidence of *Toxoplasma gondii* infection in 35 940 pregnant women in Norway and pregnancy outcome for infected women. *J. Clin. Microbiol.* **36**, 2900–2906.

Jones, C. D., Okhravi, N., Adamson, P., Tasker, S. and Lightman, S. (2000). Comparison of PCR detection methods for B1, P30, and 18S rDNA genes of *T. gondii* in aqueous humor. *Invest. Ophthalmol. Vis. Sci.* **41**, 634–644.

Kwok, S. and Higuchi, R. (1989). Avoiding false positives with PCR. *Nature* **339**, 237–238.

Landgren, M., Kyllerman, M., Bergström, T., Dotevall, L., Ljungström, L. and Ricksten, A. (1994). Diagnosis of Epstein-Barr virus-induced encephalitis by DNA amplification from cerebrospinal fluid. *Ann. Neurol.* **35**, 631–635.

Lappalainen, M., Sintonen, H., Koskiniemi, M., Hedman, K., Hiilesmaa, V., Ammala, P., Teramo, K. and Koskela, P. (1995). Cost–benefit analysis of screening for toxoplasmosis during pregnancy. *Scand. J. Infect. Dis.* **27**, 265–272.

Lebech, M. and Petersen, E. (1992). Neonatal screening for congenital toxoplasmosis in Denmark: presentation of the design of a prospective study. *Scand. J. Infect. Dis. Suppl.* **84**, 75–79.

Lin, M. H., Chen, T. C., Kuo, T. T., Tseng, C. C. and Tseng, C. P. (2000). Real-time PCR for quantitative detection of *Toxoplasma gondii*. *J. Clin. Microbiol.* **38**, 4121–4125.

Luft, B. J. and Remington, J. S. (1992). Toxoplasmic encephalitis in AIDS. *Clin. Infect. Dis.* **15**, 211–222.

Luo, W. T., Aosai, F., Ueda, M., Yamashita, K., Shimizu, K., Sekiya, S. and Yano, A. (1997). Kinetics in parasite abundance in susceptible and resistant mice infected with an avirulent strain of *Toxoplasma gondii* by using quantitative competitive pcr. *J. Parasitol.* **83**, 1070–1074.

Macpherson, J. M. and Gajadhar, A. A. (1993). Sensitive and specific polymerase chain reaction detection of *Toxoplasma gondii* for veterinary and medical diagnosis. *Can. J. Vet. Res.* **57**, 45–48.

Ostergaard, L., Nielsen, A. K. and Black, F. T. (1993). DNA amplification on cerebrospinal fluid for diagnosis of cerebral toxoplasmosis among HIV-positive patients with signs or symptoms of neurological disease. *Scand. J. Infect. Dis.* **25**, 227–237.

Ruskin, J. and Remington, J. S. (1976). Toxoplasmosis in the compromised host. *Ann. Intern. Med.* **84**, 193–199.

Savva, D., Morris, J. C., Johnson, J. D. and Holliman, R. E. (1990). Polymerase chain reaction for detection of *Toxoplasma gondii*. *J. Med. Microbiol.* **32**, 25–31.

Stray-Pedersen, B. and Jenum, P. (1992). Current status of toxoplasmosis in pregnancy in Norway. *Scand. J. Infect. Dis. Suppl.* **84**, 80–83.

Thulliez, P., Daffos, F. and Forestier, F. (1992). Diagnosis of toxoplasma infection in the pregnant woman and the unborn child: current problems. *Scand. J. Infect. Dis. Suppl.* **84**, 18–22.

van de Ven, E., Melchers, W., Galama, J., Camps, W. and Meuwissen, J. (1991). Identification of *Toxoplasma gondii* infections by B1 gene amplification. *J. Clin. Microbiol.* **29**, 2120–2124.

Wastling, J. M., Nicoll, S. and Buxton, D. (1993). Comparison of two gene amplification methods for the detection of *Toxoplasma gondii* in experimentally infected sheep. *J. Med. Microbiol.* **38**, 360–365.

Weiss, L. M., Udem, S. A., Salgo, M., Tanowitz, H. B. and Wittner, M. (1991). Sensitive and specific detection of toxoplasma DNA in an experimental murine model: use of *Toxoplasma gondii*-specific cDNA and the polymerase chain reaction. *J. Infect. Dis.* **163**, 180–186.

30 Isolation of *Toxoplasma gondii*

Maija Lappalainen

Culture of *Toxoplasma gondii* is a specific method giving an unequivocal proof of infection. Parasite isolation or identification is useful, especially in the diagnosis of foetal infection or in immuno-compromised patients.

Isolation is usually performed by inoculation of suspect material into white mice or cell cultures. *Toxoplasma gondii* can multiply in a variety of cell lines, such as VERO, MRC-5, Hep-2, L-cells, and primary monkey kidney. The human embryonic fibroblast line MRC-5 is widely used to culture *Toxoplasma* from clinical samples. *Toxoplasma gondii* has been isolated in various tissues and body fluids: blood, placenta, brain, skeletal muscle, lung biopsy, heart, cerebrospinal fluid, subretinal fluid, aqueous humor, amniotic fluid, and broncho-alveolar lavage fluid (Ho-Yen and Joss 1992). Isolation from tissues may reflect only the presence of tissue cysts and does not necessarily mean active acute infection. From lymph nodes, however, positive isolation probably indicates the presence of tachyzoites, because cysts are rarely found in nodes.

Specimens should be injected into animals and cell cultures as soon as possible after collection to prevent death of the parasite. Sample preparation before inoculation is usually needed (Table 30.1). Formalin kills the parasite, and freezing may result in the death of both the tachyzoite and the cyst forms in tissues (James *et al.* 1996). If short-term storage of specimens is necessary, refrigeration at +4°C is preferred. This maintains the encysted form for up to two months (if kept moist) and prevents death of the tachyzoite for several days. *Toxoplasma gondii* survive in blood for a week or even longer.

Table 30.1 Sample preparation for isolation of *Toxoplasma gondii*

Specimen	Management before inoculation	Inoculum[a]
Blood	Centrifugation	Clot
Body fluids	Centrifugation or none	Cells or as such
Solid tissues (muscle, placenta, etc.)	Mincing followed by trypsin[b] digestion. Filtration, centrifugation. Sediment is washed and resuspended in sterile saline	Sediment

Notes
a Inoculum volume is usually 0.5–1.0 ml.
b 0.25% trypsin treatment for 1–2 h at +37°C, if large amounts of tissues are prepared.

Mouse inoculation

Sample is inoculated intraperitoneally (ip) in *Toxoplasma* seronegative mice. In cases in which the organism is virulent for mice, *T. gondii* can be demonstrated in the peritoneal fluid after 5–10 days (examined either fresh or in Giemsa- or Wright-stained smears). If the isolate is relatively avirulent for mice, a period of 4–6 weeks is usually required for definitive demonstration of the parasite. Surviving mice are bled after six weeks from the tail vein for serologic testing. If antibodies are present, proof of infection must be obtained by demonstration of the parasite. This may be performed by examining Giemsa-stained smears of fresh brain for demonstration of cysts. If cysts are not seen, ip injection of brain tissue into fresh mice should be performed (Remington and Desmonts 1990). Animal inoculation is more sensitive than tissue culture; the sensitivity from clinical samples ranges from 60 to 80% (Derouin *et al.* 1988; Hohlfeld *et al.* 1994). The disadvantage of the animal culture is the length of time required, usually from 4 to 6 weeks.

Cell culture

Most commonly coverslip cultures in 24-well plates are used. After sample inoculation, the cells are incubated for four days. The cultures are then stained with rabbit anti-*Toxoplasma* antibody and then with fluorescent anti-rabbit IgG. Thereafter, the cultures are read by fluorescence microscopy. Cell culture is more rapid than animal culture, especially if the infected cells are identified by immunofluorescence, but cell culture appears to be less sensitive than mouse inoculation (Derouin *et al.* 1988; Hitt and Filice 1992). Polymerase chain reaction method has greatly replaced culture assays today due to its better sensitivity and its suitability to study samples with non-viable parasites.

References

Derouin, F., Thulliez, P., Candolfi, E., Daffos, F., and Forestier, F. (1988). Early prenatal diagnosis of congenital toxoplasmosis using amniotic fluid samples and tissue culture. *European Journal of Clinical Microbiology and Infectious Diseases*, **7**, 423–425.

Hitt, J. A. and Filice, G. A. (1992). Detection of *Toxoplasma gondii* parasitemia by gene amplification, cell culture, and mouse inoculation. *Journal of Clinical Microbiology*, **30**, 3181–3184.

Hohlfeld, P., Daffos, F., Costa, J.-M., Thulliez, P., Forestier, F., and Vidaud, M. (1994). Prenatal diagnosis of congenital toxoplasmosis with a polymerase-chain-reaction test on amniotic fluid. *The New England Journal of Medicine*, **331**, 695–699.

Ho-Yen, D. O., and Joss, A. W. L. (eds) (1992). *Human toxoplasmosis*, 1st edn. Oxford University Press, Oxford, UK, pp. 1–265.

James, G. S., Sintchenko, V. G., Dickeson, D. J., and Gilbert, G. L. (1996). Comparison of cell culture, mouse inoculation, and PCR for detection of *Toxoplasma gondii*: effects of storage conditions on sensitivity. *Journal of Clinical Microbiology*, **34**, 1572–1575.

Remington, J. S. and Desmonts, G. (1990). Toxoplasmosis. In: *Infectious diseases of the fetus and newborn infant* (J. S. Remington and J. O. Klein, eds), 3rd edn. WB Saunders, Philadelphia, pp. 89–195.

31 Serology – *Toxoplasma gondii*

Inger Ljungström

Various assays are available for laboratory diagnosis of toxoplasmosis including serological methods, Polymerase Chain Reaction (PCR) and isolation of the parasites in tissue culture and/in laboratory animals. PCR and *in vitro* and *in vivo* isolation of the parasite are presented in separate headings in other sections of the book. Here, we will deal with the diagnosis of toxoplasmosis using serological methods.

Reference method

The Dye test (Sabin and Feldman 1948) is the golden standard for antibody detection when *Toxoplasma* infection is suspected. This assay requires live parasites, which are kept in culture or in animals, primarily in mice. This make the method laborious and very tedious and most laboratories do not provide the method routinely. Over the years, a number of other assays have been developed and validated against the Dye test (Petithory *et al.* 1996).

Screening methods

To screen for antibodies against *T. gondii* most of the commercial and 'in house' methods can be used. These include ELISA, direct agglutination (DA), indirect immunofluorescense (IFL) and semi-automatized assays based on microparticle enzyme immunoassay (Sandin *et al.* 1991; De Champs *et al.* 1997; Cimon *et al.* 1998). The antigen used in IFL (Huldt *et al.* 1975) and DA (Desmonts and Remington 1980) is whole, fixed tachyzoites, while more or less purified homogenates are used in ELISA and semi-automatized assays. In Chapter 13 the various antigens are presented.

Methods to determine acute toxoplasmosis

On suspicion of primary infection, the specific IgM, IgA and IgG antibodies to the parasite can be determined. Again, several methods have been developed for this purpose, for example, ELISA, IFL, 'capture'-ELISA, ISAGA (Desmonts *et al.* 1981). IgM and IgG antibodies can be demonstrated from about 3–4 weeks after infection. With IFL, the specific IgM can be detectable up to 6 months after primary infection. ISAGA and semi-automatized assays seem to be more sensitive, demonstrating IgM up to a year after infection. In around 50 per cent of the cases, specific IgA can also be detected up to several months after infection. Specific IgG will persist for the rest of the infected person's life. The avidity of specific IgG antibodies can also be measured by an ELISA assay (Hedman *et al.* 1989). A recent primary infection can be indicated by a low avidity of the specific IgG antibodies.

Diagnosis of congenital infection

On suspicion of congenital infection, serum samples from the mother are first analysed, this may be followed by testing the amniotic fluid. The sero-diagnosis is primarily based on comparison between the prenatal testing serum and current serum sample from the pregnant woman. Sero-conversion from negative to positive during the pregnancy gives a definitive diagnosis. A significant increase in IgG antibodies and the presence of IgM and/or IgA is interpreted as a probable diagnosis. A single serum sample containing specific IgG, IgM and/or IgA antibodies can be an indication of primary infection of the mother (European Research Network 1995). In these cases it may be necessary to perform confirmatory tests. These include determination of IgG antibody avidity, investigation of amniotic fluid by PCR (Hohlfeld *et al.* 1994; Guy *et al.* 1996; Pelloux *et al.* 1998) and/or isolation of the parasite *in vivo* and/or *in vitro*.

In newborns suspected of having congenital infection, cord blood or/and EDTA blood could be collected and analysed for the parasite itself or its DNA. Furthermore, it is recommended that the antibody response be followed by collecting blood samples at birth and at the age of 3, 6 and 12 months to measure the IgG, IgM and IgA antibodies.

Diagnosis of cerebral toxoplasmosis

Screening for antibodies against *T. gondii* is recommended in HIV-positive persons, new cases of certain cancers, for example, hairy-cell leukaemia, and before transplantation to check whether the individual harbours the *Toxoplasma* parasite. If the person is sero-positive there is a risk of developing cerebral toxoplasmosis during the progress of the disease. Lack of specific antibodies against *Toxoplasma* in patients with CNS symptoms speak against cerebral toxoplasmosis, while the presence of these antibodies in a patient with CNS symptoms can support the diagnosis. Cerebral toxoplasmosis is rarely followed by an altered serological response, that is, an increase in IgG or IgM antibodies.

References

Cimon, B., Marty, P., Morin, O., Bessieres, M.H., Marx-Chemla, C., Gay-Andrieum, F. and Thulliez, P. (1998) Specificity of low anti-Toxoplasma IgG titers with IMx and AxSYM Toxo IgG assays. *Diagnostic Microbiology & Infectious Disease*, **32**, 65–67.

De Champs, C., Pelloux, H., Cambon, M., Fricker-Hidalgo, H., Goullier-Fleuret, A. and Ambroise-Thomas, P. (1997) Evaluation of the second generation IMx Toxo IgG antibody assay for detection of antibodies to *Toxoplasma gondii* in human sera. *Journal of Clinical Laboratory Analysis*, **11**, 214–219.

Desmonts, G. and Remington, J.S. (1980) Direct agglutination for test diagnosis of toxoplasma infection; method for increasing sensitivity and specificity. *Journal of Clinical Microbiology*, **11**, 562–568.

Desmonts, G., Naot, Y. and Remington, J.S. (1981) Immunoglobulin M-immunosorbent agglutination assay for diagnosis of infectious diseases: Diagnosis of acute congenital and acquired *Toxoplasma* infections. *Journal of Clinical Microbiology*, **14**, 486–491.

European Research Network on Congenital Toxoplasmosis (coordinator: Eskild Pedersen, Copenhagen, Denmark). Classification and Case Definition, 1995.

Guy, E.C., Pelloux, H., Lappalainen, M., Aspock, H., Hassl, A., Melby, K.K. *et al.* (1996) Interlaboratory comparison of polymerase chain reaction for the detection of *Toxoplasma gondii* DNA added to samples of amniotic fluid. *European Journal of Clinical Microbiology and Infectious Diseases*, **15**, 836–839.

Hedman, K., Lappalainen, M., Seppälä, I. and Mäkelä, O. (1989) Recent primary toxoplasma infection indicated by a low avidity of specific IgG. *Journal of Infectious Diseases*, **159**, 736–740.

Hohlfeld, P., Daffos, F., Costa, J.M., Thulliez, P., Forestier, F. and Vidaud, M. (1994) Prenatal diagnosis of congenital toxoplasmosis with a polymerase-chain-reaction test on amniotic fluid. *New England Journal of Medicine*, **331**, 695–699.

Huldt, G., Ljungström, I. and Aust-Kettis, A. (1975) Detection by immunofluorescense of antibodies to parasitic agents. Use of class-specific conjugates. *Annals of the New York Academy of Sciences*, **254**, 304–314.

Pelloux, H., Guy, E., Angelici, M.C., Aspock, H., Bessieres, M.H., Blatz, R. *et al.* (1998) A second European collaborative study on polymerase chain reaction for *Toxoplasma gondii*, involving 15 teams. *FEMS Microbiology Letters*, **165**, 231–237.

Petithory, J.C., Reiter-Owoba, I., Berthelot, F., Milgram, M., De Loye, J. and Petersen, E. (1996) Performance of European laboratories testing serum samples for *Toxoplasma gondii*. *European Journal of Clinical Microbiology and Infectious Diseases*, **15**, 45–49.

Sabin, A.D. and Feldman, H.A. (1948) Dye as microchemical indicators of a new immunity phenomenon affecting a protozoon parasite (Toxoplasma). *Science*, **108**, 660–663.

Sandin, R.L., Knapp, C.C., Hall, G.S., Washington, J.A. and Rutherford, I. (1991) Comparison of the Vitek Immunodiagnostic Assay System with an indirect immunoassay (Toxostat Test Kit) for detection of immunoglobulin G antibodies to *Toxoplasma gondii* in clinical specimens. *Journal of Clinical Microbiology*, **29**, 2763–2767.

32 Serology – other parasites

Inger Ljungström

Serology

Originally the laboratory diagnosis of parasitic infections has been based on the demonstration of the organism such as eggs, larvae, adults, cysts, and trophozoites in feacal specimens, tissue, and body fluids. To find and identify the parasite is the definitive proof of infection. In many instances it is not possible or very difficult to find the organism because of low/irregular excretion or their hiding in the tissue. To facilitate the laboratory diagnosis, indirect methods have been developed. These methods rely on detection of antibodies directed against the organism or antigens secreted by the organisms. Lately also detection of nucleic acid-based probe tests (PCR) have been developed.

It is important to be aware of sensitivity and specificity and the purpose of the investigation when deciding methods to be utilized. Under certain circumstances, a method with high specificity and less sensitivity is preferable, but opposite can be as likely. The criteria for diagnosis of individuals may differ from the criteria needed for screening of populations.

Helminths

Anisakiosis

Clinical diagnosis of anisakiosis is rarely done before surgery has been performed for suspicion of appendicitis, gastric tumour, etc. Often a biopsy is taken and the correct diagnosis made when the larvae are identified in histological sections. Skin tests, indirect immunofluorescence (IFL) and the enzyme linked immunosorbent assay (ELISA) have been tried with limited success due to cross-antigenicity with other helminths and failure to detect light infections (Ruitenberg 1970). An assay using monoclonal antibody specific for *Anisakis* larvae antigen seems, however, promising (Yagihashi *et al.* 1990).

Several methods have been developed for diagnosis of *Anisakis simplex* allergy, these include IgE immunoblotting (Garcia *et al.* 1997) and capture ELISA (Lorenzo *et al.* 1999).

Ascaris lumbricoides *infection*

The traditional diagnosis of *Ascaris* infection is the identification of eggs in faeces or the worm itself.

The serological methods available are not regularly used for diagnosis of ascariosis in patients, but, in general, been utilized for epidemiological studies in various populations.

(Haswell-Elkins *et al.* 1992; van Knapen *et al.* 1992; Lynch *et al.* 1993). On individual bases, immunodiagnosis has been of little help due to cross-reactivity with other helminthic antigens. Recently, Chatterjee *et al.* (1996) reported improved sensitivity and specificity by measuring specific IgG4 against a fractionated excretory/secretory antigen.

Echinococcus granulosus *infection*

Non-invasive methods for diagnosis of echinococcosis such as ultrasonography, computed tomography and chest radiographs have comparable sensitive. However, differential diagnosis from benign cyst and abscesses can be difficult. Ultrasonography seems to be an efficient mass-screening method in combination with serology.

A wide variety of immunological tests have been applied for detection of antibodies in patients where echinococcosis is suspected. These include complement-fixation, latex agglutination, indirect haemagglutination, immunodiffusion, immunoelectrophoresis, Casoni intradermal test, IFL, and ELISA. For the antigen crude cyst fluid appears to be the most suitable in terms of immunogenicity and availability, but the complex mixture of proteins produce varying degree of both positive and negative results (Thompson and Lymbery 1995). The use of more purified fractions of cyst fluid antigens may improve specificity. The cyst fluid contains two major lipoproteins referred to as Antigen 5 and Antigen B. Antibodies to Antigen B have been considered specific for *Echinococcus granulosus*, but it is now known that a certain proportion of cross-reactions occur with sera from persons infected with other species of *Echinococcus* (Ito *et al.* 1999).

ELISA provides a sensitive assay and IgG is the dominant class of specific antibody in *E. granulosus* infection and IgG4 seems to be of diagnostic value (Shambesh *et al.* 1997; Grimm *et al.* 1998; Sterla *et al.* 1999). Replacement of native antigens by recombinant antigen may provide a better standardised specific test (McVie *et al.* 1997). The diagnosis can also be improved by the application of at least two serological methods.

Fasciola hepatica *infection*

In *Fasciola hepatica*, infection examination of faeces for eggs are of limited value, since during the acute phase of the disease, no eggs are excreted until the larvae have matured to reproductive adults, which takes up to 4–5 months. During the chronic phase, the eggs in faeces are often scanty. A further complication is that eggs may be detected after ingestion of liver from infected animals. Thus, positive cases should be reconfirmed if liver has been eaten recently.

Over the years, most of the available immunodiagnostic methods have been used for demonstrating antibodies against *Fasciola*, thus indirectly proof of infection. Generally, the assays have shown good sensitivity, but there are problems with cross-reactivity. By using cathepsin L1 (CL 1) as antigen an increase in specificity was obtained when compared to crude antigen such as liver fluke homogenates. The excretory/secretory products (ES) had higher sensitivity than CL 1 antigen, but less specificity (O'Neill *et al.* 1998). The performance of recombinant CL 1 compared to native CL 1 showed a highly statistically significant correlation (O'Neill *et al.* 1999). The predominant isotype elicited by the infection seems to be IgG4 and determination of IgG4 seems to increase the specificity (O'Neill *et al.* 1998; Maher *et al.* 1999).

Assays for demonstration of coproantigen of *Fasciola* in faeces have mostly been utilized to identify infected animals (el-Bahi *et al.* 1992). However, some assays for coproantigen detection have been applied for human diagnosis (Espino *et al.* 1998).

Methods to detect circulating ES antigens in sera have also been developed over the years (Espino *et al.* 1990; Espino and Finlay, 1994; Shehab *et al.* 1999).

Hymenolepis nana *infection*

Discovery of eggs in the faeces provides definitive diagnosis of *Hymenolepis* infection, which is the standard procedure for diagnosis. ELISA has been developed to detect serum antibodies in humans. The assay shows good sensitivity, but low specificity (Castillo *et al.* 1991; Gomez-Priego *et al.* 1991).

Toxocariosis

Demonstration of larvae in the tissue is very difficult and seldom successful. For that reason the diagnosis is based on indirect methods such as antibody detection. The current assays mostly rely on excretory/secretory products of cultured larvae (de Savigny 1975) used in ELISA (Ljungström and van Knapen 1989; Jacquier *et al.* 1991). Cross-reactivity between *Toxocara* spp. and *Ascaris* in healthy population seems not to exist. However, in suspected patients, antibodies to both parasites are seen in almost 50% of the seropositive individuals (van Knapen *et al.* 1992). In areas endemic for both *Ascaris* and *Toxocara*, both infections in one person may take place simultaneously.

Trichinellosis

Examining a muscle biopsy constitutes the definitive test for trichinellosis. False negative biopsy results may occur if the patient is only lightly infected or the analysis might have been performed too early after infection.

The serological test available for the detection of circulating *Trichinella* antigens are not widely used for human diagnosis. In general indirect methods are applied, that is, they are geared at the detection of specific antibodies. Various assays have been applied for detection of specific antibodies such as complement-fixation, bentonite flocculation, latex agglutination indirect haemagglutination, counterimmunoelectrophoresis, IFL, and ELISA (Ljungström, 1983). The most common assays in Europe are those employing labelled antibodies which have the advantage that the various antibody isotypes can be determined and differentiated from the overall humoral response. By determination of specific IgM, IgA and IgG it is possible to distinguish between acute and past infection (Ljungström *et al.* 1988). The diagnosis is improved by the application of at least two serological diagnostic methods. The various antigens are presented in the chapter on *Trichinella*.

Protozoa

Amoebiasis

The standard method to diagnose amoebiasis is detection and identification of *Entamoeba histolytica* in faeces, biopsies, or abscess material. However, cysts of the pathogenic *E. histolytica* and the apathogenic *E. dispar* are morphological indistinguishable. Over the last years various techniques including ELISA, IFL, and PCR have been developed to differentiate between the two amoebae species in faeces (Mirelman *et al.* 1997; Tachibana *et al.* 1997; Haque *et al.* 1998).

Most of the serological methods available over the last decades have been used for antibody detection when amoebiasis, both intestinal and extra intestinal, has been suspected. The positive antibody response is in general a good predictor for *Entamoeba* infection in patients from non-endemic areas, but in endemic areas the healthy population is often seropositive (Hossain *et al.* 1983). Immunodiffusion is less sensitive compared to IFL and ELISA, but has a higher specificity and shows a close correlation with clinical disease in an area of low endemicity (Stamm *et al.* 1976).

The cysteine-rich immunodominant domain of the antigenic 170-kDa subunit of the galactose-*N*-acetylgalactosamine binding lectin expressed as a recombinant lectin seems to be able to replace the crude amoeba antigen for use in serological test showing a sensitivity and specificity over 90% (Zhang *et al.* 1992; Shenai *et al.* 1996).

Babesiosis

Diagnosis is based on microscopical identification of parasites in Giemsa-stained blood films. This kind of diagnosis is tedious and the parasite can be overlooked or be confused with other intraerythrocytic inclusions. Over the years, several other techniques have been developed such as IFL, ELISA, and PCR primarily to diagnose babesiosis in livestock (Bose *et al.* 1995).

In humans the infection is rare and the babesias are often indistinguishable from the ring stages of malaria parasites. The first case of human babesiosis in Europe was reported in 1957 and since then another 28 human cases of babesiosis have been reported in Europe (Gorenflot *et al.* 1998). For obvious reason, this has resulted in very few assays for diagnosis of human babesiosis, especially for *B. divergens* infection. When a human case is suspected, physicians tend to collaborate with the veterinaries as was done when the first case in Sweden was documented (Uhnoo *et al.* 1992).

Cryptosporidios

Microscopic identification of oocyst in faeces after Ziehl–Neelsen staining is mainly used for diagnosis. To assist the detection of oocysts in faeces, various immunoassays (Siddons *et al.* 1992; Tee *et al.* 1993; Dagan *et al.* 1995) and PCR techniques (Zhu *et al.* 1998) have been developed. Serological assays have been described but are in general not used for diagnostic purposes (Frost *et al.* 1998).

Giardiasis

Diagnosis of *Giardia* by microscopic methods is a reliable in spite of irregular excretion of cysts. However, at least three faecal samples obtained on non-consecutive days are required to overcome the disadvantage of this irregular excretion. To assist the detection of infection a variety of coproantigen detection assays based on ELISA have been developed which have been met with success (Wolfe 1992; Faubert 2000).

Serological analysis can contribute to diagnosis in non-endemic areas if the assay is performed at least three weeks after an untreated infection (Ljungström and Castor 1992). The determination of specific IgM antibodies seems to correlate with active *Giardia* in endemic areas (Sullivan *et al.* 1991; Chaudhuri *et al.* 1992).

References

Bose, R., Jorgensen, W. K., Dalgliesh, R. J., Friedhoff, K. T. and de Vos, A. J. (1995). Current state and future trends in the diagnosis of babesiosis. *Veterinary Parasitology*, **57**, 61–74.

Castillo, R. M., Grados, P., Carcamo, C., Miranda, E., Montenegro, T., Guevara, A. and Gilman, R. H. (1991). Effect of treatment on serum antibody to *Hymenolepis nana* detected by enzyme-linked immunosorbent assay. *Journal of Clinical Microbiology*, **29**, 413–414.

Chatterjee, B. P. Santra, A., Karmakar, P. R. and Mazumder, D. N. (1996). Evaluation of IgG4 response in ascariasis by ELISA for serodiagnosis. *Tropical Medicine and International Health*, **1**, 633–639.

Chaudhuri, P. P., Sengupta, K., Manna, B., Saha, M.-K., Pal, S. C. and Das P. (1992). Detection of specific anti-Giardia antibodies in the serodiagnosis of symptomatic giardiasis. *Journal of Diarrhoeal Diseases Research*, **10**, 151–155.

Dagan, R., Fraser, D., El-On, J., Kassis, I., Deckelbaum, R. and Turner, S. (1995). Evaluation of an enzyme immunoassay for the detection of *Cryptosporidium* spp. in stool specimens from infants and young children in field studies. *American Journal of Tropical Medicine and Hygiene*, **52**, 134–138.

de Savigny, D. H. (1975). *In vitro* maintenance of *Toxocara canis* larvae and a simple method for the production of *Toxocara* ES antigens for use in serodiagnostic tests for visceral larva migrans. *Journal of Parasitology*, **61**, 781–782.

el-Bahi, M. M., Malone, J. B., Todd, W. J. and Schnorr, K. L. (1992). Detection of stable diagnostic antigen from bile and faeces of *Fasciola hepatica* infected cattle. *Veterinary Parasitology*, **45**, 157–167.

Espino, A. M., Marcet, R. and Finlay, C. M. (1990). Detection of circulating excretory secretory antigens in human fascioliasis by sandwich enzyme-linked immunosorbent assay. *Journal of Clinical Microbiology*, **28**, 2637–2640.

Espino, A. M. and Finlay, C. M. (1994). Sandwich enzyme-linked immunosorbent assay for detection of excretory secretory antigens in humans with fascioliasis [published erratum appears in *Journal of Clinical Microbiology*, 1994, **32**(3), 860]. *Journal of Clinical Microbiology*, **32**, 190–193.

Espino, A. M., Diaz, A., Perez, A. and Finlay, C. M. (1998). Dynamics of antigenemia and coproantigens during a human *Fasciola hepatica* outbreak. *Journal of Clinical Microbiology*, **36**, 2723–2726.

Faubert, G. (2000). Immune response to Giardia duodenalis. *Clinical Microbiology Reviews*, **13**, 35–54.

Frost, F. J., de la Cruz, A. A., Moss, D. M., Curry, M. and Calderon, R. L. (1998). Comparisons of ELISA and Western blot assays for detection of *Cryptosporidium* antibody. *Epidemiology and Infection*, **121**, 205–211.

Garcia, M., Moneo, I., Audicana, M. T., del Pozo, M. D., Munoz, D., Fernandez, E. *et al.* (1997). The use of IgE immunoblotting as a diagnostic tool in *Anisakis simplex* allergy. *Journal of Allergy and Clinical Immunology*, **99**, 497–501.

Gomez-Priego, A., Godinez-Hana A. L. and Gutierrez-Quiroz, M. (1991). Detection of serum antibodies in human *Hymenolepis* infection by enzyme immunoassay. *Transactions of the Royal Society of Tropical Medicine and Hygiene*, **85**, 645–647.

Gorenflot, A., Moubri, K., Precigout, E., Carcy, B. and Schetters, T. P. (1998). Human babesiosis. *Annuals of Tropical Medicine and Parasitology*, **92**, 489–501.

Grimm, F., Maly, F. E., Lu, J. and Llano, R. (1998). Analysis of specific immunoglobulin G subclass antibodies for serological diagnosis of *Echinococcosis* by a standard enzyme-linked immunosorbent assay. *Clinical and Diagnostic Laboratory Immunology*, **5**, 613–616.

Haque, R., Ali, I. K., Akther, S. and Petri, W. A. Jr. (1998). Comparison of PCR, isoenzyme analysis, and antigen detection for diagnosis of *Entamoeba histolytica* infection. *Journal of Clinical Microbiology*, **36**, 449–452.

Haswell-Elkins, M. R., Leonard, H., Kennedy, M. W., Elkins, D. B. and Maizel, R. M. (1992). Immunoepidemiology of *Ascaris lumbricoides*: relationships between antibody specificities, exposure and infection in a human community. *Parasitology*, **104**, 153–159.

Hossain, M. M., Ljungström, I., Glass, R. I., Lundin, L., Stoll, B. J. and Huldt, G. (1983). Amoebiasis and giardiasis in Bangladesh: parasitological and serological studies. *Transactions of the Royal Society of Tropical Medicine and Hygiene*, **77**, 552–554.

Isaac-Renton, J., Blatherwick, J., Bowie, W. R., Fyfe, M., Khan, M., Li, A. *et al.* (1999). Epidemic and endemic seroprevalence of antibodies to Cryptosporidium and Giardia in residents of three communities with different drinking water supplies. *American Journal of Tropical Medicine and Hygiene*, **60**, 578–583.

Ito, A., Ma, L., Schant, P. M., Gottstein, B., Liu, Y. H., Chai, J. J. *et al.* (1999). Differential serodiagnosis for cystic and alveolar echinococcosis using fractions of *Echinococcus granulosus* cyst fluid (antigen B) and *E. multilocularis* protoscolex (EM18). *American Journal of Tropical Medicine and Hygiene*, **60**, 188–192.

Jacquier, P., Gottstein, B., Stingelin, Y. and Eckert, J. (1991). Immunodiagnosis of toxocarosis in humans: evaluation of a new enzyme-linked immunosorbent assay kit. *Journal of Clinical Microbiology*, **29**, 1831–1835.

Lorenzo, S., Iglesias, R., Audicana, M. T., Garcia-Villaescusa, R., Pardo, F., Sanmartin, M. L. and Ubeira, F. M. (1999). Human immunoglobulin isotype profiles produced in response to antigens recognized by monoclonal antibodies specific to *Anisakis simplex*. *Clinical and Experimental Allergy*, **29**, 1095–1101.

Ljungström, I. (1983). Immunodiagnosis in man. In: *Trichinella and Trichinellosis* (W. C. Campbell, ed.), pp. 403–429. New York : Plenum Publishing Corporation.

Ljungström, I., Hammarström, L., Kociecka, W. and Smith, C. I. (1988). The sequential appearance of IgG subclasses and IgE during the course of *Trichinella spiralis* infection. *Clinical and Experimental Immunology*, **74**, 230–235.

Ljungström, I. and van Knapen, F. (1989). An epidemiological and serological study of toxocara infection in Sweden. *Scandinavian Journal of Infectious Diseases*, **21**, 87–93.

Ljungström, I. and Castor, B. (1992). Immune response to *Giardia lamblia* in a water-borne outbreak of giardiasis in Sweden. *Journal of Medical Microbiology*, **36**, 347–352.

Lynch, N. R., Hagel, I., Vargas, V., Rotundo, A., Varela, M. C., Di Prisco, M. C. and Hodgen, A. N. (1993). Comparable seropositivity for ascariasis and toxocariasis in tropical slum children. *Parasitology Research*, **79**, 547–550.

Maher, K., El Ridi, R., Elhoda, A. N., El-Ghannam, M., Shaheen, H., Shaker, Z. and Hassanein, H. I. (1999). Parasite-specific antibody profile in human fascioliasis: application for immunodiagnosis of infection. *American Journal of Tropical Medicine and Hygiene*, **61**, 738–742.

McVie, A., Ersfeld, K., Rogan, M. T. and Craig, P. S. (1997). Expression and immunological characterisation of *Echinococcus granulosus* recombinant antigen B for IgG4 subclass detection in human cystic echinococcosis. *Acta Tropica*, **67**, 19–35.

Mirelman, D., Nuchamowitz, Y. and Stolarsky, T. (1997). Comparison of use of enzyme-linked immunosorbent assay-based kits and PCR amplification of rRNA genes for simultaneous detection of *Entamoeba histolytica* and *E. dispar*. *Journal of Clinical Microbiology*, **35**, 2405–2407.

O'Neill, S. M., Parkinson, M., Strauss. W., Angles, R. and Dalton, J. P. (1998). Immunodiagnosis of *Fasciola hepatica* infection (fascioliasis) in a human population in the Bolivian Altiplano using purified cathepsin L cysteine proteinase. *American Journal of Tropical Medicine and Hygiene*, **58**, 417–423.

O'Neill, S. M., Parkinson, M., Dowd, A. J., Strauss, W., Angles, R. and Dalton, J. P. (1999). Short report: immunodiagnosis of human fascioliasis using recombinant *Fasciola hepatica* cathepsin L1 cysteine proteinase. *American Journal of Tropical Medicine and Hygiene*, **60**, 749–751.

Ruitenberg, E. J. (1970). *Anisakis. Pathogenesis, serodiagnosis and prevention*. DVS Thesis, University of Utrecht, Utrecht, The Netherlands.

Shambesh, M. K., Craig, P. S., Wen, H., Rogan, M. T. and Paolillo, E. (1997). IgG1 and IgG4 serum antibody responses in asymptomatic and clinically expressed cystic echinococcosis patients. *Acta Tropica*, **64**, 53–63.

Shehab, A. Y., Hassan, E. M., Abou Basha, L. M., Omar, E. A., Helmy, M. H., El-Morshedy, H. N. and Farag, H. F. (1999). Detection of circulating E/S antigens in the sera of patients with fascioliasis by ELISA: a tool of serodiagnosis and assessment of cure. *Tropical Medicine and International Health*, **4**, 686–690.

Shenai, B. R., Komalam, B. L., Arvind, A. S., Krishnaswamy, P. R. and Rao P. V. (1996). Recombinant antigen-based avidin-biotin microtiter enzyme-linked immunosorbent assay for serodiagnosis of invasive amebiasis. *Journal of Clinical Microbiology*, **34**, 828–833.

Siddons, C. A., Chapman, P. A. and Rush, B. A. (1992). Evaluation of an enzyme immunoassay kit for detecting cryptosporidium in faeces and environmental samples. *Journal of Clinical Pathology*, **45**, 479–482.

Stamm, W. P., Ashley, M. J. and Bell, K. (1976). The value of amoebic serology in an area of low endemicity. *Transactions of the Royal Society of Tropical Medicine and Hygiene*, **70**, 49–53.

Sterla, S., Sato, H. and Nieto, A. (1999). *Echinococcus granulosus* human infection stimulates low avidity anticarbohydrate IgG2 and high avidity antipeptide IgG4 antibodies. *Parasite Immunology*, **21**, 27–34.

Sullivan, P. B., Neale, G., Cevallos, A. M. and Farthing, M. J. (1991). Evaluation of specific serum anti-giardia IgM antibody response in diagnosis of giardiasis in children. *Transactions of the Royal Society of Tropical Medicine and Hygiene*, **85**, 748–749.

Tachibana, H., Kobayashi, S., Kaneda, Y., Takeuchi, T. and Fujiwara, T. (1997). Preparation of a monoclonal antibody specific for *Entamoeba dispar* and its ability to distinguish *E. dispar* from *E. histolytica*. *Clinical and Diagnostic Laboratory Immunology*, **4**, 409–414.

Tee, G. H., Moody, A. H., Cooke, A. H. and Chiodini, P. L. (1993). Comparison of techniques for detecting antigens of *Giardia lamblia* and *Cryptosporidium parvum* in faeces. *Journal of Clinical Pathology*, **46**, 555–558.

Thompson, R. C. A. and Lymbery, A. J. (1995). *Echinococcus and hydatid disease*. Oxon: CAB International.

Uhnoo, I., Cars, O., Christensson, D. and Nystrom-Rosander, C. (1992). First documented case of human babesiosis in Sweden. *Scandinavian Journal of Infectious Diseases*, **24**, 541–547.

van Knapen, F., Buijs, J., Kortbeek, L. M. and Ljungström, I. (1992). Larva migrans syndrome: toxocara, ascaris, or both? *Lancet*, **340**, 550–551.

Wolfe, M. S. (1992). Giardiasis. *Clinical Microbiology Reviews*, **5**, 93–100.

Yagihashi, A., Sato, N., Takahashi, S., Ishikura, H. and Kikuchi, K. (1990). A serodiagnostic assay by microenzyme-linked immunosorbent assay for human anisakiasis using a monoclonal antibody specific for *Anisakis* larvae antigen. *Journal of Infectious Diseases*, **161**, 995–998.

Zhang, Y., Li, E., Jackson, T. F., Zhang, T., Gathiram, V. and Stanley, S. L. Jr. (1992). Use of a recombinant 170-kilodalton surface antigen of *Entamoeba histolytica* for serodiagnosis of amebiasis and identification of immunodominant domains of the native molecule. *Journal of Clinical Microbiology*, **30**, 2788–2792.

Zhu, G., Marchewka, M. J., Ennis, J. G. and Keithly, J. S. (1998). Direct isolation of DNA from patient stools for polymerase chain reaction detection of *Cryptosporidium parvum*. *Journal of Infectious Diseases*, **177**, 1443–1446.

33 Parasite detection

*Michael Willcox, Marianne Lebbad and
Jadwiga Winiecka-Krusnell*

The laboratory detection of parasites requires little specialized apparatus unless culture is done. Motile trophozoites are seen in fresh preparations, but permanent staining is necessary when there is delay in examination. Concentration methods such as formol ethyl acetate are essential for detection of cysts and ova in faeces. For certain parasites special methods are used for detection, for example, culture for *Acanthamoeba* and *Trichomonas*, fortified trichrome staining for Microsporidia, tape method for *Enterobius*.

Introduction

Definitive diagnosis of a parasitic infection is by finding and identifying the parasite. Of the newer methods for parasite detection, those using gene technology are not yet available for general use in the routine laboratory and labelled monoclonal antibody techniques are costly and available for some species only. Thus, routine investigation is still mainly dependent on microscopy. Culture is used occasionally. The proper use of both internal and external quality control programmes is vital.

Although equipment needs for basic parasitology are modest, laboratories carrying out culture, for example, for *Acanthamoeba*, or handling highly infectious material will need a Class 2 biological safety cabinet or bench. The single most important instrument in routine parasitology is a binocular microscope with high quality optics. A properly calibrated eye-piece micrometer is essential. It should also be used. Some staining methods require the use of a fluorescence microscope and for *Acanthamoeba* culture an inverted microscope is needed. As many of the reagents used are highly inflammable or toxic, an efficiently ventilated fume hood is essential for specimen preparation and staining and a fume extraction device over the centrifuge is desirable. A swing-out centrifuge is preferable to an angle model. Laboratories carrying out culture methods will need incubators.

Techniques for the detection of parasites

General techniques for the detection of parasites, are described first. Methods for some important parasites in Northern countries are discussed later. Most of the techniques described are those used at the Division of Parasitology, Swedish Institute for Infectious Disease Control.

Direct microscopy of wet mounts

Direct microscopy of fresh unfixed stools or other clinical material is done to detect motile trophozoites. These die quickly, microscopy must be done immediately the specimen is collected. If immediate examination is not possible, the specimen should be suitably fixed for permanent staining.

Reagents required

- Isotonic saline (0.9% NaCl).
- d'Antoni's iodine (stock solution): Dissolve 1 g potassium iodide and 1.5 g iodine crystals in 100 ml distilled water in a brown glass-stoppered bottle. Some crystals should remain on the bottom. Store at room temperature (RT); shelf life is 1 year.

For use, transfer some of the solution into a drop bottle. When the strong tea colour of the working solution lightens, replace it with fresh iodine.

Procedure

1   Place a drop of 0.9% saline on one end of a microscope slide, on the other a drop of d'Antoni's iodine solution.
2   Using an applicator stick, pick up a small amount of specimen and mix thoroughly with the saline. Repeat the process this time mixing with the iodine.
3   Place a coverglass on each suspension.
4   Scan both preparations with the 10× objective, the saline suspension first. The whole coverslip area should be examined. When necessary, change to the 40× objective for a more detailed study. Some of the coverslip area always should be examined with this objective.

Results and technical notes

Protozoan cysts appear as opalescent refractile bodies in the saline preparation; internal structures are easier seen in the iodine suspension. Ova are more readily identified in the saline preparation.

   Cysts are not easily detected in thick faecal suspensions. Too thin a suspension gives insufficient material for examination. The correct amount is found by experience. Amoeba trophozoites cannot be identified in fixed material but *Giardia intestinalis* may be recognized. Correct illumination is vital. Too much light causes objects to be missed. When changing to the 40× objective, the light should be increased.

Preservation of faecal specimens

Stools sent to the laboratory by post should be fixed in 10% formalin or SAF fixative (Yang and Scholten 1977) to preserve parasite morphology and prevent growth of yeast and moulds.

   Formalin (10%): mix 100 ml formaldehyde solution (usually 37–40%) with 900 ml distilled water. Store at RT. Shelf life, 2 years. For use, mix one part faeces with three parts formalin to make a homogenous suspension.

   SAF fixative: dissolve 15 g sodium acetate in some distilled water. Add 20 ml glacial acetic acid, 40 ml formaldehyde solution and distilled water to 1 l. Store at RT; shelf life, 2 years. Use as for 10% formalin.

Positive control reference specimens

Positive control specimens should always be used when available. Methods of preparation of some of these are described under techniques for specific parasites. For use with the formol–ethyl acetate method for concentration of stool specimens, Bayer's fixative is preferable to formalin for preserving cyst morphology in specimens used for quality control (Moody 1986).

Bayer's fixative (stock solution): copper chloride dihydrate ($CuCl_2 \cdot 2H_2O$), 7 g: glacial acetic acid, 70 ml: 20% v/v formalin solution, 1 000 ml (200 ml 37–40% formaldehyde mixed with 800 ml distilled water). Store at RT; stable indefinitely.

For use: dilute the stock solution 1 in 10 with distilled water.

Formol–ethyl acetate method for concentration of faecal specimens

Parasites may be scanty in stool specimens. Concentration methods remove faecal debris so that cysts or ova are more readily detected. In this method (Young *et al.* 1979) formalin or SAF-fixed faeces are filtered, shaken with ethyl acetate and centrifuged. Parasitic elements are found in the deposit. Trophozoites are not concentrated.

Method

Gloves should be used and the procedure carried out in a fume cupboard

1   With unfixed specimens add a spoonful (about 1 g) of faeces to about 7 ml 10% forma- lin in a tube. Mix well with two applicator sticks. Wait about 30 min for inactivation and fixing of specimen.
2   Filter a suitable amount of faeces suspension through one layer of dampened surgical gauze into a beaker. With very thin or mucoid specimens, see below.
3   Transfer the filtrate to a centrifuge tube. Adjust the volume to 7 ml with formalin.
4   Add 3 ml ethyl acetate and stopper the tube. Shake the tube for 15 s on a vortex mixer or 30 s by hand.
5   Centrifuge the stoppered tube at 1 000 g for 60 s or for 5 min if investigation for *Cryptosporidium* is requested. There will now be four layers, the sediment, a formalin/debris layer, a fatty plug and an ethyl acetate layer.
6   Unstopper the tube with care because of positive pressure inside. Carefully loosen the fatty plug with an applicator stick. Pour off the plug and fluid layers leaving the sediment in the tube. A little fluid will run back to the sediment.
7   Dry the inside of the tube with a cotton swab. If the sediment appears too dry add a drop or two of 0.9% saline and mix well.
8   Leave the tube in the fume cupboard for a few minutes to allow any remaining ethyl acetate to evaporate.

Notes

If, after processing, the deposit is too heavy, either too much faecal material was taken, the shaking was inadequate or centrifugation too long or too heavy.

With very fluid, thin or mucoid specimens little material is left after processing. These specimens should be centrifuged directly after fixation in formalin without filtration or shaking with ethyl acetate.

Examination of specimen

1   Mix the sediment thoroughly and make a saline and iodine preparation from each specimen as described earlier.

2   Systematically examine the whole of each preparation first with the 10× objective and the 40× when necessary. Measure and identify any parasites.
3   Record positive findings including species normally regarded as non-pathogenic.

Trichrome stain for demonstration of protozoa

Permanently stained smears of stool are essential for detection of trophozoites when there is a delay in examination. They also allow more time for observation and show better nuclear and cytoplasmic detail than wet mounts (Garcia *et al.* 1979).

Specimen collection

The specimen is fixed in SAF fixative *immediately* after collection. Note: formalin-fixed specimens are unsuitable for trichrome staining.

Reagents

- Positive control: SAF-fixed faeces containing trophozoites of any species.
- Mayer's egg albumin.
- Wheatley's (1951) trichrome stain: chromotrope 2R 0.6 g, light green SF 0.3 g, phospho-tungstic acid 0.75 g, glacial acetic acid 1 ml, distilled water 100 ml. Add acetic acid to dry components. Allow to stand 15–30 min at RT. Add the distilled water. Store at RT. The stain has a shelf life of 24 months.
- Acid–alcohol: 0.45% glacial acetic acid in 90% ethanol.
- Ethanol, 70%, 95%, and absolute.
- Xylene or substitute, for example, Histoclear (National Diagnostics).
- Mounting medium, for example, DPX (BDH).

Preparation of specimen

1   Filter some of the fixed specimen through a layer of dampened surgical gauze into a beaker.
2   Place about 3 ml of the filtered specimen into a centrifuge tube and fill the tube with 0.9% saline.
3   Centrifuge for 1 min at 800 g.
4   Pour off enough of the supernatant so that about 1 ml remains in the tube. Dilute with 0.9% saline if the specimen appears too thick.
5   Mix a drop of egg albumin and specimen on a microscope slide. Make a thin smear which varies in thickness. Prepare a positive control in the same way.
6   Allow the smear to dry until it appears 'sticky', about 10 min.

Staining procedure

Set up 10 staining (Coplin) jars as shown in Figure 33.1.

1   70% ethanol: 30 min to harden and fix smear.
2   70% ethanol: 10 min.
3   70% ethanol: 3–5 min.
4   70% ethanol: 3–5 min. Slides may remain here up to 24 h.

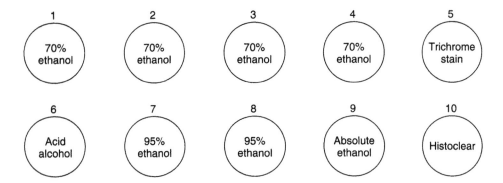

Figure 33.1 Set up of staining jars for trichrome stain for intestinal protozoa. Change contents of all jars except those with trichrome stain and Histoclear when about 10 slides have been processed. Change trichrome stain and Histoclear about every 50 slides or more often depending on the staining of the positive control.

5   Trichrome stain: 6–8 min, remove excess stain with absorbent tissue.
6   Acid–alcohol: 5–10 s; transfer to jar 7 immediately stain runs from smear.
7   95% ethanol: two dips. Change ethanol frequently.
8   95% ethanol: 5 min.
9   Absolute ethanol: 3 min.
10   Histoclear: 3 min.
11   Mount the slide with DPX under a coverglass. Allow to dry.

Examine preparation under the light microscope using 10×, 40×, and oil immersion objectives. Protozoal cysts and trophozoites are stained in different shades of red and blue-green against a green background. Chromidial bars are red.

Iron haematoxylin stain

Trichrome staining is suitable for routine use but for fine nuclear detail, iron haematoxylin staining (Spencer and Munroe 1977) is superior.

Reagents
- Iron haematoxylin.
- Solution 1: haematoxylin 10 g dissolved in 1 000 ml 99.5% ethanol. Allow the solution to stand in light for 1 week. Store in brown glass-stoppered bottle.
- Solution 2: Ferrous ammonium sulphate 10 g, Ferric ammonium sulphate 10 g, concentrated hydrochloric acid 10 ml, distilled water 100 ml. Shelf life, 6 months.

For use: mix equal parts of each solution. The mixture keeps about 1 week.

- Acid–alcohol: 1% hydrochloric acid in 70% ethanol.

Procedure

Smears are prepared as described above for trichrome staining.

Staining of smears

1   Place in 70% ethanol: 30 min.
2   Transfer to two changes of 70% ethanol: 5 min each.
3   Wash under a gently running tap: 10 min.
4   Place in working solution of haematoxylin: 5 min.
5   Wash in running tap water: 10 min.
6   Dip slide in acid–alcohol for a few seconds.
7   Wash in running tap water: 10 min.
8   Transfer to two changes of 70% ethanol: 5 min each.
9   Transfer to 95% ethanol: 5 min.
10   Place in absolute ethanol: 5 min.
11   Transfer the slide to Histoclear: 3 min.
12   Mount under a coverslip with DPX.

Examine under the light microscope using 10×, 40×, and 100× objectives for at least 100 fields. Measure objects with the calibrated eyepiece micrometer. Nuclear chromatin and chromidial bars are black, other structures, varying shades of gray to black.

Giemsa stain for protozoa

The main use for Giemsa staining in the parasitology laboratory is in the investigation of suspected malaria when both thick and thin films are examined. Giemsa is also useful for staining other protozoa, for example, *Acanthamoeba*, *Trichomonas*, etc.

Reagents

•   Absolute methanol, acetone free.
•   Giemsa stock solution. This may be prepared locally but commercially produced stain is more satisfactory: place 3.8 g Giemsa stain, 250 ml methanol and 250 ml glycerol in a glass bottle with a few beads. Mix well at intervals over a day until the stain is dissolved. Leave the stain in a warm place. Ready for use the next day but improves on keeping.
•   Phosphate buffer pH 7.2: dissolve 1 g disodium hydrogen phosphate, Na_2HPO_4 (or 2.5 g $Na_2HPO_4 \cdot 12H_2O$) and 0.7 g potassium dihydrogen phosphate, KH_2PO_4 in distilled water; make the volume up to 1 l. Check the pH.

Ready prepared buffer tablets pH 7.2 to be dissolved in distilled water can be obtained from, for example, Merck, BDH, Gurr, etc.

Staining of smears

Prepare 5% Giemsa dilution every time a batch of slides is stained. Single slides may be stained on a staining rack, several slides in a staining jar. For single slides dilute 1 ml Giemsa stock stain with 19 ml phosphate buffer pH 7.2; for several slides dilute 5 ml Giemsa stain with 95 ml buffer.

1   Fix smears in absolute methanol in a staining jar with lid for 30 s.
2   Stain the slides with Giemsa dilution for 20 min.
3   Carefully rinse off stain under running water, either from a squeeze bottle or under the tap. Never direct the water jet directly at the smear.
4   Allow the slide to dry in an upright position.

Examine under immersion oil or mount under a coverslip with DPX.

Detection of important protozoa of the North

The detection of *Toxoplasma* is discussed elsewhere.

Entamoeba histolytica

The diagnosis of *E. histolytica* infection by detection of trophozoites or cysts in clinical material requires careful attention to specimen collection, processing and examination. Different types of clinical material may be submitted to the laboratory. The majority of specimens are faecal but microscopical examination of material obtained from rectoscopy may also be requested. With suspected invasive amoebiasis, specimens may be aspirates or biopsy material.

Examination of dysenteric or diarrhoeic stools and rectal scrapes

The chances of detecting motile trophozoites are highest with this type of specimen; they should be examined without delay, preferably within 15–30 min of collection although trophozoites sometimes survive for longer periods. For direct microscopy, wet mounts are prepared by emulsifying, if present, bloody or mucoid portions of the specimen in warm (37°C) saline. Fluid specimens need less or no saline. A portion of specimen should also be fixed in SAF fixative for permanent staining.

Formed stool specimens

Cysts rather than trophozoites are found in formed stool. Specimens fixed in formalin or SAF fixative are processed by formol ethyl acetate concentration. SAF-fixed material is essential for permanent staining.

Material from aspiration

Aspirates, usually from liver abscesses, are treated as for rectal scrapes. The chances of finding amoebae are better with material from the edge of the abscess. Usually, only necrotic material with occasional leukocytes is seen. Bacteria are not present in amoebic abscesses; if they are seen together with large numbers of pus cells, the abscess is bacterial rather than amoebic in origin.

Biopsy material

This is fixed in formol saline and processed in the same way as other surgical material in the histology laboratory.

Culture of amoebae

Culture of amoebae has been long established (Robinson 1968). However, it is carried out in few laboratories and usually as part of isoenzyme typing of *E. histolytica* isolates. Nevertheless, Haque *et al.* (1995) found that culture in Robinson's medium (1968) was more reliable than microscopy in identifying *Entamoeba* species.

Triage Micro Parasite Panel (Biosite Inc., San Diego, CA) is a commercial kit detecting antigens of *G. intestinalis*, *E. histolytica/dispar* and *Cryptosporidium* in unpreserved, fresh or frozen stool samples. The test is rapid and can be used as a screen for immediate testing of specimens however, the high cost of the assay may limit its applications and this test does not differentiate between *E. histolytica* and *E. dispar* (Sharp *et al.* 2001).

Differentiation of Entamoeba histolytica *and* Entamoeba dispar

Entamoeba histolytica and the non-pathogenic *E. dispar* are morphologically identical. The finding of haematophagous trophozoites is diagnostic for *E. histolytica*. Otherwise positive findings, whether of cyst or trophozoite must be reported as *E. histolytica/E. dispar*. Antigen-detection methods that can differentiate between *E. histolytica* and *E. dispar* have been developed recently (Abd-Alla *et al.* 1993; Haque *et al.* 1994) and a commercial version of one of these, the *E. histolytica* test (TechLab, Inc., Blacksberg, VA) has been tested in the field with encouraging results (Haque *et al.* 1995). An improved, second-generation LabTech kit showed an increased sensitivity of 100% and 95.7% of antigen detection in intestinal infections and liver abscess cases respectively (Haque *et al.* 2000). Another ELISA-based commercial test that has been investigated (Ang *et al.* 1996) is the ProSpecT® (Alexon Inc, Sunnyvale, CA, USA).

Molecular methods based on detection of *E. histolytica*-specific DNA by PCR amplification have been developed in recent years. A number of primers designed for *E. histolytica*-unique genes proved to be useful for the detection of parasites in both stool samples and liver abscess material (Clark and Diamond 1991; Tachibana *et al.* 1991; Prakash *et al.* 2000; Zaman *et al.* 2000). The PCR-based diagnostics of *E. histolytica* infections although specific and sensitive are not yet widely used. One of the limiting factors is that formalin based fixatives usually used to preserve specimens could interfere with DNA recovery from the sample. Further optimization of sample processing is necessary to allow more common application of molecular techniques.

Giardia intestinalis

Diagnosis of *G. intestinalis* infection is made by detecting cysts or, less frequently, trophozoites in the stool. As cyst excretion is often irregular and of low density, formol ethyl acetate concentration of three stool specimens collected every other day is preferable. Although trophozoites are not concentrated they may be present and identifiable. Nevertheless, direct microscopy of unfixed material or trichrome staining of SAF-fixed stool is more reliable for detecting trophozoites. Trophozoites may also be detected in faecal specimens by staining thin smears of material by Giemsa stain.

Duodenal aspirates have been used to detect trophozoites. The sensitivity is lower than stool microscopy but useful as a complement to the latter (Goka *et al.* 1990).

Immunological methods of detecting Giardia intestinalis *in faecal specimens*

These are available commercially in kit form. They are of different types; faecal antigen ELISA tests (Goldin *et al.* 1992), immunofluorescence assays using monoclonal antibodies (Garcia *et al.* 1992) and Triage Micro Parasite Panel (Biosite Inc., San Diego, CA). When using these kits it is essential to follow the manufacturers' instructions. The main disadvantages are cost and the interpretation of microscopy negative, immunological test positive results.

Cryptosporidium parvum

Like *E. histolytica* and *G. intestinalis*, monoclonal antibody-based methods have been developed for the detection of *C. parvum*. Combined *Giardia/Cryptosporidium* kits are available (Garcia *et al.* 1992; Sharp *et al.* 2001). However, traditional staining methods are more commonly used for the detection of *Cryptosporidium*. The different species of coccidia that are found in the stool must be differentiated.

Modified Ziehl–Neelsen stain for the demonstration of Coccidia

Coccidia oocysts after staining with cold Ziehl–Neelsen stain are resistant to decolourization with acid–alcohol and appear bright red in colour against a green background.

Clinical material

Formalin- or SAF-fixed faeces. Staining is normally done after formol ethyl acetate concentration but unconcentrated and even unfixed material may be used. Sputum and bronchoalveolar lavage are treated as for suspected *Pneumocystis* infection.

Reagents

- Absolute methanol.
- Strong carbol fuchsin stain: Basic fuchsin 10 g; Absolute ethanol 100 ml; 5% phenol 900 ml. May be purchased ready made, for example, from Merck, BDH, Gurr, etc.
- Acid alcohol, 1% HCl in 95% ethanol.
- Malachite green stain, 10 g malachite green in 1 000 ml 10% ethanol.
- DPX mounting medium (BDH), or similar.

Positive control

Formalin-fixed faeces containing *Cryptosporidium* oocysts kept at 4°C. Valid at least 1 year. Slides ready for staining are prepared and fixed in advance.

Staining procedure

1. Place a drop of formol ethyl acetate deposit on a microscope slide and allow to airdry for at least 1 h. With unconcentrated material make a thin smear on a slide. Process a positive control at the same time as the clinical specimens.
2. Fix the slide in methanol in a staining jar for 10 min.
3. Place the slide on a staining rack and cover with carbol fuchsin. Stain for 20 min.
4. Rinse well under running tap water.
5. Decolourize with acid alcohol in a staining jar until the red colour of the smear has disappeared.
6. Rinse under running tap water.
7. Cover the slide with malachite green stain for 30 s.
8. Rinse under running tap water.
9. Air-dry the preparation.
10. Mount under a coverslip with DPX. Allow to dry.

Examine the whole preparation under the light microscope. Use the 40× objective for screening and 100× oil immersion for confirmation. Measure any oocysts.

Cryptosporidium oocysts; round, 4–6 μm in diameter, stained in varying shades of pale pink to dark red against a green background. Completely unstained oocysts may be present. Identifying sporozoites inside oocysts is diagnostic for *Cryptosporidium*.

Cyclospora oocysts; 8–10 μm in diameter and similar in colouring to *Cryptosporidium* but without a definite inner structure. Completely uncoloured oocysts with a glassy appearance are often seen.

Isospora oocysts; 20–30 × 10–19 μm containing either a granular zygote or two sporoblasts. They are coloured in different shades of red; some oocysts are unstained.

Microsporidia

Of the options available for detection of microsporidia, electron microscopy (EM) is the 'gold standard'. However, EM is not available to most laboratories and is less sensitive than other methods. Other techniques include Giemsa staining and modified trichrome stains. Screening methods employing chemofluorescent agents such as Calcofluor are sensitive but non-specific. Newer approaches include the use of poly- and monoclonal antibodies but are not yet commercially available.

Microsporidia are difficult to detect because of their small size, often against a debris-filled background. A modified trichrome stain with increased Chromotrope 2R and aniline blue as counterstain instead of light green SF (Ryan *et al.* 1993) gives good contrast between microsporidia and background.

Trichrome blue stain for the demonstration of Microsporidia spores

Reagents

- Absolute methanol.
- Trichrome blue stain: Add 6 g chromotrope 2R, 0.5 g aniline blue and 0.25 g phospho-tungstic acid to 3 ml glacial acetic acid. Wait 30 min and add 100 ml distilled water. Store at RT. Shake the stain before using. Filter if precipitates are present. The stain is available commercially (Para-Pak Trichrome Blue, Meridian Diagnostics).
- Other reagents are as for the standard trichrome method.

Positive Microsporidia control

An untreated faeces specimen containing microsporidial spores stored at −40°C. For use this is treated wih 10% formalin as for standard faeces specimens and kept at 4°C for up to 1 year.

Pre-treatment of clinical material

Faeces; the investigation is done on unconcentrated faeces fixed in formalin or SAF fixative. Very thin faeces are centrifuged for 5 min at 1 800 g and the deposit used to make a smear.

Urine; about 10 ml urine is centrifuged at 1 800 g for 10 min. Discard the supernatant and mix the deposit with a few drops of 10% formalin.

Sputum and BAL specimens; the specimens are treated as for demonstration of *Pneumocystis carinii* except inactivation is done with 10% formalin instead of methanol.

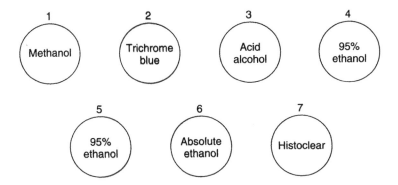

Figure 33.2 Set up of staining jars for trichrome blue stain. Change contents of jars 3–6 each time slides are processed. Change methanol once a week and Histoclear after about 30 slides. The trichrome blue can be used as long as the spores stain correctly, usually for about 30 slides.

Staining procedure

Carried out in a fume cupboard using gloves. Set up seven staining (Coplin) jars as shown in Figure 33.2.

1 Prepare a very thin smear from about 10 µl specimen; allow to dry.
2 Methanol: 5 min, to fix smear.
3 Stain in Trichrome blue for 90 min.
4 Carefully remove excess stain with absorbent tissue.
5 Decolourize smears in acid–alcohol, about 5 s.
6 Rinse quickly in 95% ethanol.
7 95% ethanol: 5 min.
8 Absolute ethanol: 10 min.
9 Histoclear: 10 min.
10 Mount the specimen under a coverglass with DPX; allow to dry.

Protect the mounted specimen against light.

Examination of smear

Observe 200 fields using 100× oil immersion objective; for conjunctival scrapes examine the whole preparation. Microsporidia spores range from about 1.5 × 0.9 mm to 3.5 × 2.0 mm in size, pinkish-red in colour against a blue background. Individual Microsporidia species cannot be identified. A negative result does not exclude infection.

Trichomonas vaginalis

Of the three trichomonads infecting humans, only *T. vaginalis* is generally regarded as pathogenic. Little is known about *Trichomonas tenax* inhabiting the oral cavity and the position is not entirely clear regarding the intestinal parasite *Trichomonas hominis*. The latter is often recovered from diarrhoeic stool but is considered non-pathogenic.

The usual method of diagnosing *T. vaginalis* infection is by direct microscopy of vaginal secretion or, in infection in the male, by microscopy of urethral secretion or urine sediment especially after prostatic massage. However, direct microscopy has a sensitivity of only between 40 and 80%. Staining, for example, with acridine orange does not increase this appreciably. Fluorescent antibody techniques have a higher sensitivity than other direct methods (Lossick and Kent 1991) but culture is most sensitive (>95%).

Method for direct microscopy

1   Mix a drop of vaginal secretion with a drop of isotonic saline on a microscope slide. Urethral secretions are treated similarly; urine specimens are first centrifuged for 10 min at 500 g and the deposit used for microscopy.
2   Place a coverslip over the preparation. Examine immediately with the 40× objective. Phase contrast may be helpful.

Trichomonas vaginalis, little larger than a leukocyte, has a jerky motility, four flagellae and an undulating membrane. Non-motile protozoa are not detectable in wet mounts.

Culture of **Trichomonas vaginalis**

Materials and reagents

- Charcoal swabs.
- Stuart's transport medium.
- Modified Diamond's medium (Fouts and Kraus 1980): dissolve 2 g trypticase, 1 g yeast extract, 0.5 g maltose, 0.1 g L-cystein hydrochloride and 0.02 g ascorbic acid in 90 ml distilled water. Adjust pH to 6.0, add 0.1 g agar and autoclave at 15 psi. for 10 min. Cool to 48°C and add 10 ml inactivated sterile horse serum, 0.6 mg penicillin G, 2 mg netilmicin and 0.2 mg amfotericin B. Dispense 9–10 ml amounts into sterile screw-capped tubes. Store at 4°C.

Specimens

Specimens are taken with a charcoal swab, inoculated directly into Diamond's or Stuart's transport medium and sent to the laboratory immediately. Allow the tubes of medium to attain RT before inoculating specimens. Urethral specimens should be taken at least 2 hours after last passing urine. For urine specimens, the first portion passed is centrifuged and the deposit inoculated into Diamond's medium.

Culture procedure

1   Incubate the inoculated Diamond's medium for up to 7 days at 37°C.
2   With a sterile Pasteur pipette take a drop from the culture deposit on days 2, 3, 6 and 7 and make a wet mount. Examine microscopically with the 40× objective.

Less than 10 organisms per specimen can be detected by this method (Fouts and Kraus 1980). It is important to adhere to the schedule for investigation as trichomonads are sensitive and die easily.

Differentiation of trichomonads

Although the different species of trichomonad are normally located in different sites, contamination of, for example, vaginal secretions or urine with faeces necessitates the differentiation of species. There are no cyst forms, only trophozoites.

Trichomonas vaginalis is pear-shaped and 7–23 μm long by 5–15 μm wide. It has four anterior flagellae of uneven length and in fresh preparations, shows a nervous jerky motility. There is a short undulating membrane with a costa.

Trichomonas hominis is a little smaller, 5–15 μm long and 7–10 μm wide. Motility, by four to five anterior flagellae, is similar to that of *T. vaginalis*. The undulating membrane and costa extend the whole length of the organism. There is a free trailing flagellum.

Acanthamoeba

Acanthamoeba spp. can sometimes be detected by microscopy of clinical material using Giemsa or trichrome staining. Cysts may be demonstrated using calcofluor white, a vital fluorescent dye (Marines *et al.* 1987). Culture is more reliable.

Demonstration of Acanthamoeba *species by culture*

Reagents

- Page's saline: dissolve 120 mg sodium chloride, 4 mg magnesium sulphate ($MgSO_4 \cdot 7H_2O$), 4 mg calcium chloride ($CaCl_2 \cdot 2H_2O$), 142 mg disodium hydrogen phosphate (Na_2HPO_4) and 136 mg potasium dihydrogen phosphate (KH_2PO_4) in distilled water and make the volume up to 1 l. Autoclave for 15 min at 15 psi. Store at 4°C. Shelf life is 6 months.
- Non-nutrient agar: Dissolve 1.5 g agar in 100 ml Page's saline with gentle heat. Stir or swirl to dissolve. Dispense 20 ml amounts into screw-capped tubes (20 × 150 mm). Autoclave for 15 min at 15 psi. Store at 4°C. The shelf life is 12 months. For use, the contents of a tube are melted in a boiling water bath and poured into a 9 cm sterile Petri dish. After setting, the prepared plates are stored at 4°C. Plates sealed with Parafilm may be stored up to 1 month.
- Blood agar plates.

Esherichia coli K12 stock culture; subcultured onto blood agar plate and incubated for 1 day at 37°C. Subculture stored at 4°C for 1 month and then re-cultured.

Positive Acanthamoeba *control*

Acanthamoeba stock culture of any species; stored on non-nutrient agar plates sealed with Parafilm for 1 month and then subcultured as described below.

Procedure

Clinical material; corneal scrapings or biopsy in Page's saline (0.5 ml); contact lens or contact lens solution, sent to the laboratory as quickly as possible to ensure survival of trophozoites.

Culture

All procedures should be carried out on a Class 2 safety bench or cabinet using gloves.

1   Use one non-nutrient agar plate for each test including a control. Allow the plates to warm up to RT.
2   Pipette about 0.5 ml Page's saline onto the blood agar plate with the *E. coli* culture. Emulsify some of the growth with a 10 μl bacteriological loop. Place three to four drops of bacterial suspension onto each non-nutrient plate. Spread out the bacterial suspension with a 10 μl loop and allow the fluid to absorb into the agar.
3   Inoculate each specimen in the centre of a separate *E. coli* – seeded plate. Fluid specimens are first centrifuged at 500 g, the supernatant removed and the deposit inoculated onto the plate. Avoid flooding the plate with liquid inoculum. Incubate the inoculated plates inverted at 30°C in a moist chamber, for example, a plastic bag containing some moistened absorbent tissue.
4   Before setting up the control strain, swab the working surfaces with 70% alcohol.
5   Culture the *Acanthamoeba* reference strain by cutting a 1 cm² piece of agar from the plate with a sterile scalpel. Place it upside down in the centre of a seeded non-nutrient agar plate. Incubate as for the clinical specimens but in a separate moist chamber.

All plates are incubated for 1 week and examined daily.

Examination

Clinical specimens are examined first and then the control plate after swabbing the working area with 70% alcohol and changing gloves.

The whole surface of each plate is scanned for trophozoites and cysts under an inverted microscope. A smear or imprint is made on a microscope slide from plates that appear positive. These are stained with Giemsa stain as described previously.

Criteria for identification of Acanthamoeba spp.

Direct microscopy; viable trophozoites contain a vacuole, which empties itself several times each minute. They are between 20–40 μm in size and assume different shapes. Cysts, 12–18 μm in diameter, polygonal with a double cyst wall, similar to talc crystals.

The cytoplasm of Giemsa-stained trophozoites, contain several vacuoles. The nucleus is centrally placed and contains a large endosome.

Negative findings do not exclude *Acanthamoeba* spp. infections. The sensitivity for culture of corneal scrapes is about 76%. Deep biopsy of the cornea gives a higher sensitivity (Auran *et al.* 1987).

Microscopical demonstration of Pneumocystis carinii in lower respiratory tract infections

Induced sputum, bronchoalveolar lavage (BAL) or biopsy specimens may be submitted to the laboratory for investigation. Smear, touch or imprint preparations are most often used when little biopsy material is obtained from transbronchial biopsy, paraffin or frozen sections when larger amounts of lung tissue are available, for example after post mortem examination. The prepared specimens are examined microscopically with a cytochemical method such as Toluidine blue O staining or an immunological procedure such as immunofluorescence.

Preparation of clinical material

This is carried out in a Class 2 safety cabinet or bench.

Sputum

1   Estimate the volume of the specimen and add a equal amount of 10% Sputolysin (6.3 mM dithiothreitol).
2   Shake the mixture, wait 5–15 min and shake the specimen again. If clumps of mucoid material are still present repeat the treatment.
3   Add an equal volume of 50% ethanol. Allow the mixture to stand for 15 min.
4   Centrifuge at 1 300 g for 5–10 min. Remove the supernatant by suction or pouring off.
5   Mix the pellet with the small amount of fluid remaining in the tube.
6   Place 25–50 μl in a well of a Teflon-coated slide for immunofluorescence and onto an ordinary microscope slide for Toluidine blue staining.

BAL specimens

If the specimen contains mucus treat it first with Sputolysin as above. Add an equal volume of 50% alcohol to the specimen. Centrifuge and make slides as for sputum specimens.

Biopsy material. Larger biopsies are fixed in formalin; material on slides is fixed in absolute ethanol or methanol.

Toluidine blue O staining

Reagents

- Fixing solution: measure 75 ml glacial acetic acid into a beaker standing in a water-bath containing ice (temperature <10°C). Slowly add 25 ml concentrated sulphuric acid while stirring the mixture constantly. When ready, the mixture is stored in a staining jar with a tight fitting lid. The shelf life is 1 week at RT.
- Toluidine blue stain: dissolve 0.45 g Toluidine blue O in 90 ml distilled water. Add 3 ml concentrated HCl and 210 ml absolute ethanol. The shelf life is 1 year at RT.
- Isopropanol.
- Xylene or substitute, for example, Histoclear (National Diagnostics).
- DPX mounting medium (BDH).

Positive Pneumocystis carinii *control*

A positive specimen is stored at −40°C in small amounts. To conserve material, strongly positive specimens may be diluted with treated negative BAL material. For use, a portion of control specimen is thawed and placed on slides as for clinical specimens.

Procedure

1   Place the slide in fixing solution for 10 min. The latter should be remixed before the slide is inserted.
2   Wash the slide in a staining jar under a running tap.
3   Stain in Toluidine blue O solution for 4–5 min. *Each slide must now be treated one at a time.*
4   Remove excess stain around the preparation and from the back of the slide with absorbent tissue.
5   Rinse away excess stain by dipping the slide three times in each of two staining jars containing isopropanol.
6   Dip the slide in Histoclear until the preparation clears.

7 Allow the slide to remain in a further jar containing Histoclear until all slides have reached point 7.
8 Wipe away excess Histoclear and mount under a coverslip with DPX whilst the preparation is still damp. Allow to dry.

Immunofluorescent staining method

Reagents

1 Fixing solution; equal parts absolute methanol and acetone.
2 Trypsin solution: 0.25% trypsin in phosphate buffered saline (PBS) pH 7.1.
3 Monoclonal mouse anti-*Pneumocystis* specific antibody (Dako) optimally diluted.
4 FITC-conjugated sheep anti-mouse immunoglobulin (Dako) optimally diluted in 1 in 10 000 Evans blue solution.
5 Buffered glycerine mounting medium.

Positive Pneumocystis carinii *control specimen*

Control specimens are prepared as above. A series of control slides may be prepared, fixed in equal parts methanol and acetone and stored at 4°C until used.

Procedure

1 Fix the slide in equal parts methanol and acetone. Store at 4°C until stained.
2 Treat the slide with trypsin solution for 30 min at 37°C to remove interfering immunoglobulins. Paraffin sections do not need this treatment.
3 Transfer the slide to three changes of PBS, 1 min each.
4 Incubate with monoclonal antibody in a moist chamber for 30 min at RT.
5 Transfer to three changes of PBS for 1 min each.
6 Incubate with diluted class specific monoclonal anti-mouse FITC conjugate for 30 min at RT in a moist chamber.
7 Rinse in three changes of PBS for 1 min each.
8 Mount under a coverslip with buffered glycerine.

Examination and assessment of specimens

Morphologically identifiable *P. carinii* cysts (about 7 mm) in groups or bunches should be seen for a positive result. Large numbers of trophozoites are observed in some preparations. To be judged as positive, these should have typical morphology or, as is often the case, isolated cysts should be present also.

If the specimen does not contain alveolar macrophages and there are many epithelial cells, the material is not representative and a new specimen should be requested.

Detection of some important helminths of the North

Enterobius vermicularis

Adult worms are small. The female measures 8–12 mm in length and may be filled with ova. The male is much smaller and seldom seen. Both are white in colour. The male worm's tail is curved with a single spicule, the female's is long and pointed.

Adhesive tape test for demonstration of Enterobius vermicularis *ova*

Microscopy of stool is not efficient for diagnosing *E. vermicularis* infection. Instead, ova and sometimes adult worms can be captured by pressing a strip of clear adhesive tape around the patient's anus early in the morning before the patient has bathed.

Specimen collection

1   Press the adhesive side of a 10 cm length of transparent adhesive tape (Sellotape, Scotch tape, etc.) on the skin in several places around the anal opening. The non-sticky side of the tape can be folded around the end of a tongue depressor or spoon handle for support.
2   Stick the tape to a microscope slide.
3   Place the slide in a slide tray for transport to the laboratory. If the preparation cannot be examined immediately it may be stored at 4°C for a few days without alteration of morphology.

Examination

Scan the whole preparation under a light microscope with the 10× objective and the 40× objective for verification.

Both ova and adult worms may be recovered by this procedure. The preparation should be inspected carefully with the naked eye. The ova contain infective larvae; gloves should be used as appropriate. *Taenia* ova may sometimes be captured.

Microscopical examination of the contents of hydatid cysts for Echinococcus scolices

Aspiration of cyst contents is a dangerous procedure. Escape of the fluid into the body cavity during aspiration may cause anaphylactic shock and the scolices may develop into further cysts. Nevertheless, cyst contents do sometimes arrive at the laboratory with the request to exclude hydatid disease.

Procedure

1   Record the macroscopical appearance. With unfixed specimens add an equal amount of formalin solution. Allow to react for 30 min.
2   Mix the specimen. Transfer 10 ml to a centrifuge tube. Centrifuge 10 min at 800 g.
3   Discard the supernatant.
4   Examine the whole deposit under the microscope using the 10× objective and the 40× objective where necessary.

Record the presence of protoscolices and hooklets. If necessary, compare with a reference specimen. A negative result does not exclude hydatid disease.

Trichinella *species*

The laboratory diagnosis of trichinosis is usually made by immunological methods. Adult worms and larvae have been identified in the faeces and larvae isolated from peripheral blood in the early stages of the migration phase (Gilles 1996). *Trichinella* larvae may also be detected in striated muscle.

Demonstration of Trichinella *spp. larvae in muscle*

Larvae are present in muscle tissue from about two to three weeks after infection.

Specimen

A biopsy of striated muscle is required. A 1 cm^2 piece of deltoid biceps, gastroenemius or pectoralis major muscles collected under local anaesthesia is suitable.

Direct Slide Technique:

Gloves are used at all stages

1   With a sterile scalpel cut the tissue into thin slices. Use forceps to hold the tissue. Place the tissue onto a glass slide and squash the pieces with another slide.
2   Hold the slides together with adhesive tape applied at the ends.
3   Examine the preparation under the microscope using the 10× objective and reduced light.
4   If no larvae are seen, a drop of glycerine applied between the slides clears the tissue making any larvae present easier to observe.

Digestion Technique:

Acid pepsin digests the tissue and any larvae are concentrated by centrifugation.

   Acid pepsin reagent: Dissolve 1 g pepsin in about 50 ml distilled water. Add 1 ml concentrated HCl and make up to 100 ml with distilled water. Mix well. Store at 4°C. The shelf life is several months.

Procedure

Gloves are worn and forceps used to handle the specimen.

1   Place small pieces of tissue into 20 ml of acid pepsin reagent. Incubate overnight at 37°C.
2   Transfer the contents of the bottle to two centrifuge tubes. Centrifuge at 500 g for 2–5 min.
3   Discard the supernatant.
4   Add a small amount of 10% formalin, mix, resuspend the pellet and fill the tubes with fixative.
5   Centrifuge the tubes again and discard the supernatant.
6   Transfer the deposits to glass slides. Cover with coverslips and examine each preparation under the microscope using the 10× objective and reduced light for *Trichinella* larvae.

Other helminths

Of the other important helminths, *Dyphyllobothrium latum* ova are readily detected using standard methods; worm segments are seldom seen. *Toxocara* and *Anisakis* larvae are not detected by standard parasitological methods although they may sometimes be seen in histopathological preparations. *Anisakis* larvae may sometimes be seen by fiber optic endoscopy when infecting the stomach mucosa. They are milky white in colour, 19–36 mm long by 0.3–6 mm broad with a blunt tail.

References

Abd-Alla, M. D., Jackson, T. F. H. G., Gathiram, V., El-Hawey, E. M. and Ravdin, J. L. (1993). Differentiation of pathogenic *Entamoeba histolytica* infections from nonpathogenic infections by detection of galactable–inhibitable adherence protein antigen in sera and faeces. *Journal of Clinical Microbiology*, **31**: 2843–2850.

Ang, S. J., Cheng, M. G., Liu, K. H. and Horng, C.-B. (1996). Use of the ProSpecT® microplate enzyme immunoassay for the detection of pathogenic and non-pathogenic *Entamoeba histolytica* in faecal specimens. *Transactions of the Royal Society of Tropical Medicine and Hygiene*, **90**: 248.

Auran, J. D., Starr, M. B. and Jakobiec, F. A. (1987). *Acanthamoeba* keratitis. A review of the literature. *Cornea*, **6**: 2–26.

Clark, C. G. and Diamond, L. S. (1991). Ribosomal RNA genes of 'pathogenic' and 'nonpathogenic' *Entamoeba histolytica* are distinct. *Molecular and Biochemical Parasitology*, **49**: 297–302.

Fouts, A. C. and Kraus, S. J. (1980). *Trichomonas vaginalis*: Reevaluation of its clinical presentation and laboratory diagnosis. *Journal of Infectious Diseases*, **141**: 137.

Garcia, L. S., Brewer, T. C. and Bruckner, D. A. (1979). A comparison of the formalin–ether concentration and trichrome stained smear methods for the recovery and identification of intestinal protozoa. *American Journal of Medical Technology*, **44**: 932–935.

Garcia, L. S., Shum, A. C. and Bruckner, D. A. (1992). Evaluation of a new monoclonal antibody combination reagent for direct fluorescence detection of *Giardia* cysts and *Cryptosporidium* oocysts in human fecal samples. *Journal of Clinical Microbiology*, **30**: 3255–3257.

Gilles, H. M. (1996). Soil transmitted helminths (Geohelminths). In: *Manson's Tropical Diseases*, 20th edn, G. C. Cook (ed.), p. 1406. London: Saunders.

Goka, A. K. J., Rolston, D. D. K., Mathan, V. I. and Farthing, M. J. G. (1990). The relative merits of faecal and duodenal juice microscopy in the diagnosis of Giardiasis. *Transactions of the Royal Society of Tropical Medicine and Hygiene*, **84**: 66–67.

Goldin, A. J., Apt, W., Aquilera, X., Zulantay, I., Warhurst, D. C. and Miles, M. A. (1992). A capture ELISA detects *Giardia lamblia* antigens in formalin-fixed faecal samples. *Transactions of the Royal Society of Tropical Medicine and Hygiene*, **86**: 164–165.

Haque, R., Lyerly, D., Wood, S. and Petri, Jr., W. A. (1994). Detection of *Entamoeba histolytica* and *Entamoeba dispar* directly in stool. *American Journal of Tropical Medicine and Hygiene*, **50**: 595–596.

Haque, R., Neville, L. M., Hahn, P. and Petri, Jr., W. A. (1995). Rapid diagnosis of *Entamoeba* infection by using *Entamoeba* and *Entamoeba histolytica* stool antigen detection kits. *Journal of Clinical Microbiology*, **33**: 2558–2561.

Haque, R., Mollah, N. U., Ali, I. K., Alam, K., Eubanks, A., Lyerly, D. and Petri, Jr., W. A. (2000). Diagnosis of amebic liver abscess and intestinal infection with the TechLab *Entamoeba histolytica* II antigen detection and antibody tests. *Journal of Clinical Microbiology*, **38**: 3235–3239.

Lossick, J. G. and Kent, H. L. (1991). Trichomoniasis. Trends in diagnosis and management. *American Journal of Obstetrics and Gynaecology*, **165**: 1217–1222.

Marines, H. M., Osato, M. S. and Font, R. L. (1987). The value of calcofluor white in the diagnosis of mycotic and *Acanthamoeba* infections of the eye and ocular adnexa. *Opthalmology*, **94**: 23–26.

Moody, A. (1986). The development of internal quality control and external quality assessment in parasitology. In: *Laboratory Health Care. Developing Country Proceedings*, M. Cheesbrough (ed.), p. 76. Cambridge: Tropical Health Technology.

Prakash, A., Chakraborti, A., Mahajan, R. C., Ganguly, N. K. (2000). *Entamoeba histolytica*: rapid detection of Indian isolates by cysteine proteinase gene-specific polymerase chain reaction. *Experimental Parasitology*, **95**: 285–287.

Robinson, G. L. (1968). The laboratory diagnosis of human parasitic amoebae. *Transactions of the Royal Society of Tropical Medicine and Hygiene*, **62**: 285–294.

Ryan, N. J. *et al.* (1993). A new trichrome blue stain for detection of microsporidial species in urine, stool and nasopharyngeal specimens. *Journal of Clinical Microbiology*, **31**: 3264–3269.

Sharp, S. E., Suarez, C. A., Duran, Y. and Poppiti, R. J. (2001). Evaluation of the Triage Micro Parasite Panel for detection of *Giardia lamblia, Entamoeba histolytica/Entamoeba dispar*, and *Cryptosporidium parvum* in patient stool specimens. *Journal of Clinical Microbiology*, **39**: 332–334.

Spencer, F. M. and Munroe, L. S. (1977). *The Color Atlas of Intestinal Parasites*, 7th printing, pp. 17–19. Springfield: Thomas.

Tachibana, H., Kobayashi, S., Takekoshi, M. and Ihara, S. (1991). Distinguishing pathogenic isolates of *Entamoeba histolytica* by polymerase chain reaction. *Journal of Infectious Diseases* **164**: 825–826.

Tachibana, H., Kobayashi, S., Okuzawa, E. and Masuda, G. (1992). Detection of pathogenic *Entamoeba histolytica* DNA in liver abscess fluid by polymerase chain reaction. *Internationl Journal of Parasitology*, **22**: 1193–1196.

Wheatley, W. B. (1951). A rapid staining procedure for intestinal amoebae and flagellates. *American Journal of Clinical Pathology*, **21**: 990–991.

Young, K. H., Bullock, S. L., Melvin, D. M. and Spruill, C. L. (1979). Ethyl acetate as a substitute for diethyl ether in the formalin-ether sedimentation technique. *Journal of Clinical Microbiology*, **10**: 852–853.

Yang, J. and Scholten, T. (1977). A fixative for intestinal parasites permitting the use of concentration and permanent-staining procedure. *American Journal of Clinical Pathology*, **67**: 300–304.

Zaman, S., Khoo, J., Ng, S. W., Ahmed, R., Khan, M. A., Hussain, R. and Zaman, V. (2000). Direct amplification of *Entamoeba histolytica* DNA from amoebic liver abscess pus using polymerase chain reaction. *Parasitology Research*, **86**: 724–728.

34 DNA detection

Bettina Lundgren and Mats Olsson

The current laboratory diagnosis for many parasites is microscopical detection of the organism or specific antigenic epitopes (by immunofluorescence, IFL). Increased sensitivity is obtained with DNA amplification by using the polymerase chain reaction (PCR), targetted to specific gene fragments of parasites. The diagnostic and clinical significance of the method, in relation to different sample material, is evaluated for the following parasite organisms/diseases: *Pneumocystis carinii, Trichomonas vaginalis, Giardia lamblia,* Microsporidia spp., *Cryptosporidium parvum, Entamoeba histolytica* and *dispar,* and *Plasmodium* spp. causing human malaria-infection.

Introduction

Molecular diagnostic methods are gaining acceptance in clinical microbiological laboratories. Development of the PCR technique initiated a large number of reports on the use of DNA-based molecular methodology. These methods have proven to be useful for microbiological detection, strain typing, drug resistance, and epidemiology.

Currently, diagnosis of parasitic infections rely on clinical symptoms, travel history, geographical location of the patient, and several detection methods depending on the parasite or the specimen being tested. If culture or animal inoculation is required for identifying the parasite and introducing therapy, PCR may offer an advantage. When direct microscopy is sufficiently good to detect and species differentiate the parasites, PCR only offers an advantage by being able to process a large number of samples with an automated assay. However, when the parasite load is low, PCR may increase the sensitivity above that of microscopy.

For diagnostic purposes, a number of commercial kits for viral and bacterial detection by molecular methods are available, however, no kit for detection of parasites has been developed. This is, in part, due to the expense of new technology as well as a scarcity of these parasites in countries where this research is ongoing.

The PCR method requires the presence of specific key (primer) DNA sequences, complementary to DNA sequences flanking a DNA fragment of interest (target DNA) and a thermostable DNA polymerase. Following sample preparation, PCR amplification itself is a fast and automated method, simply requiring a programmable heating-block and addition of a standard reaction mixture. In general, detection of microbial DNA requires three procedure steps: (i) target DNA preparation; (ii) PCR amplification (subsequently re-amplification, if nested or semi-nested PCR is used); and (iii) identification of the amplified DNA fragments (amplimers or amplicons), usually by gel-electrophoresis.

Two main approaches for parasite diagnostic PCR assays are conducted: either a one-step species-specific (single) PCR, or a two-step procedure by a generic PCR, followed by DNA hybridization with a species-specific probe, a species-specific nested PCR or a species-specific detection by endonuclease restriction digest of the amplified PCR product.

Pneumocystis carinii

Since *Pneumocystis carinii* cannot be effectively cultured in vitro, the current laboratory diagnosis relies on direct microscopic demonstration of the pathogen. This is usually performed in broncho-alveolar lavage (BAL) by histochemical or immunohistochemical staining (Montaner and Zala 1995). Wakefield *et al.* (1990) presented a clinical application of a PCR assay, based on sequences coding for mitochondrial rRNA (mt rRNA) of *P. carinii*. Several other gene targets of the *P. carinii* genome have been used in PCR assays (Table 34.1). In a comparative study, Lu *et al.* (1995) found this assay to be the most sensitive single PCR for detection of *P. carinii* in BAL samples. The nested PCR methods, based on other *P. carinii* gene targets (Lu *et al.* 1995), performed better, as expected, than single PCR methods with regard to sensitivity and specificity.

Although the sensitivity of histochemical stains for detection of *P. carinii* in BAL specimen is high, up to 98% (Lipschik *et al.* 1992; Tamburrini *et al.* 1993; Roux *et al.* 1994), patients with *Pneumocystis carinii* pneumonia (PCP) may be severely ill and unable to undergo invasive diagnostic procedures. Less invasive sampling for the diagnosis of PCP is needed. The advantages with PCR concerns these less invasive sampling methods, mainly induced sputum samples, which in general contain fewer infectious organisms (Wakefield *et al.* 1991; Lipschik *et al.* 1992; Olsson *et al.* 1993; Cartwright *et al.* 1994; Roux *et al.* 1994; Leibovitz *et al.* 1995) (Figure 34.1). A forbearing alternative sampling procedure to BAL and sputum is oropharyngeal washes, which can be performed on patients with tendencies of bleeding and respiratory failure. The sampling technique does not require special equipment and no specially trained staff are required. The method has been used in HIV-infected patients with a reported sensitivity of 78% (Wakefield *et al.* 1993) and in a prospective study using the mt RNA single round PCR. The latter study showed a sensitivity of 72%, a specificity of 96%, a positive predictive value of 95% and a negative predictive value of 77% in diagnosing PCP among 49 HIV infected patients (Lundgren *et al.* 1996). Recently, the method has proven useful in diagnosing PCP among patients with haemotological malignancies (Helweg-Larssen *et al.* 1997).

PCR on serum or blood samples has shown conflicting results (Lipschik *et al.* 1992; Schluger *et al.* 1992; Contini *et al.* 1993; Roux *et al.* 1994; Atzori *et al.* 1995; Evans *et al.* 1995; Tamburrini *et al.* 1996) (see also Table 34.2). Thus, the use of blood products as a non-invasive specimen in the diagnosis of PCP remains to be established, although in rare cases of disseminated *P. carinii* infection, PCR on blood samples may consistently show *P. carinii* DNA (Lipschik *et al.* 1992; Roux *et al.* 1994).

Table 34.1 Gene targets for diagnostic detection of *P. carinii* with PCR amplification

Single copy genes	Reference
Enzymes	
Thymidylate synthase (TS)	(Olsson *et al.* 1993)
Dihydrofolate (DFHR) synthase	(Schluger *et al.* 1992)
Nuclear genes	
Internal trancribed spacers (ITS)	(Lu *et al.* 1995)
18S RNA	(Lipschik *et al.* 1992)
18S RNA	(Kitada *et al.* 1991)
Multi-copy genes	
Mitochondrial LSU rRNA (mt LSU rRNA)	(Wakefield *et al.* 1990)

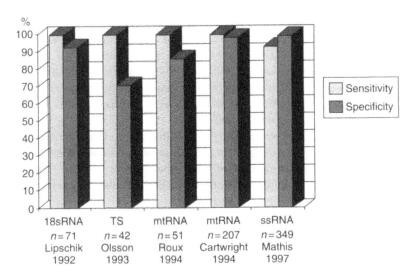

Figure 34.1 Sensitivity and specificity of PCR on induced sputum for diagnosing *P. carinii* pneumonia

Table 34.2 Serum/blood as non-invasive specimen for diagnosing PCP

Specimen	Target gene	Sensitivity (%)	Reference
Serum	DHFR	86	(Schluger *et al.* 1992)
Blood	18sRNA	10	(Lipschik *et al.* 1992)
PBMC	18sRNA	95	(Contini *et al.* 1995)
Blood	mt LSU rRNA	0	(Roux *et al.* 1994)
Serum			
PBMC			
Serum	ITS	100	(Atzori *et al.* 1995)
	DHFR	10	
Serum	mt LSU rRNA	0	(Tamburini *et al.* 1996)
PBMC		7	
PMNC		33	

The clinical significance of PCR for detection of *P. carinii* has been questioned since positive findings of *P. carinii* in immunosuppressed patients not always progressed to PCP (Leigh *et al.* 1992; Lipschik *et al.* 1992; Tamburrini *et al.* 1993; Weig *et al.* 1997). This has been interpreted as temporarly colonization or subclinical infection of *P. carinii* in susceptible individuals. On the contrary, Elvin *et al.* (1996) showed a predictive value of PCR in a study, where asymptomatic HIV infected patients without primary prophylaxis were followed during 3 years. Similar results have also been shown by others, mainly in HIV positive patients (Wakefield *et al.* 1991; Leigh *et al.* 1992, 1993; Lipschik *et al.* 1992).

In summary, laboratory diagnosis of *P. carinii* still relies on morphological identification of the organism in specimens obtained from the lung, although PCR is recommended as a diagnostic complement on negative stained induced sputum or other less invasive samples.

Trichomonas vaginalis

Trichomonas vaginalis is a common sexually transmitted pathogen causing vaginitis, exocervicitis, and ureteritis in women (Fauts and Kraus 1980). *Trichomonas vaginalis* infections has been suggested to play a role in the pathogenesis of pre-term birth, pre-term rupture of membranes and delivery of low-birth-weight infants (Cotch *et al.* 1991; Read and Klebanoff 1993). Recently *T. vaginalis* infections has been implicated as a cofactor in the transmission of HIV (Laga *et al.* 1993). *Trichomonas vaginalis* infection are frequently asymptomatic, and early, accurate diagnosis are required for specific treatment.

Routine diagnosis of *T. vaginalis* usually depends on direct microscopic identification of the parasite in wet mount preparations. However, wet mount examinations detects only 60% and the direct immunofluorescense using monoclonal antibodies detect 86% of culture positive cases in women. Although culture is considered the most reliable diagnostic method, with a sensitivity of >90% for detecting *T. vaginalis* (Spence *et al.* 1990), it requires complex media and is time-consuming (daily examination from 2 to 7 days). Furthermore, even redundant culture techniques may miss parasites in low number, defective parasites and organisms not surviving the transfer to cultural medium.

Molecular approaches to diagnose *T. vaginalis* have been reported (Riley *et al.* 1992; Briselden and Hiller 1994; Jeremias *et al.* 1994; Kenge *et al.* 1994; Lin and Shaio 1997). PCR detection of *T. vaginalis* in clinical samples was first reported by Riley *et al.* (1992), who found that primers amplifying a 102-bp genomic sequence, termed as A6, detected *T. vaginalis* in all 24 clinical isolates tested. Using the same primer pairs another study conclude that PCR analysis may be useful in women who were negative by wet mount but whose symptomatic vaginitis remained unexplained (Jeremias *et al.* 1994). Recently, the target of a family of 650-bp repeat of the *T. vaginalis* genome used in a single tube nested PCR has been evaluated in 378 clinical vaginal discharges (Shaio *et al.* 1997). For symptomatic patients, the PCR was as sensitive as culture for detection of *T. vaginalis* in vaginal discharge. The test was consistently negative in patients whose symptoms were due to bacterial vaginosis and vaginal candidiasis, and for asymptomatic women nested PCR was more sensitive than culture for detection of *T. vaginalis*.

To date, PCR is not a routine detection method in diagnosing *T. vaginalis*.

Giardia lamblia

Diagnosis of giardiasis relay on direct visualization of *G. lamblia* trophozoite cysts by direct microscopy of faecal specimens, which has a sensitivity between 50 and 70% (Buke, 1975). PCRs have had difficulties in detecting the organisms due to problems in lysis of the cysts and large amounts of inhibitory substances present in the faecal specimens (Butcher and Farthing 1989; Lewis *et al.* 1990). Only one study have applied PCR to the evaluation of human faecal specimens (Weiss *et al.* 1992). Using a small sub-unit rRNA PCR assay on formalin-fixed faecal specimens, the test found both false positive and false negative compared to microscopy. The PCR technique has been used to genomically differentiate *G. lamblia* and have found that there was genetic variation among the human isolates (Mahbubani *et al.* 1992).

In recent years, there has been an increase in the incidence of waterborne disease outbreaks caused by *G. lamblia* (Ongerth *et al.* 1995). The PCR test has been used to screen environmental water samples (Mahbubani *et al.* 1992; Rochelle *et al.* 1997), and it was also possible to distinguish live from dead cysts which is of great importance for the surveillance of the water supplies (Mahbubani *et al.* 1991). This environmental screening will probably be the first place, where the PCR assay will have a place in detection of *G. lamblia*.

Microsporidia

Six genera of microsporidian organisms have been associated with human microsporidiosis (*Enachiola, Enterocytozoon, Septata, Pleistophora, Trachipleistophora,* and *Vittaforma*). Currently, diagnosis of microsporidiosis depends on direct visualization of the parasites by light and transmission electron microscopy (TEM). For exact species differentiation, ultrastructural analysis of spores and tissue stages has been necessary (Weber *et al.* 1992; Cali *et al.* 1993). Serology has not proved useful in the diagnosis of human microsporidial infections (Weber *et al.* 1994).

Using primers targeted to the small sub-unit (SSU) and internal transcribed spacers (ITS) of the rRNA gene of microsporidia, amplified DNA of *Enterocytozoon bieneusi* and *Septata intestinalis* in intestinal biopsies (Zhu *et al.* 1993; Franzen *et al.* 1995, 1996; Velasquez *et al.* 1996) and of *E. bieneusi, E. cuniculi,* and *S. intestinalis* in stool specimens, have been detected (Fedorko *et al.* 1995; Katzwinkel-Wladarsch *et al.* 1996; Velasquez *et al.* 1996; Ombrouck *et al.* 1997). Subsequently, species identification by Southern blot or endonuclease restriction digest is also reported (Fedorko *et al.* 1995; Franzen *et al.* 1996).

The resistant spores of microsporidia and the risk of co-extracting PCR inhibitors from stool specimens, have required laborious and time-consuming methods for extracting DNA from microsporidia, particularly from faecal samples (Fedorko and Hijazi 1996). However, Kock *et al.* (1997) presented recently two DNA preparation methods, facilitating a fast and simple DNA isolation from stool and tissue, not requiring toxic reagents. The potential for PCR to identify species of microsporidia, mainly from non-invasively acquired specimens, such as sputum and stool specimen, makes it an attractive diagnostic option. A recent comparison between microscopy, assuming to be the correct diagnostic tool, and PCR for detection of microsporidia, showed 88% sensitivity and 78% specificity for PCR. However, a previous study showed that sensitivity of microscopical detection of microsporidia was less than 100% (Beauvais *et al.* 1993).

Considering the pathognomonic clinical features of microsporidiosis, the true status of the PCR positive cases cannot yet be resolved (Katzwinkel-Wladarsch *et al.* 1997). Therefore, the paradigm for a laboratory diagnosis of microsporidiosis still involves screening for microsporidial organisms by morphological procedures, followed by confirmation and speciation with PCR (Federko *et al.* 1995; Ombrouck *et al.* 1997).

Cryptosporidium parvum

Cryptosporidium parvum is a coccidian protozoan that causes diarrhoea in human, usually self-limited in immunocompetent hosts but severe and debilitating in immunocompromised hosts (Guerrent 1997). Cryptosporidiosis is very common in patients with AIDS (Balatbat *et al.* 1996). Commonly utilized coprodiagnostic methods involving stool concentration and vizualization using acid-fast (AF) or immunofluorescent staining (IF) have a low sensitivity, especially in formed stools, in which the threshold for detection may require 50 000 (IF) to 500 000 (AF) oocysts per gram of stool (Weber *et al.* 1991). Development of more sensitive methods would, therefore, have significant clinical implications.

Laxer *et al.* (1991) reported a *C. parvum* specific PCR, which was evaluated on human stool samples and on fixed, paraffin-embedded tissue samples (Laxer *et al.* 1992). The assay was capable to detect only one parasite. The same primer sequences were used by Gobet *et al.* (1997), who introduced hypochlorite inactivation of the samples. A more simplified oocyst preparation procedure, combined with a nested PCR, detected 500 oocysts per gram of stool or 500 ng *C. parvum* DNA (Balatbat *et al.* 1996). Diagnostic primers for *Cryptosporidium* spp.

have also been developed utilizing the random amplified polymorphic DNA (RAPD) analysis, which also have been used to genetically differentiate isolates of *Cryptosporidium* spp. (Morgan *et al.* 1995). In addition, the 18S sequence of rRNA genes of *C. parvum* and *C. muris* have been used in an assay, based on PCR and endonuclease restriction, for detection and species identification (Awad-el-Kariem *et al.* 1994). The increased sensitivity of the PCR assays, compared to current methods, may also contribute to identification of carriers of *C. parvum*, with or without active diseases, at a threshold below what is detectable by the current methods (Balatbat *et al.* 1996).

Another application of *Cryptosporidia*-PCR is to investigate the presence of *Cryptosporidia*-species in environmental water (surface, drinking water, and sewage samples) (Johnson *et al.* 1995; Stinear *et al.* 1996; Rochelle *et al.* 1997).

Entamoeba histolytica and *Entamoeba dispar*

It has now been established that two distinct species, although morphologically identical, exist within what was originally known as *Entamoeba histolytica: E. dispar* and *E. histolytica*, for the nonpathogenic and pathogenic forms, respectively (Diamond and Clark 1993). For both diagnostic as well as epidemiological reasons, it is necessary to establish markers to differentiate between the two *Entamoeba* species. The PCR has shown to be a promising tool for this purpose, since no time-consuming cultivation of the organism is needed.

Tachibani *et al.* (1991) clarified DNA differences between pathogenic and non-pathogenic isolates of *E. histolytica* (Tachibani 1991), which has been used to design PCR assays for diagnosis of stool specimen (Rivera *et al.* 1996; Sanuki *et al.* 1997). Other gene targets for *Entamoeba* species-specific single or nested PCR, are the genes for the small subunit rRNA (Clark and Diamond 1991; Katzwinkel-Wladarsch *et al.* 1994) and the 16S rRNA (Troll *et al.* 1997). PCR based on extra-chromosomal circular DNA from *E. histolytica* has also been used to differentiate between the two *Entamoeba*-species (Acuna-Soto *et al.* 1993; Aguirre *et al.* 1995). The latter assay was modified and used in a PCR assay, which was preceded by a rapid extraction procedure and followed by colorimetric product detection, PCR-SHELA (Britten *et al.* 1997).

Although microscopy remains the method of choice for *Entamoeba*-detection, specific PCR assays for differentiation between *E. dispar* and *E. histolytica* will greatly help the physicians determine whether they must treat the patients or not.

Malaria

We like to mention malaria in this book due to the frequency of specimens tested in Nordic laboratories. Accurate diagnosis of *Plasmodium* spp. is essential for successful treatment. Generally microscopic examination of smears is adequate for diagnosis of the majority of cases. The species do differ morphologically; which can be detected by microscopy, but it suffers from poor sensitivity in cases with low-level parasitemiae. PCR has been developed both for diagnostic purposes as well as for the possibility to detect drug resistant *P. falciparum* (Weiss, 1995).

Various assays based on the PCR have been described in the diagnosis of malaria, which were specific either for *P. falciparum* (Zalis *et al.* 1996) or *Plasmodium vivax* (Kain *et al.* 1993). A PCR based on the small subunit rRNA has been described, which enables detection of all four human *Plasmodium* species (Waters 1988; Snounou *et al.* 1993). In field studies, sampling of blood direct to membranes followed by PCR detection has been shown to be the best and easiest way of testing large number of samples (McLaughlin *et al.* 1991). In most of the reports, PCR detection was reported to be more sensitive than microscopic evaluation of blood smears, but did, due to PCR inhibitory components, sometimes fail to detect *Plasmodium* in specimens known to be positive (Walliker 1994; Oliveira *et al.* 1996). However, generally microscopy produced adequate sensitivity and specificity (Payne 1988; Weiss 1995).

The previous demonstrated sensitivity and specificity of PCR in the diagnosis of malaria may be preferable to that of microscopy in evaluating new diagnostic tests. Recently the diagnosis of *P. falciparum* was evaluated in 151 travellers comparing the dipstick test 'Parasight F', PCR, and direct microscopy (Humar *et al.* 1997). Using PCR as a gold standard, both the dipstick test and microscopy, had a sensitivity of 88%. The PCR was superior in cases of low parasitemiae.

Antifolate drugs such as pyrmethamine and proguanile are widely used for treatment and prophylaxis of malaria. Antifolate resistance is increasingly common in areas where malaria is endemic. Point mutations in the dihydrofolate gene are responsible for developing resistance. Two reports employed mutation specific primers and PCR followed by electrophoretic analysis to detect the presence or absence of specific products to distinguish pyrmethamine-susceptible from drug resistant as well as proguanile resistant isolates (Peterson *et al.* 1991; Plowe *et al.* 1995).

PCR diagnosis or detection of resistance will probably be a test that need to be centralized in reference laboratories, due to the low number of samples.

General conclusions

In conclusion, PCR for diagnosing parasitic infections is a promising approach, although its current role in routine diagnosis for most pathogens remains undefined. With our current knowledge, routine use of PCR only seems reasonable in the primary detection of *T. gondii*, *P. carinii*, differentiation of pathogenic/nonpathogenic. *Entamoeba* species and detection of drug-resistant *Plasmodium* spp. Easy sample extraction methods, reproducible results and the use of internal controls to validate the PCR result are needed.

Each clinical laboratory have to evaluate their needs, counting the number of specimens to be tested, the duration of time of conventional detection methods, sensitivity and specificity has to be critically evaluated and compared to the expensive PCR technique, before deciding to set up diagnostic PCR assays. Furthermore, PCR diagnosis of infrequent present parasites probably should be centralized and performed in national reference laboratories.

References

Acuna-Soto, R. *et al.* (1993). Application of the polymerase chain reaction to the epidemiology of pathogenic and nonpathogenic *Entamoeba histolytica*. *American Journal of Tropical Medicine and Hygiene*, **48**, 58–70.

Aguirre, A., Warhurst, D. C., Guhl, F., and Frame, I. A. (1995). Polymerase chain reaction-solution hybridization enzyme-linked immunoassay (PCR-SHELA) for the differential diagnosis of pathogenic and non-pathogenic *Entamoeba histolytica*. *Transactions of the Royal Society of Tropical Medicine and Hygiene*, **89**, 187–188.

Atzori, C. *et al.* (1995). Diagnosis of *Pneumocystis carinii* pneumonia in AIDS patients by using polymerase chain reactions on serum specimens. *Journal of Infectious Diseases*, **172**, 1623–1626.

Awad-el-Kariem, F. M., Warhurst, D. C., and McDonald, V. (1994). Detection and species identification of *Cryptosporidium* oocysts using a system based on PCR and endonuclease restriction *Parasitology*, **109**, 19–22.

Balatbat, A. B., Jordan, G. W., Tang, Y. J., and Silva, Jr, J. (1996). Detection of *Cryptosporidium parvum* DNA in human feces by nested PCR. *Journal of Clinical Microbiology*, **34**, 1769–1772.

Beauvais, B., Sarfati, C., Molina, J. M., Lesourd, A., Lariviere, M., and Derouin, F. (1993). Comparative evaluation of five diagnostic methods for demonstrating microsporidia in stool and intestinal biopsy specimens. *Annals of Tropical Medicine and Parasitology*, **87**, 99–102.

Briselden, A. M. and Hiller, S. L. (1994). Evaluation of Affirm VP Microbial Identification test for *Gardnerella* and *Trichimonas vaginalis*. *Journal of Microbiology*, **32**, 148–152.

Buke J. A. (1975). Giardiasis in childhood. *American Journal of Diseases in Children*, **129**, 1304–1310.

Britten, D., Wilson, S. M., McNerney, Moody, A. H., Chiodini, P. L., and Ackers, J. P. (1997). An improved colorimetric PCR-based method for detection and differentiation of *Entamoeba histolytica* and *Entamoeba dispar* in feces. *Journal of Clinical Microbiology*, **35**, 1108–1111.

Butcher, P. D. and Farthing, M. J. G. (1989). DNA probes for the faecal diagnosis of *Giardia lamblia* infections in man. *Biochemical Society Transactions*. **17**, 363–364.

Cali, A., Kotler, D. P., and Orestein, J. M. (1993). n.g, n.sp., an intestinal microsporidian associated with chronic diarrhea and dissemination in AIDS patients. *Journal of Eucaryotic Microbiology*, **40**, 101–112.

Cartwright, C. P., Nelson, N. A., and Gill, V. J. (1994). Development and evaluation of a rapid and simple procedure for detection of *Pneumocystis carinii* by PCR. *Journal of Clinical Microbiology*, **32**, 1634–1638.

Clark, C. G. and Diamond, L. S. (1991). Ribosomal RNA genes of "pathogenic" and "nonpathogenic" *Entamoeba histolytica* are distinct. *Molecular and Biochemical Parasitology*, **49**, 297–302.

Contini, C., Romani, R., Manganaro, M., Sorice, F., and Delia, S. (1993). Tissue Culture isolation of *Pneumocystis carinii* from peripheral blood mononuclear cells of AIDS patients with PCP. *AIDS*, **7**, 1137–1138.

Cotch, M. F., Pastorek, J. G., Nugent, R. P., Yerg, D. E., Martin, D. H., and Eschenbach, D. A. (1991). Demographic and behavioural predictors of *Trichomonas vaginalis* infection among pregnant women. *Obstetrics and Gynecology*, **78**, 1087–1092.

Diamond, L. S. and Clark, C. G. (1993). A redescription of *Entamoeba histolytica* Schaudinn, 1903 (Emended Walker, 1911) separating it from *Entamoeba dispar* Brumpt 1925. *Journal of Eukaryotic Microbiology*, **40**, 340–344.

Elvin, K., Olsson, M., Lidman, C., and Björkman, A. (1996). Detection of asymptomatic *Pneumocystis carinii* infection by PCR – predictive for subsequent pneumonia, *AIDS*, **10**, 1296–1297.

Evans, R., Joss, A. W. L., Pennington, T. H., and Ho-Yen, D. O. (1995). The use of a nested polymerase chain reaction for detecting *Pneumocystis carinii* from lung and blood in rat and human infection. *Journal of Medical Microbiology*, **42**, 209–213.

Fauts, A. C. and Kraus, S. J. (1980). *Trichomonas vaginalis*: reevaluation of its clinical presentation and laboratory diagnosis. *Journal of Infectious Diseases*, **141**, 137–143.

Fedorko, D. P., Nelson, N. A., and Cartwright, C. P. (1995). Identification of Microsporidia in stool specimens by using PCR and restriction endonucleasis. *Journal of Clinical Microbiology*, **33**, 17–39.

Fedorko, D. P. and Hijazi, Y. M. (1996). Application of molecular techniques to the diagnosis of microsporidial infection. *Emerging Infectious Diseases*, **2**, 183–191.

Franzen, C. *et al.* (1995). Detection of Microsporidia (*Enterocytozoon bieneusi*) in intestinal biopsy specimens from human immunodeficiency virus-infected patients by PCR. *Journal of Clinical Microbiology*, **33**, 2294–2296.

Franzen, C. *et al.* (1996). Genetic evidence for latent *Septata intestinalis* infection in human immunodeficiency virus-infected patients with intestinal microsporidiosis. *Journal of Infections Diseases*, **173**, 1038–1040.

Guerrent, R. L. (1997). Cryptosporidiosis: an emerging, highly infectious threat. *Emerging Infectious Diseases*, **3**, 51–57.

Gobet, P. *et al.* (1997). Detection of *Cryptosporidium parvum* DNA in formed human feces by a sensitive PCR-based assay including uracil-*N*-glycosylase inactivation. *Journal of Clinical Microbiology*, **35**, 254–256.

Helweg-Larsen, J., Jensen, J. S., and Lundgren, B. (1997). Non-invasive diagnosis of *Pneumocystis carinii* pneumonia by PCR and oral washes. *The Lancet*, **350**, 1363.

Humar, A., Ohrt, C., Harrington, M. A., Pillai, D., and Kain, K. C. (1997). Parasight F test compared with polymerase chain reaction and microscopy for the diagnosis of *Plasmodium falciparum* malaria in travelers. *American Journal of Tropical Medicine and Hygiene*, **56**, 44–48.

Jeremias, J. *et al.* (1994). Detection of Trichomonas vaginalis using the polymerase chain reaction in pregnant and non-pregnant woman. *Infectious Diseases in Obstetrics and Gynecology*, **2**, 16–19.

Johnson, D. W., Pienazek, N. J., Griffin, D. W., Misener, L., and Rose, J. B. (1995). Development of a PCR protocol for sensitive detection of *Cryptosporidium* oocysts in water supplies. *Applied and Environmental Microbiology*, **61**, 3849–3855.

Kain, K. C., Brown, A. E., Mirabelli, L., and Webster, H. K. (1993). Detection of *Plasmodium vivax* by polymerase chain reaction in a field study. *Journal of Infectious Diseases*, **168**, 1323–1326.

Katzwinkel-Wladarsch, S., Loscher, T., and Rinder, H. (1994). Direct amplification and differentiation of pathogenic and nonpathogenic *Entamoeba histolytica* DNA from stool specimen. *American Journal of Tropical Medicine and Hygiene*, **51**, 115–118.

Katzwinkel-Wladarsch, S., Lieb, M., Heise, W., Löscher, T., and Rinder, H. (1996). Direct amplification and species determination of microsporidian DNA from stool specimens. *Tropical Medicine and International Health*, **1**, 373–378.

Katzwinkel-Wladarsch, S., Deplazes, P., Weber, R., Löscher, T., and Rinder, H. (1997). Comparision of polymerase chain reaction with light microscopy for detection of microsporidia in clinical specimens. *European Journal of Clinical Microbiology and Infectious Diseases*, **16**, 7–10.

Kenge, P., Veas, F., Vidal, N., Rey, J.-L., and Cuny, G. (1994). *Trichomonas vaginalis*: repeated DNA target for highly sensitive and specific polymerase chain reaction. *Cellular and Molecular Biology*, **40**, 819–831.

Kitada, K. *et al.* (1991). Detection of *Pneumocystis carinii* sequences by polymerase chain reaction: animal models and clinical application to noninvasive specimen. *Journal of Clinical Microbiology*, **29**, 1985–1990.

Kock, N. P. *et al.* (1997). Species-specific identification Microsporidia in stool and intestinal biopsy specimens by the polymerase chain reaction. *European Journal of Microbiology and Infectious Diseases*, **16**, 369–376.

Laga, M. *et al.* (1993). Nonulcerative sexually transmitted diseases as risk factors for HIV-1 transmission in woman: results from a cohort study. *AIDS*, **7**, 95–102.

Laxer, M. A., Timblin, B. K., and Patel, R. J. (1991). DNA sequences for the specific detection of *Cryptosporidium parvum* by the polymerase chain reaction. *American Journal of Tropical Medicine and Hygiene*, **41**, 688–694.

Laxer, M. A., D'Nicuola, M. E., and Patel, R. (1992). Detection of *Cryptosporidium parvum* DNA in fixed, paraffin-embedded tissue by the polymerase chain reaction. *American Journal of Tropical Medicine and Hygiene*, **47**, 450–455.

Leibowitz, E. *et al.* (1995). Comparison of PCR and standard cytological staining for detection of *Pneumocystis carinii* from respiratory specimen from patients with or at high risk for infection by human immunodeficiency virus. *Journal of Clinical Microbiology*, **33**, 3004–3007.

Leigh, T. R., Wakefield, A. E., Peters, S. E., Hopkin, J. M., and Collins, J. V. (1992). Comparison of DNA amplification and immunofluorescence for detecting *Pneumocystis carinii* in patients receiving immunosuppressive therapy *Transplantation*, **54**, 468–470.

Leigh, T. R., Kangro, H. O., Gazzard, B. G., Jeffries, D. J., and Collins, J. V. (1993). DNA amplification by the polymerase chain reaction to detect subclinical *Pneumocystis carinii* colonization in HIV-positive and HIV-negative male homosexuals with and without respiratory symptoms. *Respiratory Medicine*, **87**, 525–529.

Lewis, D. J. M., Green, E. L., and Ashall, F. (1990). Total genomic DNA probe to detect *Giardia lamblia*. *The Lancet*, **336**, 257.

Lin, P. R. and Shaio, M. F. (1997). One-tube, nested-PCR assay for the detection of *Trichomonas vaginalis* in vaginal discharges. *Annual Tropical and Medical Parasitology*, **91**, 61–65.

Lipschik, G. Y. *et al.* (1992). Improved diagnosis of Pneumocystis carinii infection by polymerase chainreaction on induced sputum and blood. PCR. *The Lancet*, **340**, 203–206.

Lu, J.-J., Chen, C.-H., Bartlett, M. S., Smith, J. W., and Lee, C.-H. (1995). Comparison of six different PCR methods for detection of *Pneumocystis carinii.. Journal of Clinical Microbiology*, **33**, 2785–2788.

Lundgren, B., Benfield, T., and Lundgren, J. D. (1996). Evaluation of PCR technique for diagnosing *Pneumocystis carinii* pneumonia in HIV positive patients using oropharyngeal washings. *Journal of Eukaryotic Microbiology*, **43**, 9S.

Mahbubani, M. H., Bej, A. K., Perlin, M. H., Schaefer III, F. W., Jakubowski, W., and Atlas, R. M. (1991). Detection of Giardia cysts by using the polymerase chain reaction and distinguishing live from dead cysts. *Applied Environmental Microbiology*, **57**, 3456–3461.

Mahbubani, M. H., Bej, A. K., Perlin, M. H., Schaefer III, F. W., Jakubowski, W., and Atlas, R. M. (1992). Differentiation of *Giardia duodenalis* from other *Giardia* spp. by using polymerase chain reaction and gene probes. *Journal of Clinical Microbiology*, **30**, 74–78.

McLaughlin, G. L., Decrind, C., Deyal-Drager, R., Hassan-King, M., Subramanian, S., and Greenwood, B. M. (1991). Optimization of a rapid nonisotopic DNA probe assay for *Plasmodium falciparum* in the Gambia. *Journal of Clinical Microbiology*, **29**, 1517–1519.

Montaner, J. S. G. and Zala, C. (1995). The role of the laboratory in the diagnosis and management of AIDS-related *Pneumocystis carinii* pneumonia. In: *Clinical Infectious Deseases, International Practice and Research, Pneumocystis carinii*, vol 2, No. 3, F. R. Sattler and P. D. Walzer, eds, pp. 471–485. London: Baillíere Tindall.

Morgan, U. M., Constantine, C. C., O'Donoghue, P., Meloni, B. P., O'Brien, P. A., and Thompson, C. A. (1995). Molecular characterization of *Cryptosporidium* isolates from human and other animals using random amplified polykorphic DNA analysis. *American Journal of Tropical Medicine and Hygiene*, **52**, 559–564.

Oliveira, D. A. *et al.* (1996). Field evaluation of a polymerase chain reaction-based nonisotopic liquid hybridization assay for malaria diagnosis. *Journal of Infectious Diseases*, **173**, 1284–1287.

Olsson, M., Elvin, K., Löfdahl, S., and Linder, E. (1993). Detection of *Pneumocystis carinii* DNA in sputum and bronchoalveolar lavage samples by polymerase chain reaction. *Journal of Clinical Microbiology*, **31**, 221–226.

Ombrouck, C. *et al.* (1997). Specific PCR assay for direct detection of intestinal Microsporidia *Enterocytozoon bieneusi* and *Enterocytozoon intestinalis* in fecal specimens from human immunodeficiency virus-infected patients. *Journal of Clinical Microbiology*, **35**, 652–655.

Ongerth, J. E., Hunter, G. D., and DeWalle, F. B. (1995). Watershed use and *Giardia* cysts presence. *Water Research*, **29**, 1295–1299.

Payne, D. (1988). Use and limitations of light microscopy for diagnosing malaria at the primary health care level. *Bulletin of the World Health Organization*, **66**, 621–626.

Peterson, D. S., Di Santi, S. M., Povoa, M., Calvosa, V. S., Do Rosario, V. E., and Wellems, T. E. (1991). Prevalence of the dihydrofolate reductase Asn-108 Mutation, as the basis for pyrmethamine-resistant falcparum malaria in the Brazilan Amazon. *American Journal of Tropical Medicine and Hygiene*, **45**, 492–497.

Plowe, C. V., Djimde, A., Bouare, M., Doumbo, O., and Wellems, T. (1995). Pyrimethamine and proguanil resistance-conferring mutations in *Plasmodium falciparum* dihydrofolate reductase: polymerase chain reaction methods for surveillance in Africa. *Tropical Medicine and Hygiene*, **52**, 565–568.

Read, J. S. and Klebanoff, M. A. (1993). Sexual intercourse during pregnancy and preterm delivery: effects of vaginal microorganisms. *American Journal of Obstetric Gynecology*, **168**, 514–519.

Riley, D. E., Roberts, M. C., Takayama, T., and Krieger, J. N. (1992). Development of a polymerase chain reaction-based diagnosis of *Trichomonas vaginalis. Journal of Clinical Microbiology*, **30**, 465–472.

Rivera, W. R., Tachibana, H., Silva-Tahat, M. R. A., Uemura, H., and Kanbara, H. (1996). Differentiation of *Entamoeba histolytica* and *E. dispar* DNA from cysts present in stool specimens by polymerase chain reaction: its field application in the Philippines. *Parasitology Research*, **82**, 585–589.

Rochelle, P. A., de Leon, R., Stewart, M. H., and Wolfe, R. L. (1997). Comparison of primers and optimization of PCR conditions for detection of *Cryptosporidium parvum* and *Giardia lamblia* in water. *Applied and Environmental Microbiology*, **63**, 106–114.

Roux, P. *et al.* (1994). Usefulness of PCR for detection of *Pneumocystis carinii* DNA. *Journal of Clinical Microbiology*, **32**, 2324–2326.

Sanuki, J.-I., Asai, T., Okuzawa, E., Kobayashi, S., and Takeuchi, T. (1997). Identification of *Entamoeba histolytica* and *E. dispar* cysts in stool by polymerase chain reaction. *Parasitology Research*, **83**, 96–98.

Schluger, N. *et al.* (1992). Application of DNA amplification to pneumocystosis: presence of serum *Pneumocystis carinii* DNA during human and experimentally induced *Pneumocystis carinii* pneumonia. *Journal of Experimental Medicine*, **176**, 1327–1333.

Shaio, M.-F., Lin, P.-R., and Liu, J.-Y. (1997). Colimetric one-tube nested PCR for detection of *Trichomonas vaginalis* in vaginal discharge. *Journal of Clinical Microbiology*, **35**, 132–138.

Snounou, G., Viriyakosol, S., Jarra, W., Thaithong, S., and Brown, K. N. (1993). Identification of the four human malaria parasites species in field samples by polymerase chain reaction and detection of a high prevalence of mixed infections. *Molecular and Biochemical Parasitology*, **58**, 283–292.

Spence, D. E., Hollander, D. H., Smith, J., McCaig, L., Swell, D., and Brockman, M. (1990). The clinical and laboratory diagnosis of *Trichomonas vaginalis* infection. *Sexual Transmitted Disease*, **7**, 168–171.

Stinear, T., Matusan, A., Hines, K., and Sandery, M. (1996). Detection of a single viable *Cryptosporidium parvum* oocyst in environmental water concentrates by reverse transcription-PCR. *Applied and Environmental Microbiology*, **62**, 3385–3390.

Tachibani, H., Ihara, S., Kobayashi, S., Kaneda, Y., Takeuchi, T., and Watanabe, Y. (1991). Differences in genomic DNA sequences between pathogenic and nonpathogenic isolates of *Entamoeba histolytica* identified by polymerase chain reaction. *Journal of Clinical Microbiology*, **29**, 2234–2239.

Tamburrini, E. *et al.* (1993). Diagnosis of *Pneumocystis carinii*. pneumonia: specificity and sensitivity of polymerase chain reaction in comparison with immunofluorescence in bronchoalveolar lavage specimen. *Journal of Medical Microbiology*, **38**, 2788–2789.

Tamburrini, E. *et al.* (1996). Detection of *Pneumocystis carinii* DNA in blood by PCR is not of value for diagnosis of *P. carinii* pneumonia. *Journal of Medical Microbiology*, **34**, 1568–1588.

Troll, H., Marti, H., and Wiess, N. (1997). Simple differential detection of *Entamoeba histolytica* and *Entamoeba dispar* in fresh stool specimens by sodium acetate–acetic acid–formalin concentration and PCR. *Journal of Clinical Microbiology*, **35**, 1701–1705.

Velasquez, J. N. *et al.* (1996). Detection of the microsporidian parasite *Enterocytozoon bieneusi* in specimen from patients with AIDS by PCR. *Journal of Clinical Microbiology*, **34**, 3230–3232.

Wakefield, A. E. *et al.* (1990). Detection of *Pneumocystis carinii* with DNA amplification. *The Lancet*, **336**, 451–453.

Wakefield, A. E., Guiver, L., Miller, R. F., and Hopkin, J. M. (1991). DNA amplification on induced sputum samples for diagnosis of *Pneumocystis carinii* pneumonia. *The Lancet*, **337**, 1378–1379.

Wakefield, A. E., Miller, R. F., Guiver, L. A., and Hopkin, J. M. (1993). Oropharyngeal samples for detection of Pneumocystis carinii by DNA amplification. *Quaterly Journal of Medicine*, **86**, 401–406.

Walliker, D. (1994). The role of molecular genetics in field studies on malaria parasites. *International Journal of Parasitology*, **24**, 799–808.

Waters, A. P. (1988). *Reviews*, **8**, 113–130.

Weber, R., Bryan, R. T., Bishop, H. S., Wahlquist, S. P., Sullivan, J. J., and Juranek, D. D. (1991). Threshold of detection of *Cryptosporidium* oocysts in human stool specimens: evidence for low sensitivity of current diagnostic methods. *Journal of Clinical Microbiology*, **29**, 1323–1327.

Weber, R. *et al.* (1992). Improved light-microscopical detection microsporidia spores in stool and duodenal aspirates. *New England Journal of Medicine*, **326**, 161–166.

Weber, R., Bryan, R. T., Schwartz, D. A., and Owen, R. L. (1994). Human microsporidial infections. *Clinical Microbiology Review*, **7**, 426–461.

Weig, M., Klinker, H., Bögner, B. H., Meier, A., and Gross, U. (1997). Usefulness of PCR for diagnosis of *Pneumocystis carinii* pneumonia in different patient groups. *Journal of Clinical Microbiology*, **35**, 1445–1449.

Weiss, J. B. (1995). DNA probes and PCR for diagnosis of parasitic infections. *Clinical Microbiology*.

Weiss, J. B., van Keulen, H., and Nash, T. E. (1992). Classification of subgroups of *Giardia lamblia* based upon ribosomal RNA gene sequence using the polymerase chain reaction. *Molecular and Biochemical Parasitology*, **54**, 73–86.

Zalis, M. G. *et al.* (1996). Malaria diagnosis: standardization of a polymerase chain reaction for detection of *Plasmodium falciparum* in individuals with low-grade parasitemia. *Parasitology Research*, **82**, 612–616.

Zhu, X., Wittner, M., Tanowitz, H. B., Kotler, D., Cali, A., and Weiss, L. M. (1993). Small subunit rRNA sequence of *Enterocytozoon bieneusi* and its potential diagnostic role with use of the polymerase chain reaction. *Journal of Infectious Diseases*, **168**, 1570–1575.

Part 6

Historical review

35 Linnaeus, Armauer Hansen, and Nordic research on some neglected diseases

Sven F. F. Britton

In memory of Elias Bengtsson

Parasitological research in the North is like the prevalence of parasites in the area – scarce. Otherwise, it has the feature of Scandinavian research in general. It is methodological in nature, thereby reflecting the mind of its inhabitants.

On the other hand, it has a different origin than parasitological research in Europe and North America at large. Whilst the latter is tightly linked to military medicine and colonization, parasitological research in this northern part of Europe comes from missionary medicine and foreign aid, activities that may be coined spiritual colonization by some. The scientist to whom this chapter is dedicated exemplifies this. Elias Bengtsson who died in 1998 was the second professor of infectious diseases at the Karolinska Institute, Stockholm where this discipline was not academically recognized as an entity of its own until 1956 up until which time it was comprised within paediatric medicine.

Elias Bengtsson's interest in parasitology and tropical medicine came both from his close links with returning missionaries from the warm continents carrying exotic diseases and also from his service as a physician in Swedish peace keeping missions under the UN in the then Congo and in Gaza in the Middle East. Through these experiences he was confronted with parasitologic infections of exotic origin and this came to be his research interest and that of his many disciples for several decades. He tried, and was very close to succeed in creating a special professorship of Tropical Medicine and Parasitology at the Karolinska Institute. However, a chair in pure Parasitology (without clinical affiliation) was not finally installed at the Karolinska Institute until 1993 and its first holder is a pupil of Elias Bengtsson. This is the first and so far only professorship of human parasitology in the Nordic countries.

However, parasitological research has an older date than this century. The most famous natural scientist until now in this part of the world (and here I believe my fellow neighbours in Denmark, Finland, and Norway agree with me) is the Swedish scientist Carl Linnaeus. He was later knighted and got the more German sounding name von Linné. He made a short, but as always with him, important contribution to parasitology. He was primarily a botanist, although professor of practical medicine at Uppsala University. Medicine was more permissive at the time. Linnaeus was a taxonomist in good Scandinavian tradition but he had that extra gift which is rare among us, to link methodological examination to function in its widest sense. In his 24 pages long doctoral thesis that he defended in Holland 1736, he discusses the relation between the marshland biotope and the appearance of a febrile disease then endemic even in northern Europe, that is, what we understand now as malaria (Linnaeus 1736). (He later had 180 PhD students of his own). It is not wrong to characterize

Linnaeus as one of the first ecologists and in this regard he studied plants, parasites, insects, and mammals including man. Another of his observations was to describe and characterize a cestode *Diphyllobotrium latum* not reported of before (Plater *et al.* 1978). His main contributions though are, of course, the systematization of plants and animals (including *Homo sapiens*) and the introduction of the binomial nomenclature.

Perhaps, the most famous Nordic research linked to a parasitological disease, although admittedly taxonomically outside of classical parasites, is the one conducted by Gerhard Armauer Hansen in the late nineteenth century in Bergen, Norway. It concerns leprosy, the causative agent of which, *Mycobacteria leprae*, parasitizes the host macrophage. Leprosy is a bacterial disease that is coined by WHO as one of the six neglected diseases. The other diseases (malaria, schistosomiasis, leishmaniasis, filariasis, and trypanosomiasis) are strictly parasitological in nature. Hansen, through careful observation of transmission chains refuted the then dogma that leprosy was an inherited disease. It has to be acknowledged that transmission studies up until today are very difficult to perform. The route of transmission of *M. leprae* is still not known, where disease penetrance is very low (~1% of exposed individuals) and where the incubation time, from time of infection to occurrence of disease, is on average 10 years. Hansen managed to show in the microscope *M. leprae* bacilli from the skin of lepromatous leprosy patients without knowing the Ziehl–Neelsen method for staining of acid-fast bacilli, simply because of the extraordinary abundance of bacilli in the affected organs of lepromatous leprosy cases. Through this observation, Hansen, who was an autodidact scientist working as a country doctor, was one of the first scientist to link a disease to a microbial pathogen (Hansen 1874). This was a couple of years before Koch linked *M. tuberculosis* to the clinical entity of tuberculosis (Koch 1882). This is by far the most important single scientific contributions by Nordic scientists to parasitological research and it also serves to demonstrate that many of the so-called tropical diseases for which a better term is 'diseases of poverty' – were present in Scandinavia up until and within the twentieth century. This includes not only leprosy where the last domestic case died in the 1950s, but also malaria where the last (latest) indigenous cases of *Plasmodium vivax* malaria occurred in Finland in the 1950s. Although some parasitological infections clearly are latitude confined, the main denomination for their spread appears to be poverty, a state of affairs that Nordic countries on the Scandinavian peninsula have successfully conquered as some of the first nations of the world. Hence, the relatively low prevalence of parasite driven infections in this part of the world today.

The discovery of 'Hansen bacilli' or *M. leprae* and the studies of Linnaeus have had profound influence of host parasite research in general and of such research in Scandinavia as well. A direct spin-off is the foundation of a research institute for the study of leprosy and leishmaniasis in Addis Ababa, Ethiopia. This Institute called The Armauer Hansen Research Institute came to be through initiative from Norwegian and Swedish scientists of which Elias Bengtsson was one. The idea was to conduct research on diseases of poverty at the site where these diseases appeared and at the same time provide scientific training of local science students. Thus, this Institute was erected 30 years ago with support from the Norwegian and Swedish governments in direct association with a large leprosy hospital founded by the Swedish and Norwegian Save the Children organizations in the mid-1930s. The Institute was started in the mid-1960s when modern immunology had its most blooming period and hence its emphasis has been biomedical research on leprosy and leishmaniasis with an immunological angle to it. Its first director was Morten Harboe, who has remained an international nester of leprosy research ever since. He has used counter electrophoresis-methodology for characterizing the humoral immune response to *M. leprae* bacilli (Harboe *et al.* 1978) thus applying the original method of protein separation, through electrophoresis of Arne Tiselius (Tiselius *et al.* 1965) and its application to antibody mediated precipitation by Örjan Ouchterlony (Navalkar *et al.* 1965), to leprosy research.

Harboe, who retired as professor of immunology at Oslo University in 1998, was succeeded by the Director of AHRI, Tore Godal, a Norwegian MD with interest in disorders of immunohaematological origin. At AHRI he made some fundamental discoveries in collaboration with local scientists, on the nature of the specific immunological unresponsiveness in lepromatous leprosy (Godal *et al.* 1971) and its putative correction through immunological interventions. Godal, later on came to be a very important figure in modern parasitological research in general. In collaboration with, among others, Kenneth Warren and Barry Bloom, he alerted WHO on the need for research on the six great neglected diseases which resulted in the creation of a new department within WHO in 1984, coined Tropical Diseases Research (TDR) of which Tore Godal became the second director. Through the initiative of Ken Warren, a famous American scientist, on the immunopathology of *Schistosoma haematobium* resources were generated from the Rockefeller Foundation for research in parasitology which resulted in a world wide upsurge of such research with participation of scientists from developing countries. This development is still ongoing albeit the initial enthusiasm and resource generation may have subsided somewhat, to which the premature death of Ken Warren in 1996 may have contributed.

Nordic scientists at AHRI have continued to contribute to the progress in understanding and treatment of leprosy, notably through work of the Norwegian Gunnar Bjune on the immunopathological origin of reversal nerve reactions in leprosy and their suppression by high doses of steroids (Bjune *et al.* 1976). The findings of one of the later AHRI directors, Rolf Kiessling described the cellular nature and specificity of cytotoxic reactions in leprosy (Kaleab *et al.* 1990). Hannah Akuffo described through work at AHRI the putative differences in parasite strains causing diffuse versus local cutaneous leishmaniasis. She, later on in Sweden, described an innate responsiveness to *Leishmania* antigens even in a fraction of non-exposed Swedes (Akuffo and Britton 1992), perhaps the cellular basis for susceptibility versus resistance to this infection.

Danish parasitological research had a flashing start when Jens Fibiger in 1927 was awarded the Nobel prize in medicine for his demonstration of the minute nematode *Spiroptera neoplastica* and its role as a causative agent in gastric cancers of rats. However, these findings could never be repeated outside the laboratory of Fibiger and much to the disgrace of Fibiger himself, but even more so to the Nobel committee. It turned out that what Fiber saw in his microscope most probably were artefacts or contaminating impurities and the cancers were most likely induced by bad breeding conditions of the rats (vitamin deficiencies). However, after this debacle Danish parasitological research proved itself of good international standard when J. C. Siim, later on director of the Statens Serum Institut, described an acquired lymphoglandular disease caused by *Toxoplasma gondii*. This protozoa with world-wide distribution, had first been demonstrated by the discoverer of the malaria *Plasmodium*, namely the Frenchman, Laveran. Twenty years after his major finding (malaria), he described the *Toxoplasma gondii* in Java sparrows in 1900. Whereas congenital toxoplasmosis had been described already in the 1930s by Wolf, Cohen Siim was the first to identify the acquired febrile lymphoglandular form of this normally asymptotic infection, a very elegant and clinically important achievement (Siim 1961) first presented in 1950. Two decades later, the biologist Mandahl-Barth, disclosed parts of the intermediate snail-worm-relations in *Schistosoma hematobium* infection. He later on established the Danish Bilharzia laboratory (DBL) which has been a training and research institute for waterborne parasitological disease ever since, although lately it has been engaged in general nutritional health problems as well.

Finnish parasitology has shown fidelity to the parasites of the North, reflecting their more insulated character and Spöring 1747 published the first paper on the occurrence of *D. latum* in Finland. J. W. Runeberg already 1886 described the clinical entity of anaemia and weight loss resulting from consumption of raw fish infected with *D. latum* larvae. The habit of eating raw fish was quite common in Finland explaining the high incidence of this infection, reflecting the similarity of both the lake districts and eating habits in Finland and the former Soviet Karelia. The *D. latum* prevalence in Finland was estimated to be as high as 20% in the late 1940ies. A rapid decline was seen as the result of intense chemotherapy and information campaigns. Parasite prevalence figures in Finland have been different from those of the other Nordic countries and malaria was seasonal there, many years after it was eradicated in the rest of Scandinavia. Finnish clinical parasitology research has centered around the pathogenesis of tapeworm anaemia by B. von Bonsdorff and his tapeworm team, whereas active research on fish parasites and parasites of veterinary importance has been performed at the Parasitological institute at the Academy of Åbo (Åbo Akademi) established 1963 by the Societas Scientarum Fennica. B-J. Wikgren was the head of the Institute, followed by his student G. Bylund.

In recent years E. Linder has contributed to the development of improved diagnostic methods, notably in *Schistosoma* and *Pneumocystis carinii* infections in Sweden.

Parasitological research in Scandinavia today has centres in the Swedish Institute of Infectious Disease Control, Karolinska Institutet and Department of Immunology, University of Stockholm and in the Panum Institute of Copenhagen University with its connections with Statens Serum Institut and the Danish Bilharzia Institute. In Norway, it is Department of Infectious Diseases, Oslo University, and the Center for International Health, University of Bergen.

At the Karolinska Institutet, the first professor of human parasitology in Scandinavia, Mats Wahlgren is heading an internationally composed group exploring a phenomenon first described by Wahlgren in 1989 (Udomsangpetch *et al.* 1989), that is, red cell rosetting in cerebral malaria. Wahlgren observed that *Plasmodium falciparum* infected red cells formed rosetting agglutinates with uninfected red cells and through epidemiological field studies in West Africa, his group described that *P. falciparum* isolated from patients with cerebral malaria had significantly higher tendency to create rosettes than strains from patients without cerebral malaria. His group is now working on which receptors that are involved in this process and this interaction can be interfered with as a mode of treating or preventing cerebral malaria rather than the infection as such.

The other group in Stockholm working on malaria is from Stockholm University where Peter Perlmann, who some 20 years ago started a new career as a malaria immunologist from having been a very reputed cellular immunologist with particular interest in tumour immunology and auto-immunity. His group has mainly focused on assessing *P. falciparum* specific blood stage antigens that give protective immunity and can be used as components in a merozoite antigen-based vaccine (Berzins and Perlmann 1996). His group has also studied the immunity to *P. falciparum* and lately found that a Th2 driven IgE response can be linked to severe, for example, cerebral forms of malaria (Perlmann *et al.* 1997). At the Karolinska Institute Anders Bjorkman, yet another pupil of Elias Bengtsson has followed up the early malaria epidemiological studies from Liberia where they restored proguanil as an efficient plasmodiestatic drug when used in combination with Chloroquine. His later work with Ingegerd Rooth in Tanzania is based on a very close observation of a small village regarding malaria, measles, and HIV. An interesting observation from this work was the finding of a positive interaction between measles and malaria where parasitemia is considerably reduced for rather long time period in relation to measles similar in duration to the eliminated Purified Protein Derivative (PPD) skin reactivity (Rooth and Bjorkman 1992) following measles.

In Gothenburg, parasitological research was initiated in the 1970s when Gunnel Huldt and Inger Ljungström from the National Bacteriological Laboratory wanted to test the immuno-suppressive effects of *Toxoplasma gondii* and *Trichnella spiralis* respectively on mucosal immunity. Gothenburg and the departments under Jan Holmgren and Lars Åke Hansson of Medical Microbiology have become the Mecca of mucosal immunity reseach. Indeed such immuno-suppression did occur. Earlier on, in 1970, a famous infectious disease professor of Sweden, Ragnar Norrby had made his PhD thesis in Gothenburg on sub-cellular interactions between *Toxoplasma* and its target cell (Norrby 1970). Presently, studies from Gothenburg focus on schistosomiasis. These studies are both from a diagnostic point of view, where Örjan Ouchterlony's methodological skills are still used, and also through epidemiological studies. One study on the island of Kome in lake Victoria in Tanzania where different methods of intervention are tried and ensuing infections are measured with the immunologically based techniques that have been developed by the group together with the group of Deelder in Leiden, Holland (Jamaly *et al.* 1997).

Danish malaria research today is focused around Ib Bybjerg at the parasitological unit of the Rigshospitalet. He was recently appointed the first Danish professor of International Health. Their work concerns the epidemiology of chloroquin and mefloquin resistant malaria particularly through studies in East Africa (Bygbjerg *et al.* 1983). Another unique contribution comes from Harald Fugelsang who during 10 years in West Africa studied the natural course of blinding *Oncocerca volvolus* infection (Fuglsang *et al.* 1979). At the Pannun Institute the group around Thor Teander and Lars Hviid study visceral leishmaniasis in Sudan with particular reference to the immunological response profiles. The long-term aim is to produce a vaccine and also, like the groups of Perlmann and Wahlgren in Sweden to understand the immunology of severe malaria and how it can be interfered with.

A similar intervention study is done by Sven Gunnar Gundersen (Gundersen 1992) from the Center for International Health in Oslo using the local natural product Endod developed in Ethiopia by the late Prof. Aklilu Lemma. Endod is the product of the berry of the Endod plant, which grows along the riverbanks in Ethiopia, as well as the rest of Africa. It has been used as a washing powder since long in Ethiopia and the original observation was that the *Schistosome* transmitting snails died in the wash water downstream. A major study under Sven Gundersen using three different rivers where one received a calibrated regular dosage of Endod powder, one just Endod for river water washing and one without Endod treatment will be the study areas. The appearance of *Schistosoma* infection along these three different rivers will be followed and they all appear in a similar biotope with ethnically similar populations. This major study of interfering with schistosomiasis with a natural plant is soon to be completed and it is done in good Linnaeus tradition.

At the Center for International Health in Bergen, its first director and the first professor of International Health in the North, Prof. Bjarne Bjorvatn, was the first to establish an animal model for the study of immunity to *Leishmania major*. This model was later used to reveal the relevance of Th1/Th2 immunity in an infectious disease (Bjorvatn and Neva 1979). At that centre a broad activity of research pertaining to international health goes on and in the parasitological field can be mentioned studies on treatment and serological diagnosis of *Entamoeba histolytica* infections and the genetics of *P. falciparum*.

I hope that the reader has now appreciated that parasitological research in Scandinavia is considerably richer than the parasitic infections of the area. Parasite in Greek means sharing the food of another's table and scientists of the North have gone overseas to share the rich parasitological table of, above all, the African continent. They have made fairly substantial contributions in this field besides creating contacts with scientists of that continent of whom one is an editor of this book.

References

Akuffo, H. O. and Britton, S. F. F. (1992). The contribution of innate or non-*Leishmania* specific immunity to resistance to *Leishmania* infection in man. *Clin. Exp. Immunol.* **87**: 58–64.

Berzins, K. and Perlmann, P. (1996). Malaria vaccines: attacking infected erythrocytes. In: *Malaria vaccine development: a multi-immune approach*, S. L. Hoffman, ed. pp. 105–143.

Bjorvatn, B. and Neva, F. A. (1979). Experimental therapy of mice infected with *Leishmania tropica*. *Am. J. Trop. Med. Hyg.* **28**: 480–485.

Bjune, G., Barnetson, R. S., Ridley, D.S., and Kronvall, G. (1976). Lymphocyte transformation test in leprosy; correlation of the response with inflammation of lesions. *Clin. Exp. Immunol.* **25**: 85–94.

Bygbjerg, I. C., Shapira, A., Flachs, H., Gomme, G., and Jepsen, S. (1983). Mefloquine resistance of *falciparum* malaria from Tanzania enhanced by treatment. *Lancet* **2**;1: 774–775.

Fuglsang, H., Anderson, J., and Marshall, T. F. (1979). Studies on onchocerciasis in the United Cameroon Republic. V. A four year follow-up of 6 rain-forest and 6-Sudan-savanna villages. Some changes in skin and lymph nodes. *Trans. R. Soc. Trop. Med. Hyg.* **73**: 118–119.

Godal, T., Myklestad, B., Samuel, D. R., and Myrvang, B. (1971). Characterization of the cellular immune defect in lepromatous leprosy: a specific lack of circulating *Mycobacterium leprae*-reactive lymphocytes. *Clin. Exp. Immunol.* **9**: 821–831

Gundersen, S. G. (1992). *Studies on schistosomiasis and other infections in the Ethiopia Blue Nile valley with special reference to diagnostic improvements*. MD/PhD Thesis, Oslo University.

Hansen, G. H. A. (1874). Norsk Mag. F. Lægevidensk. *Tillægshefte Arch. F. Path. Anat. u. f. Klin. Med.* 1880, 79; 1882, **90**: 542.

Harboe, M., Closs, O., Rees, R. J., and Walsh, G. P. (1978). Formation of antibody against *Mycobacterium leprae* antigen 7 in armadillos. *J. Med. Microbiol.* **11**(4): 525–535

Jamaly, S., Chihani, T., Deelder, A. M., Gabone, R., Nilsson, L. A., and Ouchterlony, O. (1997). Polypropylene fibre web, a new matrix for sampling blood for immuno diagnosis of schistosomiasis. *Trans. R. Soc. Trop. Med. Hyg.*, **91**: 412–415.

Koch, R. (1882). Die Aetiologie der Tuberkulose. *Berl. Klin. Wchnschr.* **19**: 221–230.

Kaleab, B., Ottenoff, T., Converse, P., Halapi, E., Tadesse, G., Rottenberg, M., and Kiessling, R. (1990). Mycobacterial-induced cytotoxic T cells as well as nonspecific killer cells derived from healthy individuals and leprosy patients. *Eur. J. Immunol.* **20**(12): 2651–2659.

Linnaeus, C. (1736). Disseratio medica imangualis in qua exhibetur hypothesis nova de febrium intermittentium causa. Harderovici.

Navalkar, R.G., Norlin, M., and Ouchterlony, O. (1965). Characterization of leprosy sera with various mycobacterial antigens using double diffusion-in-gel analysis – II. *Int. Arch. Allergy Appl. Immunol.* **28**: 250–260.

Norrby, R. (1970). *Studies on the host cell penetration of Toxoplasma gondii*. MD/PhD Thesis, Göteborg University.

Perlmann, P., Perlmann, H., Flyg, B.W., Hagstedt, M., Elghazali, G., Worku, S., Fernandez, V., Rutta, A. S., and Troye-Blomberg, M. (1997). Immunoglobulin E, a pathogenic factor in *Plasmodium falciparum* malaria. *Infec. Immun.* **65**: 116–121.

Plater, F., Linneus, C., Bonne, C. *et al.* (1978). In: *Diphyllobotrisis Tropical Medicine and Parasitology, Classical Investigations*, vol. II, B. H. Kean, K. E. Mott and A. J. Russell, eds, Chap. 37, pp. 653–671.

Rooth, I. B. and Bjorkman, A. (1992). Suppression of *Plasmodium falciparum* infections during concomitant measles or influenza but not during pertussis. *Am. J. Trop. Med. Hyg.* **47**: 675–681.

Siim, J. C. (1961). Acquired toxoplasmosis. *J. Am. Med. Assoc.* (JAMA) **147**: 1641–1645.

Svensson, R. (1935). Studies on intestinal protozooa especially with regard to their demonstrability and the connection between their distribution and hygienic conditions. *Acta Med. Scand.*, *Suppl.* 70.

Tiselius, A., Hjerten, S., and Jerstedt, S. (1965). 'Particle-sieve' electrophoresis of viruses in polyacrylamide gels. *Arch. Gesamte Virusforsch.* **17**: 512–521. (Exemplified by purification of turnip yellow mosaic virus.)

Udomsangpetch, R., Wahlin, B., Carlson, J., Berzins, K., Torii, M., Aikawa, M., Perlmann, P., and Wahlgren, M. (1989). *Plasmodium falciparum*-infected erythrocytes form spontaneous erythrocyte rosettes. *J. Exp. Med.* **169**: 1835–1840.

von Bonsdorff, B. (1978). Läkare och läkekonst i Finland 1840–1940. (in Swedish). Ekenäs, Finland.

Index

Printed and bound by CPI Group (UK) Ltd, Croydon, CR0 4YY

23/10/2024

01778226-0006